D0222080

SOLUTIONS OF SELECTED PROBLEMS

for

MATHEMATICAL METHODS IN THE PHYSICAL SCIENCES

SECOND EDITION

MARY L. BOAS
DePaul University

JOHN WILEY & SONS
NEW YORK CHICHESTER BRISBANE TORONTO SINGAPORE

Preface

This manual is intended for students, to assist them in learning good problem solving techniques. Solutions are given for approximately one-quarter of the problems in the text. I have tried to select at least one example of each basic kind of problem as well as a few of the more difficult problems, derivations, etc. The routine problems are discussed in detail with special attention given to efficient methods and to points which cause difficulty for students. Hints are given for some problems, and, where needed, there are general outlines of suggestions for a particular set of problems.

Since the Solutions are closely keyed to the text, you can find material in the Solutions by using the text Index or Table of Contents. Note that the corresponding text Chapter and Section are given at the top of each facing pair of pages in the Solutions. As in the text, Problem 5.7 means Problem 7 in Section 5, but within Section 5 it is called just Problem 7. For reference, note the following items: Appendices (pages 603 ff) on Integrals and Integration, Synthetic Division, and Partial Fractions; table and graph of the gamma function (pages 346-347); graphs of Legendre polynomials (page 367) and Bessel functions (page 390).

If you are a student, let me offer you the advice I give to students in my classes about making efficient use of these Solutions and in general about studying from the text. You learn to solve problems by solving them yourself, not by reading a solution. If you can't do a problem, look for it or a similar problem in these Solutions, but read only enough to get you started--then try to continue on your own. When you have finished a problem, be sure to <u>check</u> <u>your</u> <u>answer</u> if it is given in the text (pages 747-774), or compare your work with the solution if it is given. If you are working independently or if you are having trouble with a particular topic, a good way to proceed is to try first the problems for which you have solutions. If you are not having any difficulty solving the problems, it is still a good idea to look through the solved problems for hints and suggestions about efficient methods which may save you time. Also please read my comments "To the Student" in the text, pages xiii-xiv.

If you are an instructor, please note that for each problem in the Solutions, the text includes (in most cases) several similar problems, some with answers at the end of the text. Thus you can choose the mixture you find best from problems to which your students have solutions, answers, or neither. I like to use a few of the solved problems for supervised study periods during a class. Students can at least get started on some representative new problems and if they don't finish a problem they have the solution to consult at home. May I suggest that you walk around your class, look over shoulders, and answer individual questions during a study period. You may find the experience eye-opening; we instructors tend to assume that students remember more than they do of previous courses. I hope that you will reinforce my admonitions to the students (above and in the text) about the importance of using efficient problem solving methods and of prac- tice in problem solving.

If you find errors in either the Solutions or the text, or have comments, please send them to me at the Department of Physics, DePaul University, 2219 North Kenmore, Chicago, Illinois 60614.

May 1984   *Mary L. Boas*

Chapter 1

Section 1

2. We shall do one special case of this problem and leave the general problem to you. For $n = 5$, we write S_5, multiply it by r, and subtract:

$$S_5 = a + ar + ar^2 + ar^3 + ar^4$$

$$rS_5 = ar + ar^2 + ar^3 + ar^4 + ar^5$$

$$S_5 - rS_5 = a + 0 + 0 + 0 + 0 - ar^5 = a - ar^5$$
$$S_5(1 - r) = a(1 - r^5)$$
$$S_5 = \frac{a(1 - r^5)}{1 - r} \ , \ r \neq 1.$$

In the same way, we can find S_n for any n.

For the convergence proof, write S_n in two parts:

$$S_n = \frac{a(1 - r^n)}{1 - r} = \frac{a}{1 - r} - \frac{ar^n}{1 - r} \ .$$

Convince yourself that $\lim_{n \to \infty} r^n = 0$ if $|r| < 1$. Then

$$S = \lim_{n \to \infty} S_n = \frac{a}{1 - r} - 0 = \frac{a}{1 - r} \quad \text{if } |r| < 1.$$

If $|r| > 1$, $|r^n| \to \infty$. If $r = 1$, $S_n = a + a + a + \ldots = na \to \infty$ as $n \to \infty$. If $r = -1$, then the series is $a - a + a - a + \ldots$, which does not converge. Thus a geometric series converges only when $|r| < 1$.

5. We write $0.583333\cdots = 58/100 + (3/1000 + 3/10^4 + 3/10^5 + \cdots)$. The series in parentheses is a geometric series with $a = 3/1000$ and $r = 1/10$, so by text equation (1.8) its sum is

$$\frac{a}{1 - r} = \frac{3/1000}{1 - (1/10)} = \frac{3/1000}{9/10} = \frac{3}{1000} \cdot \frac{10}{9} = \frac{1}{300} \ .$$

Then $0.58333\cdots = \frac{58}{100} + \frac{1}{300} = \frac{174 + 1}{300} = \frac{175}{300} = \frac{7}{12} \ .$

1

9. We write $0.185185 \cdots = 185/10^3 + 185/10^6 + \cdots$. This is a geometric series with $a = 185/10^3$ and $r = 1/10^3$ so its sum is

$$\frac{a}{1-r} = \frac{185/1000}{1 - (1/1000)} = \frac{185/1000}{999/1000} = \frac{185}{999} = \frac{5}{27} .$$

12. If $n = 2$, the fraction of the impurity removed in the first stage is 1/2, in the second stage 1/4, in the third stage 1/8, and so on. The sum of these fractions is a geometric series with $a = 1/2$ and $r = 1/2$, so by text equation (1.8) its sum is

$$\frac{a}{1-r} = \frac{1/2}{1 - (1/2)} = 1.$$

Thus the fraction of the impurity removed can be as near 1 as we like by using enough stages.

Similarly if $n = 3$, find

$$\frac{1}{3} + \frac{1}{9} + \frac{1}{27} + \cdots = \frac{1/3}{1 - 1/3} = \frac{1/3}{2/3} = \frac{1}{2} .$$

Thus we can come as close as we like to removing half the impurity, but half will always remain.

You might like to show in general that for any given n, the fraction which you can remove (almost) is $1/(n-1)$, so the fraction always remaining is

$$1 - \frac{1}{n-1} = \frac{n-2}{n-1} .$$

Section 2

6.   We evaluate the numerator and denominator for several n values:

n	n!	$(n!)^2$	$(2n)!$
1	1	1	$(2 \cdot 1)! = 2! = 2$
2	2	4	$(2 \cdot 2)! = 4! = 4 \cdot 3 \cdot 2 = 24$
3	$3! = 3 \cdot 2 = 6$	36	$(2 \cdot 3)! = 6! = 6 \cdot 5 \cdot 4 \cdot 3 \cdot 2 = 720$

(You might verify the next two terms.) Note carefully that
$(2n)!$ and $2(n!)$ are very different; for example, if $n = 5$, then
$(2n)! = 10! = 3.6 \times 10^6$, but $2(n!) = 2(5!) = 2(120) = 240$.   Then

$$\sum_{n=1}^{\infty} \frac{(n!)^2}{(2n)!} = \frac{1}{2} + \frac{2^2}{24} + \frac{6^2}{720} + \frac{(4!)^2}{8!} + \frac{(5!)^2}{10!} + \ldots$$

$$= 1/2 + 1/6 + 1/20 + 1/70 + 1/252 + \ldots .$$

It is not easy to recognize these numbers, so in this case it
is better to have the formula for the nth term.

7.   The numerators are 2^0, 2^1, 2^2, 2^3, etc., in other words 2^n
starting with $n = 0$. The denominators are odd numbers which we
can write as $2n + 3$ so that, when $n = 0$, the denominator is 3,
and when $n = 1$, the denominator is 5, etc. Then the series can
be written as $\sum_{n=0}^{\infty} \frac{2^n}{2n + 3}$. If we prefer to have $n = 1$ for the
first term, we can write $\sum_{n=1}^{\infty} \frac{2^{n-1}}{2n + 1}$.

Section 4

1. $S - S_n = 1/2^{n+1} + 1/2^{n+2} + \cdots$; this is a geometric series with
    $a = 1/2^{n+1}$ and $r = 1/2$. Thus by text equation (1.8),
    $S - S_n = (1/2^{n+1}) \div (1 - \frac{1}{2}) = (1/2^{n+1}) \cdot 2 = 1/2^n$. Now $\frac{1}{2^4} = \frac{1}{16} < \frac{1}{10}$,
    $\frac{1}{2^7} = \frac{1}{128} < \frac{1}{100}$, $\frac{1}{2^{10}} = \frac{1}{1024} < \frac{1}{1000}$, so some ε's and the corres-
    ponding N's are:

    $\varepsilon = 0.1$, $N = 4$, since $|S - S_n| = 1/2^n < 0.1$ for $n \geqslant N = 4$;

    $\varepsilon = 0.01$, $N = 7$, since $1/2^n < 0.01$ for $n \geqslant N = 7$;

    $\varepsilon = 0.001$, $N = 10$, since $1/2^n < 0.001$ for $n \geqslant N = 10$.

    Since $1/2^n \to 0$ as $n \to \infty$, given any ε no matter how small, we
    can choose N so that $1/2^n < \varepsilon$ for every $n \geqslant N$.

3. Using the hint, we can write the series as
    $S = 1 - \frac{1}{2} + \frac{1}{2} - \frac{1}{3} + \frac{1}{3} - \frac{1}{4} + \cdots = 1$. Similarly we can write the
    rest of the series after the nth term as
    $S - S_n = \frac{1}{n+1} - \frac{1}{n+2} + \frac{1}{n+2} - \frac{1}{n+3} + \cdots = \frac{1}{n+1}$.
    Then if $n \geqslant N = 10$, $|S - S_n| = \frac{1}{n+1} < \varepsilon = 0.1$;
        if $n \geqslant N = 100$, $|S - S_n| < \varepsilon = 0.01$, etc .

8. Using the hint, we write $\frac{1}{n^2 - 1} = \frac{1}{(n+1)(n-1)} < \frac{1}{n(n-1)}$. Then
    $\sum_{n=2}^{\infty} \frac{1}{n^2 - 1} < \sum_{n=2}^{\infty} \frac{1}{n(n-1)} = \sum_{m=1}^{\infty} \frac{1}{(m+1)m}$, where, in the last step,
    we have replaced n by $m+1$ so that the sum on m goes from 1 to
    ∞ . This is the same series as in Problem 3. Adjusting the
    results found there, we get for $\sum_{n=2}^{\infty} \frac{1}{n^2 - 1}$: If $n \geqslant N = 11$, then
    $|S - S_n| < \varepsilon = 0.1$, etc.

Section 5

2.   The general term is $\frac{\sqrt{n+1}}{n}$. For very large n, this is nearly $\frac{\sqrt{n}}{n} = \frac{1}{\sqrt{n}}$ which $\to$ 0 as $n \to \infty$. Thus we must test this series further.

4.   For very large n, $\frac{n^2}{(n+1)^2}$ is nearly $\frac{n^2}{n^2} = 1$. As $n \to \infty$, the terms alternate between nearly +1 and nearly -1. Since the terms do <u>not</u> tend to zero, the series diverges.

11.

$$S_n = a_0 + a_1 + a_2 + \cdots + a_{n-1} + a_n$$
$$S_{n-1} = a_0 + a_1 + a_2 + \cdots + a_{n-1}$$
$$\overline{\phantom{S_n = a_0 + a_1 + a_2 + \cdots + a_{n-1} + a_n}}$$
$$S_n - S_{n-1} = a_n$$

As $n \to \infty$, n - 1 also $\to \infty$ so, if the series converges, $\lim S_n = S$ and $\lim S_{n-1} = S$. Thus

$$\lim a_n = \lim(S_n - S_{n-1}) = S - S = 0.$$

Section 6

3. $\displaystyle\sum_{n=1}^{\infty} \frac{1}{n^2} = \frac{1}{1^2} + \left(\frac{1}{2^2} + \frac{1}{3^2}\right) + \left(\frac{1}{4^2} + \frac{1}{5^2} + \frac{1}{6^2} + \frac{1}{7^2}\right) + \left(\frac{1}{8^2} + \cdots + \frac{1}{15^2}\right) + \cdots$.

We compare this with the series

$\frac{1}{1^2} + \left(\frac{1}{2^2} + \frac{1}{2^2}\right) + \left(\frac{1}{4^2} + \frac{1}{4^2} + \frac{1}{4^2} + \frac{1}{4^2}\right) + \left(8 \text{ terms each equal to } \frac{1}{8^2}\right) + \cdots$

$= 1 + \frac{2}{2^2} + \frac{4}{4^2} + \frac{8}{8^2} + \cdots = 1 + \frac{1}{2} + \frac{1}{4} + \frac{1}{8} + \cdots$;   this is a geometric series with r = 1/2 and therefore converges. The terms in $\sum \frac{1}{n^2}$ are less than or equal to the corresponding terms of the geometric series; therefore $\sum \frac{1}{n^2}$ is also convergent.

5. (b) Since $\ell n\, n < n$, then $\frac{1}{\ell n\, n} > \frac{1}{n}$. The series $\sum \frac{1}{n}$ is divergent (harmonic series), so by comparison, the series $\sum \frac{1}{\ell n\, n}$ is divergent.

7. We evaluate $\int^{\infty} \frac{dn}{n\, \ell n\, n}$. (Note: no lower limit.) Let $u = \ell n\, n$; then $du = \frac{1}{n}\, dn$, $u = \infty$ when $n = \infty$.

$$\int^{\infty} \frac{dn}{n\, \ell n\, n} = \int^{\infty} \frac{du}{u} = \ell n\, u \, \Big|^{\infty} = \ell n\, (\ell n\, n) \, \Big|^{\infty} = \infty \, ,$$

so the series is divergent.

10. We want $\int^{\infty} \frac{e^n dn}{e^{2n} + 9}$. Let $u = e^n$; then $du = e^n dn$, $e^{2n} = u^2$, and $u = \infty$ when $n = \infty$. We use integral tables to evaluate

$$\int^{\infty} \frac{du}{u^2 + 9} = \frac{1}{3} \arctan \frac{u}{3} \, \Big|^{\infty} = \frac{1}{3} \frac{\pi}{2} = \frac{\pi}{6} \, .$$ Since the integral is finite, we conclude that the series converges.

15. When $p \neq 1$, $\int^{\infty} \frac{dn}{n^p} = \int^{\infty} n^{-p} dn = \frac{n^{-p+1}}{1-p} \, \Big|^{\infty} = \begin{cases} 0, & p > 1, \\ \infty, & p < 1. \end{cases}$

Thus $\sum \frac{1}{n^p}$ converges if $p > 1$ and diverges if $p < 1$. If $p = 1$,

$\int^{\infty} \frac{dn}{n} = \ell n\, n \, \Big|^{\infty} = \infty$, so the series diverges. Alternatively, we could have observed that when $p = 1$, we have the harmonic series which diverges.

16. Sketch the graph of $y = \frac{1}{x^2}$, $x = 0$ to ∞. What is the area under this graph from 0 to 1? (Answer: ∞). Look at text Figure 6.2 to see that this infinite area has nothing to do with the sum of the series.

18.
$$\rho_n = \left| \frac{2^{n+1}}{(n+1)^2} \div \frac{2^n}{n^2} \right| = \frac{2^{n+1}}{2^n} \frac{n^2}{(n+1)^2} = \frac{2n^2}{(n+1)^2} \, .$$

Divide numerator and denominator by n^2 and let $n \to \infty$:

$$\rho = \lim_{n \to \infty} \rho_n = \lim_{n \to \infty} \frac{2}{\left(1 + \frac{1}{n}\right)^2} = \frac{2}{1} = 2 > 1$$

so the series is divergent.

20.   Here $a_n = \frac{n!}{(2n)!}$.   Be careful about a_{n+1} ; when n is changed to

$n+1$, $2n$ becomes $2(n+1) = 2n+2$ (<u>not</u> $2n+1$).   Then $a_{n+1} = \frac{(n+1)!}{(2n+2)!}$

so

$$\rho_n = \left| \frac{(n+1)!}{(2n+2)!} \div \frac{n!}{(2n)!} \right|$$

$$= \frac{(n+1)\,\cancel{(n)}\,\cancel{(n-1)}\cdots\cancel{2}\cdot\cancel{1}}{\cancel{n}\cancel{(n-1)}\cdots\cancel{2}\cdot\cancel{1}} \cdot \frac{\cancel{(2n)}\,\cancel{(2n-1)}\cdots\cancel{2}\cdot\cancel{1}}{(2n+2)(2n+1)\cancel{(2n)}\,\cancel{(2n-1)}\cdots\cancel{2}\cdot\cancel{1}}$$

$$= \frac{n+1}{(2n+2)(2n+1)} = \frac{1}{2(2n+1)} \text{ since } \frac{n+1}{2n+2} = \frac{n+1}{2(n+1)} = \frac{1}{2} \, .$$

$$\rho = \lim_{n \to \infty} \frac{1}{2(2n+1)} = 0 < 1, \text{ so the series converges.}$$

23.
$$\rho_n = \left| \frac{(n+1)!}{100^{n+1}} \div \frac{n!}{100^n} \right| = \frac{n+1}{100}$$

$\lim_{n \to \infty} \rho_n = \infty$;   the series diverges.   In fact, it is so badly

divergent that we <u>could</u> have used the preliminary test.   How-

ever, in this case it is easier to find $\lim_{n \to \infty} \rho_n = \lim \frac{n+1}{100}$ than

it is to find $\lim_{n \to \infty} |a_n| = \lim_{n \to \infty} \frac{n!}{100^n}$.   In fact, if you used your

calculator to find the values of the terms $\frac{n!}{100^n}$ up to $n = 49$

$(100^{49} = 10^{98}$ is probably as far as your calculator will go),

23.   (continued)

you might become convinced that a_n was tending to zero. To see
that this is wrong as $n \to \infty$, consider $n > 100$. To go from one
term to the next, we multiply by $\frac{n}{100}$, so the terms become
larger and larger rather than tending to zero as $n \to \infty$. But
$\frac{n}{100}$ is just ρ_{n-1}; we are essentially using the ratio test!
Moral: Use the preliminary test first if you can easily find
$\lim |a_n|$; otherwise go on to another test.

30.   Let $\rho = \lim\limits_{n \to \infty} |a_{n+1}/a_n|$ with $\rho < 1$. Choose a number σ so that
$\rho < \sigma < 1$. Then for all sufficiently large n, $|a_{n+1}/a_n|$ is as
close as we like to ρ (this is what it means to say that
$\lim |a_{n+1}/a_n| = \rho$). But if $|a_{n+1}/a_n|$ is close enough to ρ, then
it is less than σ, or $|a_{n+1}| < \sigma |a_n|$ for all $n \geq N$. Then

$$|a_{N+1}| < \sigma |a_N|, \quad |a_{N+2}| < \sigma |a_{N+1}| < \sigma^2 |a_N|, \quad \text{etc. Thus}$$

$$\sum_{n=N+1}^{\infty} |a_n| < \sum_{n=1}^{\infty} \sigma^n |a_N| = |a_N|(\sigma + \sigma^2 + \sigma^3 + \cdots).$$

Since $\sigma < 1$, the geometric series converges, and therefore the
series $\sum |a_n|$ converges. Also see Problem 7.9.

If $\rho > 1$, then choose σ so that $\rho > \sigma > 1$. Then $|a_{n+1}| > \sigma |a_n|$
for all $n \geq N$. Thus a_n does not tend to zero, so the series
diverges by the preliminary test.

31.   For large n, the nth term of the series is approximately
      $(2n)(3n)/\sqrt{n^2} = 6n^2/n = 6n$.   We use as a comparison series $\sum n$,
      which is divergent.   Then by the special comparison test (text,
      Section 6D)

$$\lim_{n\to\infty} \frac{(2n+1)(3n-5)}{\sqrt{n^2-73}} \div n = \lim_{n\to\infty} \frac{(2n+1)(3n-5)}{n\sqrt{n^2-73}} = 6 > 0,$$

      so the given series diverges.   We could have seen this using
      the preliminary test!   The nth term $\to\infty$ so the series diverges.

35.   First note that for large n, n is very much larger than $\ell n\, n$.
      (For example if $n = 10^6$, $\ell n\, n \cong 14$).   Thus the numerator is about
      n^2 for large n, and the denominator is about $5n^4$.   We use as a
      comparison series $\sum \frac{n^2}{n^4} = \sum \frac{1}{n^2}$,   which is convergent.   Then by
      the special comparison test (text, Section 6D)

$$\lim_{n\to\infty} \frac{(n - \ell n\, n)^2}{5n^4 - 3n^2 + 1} \div \frac{1}{n^2} = \lim_{n\to\infty} \frac{n^2(n - \ell n\, n)^2}{5n^4 - 3n^2 + 1}$$

$$= \lim_{n\to\infty} \frac{\left(1 - \frac{\ell n\, n}{n}\right)^2}{5 - \frac{3}{n^2} + \frac{1}{n^4}} \qquad \text{(Divide numerator and denominator by } n^4.)$$

$$= \frac{1}{5} \quad \text{since } \lim_{n\to\infty} \frac{\ell n\, n}{n} = 0 .$$

Since $\frac{1}{5}$ is finite, the series converges.

Section 7

2.   We need to compare 2^n and n^2:

$$
\begin{array}{ccccccccc}
n & 1 & 2 & 3 & 4 & 5 & 6 & 7 & \ldots \\
2^n & 2 & 4 & 8 & 16 & 32 & 64 & 128 & \ldots \\
n^2 & 1 & 4 & 9 & 16 & 25 & 36 & 49 & \ldots
\end{array}
$$

For $n > 4$, we see that $2^n > n^2$. We could also see this by comparing $\ln 2^n = n \ln 2$ and $\ln n^2 = 2 \ln n$; since $n > \ln n$ for large n, $2^n > n^2$ for large n. Thus $|a_n| = 2^n/n^2$ does not tend to zero, so the series diverges by the preliminary test.

4.   This is an alternating series. However, it is probably easier to use the ratio test than to use the alternating series test. (See Problem 6.23.) We find

$$
\rho_n = \left| \frac{(-3)^{n+1}}{(n+1)!} \div \frac{(-3)^n}{n!} \right| = \frac{3}{n+1}
$$

$$
\rho = \lim_{n \to \infty} \rho_n = 0.
$$

Thus the series is absolutely convergent and so convergent (see Problem 9 below).

To use the alternating series test, we must see that, for large n, the terms $\frac{3^n}{n!}$ decrease steadily to zero. Observe that $\rho_n = \frac{3}{n+1}$ (which we found above) is what we multiply $|a_n|$ by to get $|a_{n+1}|$. Thus $|a_n|$ decreases steadily to zero and the alternating series converges.

8. For large n, $|a_n|$ is approximately $\frac{\sqrt{10n}}{n} = \sqrt{\frac{10}{n}}$; thus $|a_n|$ decreases steadily to zero. Thus by the alternating series test, the series converges.

9. We have discussed a number of tests for convergence of series with positive terms; we can use these to test series for absolute convergence. If the series of absolute values of the terms converges, we would like to know that the series still converges if we put back the minus signs; let us prove this. Suppose $\sum |a_n|$ is convergent. Let $b_n = a_n + |a_n|$; then $b_n \geq 0$ and $b_n \leq 2|a_n|$, so $\sum b_n \leq 2 \sum |a_n|$ is a convergent series. Let A_n, B_n, and C_n be the partial sums [see text equations (4.5) and (4.6)] of the a_n, b_n, and $|a_n|$ series. Then since $a_n = b_n - |a_n|$, $A_n = B_n - C_n$, and $\lim A_n = \lim B_n - \lim C_n$ which is finite because the b_n and $|a_n|$ series converge. Thus $\sum a_n$ converges.

10. (b) Both $1/n$ and $1/\sqrt{n}$ tend to zero as $n \to \infty$, and the series is alternating. However, the alternating series test also requires that $|a_{n+1}| < |a_n|$. This requirement is not satisfied, as we can easily see by writing the terms as decimals:

$$0.707 - 0.5 + 0.577 - 0.333 + 0.5 - 0.25 + 0.477 \ldots,$$

and in general $1/\sqrt{n+1} > 1/n$ for $n > 1$.

Section 9

Some general hints about testing series for convergence.

(A) Always take a minute or two on the preliminary test. However, if you cannot find the limit of the nth term easily and quickly, go on to another test.

(B) Suggestions about the kind of problems that are easily done using the various tests:

Integral test: Use this if you can integrate the nth term.

Ratio test: Problems containing factorials, and nth powers (2^n, 3^n, etc.).

Special comparison test: Problems with polynomials in numerator and denominator.

3.   Since $\ln 3 > 1$, this is a p-series with $p > 1$ (see Problem 6.15). Thus it is convergent.

10. Since $\lim_{n \to \infty} |n/(n-1)| = 1 \neq 0$, the series diverges by the preliminary test.

13. We evaluate $\int^{\infty} \dfrac{n \, dn}{(n^2+4)^{3/2}}$.   Let $u = n^2 + 4$; then $du = 2n \, dn$, and $u = \infty$ when $n = \infty$.   Thus we want

$$\frac{1}{2} \int^{\infty} \frac{du}{u^{3/2}} = \frac{1}{2} \int^{\infty} u^{-3/2} du = -u^{-1/2} \Big|^{\infty} = 0.$$

Since the integral is finite, the series converges.

We could also use the special comparison test (text, Section 6D).   For large n, $n/(n^2+4)^{3/2}$ is nearly $n/(n^2)^{3/2} = n/n^3 = 1/n^2$. We find

13.   (continued)

$$\lim_{n\to\infty}\left[\frac{n}{(n^2+4)^{3/2}}\div\frac{1}{n^2}\right]=\lim_{n\to\infty}\frac{n^3}{(n^2+4)^{3/2}}=1.$$

Since $\sum\frac{1}{n^2}$ is a convergent series (p-series or integral test),

then $\sum\frac{n}{(n^2+4)^{3/2}}$ is also a convergent series.

16.   Since $\sum 1/n^2$ is convergent (integral test or p-series test),

so is $\sum 1/(n^2+7)$ by comparison $[1/(n^2+7)<1/n^2]$.   Then both

the series $\sum 2/(n^2+7)$ and $\sum(-1)^n/(n^2+7)$ are convergent (see

Fact 1, Section 9, and Problem 7.9).   By Fact 2, Section 9, the

sum of the two series is convergent.

20.   This is a sum term-by-term of two convergent series: the

alternating harmonic series $\frac{1}{2}-\frac{1}{3}+\frac{1}{4}\cdots$ (see Section 7), and

$\sum(-1)^n/n^2$ (see Problem 16).   By Fact 2, Section 9, the sum of

the two series converges.

Section 10

1. We use the ratio test to find the interval of convergence.
    Always be careful to put in the absolute value signs in ρ_n
    and ρ since x can be negative. We find

    $$\rho_n = |(-x)^{n+1} \div (-x)^n| = |x|;$$

    $$\rho = \lim_{n \to \infty} \rho_n = |x|. \quad \text{The series converges if } |x| < 1.$$

    Endpoint tests:

    If $x = 1$, the series is $1 - 1 + 1 - 1 + \dots$.

    If $x = -1$, the series is $1 + 1 + 1 + \dots$.

    These series both diverge (preliminary test), so the given
    series converges for $-1 < x < 1$.

3. $$\rho_n = \left| \frac{(-1)^{n+1} x^{n+1}}{(n+1)(n+2)} \div \frac{(-1)^n x^n}{n(n+1)} \right| = \frac{n(n+1)}{(n+1)(n+2)} |x|.$$

    Since $\lim_{n \to \infty} \frac{n}{n+2} = 1$, $\rho = |x|$; the series converges for $|x| < 1$. For
    $x = 1$, the series is $\sum_1^\infty \frac{(-1)^n}{n(n+1)}$; for $x = -1$, it is $\sum_1^\infty \frac{1}{n(n+1)}$.
    These are both convergent (that is, $\sum_1^\infty \frac{(-1)^n}{n(n+1)}$ is absolutely
    convergent) by comparison with $\sum_1^\infty \frac{1}{n^2}$. Thus the given series
    converges for $-1 \leqslant x \leqslant 1$.

5.
$$\rho_n = \left| \frac{x^{n+1}}{[(n+1)!]^2} \div \frac{x^n}{(n!)^2} \right| = \left| \frac{x^{n+1}}{x^n} \right| \left[\frac{n!}{(n+1)!} \right]^2 = \frac{|x|}{(n+1)^2} \quad ,$$

$$\rho = |x| \lim_{n \to \infty} \frac{1}{(n+1)^2} = 0 \qquad \text{for all } x.$$

Therefore the series converges for all values of x.

11.
$$\rho_n = \left| \frac{1}{n+1} \left(\frac{x}{5} \right)^{n+1} \div \frac{1}{n} \left(\frac{x}{5} \right)^n \right| = \frac{n}{n+1} \left| \frac{x}{5} \right| .$$

$$\rho = \left| \frac{x}{5} \right| < 1 \qquad \text{if} \qquad |x| < 5.$$

If $x = 5$, the series is $\sum_1^\infty \frac{1}{n}$, divergent.

If $x = -5$, the series is $\sum_1^\infty \frac{(-1)^n}{n}$, convergent.

Then the given series converges if $-5 \leqslant x < 5$.

17.
$$\rho_n = \left| \frac{(-1)^{n+1}(x+1)^{n+1}}{n+1} \div \frac{(-1)^n x^n}{n} \right| = |x+1| \frac{n}{n+1} \quad ,$$

$$\rho = |x+1| \lim_{n \to \infty} \frac{n}{n+1} = |x+1| .$$

The series converges if $|x+1| < 1$, that is if $-1 < x+1 < 1$ or $-2 < x < 0$. If $x = -2$, the series is $\sum 1/n$ (harmonic series, divergent) and if $x = 0$, the series is $\sum (-1)^n/n$ (alternating harmonic, convergent). Thus the series converges in the interval $-2 < x \leqslant 0$.

22.
$$\rho_n = \left| \frac{(n+1)!(-1)^{n+1}}{x^{n+1}} \div \frac{n!(-1)^n}{x^n} \right| = \left| \frac{n+1}{x} \right| .$$

$$\rho = \lim_{n \to \infty} \left| \frac{n+1}{x} \right| = \infty \qquad \text{for any } x.$$

Thus the series diverges for all x.

24.   Let $y = \sqrt{x^2 + 1}$, and consider $\sum\limits_{0}^{\infty} \dfrac{2^n y^n}{3^n + n^3}$.

$$\rho_n = \left| \frac{2^{n+1} y^{n+1}}{3^{n+1} + (n+1)^3} \div \frac{2^n y^n}{3^n + n^3} \right| = 2|y| \frac{3^n + n^3}{3^{n+1} + (n+1)^3} \; .$$

Now for large n, $3^n \gg n^3$ (for example, $3^{1000} \cong 5 \times 10^{47}$,

$1000^3 = 10^9$). Thus

$$\lim_{n \to \infty} \frac{3^n + n^3}{3^{n+1} + (n+1)^3} = \lim_{n \to \infty} \frac{3^n}{3^{n+1}} = \frac{1}{3} \; ,$$

so $\rho = \frac{2}{3}|y|$. The series converges for $\frac{2}{3}|y| < 1$, or $|y| < \frac{3}{2}$.

If $y = \pm\frac{3}{2}$, the series is $\sum\limits_{0}^{\infty} \dfrac{(\pm 3)^n}{3^n + n^3}$, divergent by the

preliminary test.

Now $y = \sqrt{x^2 + 1} < \frac{3}{2}$ or $x^2 + 1 < \frac{9}{4}$, $x^2 < \frac{5}{4}$.   The given series

converges if $|x| < \frac{1}{2}\sqrt{5}$.

Section 12

1.   Binomial series. We write

(1) $(1+x)^P = a_0 + a_1 x + a_2 x^2 + a_3 x^3 + \dots$

and try to find the values of the coefficients a_n . Let $x = 0$

in (1), to find $1 = a_0 + 0$, so $a_0 = 1$. Next differentiate (1)

repeatedly:

(2) $p(1+x)^{p-1} = a_1 + 2a_2 x + 3a_3 x^2 + \dots$,

           $p(p-1)(1+x)^{p-2} = 2a_2 + 3!a_3 x + \dots$,

           $p(p-1)(p-2)(1+x)^{p-3} = 3!a_3 + \dots$.

Letting $x = 0$ in each of these equations gives

1.   (continued)

$$p = a_1 + 0 \ , \qquad\qquad a_1 = p \ ,$$

$$p(p - 1) = 2a_2 \ , \qquad\qquad a_2 = \frac{p(p-1)}{2} \ ,$$

$$p(p - 1)(p - 2) = 3!a_3 \ , \qquad a_3 = \frac{p(p-1)(p-2)}{3!} \ , \quad \text{etc.}$$

Section 13

3.   We write the $\tan x$ series from Example 3 in text Section 13B
      and square it. Recall that the square of a polynomial is
      $(a + b + c + \ldots)^2 = a^2 + b^2 + c^2 + \ldots + 2ab + 2ac + 2bc + \ldots$.   Let
      us decide to stop with the x^6 term in the series for $\tan^2 x$.
      Then we tabulate the result as follows:

$$\tan^2 x = \left(x + \frac{x^3}{3} + \frac{2x^5}{15} \cdots \right)^2$$

$$= x^2 + 2x \cdot \frac{x^3}{3} + \frac{x^6}{9} \cdots$$

$$+ 2x \cdot \frac{2x^5}{15} \cdots$$

$$\overline{}$$

$$= x^2 + 2x^4/3 + 17x^6/45 \cdots \ .$$

6.   We write the series for $(1-x)^{-1}$, multiply it times the e^x series, and tabulate the results as far as the x^4 terms:

$$\frac{e^x}{1-x} = \left(1+x+x^2+x^3+x^4+\cdots\right)\left(1+x+\frac{x^2}{2}+\frac{x^3}{3!}+\frac{x^4}{4!}+\cdots\right)$$

$$= 1 + x + \frac{x^2}{2} + \frac{x^3}{6} + \frac{x^4}{24} + \cdots$$

$$x + x^2 + \frac{x^3}{2} + \frac{x^4}{6} + \cdots$$

$$x^2 + x^3 + \frac{x^4}{2} + \cdots$$

$$x^3 + x^4 + \cdots$$

$$x^4 + \cdots$$

$$\overline{1 + 2x + \frac{5x^2}{2} + \frac{8x^3}{3} + \frac{65x^4}{24} + \cdots}$$

Another method is to divide the e^x series by $(1-x)$:

$$1 + 2x + \frac{5}{2}x^2 + \frac{8}{3}x^3 + \frac{65}{24}x^4 + \cdots$$

$$1-x\overline{\smash{\big)}\ 1 + x + \frac{1}{2}x^2 + \frac{1}{3!}x^3 + \frac{1}{4!}x^4 + \cdots}$$

$$\underline{1 - x}$$

$$2x$$

$$\underline{2x - 2x^2}$$

$$\frac{5}{2}x^2$$

$$\underline{\frac{5}{2}x^2 - \frac{5}{2}x^3}$$

$$\frac{8}{3}x^3$$

$$\underline{\frac{8}{3}x^3 - \frac{8}{3}x^4}$$

$$\frac{65}{24}x^4 \quad \text{etc.}$$

11.   Using text equation (13.3) with x replaced by 2x, we write

$$\frac{2x}{e^{2x}-1} = \frac{2x}{\left(1+2x+\frac{(2x)^2}{2!}+\frac{(2x)^3}{3!}+\frac{(2x)^4}{4!}+\frac{(2x)^5}{5!}+\cdots\right)-1}$$

$$= \frac{2x}{2x\left(1+x+\frac{2x^2}{3}+\frac{x^3}{3}+\frac{2x^4}{15}\cdots\right)} = \left(1+x+\frac{2x^2}{3}+\frac{x^3}{3}+\frac{2x^4}{15}\cdots\right)^{-1}.$$

To simplify our work, we write this as $(1+u)^{-1}$, where

$$u = x+\frac{2x^2}{3}+\frac{x^3}{3}+\frac{2x^4}{15}+\cdots = x\left(1+\frac{2}{3}x+\frac{x^2}{3}+\frac{2x^3}{15}+\cdots\right).$$

In order to find $(1+u)^{-1}$, we will need the powers of u, so we compute these first, keeping terms as far as x^4.

$$u^2 = x^2\left(1+2\cdot\frac{2}{3}x+\frac{4}{9}x^2+2\cdot\frac{x^2}{3}+\cdots\right)$$

$$= x^2+\frac{4}{3}x^3+\frac{10}{9}x^4+\cdots \quad \text{(See Problem 3)};$$

$$u^3 = x^3\left(1+3\cdot\frac{2}{3}x+\cdots\right) = x^3+2x^4+\cdots \quad \left[\text{See text equation}\right.$$

$$(13.5): \quad \left(1+\frac{2}{3}x\cdots\right)^3 = \left(1+3\cdot\frac{2}{3}x\cdots\right)\Big];$$

$$u^4 = x^4+\cdots.$$

We write the binomial series for $(1+u)^{-1}$, substitute the powers of u, and tabulate the results:

$$(1+u)^{-1} = \left(1-u+u^2-u^3+u^4\cdots\right)$$

$$= 1-x-\frac{2x^2}{3}-\frac{x^3}{3}-\frac{2x^4}{15}\cdots$$

$$+ x^2+\frac{4}{3}x^3+\frac{10}{9}x^4\cdots$$

$$- x^3-2x^4\cdots$$

$$+ x^4\cdots$$

$$\overline{\quad 1-x+\frac{x^2}{3}+0-\frac{1}{45}x^4\cdots.\quad}$$

14.   Change x to $\sqrt{x}$ in text equation (13.1), and divide by $\sqrt{x}$.

$$\frac{\sin\sqrt{x}}{\sqrt{x}} = \frac{1}{\sqrt{x}}\left(\sqrt{x} - \frac{(\sqrt{x})^3}{3!} + \frac{(\sqrt{x})^5}{5!} \cdots\right)$$

$$= 1 - x/3! + x^2/5! \cdots .$$

16.   We write $\sin[\ln(1+x)] = \sin u$, where $u = \ln(1+x)$. Using text equation (13.4), we write

$$u = \ln(1+x) = x - \frac{x^2}{2} + \frac{x^3}{3} - \frac{x^4}{4} + \frac{x^5}{5} \cdots$$

$$= x\left(1 - \frac{x}{2} + \frac{x^2}{3} - \frac{x^3}{4} + \frac{x^4}{5} \cdots\right).$$

Then

$$u^3 = x^3\left(1 - \frac{x}{2} + \frac{x^2}{3} \cdots\right)^3.$$

To find the cube of the parenthesis, we use text equation (13.5). Then

$$u^3 = x^3\left[1 + 3\left(-\frac{x}{2} + \frac{x^2}{3} \cdots\right) + 3\left(-\frac{x}{2} \cdots\right)^2 \cdots\right]$$

$$= x^3 - 3x^4/2 + 7x^5/4 \cdots ;$$

$$u^5 = x^5(1 + \cdots).$$

We substitute the powers of u into the series for $\sin u$ and tabulate the results:

$$\sin u = u - u^3/3! + u^5/5! \quad \cdots$$

$$= x - \frac{x^2}{2} + \frac{x^3}{3} - \frac{x^4}{4} + \frac{x^5}{5} \quad \cdots$$

$$- \frac{x^3}{3!} + \frac{x^4}{4} - \frac{7}{24}x^5 \cdots$$

$$\frac{x^5}{5!}$$

$$\overline{\qquad\qquad\qquad\qquad\qquad}$$

$$= x - \frac{x^2}{2} + \frac{x^3}{6} + 0 - \frac{x^5}{12} = x - x^2/2 + x^3/6 - x^5/12 \cdots .$$

19. Expand $(1 - t^2)^{-1}$ using text equation (13.5) and integrate term by term.

$$\ell n \sqrt{\frac{1+x}{1-x}} = \int_0^x \frac{dt}{1-t^2} = \int_0^x (1 + t^2 + t^4 + t^6 \cdots) dt$$

$$= x + x^3/3 + x^5/5 + x^7/7 \cdots .$$

Another way to find this series is to write

$$\ell n \sqrt{\frac{1+x}{1-x}} = \frac{1}{2} \ell n \frac{1+x}{1-x} = \frac{1}{2} [\ell n (1+x) - \ell n (1-x)]$$

$$= \frac{1}{2} \left[\left(x - \frac{x^2}{2} + \frac{x^3}{3} - \frac{x^4}{4} + \frac{x^5}{5} \cdots \right) - \left(-x - \frac{x^2}{2} - \frac{x^3}{3} - \frac{x^4}{4} - \frac{x^5}{5} \cdots \right) \right]$$

$$= \frac{1}{2} (2x + 2x^3/3 + 2x^5/5 \cdots) = x + x^3/3 + x^5/5 \cdots .$$

28. Using text equation (13.3), we find

$$2 - e^{-x} = 2 - (1 - x + x^2/2! - x^3/3! + x^4/4! - x^5/5! \cdots)$$

$$= 1 + x(1 - x/2 + x^2/6 - x^3/4! + x^4/5! \cdots).$$

Then we write $\ell n(2 - e^{-x}) = \ell n(1 + u)$ where

$u = x(1 - x/2 + x^2/6 - x^3/4! + x^4/5! \cdots)$;

$u^2 = x^2 [1 + 2(-x/2 + x^2/6 - x^3/4! \cdots) + (-x/2 + x^2/6 \cdots)^2]$

$\quad = x^2 (1 - x + 7x^2/12 - x^3/4 \cdots)$;

$u^3 = x^3 [1 + 3(-x/2 + x^2/6 \cdots) + 3(-x/2 \cdots)^2 \cdots]$

$\quad = x^3 (1 - 3x/2 + 5x^2/4 \cdots)$;

$u^4 = x^4 (1 - 4x/2 \cdots)$;

$u^5 = x^5 (1 \cdots).$

We substitute the powers of u into the series for $\ell n(1 + u)$ and tabulate the results:

28.   (continued)

$$ln(1+u) = u - u^2/2 + u^3/3 - \quad u^4/4 \quad + \quad u^5/5 \;\cdots$$

$$= x - x^2/2 + x^3/6 - \quad x^4/24 \quad + \quad x^5/5! \;\cdots$$

$$- x^2/2 + x^3/2 - 7x^4/24 \quad + \quad x^5/8 \;\cdots$$

$$x^3/3 - \quad x^4/2 \quad + 5x^5/12 \cdots$$

$$- \quad x^4/4 \quad + \quad x^5/2 \;\cdots$$

$$x^5/5 \;\cdots$$

$$\overline{\quad x - \quad x^2 \quad + \quad x^3 \quad - 13x^4/12 \; + \; 5x^5/4 \;\cdots \;.}$$

29.   By text equation (13.1), we write

$$\frac{x}{\sin x} = \frac{x}{x - x^3/3! + x^5/5! - x^7/7! \;\cdots}$$

$$= \frac{1}{1 - x^2/3! + x^4/5! - x^6/7! \;\cdots} = \frac{1}{1-u} \;,$$

where $u = x^2/3! - x^4/5! + x^6/7! \;\cdots \;,$

$$u^2 = x^4/6^2 + 2(x^2/3!)(-x^4/5!) \;\cdots$$

$$= x^4/6^2 - x^6/(3\cdot5!) \;\cdots$$

$$u^3 = x^6/6^3 \;\cdots \;.$$

We substitute these powers of u into

$$(1-u)^{-1} = 1 + u + u^2 + u^3 + \cdots \qquad \text{to get}$$

$$\frac{x}{\sin x} = 1 + x^2/3! - \quad x^4/5! \quad + \quad x^6/7! \quad \cdots$$

$$x^4/6^2 \quad - \quad x^6/(3\cdot5!) \quad \cdots$$

$$x^6/6^3 \quad \cdots$$

$$\overline{\; = \; 1 + x^2/3! + 7x^4/(3\cdot5!) + 31x^6/(3\cdot7!) \;\cdots \;.}$$

37.   We first write $e^x = e^3 e^{x-3}$. Then we use text equation (13.3)

with x replaced by (x - 3) to get

$$e^{x-3} = 1 + (x - 3) + (x - 3)^2/2! + (x - 3)^3/3! \cdots .$$

Thus $e^x = e^3 e^{x-3} = e^3[1 + (x - 3) + (x - 3)^2/2! + (x - 3)^3/3! \cdots]$.

40. $\sqrt{x} = \sqrt{25 + (x - 25)} = 5\sqrt{1 + (x - 25)/25}$.

We use text equation (13.5) to find

$\sqrt{1 + u} = 1 + \frac{1}{2}u - \frac{1}{8}u^2 + \frac{1}{16}u^3 \cdots$, and set $u = (x - 25)/25$. Then

$\sqrt{x} = 5[1 + (x - 25)/(2 \cdot 25) - (x - 25)^2/(8 \cdot 25^2) + (x - 25)^3/(16 \cdot 25^3) \cdots]$

$= 5 + (x - 25)/10 - (x - 25)^2/1000 + (x - 25)^3/50,000 \cdots$.

Section 14

2.   Error $= |a_{N+1}x^{N+1} + a_{N+2}x^{N+2} + a_{N+3}x^{N+3} + \cdots|$

$\leq |a_{N+1}x^{N+1}| + |a_{N+2}x^{N+2}| + |a_{N+3}x^{N+3}| + \cdots$

$\leq |a_{N+1}x^{N+1}| + |a_{N+1}x^{N+2}| + |a_{N+1}x^{N+3}| + \cdots$

$\leq |a_{N+1}x^{N+1}|(1 + |x| + |x|^2 + \cdots) = |a_{N+1}x^{N+1}| \cdot \frac{1}{1 - |x|}$.

5.   By text equation (13.2)

$$\cos x = 1 - x^2/2! + x^4/4! - x^6/6! \cdots \qquad \text{or}$$

$$1 - \cos x = x^2/2 - x^4/4! + x^6/6! \cdots .$$

This is a convergent alternating series with terms of decreasing

absolute value. Thus, by Theorem (14.3), the error is less than

the first neglected term. Then $1 - \cos x = x^2/2$ with an error

$< x^4/4!$ which is $\leq (1/2)^4/4! = 0.0026 < 0.003$ for $|x| < 1/2$.

7.   Using text equation (13.5), we find

$$\frac{2}{\sqrt{4-x}} = \frac{1}{\sqrt{1-x/4}} = \left(1-\frac{x}{4}\right)^{-1/2} = 1 + \frac{1}{2}\left(\frac{x}{4}\right) + \frac{3}{8}\left(\frac{x}{4}\right)^2 \cdots \ .$$

Let $y = x/4$; if $0 < x < 1$, then $0 < y < 1/4$.   Then

$$\left(1-\frac{x}{4}\right)^{-1/2} = (1-y)^{-1/2} = 1 + \frac{1}{2}y + \frac{3}{8}y^2 \cdots \ .$$

By theorem (14.4), the error after 2 terms is less than $\frac{3}{8}y^2 \div (1 - |y|)$; for $0 < y < \frac{1}{4}$,   this is less than

$$\frac{3}{8}\left(\frac{1}{4}\right)^2 \div \left(1 - \frac{1}{4}\right) = \frac{3}{8}\left(\frac{1}{4}\right)^2 \cdot \frac{4}{3} = \frac{1}{32} \ . \quad \text{Thus } \frac{2}{\sqrt{4-x}} = 1 + \frac{1}{2}\left(\frac{x}{4}\right) = 1 + x/8$$

with an error $< \frac{1}{32}$,   for $0 < x < 1$.

11.   If $f(x) = \sin x$, all derivatives of $f(x)$ are $\pm \sin x$ or $\pm \cos x$, so $|f^{(n+1)}(x)| \leq 1$ for any x.   Then, by text equation (14.2) with $a = 0$ (Maclaurin series),

$$|R_n(x)| = \left|\frac{x^{n+1}f^{n+1}(c)}{(n+1)!}\right| < \frac{|x|^{n+1}}{(n+1)!} \ .$$

Then $\lim |R_n(x)| = 0$ for any fixed x since, as soon as n is greater than $|x|$, all the factors $|x|/n$ are less than 1.   Thus the Maclaurin series for $\sin x$ converges to $\sin x$.

Section 15

3.   We use the answer to Problem 13.32, and text equation (13.1):

$$\ell n\left(x+\sqrt{1+x^2}\right) = x - x^3/6 + 3x^5/40 - 5x^7/112 \cdots$$

$$\sin x = x - x^3/3! + x^5/5! - x^7/7! \cdots$$

$$\ell n\left(x+\sqrt{1+x^2}\right) - \sin x = x^5/15 - 2x^7/45 \cdots$$

For $x = 0.005$, $x^5/15 = 2.08 \times 10^{-13}$; the error in this approximation is less than the absolute value of the next term, which is
$2(0.005)^7/45 = 3.5 \times 10^{-18}$.

7.   From Example 3, Section 13B of the text, we find

$$\tan x = x + x^3/3 \cdots .$$

Then $x^8 \tan^2 x = x^{10}$ + higher powers of x. If we differentiate
x^{10} ten times, we get $10 \cdot 9 \cdot 8 \cdots 2 \cdot 1 = 10!$; if we differentiate
any higher power of x, we are left with a power of x which
equals 0 when $x = 0$.   Thus

$$\left.\frac{d^{10}}{dx^{10}}\left(x^8 \tan^2 x\right)\right|_{x=0} = 10!$$

10.
$$\int_0^1 \frac{e^x - 1}{x}\, dx = \int_0^1 \frac{\left(1+x+\frac{x^2}{2!}+\frac{x^3}{3!}+\cdots\right)-1}{x}\, dx$$

$$= \int_0^1 \left(1+\frac{x}{2}+\frac{x^2}{3!}+\frac{x^3}{4!}+\cdots\right)dx = x+\frac{x^2}{4}+\frac{x^3}{3\cdot3!}+\frac{x^4}{4\cdot4!}+\cdots\Bigg|_0^1$$

$$= 1+\frac{1}{4}+\frac{1}{18}+\frac{1}{96}+\cdots \cong 1.32.$$

12.
$$\lim_{x\to 0}\frac{1-\cos x}{x^2} = \lim_{x\to 0}\frac{1-\left(1-\frac{x^2}{2!}+\frac{x^4}{4!}\cdots\right)}{x^2}$$

$$= \lim_{x\to 0}\left(\frac{1}{2!}-\frac{x^2}{4!}\cdots\right)= \frac{1}{2}\ .$$

12'. Since $\frac{1-\cos 0}{0^2}$ is of the form $\frac{0}{0}$, L'Hôpital's rule applies.

$$\lim_{x\to 0}\frac{1-\cos x}{x^2} = \lim_{x\to 0}\frac{\sin x}{2x} = \lim_{x\to 0}\frac{\cos x}{2} = \frac{1}{2}\ .$$

17.
$$\lim_{x\to 0}\frac{\ell n(1-x)}{x} = \lim_{x\to 0}\frac{-x-x^2/2\cdots}{x} = \left. -1-x/2\right|_{x=0} = -1.$$

17'. Since $\frac{\ell n(1-0)}{0}$ is of the form $\frac{0}{0}$, L'Hôpital's rule applies.

$$\lim_{x\to 0}\frac{\ell n(1-x)}{x} = \lim_{x\to 0}\frac{-1/(1-x)}{1} = -1.$$

20.
$$\lim_{x\to 0}\left(\csc^2 x - \frac{1}{x^2}\right) = \lim_{x\to 0}\left(\frac{1}{\sin^2 x}-\frac{1}{x^2}\right) = \lim_{x\to 0}\left(\frac{x^2-\sin^2 x}{x^2\sin^2 x}\right)$$

$$= \lim_{x\to 0}\frac{x^2-\left(x-\frac{x^3}{6}\cdots\right)^2}{x^4+\cdots} = \lim_{x\to 0}\frac{x^2-\left(x^2-2x\cdot\frac{x^3}{6}\cdots\right)}{x^4+\cdots}$$

$$= \lim_{x\to 0}\frac{x^4/3}{x^4} = \frac{1}{3}\ .$$

Although this problem can be done by L'Hôpital's rule it is much easier using series.

23.   (b) By L'Hôpital's rule

$$\lim_{x \to \pi} \frac{x \sin x}{x - \pi} = \lim_{x \to \pi} \frac{x \cos x + \sin x}{1} = -\pi.$$

24.   (c) Since $\ell n(1-1) = \ell n\, 0$ is infinite, this is an indeterminate

form of the $\frac{\infty}{\infty}$ type. By L'Hôpital's rule,

$$\lim_{x \to 1} \frac{\ell n(1-x)^2 + x^2}{\ell n(1-x^2) + e^x} = \lim_{x \to 1} \frac{-2/(1-x) + 2x}{-2x/(1-x^2) + e^x}$$

$$= \lim_{x \to 1} \frac{-2 + 2x(1-x)}{-2x/(1+x) + (1-x)e^x} = 2.$$

24.   (e) Write $x^n e^{-x}$ as $\frac{x^n}{e^x}$ which is of the form $\frac{\infty}{\infty}$. Then

$$\lim_{x \to \infty} \frac{x^n}{e^x} = \lim_{x \to \infty} \frac{nx^{n-1}}{e^x} = \lim_{x \to \infty} \frac{n(n-1)x^{n-2}}{e^x} \, ,$$

etc. until finally we reach

$$\lim_{x \to \infty} \frac{n(n-1)(n-2)\cdots 2 \cdot 1}{e^x} = 0.$$

26.   If $f(t) = e^{-1/t^2}$, then

$$f'(t) = 2t^{-3}e^{-1/t^2}$$

$$f''(t) = -6t^{-4}e^{-1/t^2} + 4t^{-6}e^{-1/t^2} \text{ and so on.}$$

We substitute $t = 1/x$, and let $t \to 0$; then $x \to \infty$. Thus

$$\lim_{t \to 0} f(t) = \lim_{x \to \infty} e^{-x^2} = 0$$

$$\lim_{t \to 0} f'(t) = \lim_{x \to \infty} 2x^3 e^{-x^2} = 0 \text{ as in Problem 24e. Similarly}$$

26.   (continued)

f''(0) = 0, and so on for all derivatives. Thus the formal
Maclaurin series is $0 + 0 + 0 + \cdots$ which does not converge to
$f(t) = e^{-1/t^2}$.

You may ask about the series

$$e^{-1/t^2} = 1 - \frac{1}{t^2} + \frac{1}{2t^4} - \frac{1}{3!t^6} + \cdots \quad .$$

This series is convergent for $|t| > 0$ and converges to $f(t)$,
but it is not a power series in t; a power series or Maclaurin
series in t means a series of positive powers of t. A series
which includes negative powers is called a Laurent series (see
text, Chapter 14, Section 4).

27.   (b) V = 500   (Note carefully: V = number of million volts).

$$1 - \frac{v}{c} = 1 - \sqrt{1 - \frac{1}{4(500)^2}} = 1 - \left(1 - \frac{1}{10^6}\right)^{1/2}$$

$$= 1 - \left(1 - \frac{1}{2}\cdot 10^{-6}\right) = 5 \times 10^{-7}.$$

Then v = 0.9999995c.

30.   (b) $T = \dfrac{F}{2 \sin \theta}$

$$= \frac{F}{2\left(\theta - \theta^3/3! + \theta^5/5! \cdots\right)} = \frac{F}{2\theta}\,\frac{1}{1 - \theta^2/3! + \theta^4/5! \cdots}$$

$$= \frac{F}{2\theta}\left(1 + \theta^2/6 + 7\theta^4/360 \cdots\right). \quad \text{(See Problem 13.29.)}$$

Note: For a given F, T can be made as large as we like by
making x and θ very small.

31.   (a) Area $= \pi \sum_{n=2}^{\infty} \dfrac{1}{n^2 (\ell n\ n)^2}$;   since $\dfrac{1}{n^2 (\ell n\ n)^2} < \dfrac{1}{n^2}$ and $\sum \dfrac{1}{n^2}$ con-

verges, the area is finite.

(b) Total length $= 2\pi \sum_{n=2}^{\infty} \dfrac{1}{n\ \ell n\ n}$;   this series diverges (use

integral test), so the total length is infinite.

(c) We can't compare length and area. Suppose we decide to
cut off little strips of width 10^{-3} units; then we can compare
the area cut off with the total area. But when $n\ \ell n\ n > 1000$
($n > 190$), the width of the strip you want to cut off (10^{-3}
units) is greater than the radius $1/(n\ \ell n\ n)$ of the circular disk
from which you want to cut it. No matter how narrow a strip
you decide to cut off, after a while the disks will be so small
that the needed area is greater than the area of the disks.

Section 16

1.   Number the
books down
from the top.
Let each book
have length
2ℓ and weight
w.   Look at

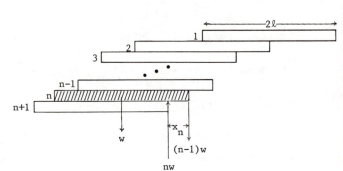

book n.   Three forces act on it (see diagram): its own weight w
at its center; the weight of the (n - 1) books above it acting
at its right-hand end; the upward force at the right-hand end
of book n + 1 which supports n books.   We take torques about
the right-hand end of book n.   Force w has lever arm ℓ; force
nw has lever arm x_n , and force (n - 1)w has lever arm zero.   The
two non-zero torques tend to rotate the book in opposite directions; if they balance, book n is in equilibrium.   Thus

$$w\ell = nwx_n ,$$

so $x_n = \frac{\ell}{n}$.   The series of setbacks x_n is then

$$x_1 + x_2 + x_3 + \cdots = \ell\left(1 + \frac{1}{2} + \frac{1}{3} + \cdots\right) = \ell\sum\frac{1}{n} .$$

This is the harmonic series; since it diverges, it is possible
by using enough terms to make the sum as large as desired.   It
is interesting to see how many books are necessary in order to
have a given overhang.   As you can see from the diagram in the
text, 5 books is enough so that the top book is not over the

1.   (continued)

bottom book at all, that is, so that the overhang is (slightly

over) one book.   For an

<table>
<tr><td>overhang of</td><td>the total number required is</td></tr>
<tr><td>2 books</td><td>32 books</td></tr>
<tr><td>3 books</td><td>228 books</td></tr>
<tr><td>5 books</td><td>over 12 thousand books</td></tr>
<tr><td>10 books</td><td>over 272 million books (a stack of height greater than the radius of the earth if the books are one inch thick)</td></tr>
<tr><td>50 books</td><td>1.5×10^{43} books</td></tr>
</table>

6.   For large n, $\dfrac{\sqrt{n+1}}{(n+1)^2 - 1}$ is nearly $\dfrac{\sqrt{n}}{n^2} = \dfrac{1}{n^{3/2}}$;   the series $\sum \dfrac{1}{n^{3/2}}$

is convergent by the integral test (or p-series, see Problem

6.15).   We use the special comparison test (Section 6, test D)

with $a_n = \dfrac{\sqrt{n+1}}{(n+1)^2 - 1}$ and $b_n = \dfrac{1}{n^{3/2}}$.   Then

$$\lim_{n \to \infty} \frac{\sqrt{n+1}}{(n+1)^2 - 1} \div \frac{1}{n^{3/2}} = \lim_{n \to \infty} \frac{n^{3/2}\sqrt{n+1}}{(n+1)^2 - 1} = \lim_{n \to \infty} \frac{\sqrt{1+\dfrac{1}{n}}}{(1+\dfrac{1}{n})^2 - \dfrac{1}{n^2}} = 1.$$

(In the next to the last step, divide numerator and denominator

by n^2.)   Thus the given series converges.

10.   Using the ratio test, we find

$$\rho_n = \left| \frac{[(n+1)!]^2 x^{n+1}}{[2(n+1)]!} \div \frac{(n!)^2 x^n}{(2n)!} \right| = |x| \left(\frac{(n+1)!}{n!} \right)^2 \frac{(2n)!}{(2n+2)!}$$

$$= |x| \frac{(n+1)^2}{(2n+2)(2n+1)} = |x| \frac{n+1}{2(2n+1)} \;;$$

$\rho = \lim\limits_{n \to \infty} \rho_n = |x|/4$, so the series converges for $|x| < 4$.

Endpoint tests: When $|x| = 4$,

$$\rho_n = \frac{4(n+1)^2}{2(n+1)(2n+1)} = \frac{2n+2}{2n+1} > 1.$$

Then the terms are increasing in absolute value, so the limit of the n th term is not zero.   Thus the series diverges by the preliminary test when $|x| = 4$.   Answer: Series converges for $|x| < 4$.

15.   We divide text equation (13.1) by x to get

$$\frac{\sin x}{x} = 1 - x^2/3! + x^4/5! - x^6/7! \cdots .$$

We call $(\sin x)/x = 1 + u$, and find the powers of u as far as terms in x^6 :

$u = -x^2/3! + x^4/5! - x^6/7! \cdots;$

$u^2 = (-x^2/3!)^2 + 2(-x^2/3!)(x^4/5!) \cdots = x^4/36 - 2x^6/(3!5!) \cdots$

$u^3 = -x^6/6^3 \cdots .$

Now we substitute these powers of x into the series for $\ell n(1+u)$ and tabulate the results:

15.   (continued)

$$\ln\left(\frac{\sin x}{x}\right) = \ln(1+u) = u - u^2/2 + u^3/3 \cdots$$

$$= -x^2/3! + \quad x^4/5! \quad - \quad x^6/7! \qquad \cdots$$

$$- x^4/(2 \cdot 36) + \quad x^6/(3!5!) \quad \cdots$$

$$- \quad x^6/(3 \cdot 6^3) \quad \cdots$$

$$\overline{\quad - x^2/6 \quad - \quad x^4/180 \quad - x^6/(3^4 \cdot 5 \cdot 7) \quad \cdots \quad .}$$

20.   We write $\sqrt[3]{x} = \sqrt[3]{8 + (x - 8)} = 2\sqrt[3]{1 + (x - 8)/8}$ and use text equation
       (13.5) with $p = 1/3$, and x replaced by $(x - 8)/8$.

$$\sqrt[3]{1 + (x - 8)/8} = [1 + (x - 8)/8]^{1/3}$$

$$= 1 + \frac{1}{3}\frac{(x - 8)}{8} + \frac{(1/3)(-2/3)}{2}\left(\frac{x - 8}{8}\right)^2 + \frac{(1/3)(-2/3)(-5/3)}{3!}\left(\frac{x - 8}{8}\right)^3 \cdots .$$

Then

$$\sqrt[3]{x} = 2\sqrt[3]{1 + (x-8)/8} = 2 + (x-8)/12 - (x-8)^2/(3^2 \cdot 2^5) + 5(x-8)^3/(3^4 \cdot 2^8) \cdots .$$

24.   By text equation (13.3), $e^x - 1 = x + x^2/2! + x^3/3! \cdots$.
       With $x = \ln 3$, we have

$$\ln 3 + (\ln 3)^2/2! + (\ln 3)^3/3! + \cdots = e^{\ln 3} - 1 = 3 - 1 = 2.$$

26.   We first replace $1 - \cos^2 x$ by $\sin^2 x$ and combine the two fractions
       to get

$$\frac{1}{x^2} - \frac{1}{\sin^2 x} = \frac{\sin^2 x - x^2}{x^2 \sin^2 x} \ .$$

By text equation (13.1), we find

26.   (continued)

$$\sin^2 x = (x - x^3/6 \cdots)^2 = x^2 - 2x^4/6 \cdots = x^2 - x^4/3 \cdots.$$

Then

$$\lim_{x \to 0}\left(\frac{1}{x^2} - \frac{1}{1 - \cos^2 x}\right) = \lim_{x \to 0} \frac{\sin^2 x - x^2}{x^2 \sin^2 x}$$

$$= \lim_{x \to 0} \frac{x^2 - x^4/3 - x^2}{x^2(x^2 \cdots)} = \lim_{x \to 0} \frac{-x^4/3 \cdots}{x^4 \cdots} = -\frac{1}{3}.$$

31.   By text equation (13.5),

$$x/(1+x) = x(1+x)^{-1} = x(1 - x + x^2 - x^3 + x^4 \cdots) = x - x^2 + x^3 - x^4 + x^5 \cdots$$

By Problem 13.28,

$$\ln(2 - e^{-x}) = x - x^2 + x^3 - 13x^4/12 + 5x^5/4 \cdots.$$

We subtract these two equations to get

$$x/(1+x) - \ln(2 - e^{-x}) = x^4/12 - x^5/4 \cdots.$$

For $x = 0.0012$, this gives 1.73×10^{-13} with an error of the order of 10^{-15}.

35.   By text equation (13.3),

$$(e^x - e^{-x})/2 = \frac{1}{2}(1 + x + x^2/2! + x^3/3! \cdots) - \frac{1}{2}(1 - x + x^2/2! - x^3/3! \cdots$$

$$= x + x^3/3! + x^5/5! \cdots.$$

For $x = 1$, we have

$$1 + 1/3! + 1/5! \cdots = (e - e^{-1})/2 = 1.18 \text{ by calculator.}$$

[Comment: This is $\sinh 1$; see text, page 69, equation (12.2).]

Chapter 2

Section 4

1. $1+i$ is $x+iy$ with $x = 1$, $y = 1$; we plot the
    point $(1,1)$ and find from the figure $r = \sqrt{2}$,
    $\theta = \pi/4$. Then the 5 ways to label the point
    are: $(1,1)$, $1+i$, $(\sqrt{2}, \pi/4)$, $\sqrt{2}\left(\cos\frac{\pi}{4} + i \sin\frac{\pi}{4}\right)$,
    $\sqrt{2}e^{i\pi/4}$. The complex conjugate of $1+i$ is
    $1-i$; this point may be labeled: $(1,-1)$, $1-i$,
    $(\sqrt{2}, -\pi/4)$ or $(\sqrt{2}, 7\pi/4)$, $\sqrt{2}\left(\cos\frac{\pi}{4} - i \sin\frac{\pi}{4}\right)$ or
    $\sqrt{2}\left(\cos\frac{7\pi}{4} + i \sin\frac{7\pi}{4}\right)$, $\sqrt{2}e^{-i\pi/4}$ or $\sqrt{2}e^{7i\pi/4}$.

7. $-1 = x + iy$ with $x = -1$, $y = 0$. We plot the point $(-1,0)$. The 5
    ways of labeling the point are: $(-1,0)$, -1,
    $(1,\pi)$, $\cos\pi + i \sin\pi$, $e^{i\pi}$. The complex
    conjugate of -1 is -1; any real number is
    its own complex conjugate. If we take the
    complex conjugate of $e^{i\pi}$, we get $e^{-i\pi}$;
    from the figure we see that these are equal since the point
    $r = 1$, $\theta = -\pi$ is the same as the point $r = 1$, $\theta = \pi$, so $e^{-i\pi} = -1 = e^{i\pi}$.
    Similarly $\cos\pi + i \sin\pi = \cos\pi - i \sin\pi = -1$ since $\sin\pi = 0$.

11. We plot the point with polar coordinates $r = 2$, $\theta = \pi/6$, and find

$$x = r \cos \theta = 2 \cos \frac{\pi}{6} = \sqrt{3},$$

$$y = r \sin \theta = 2 \sin \frac{\pi}{6} = 1.$$

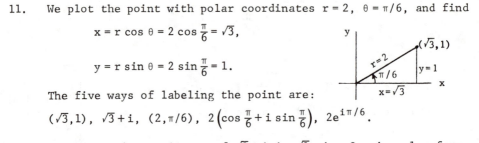

The five ways of labeling the point are:

$(\sqrt{3},1)$, $\sqrt{3}+i$, $(2,\pi/6)$, $2\left(\cos \frac{\pi}{6} + i \sin \frac{\pi}{6}\right)$, $2e^{i\pi/6}$.

The complex conjugate of $\sqrt{3}+i$ is $\sqrt{3}-i$. Or, in polar form, the complex conjugate of $2\left(\cos \frac{\pi}{6} + i \sin \frac{\pi}{6}\right)$ is

$2\left(\cos \frac{\pi}{6} - i \sin \frac{\pi}{6}\right) = 2\left[\cos\left(-\frac{\pi}{6}\right) + i \sin\left(-\frac{\pi}{6}\right)\right]$.

Thus taking the complex conjugate of a number in polar form does not change r but replaces θ by $-\theta$ (see sketch). We see this easily in the $re^{i\theta}$ form; the complex conjugate of $2e^{i\pi/6}$ is $2e^{-i\pi/6}$.

17. See Problem 1; the complex conjugate of $\sqrt{2}e^{-i\pi/4}$ is $\sqrt{2}e^{i\pi/4} = 1 + i$ as in Problem 1.

20. We write $7(\cos 110° - i \sin 110°) = 7[\cos(-110°) + i \sin(-110°)]$, and plot the point $r = 7$, $\theta = -110° = -1.92$ radians. Using a calcula-tor, we find $x = 7 \cos 110° = -2.39$,

$y = -7 \sin 110° = -6.58$. The point may be labeled: $(-2.39,-6.58)$, $-2.39 - 6.58i$, $(7,-110°)$, $7(\cos 110° - i \sin 110°) =$ $7(\cos 1.92 - i \sin 1.92)$, $e^{-1.92i}$.

Section 5

2. See text Example 2, Section 5A. We multiply numerator and denominator by the conjugate of $i - 1$ which is $-i - 1$:

$$\frac{1}{i-1} \cdot \frac{-i-1}{-i-1} = \frac{-i-1}{2} = -\frac{1}{2} - \frac{1}{2}i.$$

We plot the point $(-1/2, -1/2)$, and find

$$r = \sqrt{(-1/2)^2 + (-1/2)^2} = \sqrt{1/2}, \quad \theta = -3\pi/4 \ (\text{or } 5\pi/4).$$

The 5 ways of labeling the point are $(-1/2, -1/2)$, $-1/2 - i/2$,

$(1/\sqrt{2}, -3\pi/4)$, $\frac{1}{\sqrt{2}}\left(\cos\frac{3\pi}{4} - i\sin\frac{3\pi}{4}\right)$, $\frac{1}{\sqrt{2}}e^{-3\pi i/4}$.

   We could also do this problem by writing $i - 1$ in polar form. We sketch (or picture mentally), the point $(-1, 1)$ (see text Figure 9.6) and find $i - 1 = \sqrt{2}e^{3\pi i/4}$. Then

$$\frac{1}{i-1} = \frac{1}{\sqrt{2}e^{3\pi i/4}} = \frac{1}{\sqrt{2}}e^{-3\pi i/4} \text{ as above.}$$

6. Using polar coordinates, we write $1 + i = \sqrt{2}e^{i\pi/4}$ (visualize the sketch or see Problem 4.1) and $1 - i = \sqrt{2}e^{-i\pi/4}$. Then

$$\left(\frac{1+i}{1-i}\right)^2 = \left(\frac{\sqrt{2}e^{i\pi/4}}{\sqrt{2}e^{-i\pi/4}}\right)^2 = \left(e^{i\pi/2}\right)^2 = e^{i\pi} = -1.$$

The ways of labeling this point are given in Problem 4.7.

16.   We can write

$$\frac{1}{0.5(\cos 40° + i \sin 40°)} = \frac{1}{0.5e^{i(40°)}} = 2e^{-i(40°)}.$$

Then we can plot the point with polar

coordinates $(2,-40°)$ and find

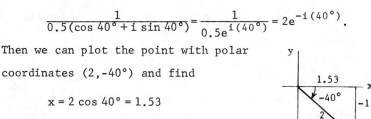

$$x = 2 \cos 40° = 1.53$$

$$y = -2 \sin 40° = -1.29.$$

The five ways of labeling the point are

$(1.53,-1.29)$, $1.53 - 1.29i$, $(2,-40°)$, $2(\cos 40° - i \sin 40°)$ or

$2(\cos 0.7 - i \sin 0.7)$, $2e^{-0.7i}$.

(Note that if we mean degrees we must say so; if an angle is not

labeled as degrees, then it is understood to be in radians.)

17.   By calculator, we find for the complex number $1.7 - 3.2i$,

$r = \sqrt{(1.7)^2 + (3.2)^2} = 3.62$, and $\theta = -\arctan \frac{3.2}{1.7} = -1.08$. Then

$1.7 - 3.2i = 3.62e^{-1.08i}$ so $(1.7 - 3.2i)^2 = (3.62)^2(e^{-1.08i})^2$

$= 13.1e^{-2.16i}$.   Thus for the complex number $(1.7 - 3.2i)^2$, we

have $r = 13.1$, and $\theta = -2.16$ or $-124°$. We plot this point and

find $x = 13.1 \cos 2.16 = -7.35$,

y $= -13.1 \sin 2.16 = -10.9$. The point

$(1.7 - 3.2i)^2$ may be labeled:

$(-7.35,-10.9)$, $-7.3 - 10.9i$,

$(13.1,-124°)$, $13.1(\cos 124° - i \sin 124°)$,

$13.1e^{-2.16i}$.

19. If $z = 2 - 3i$, $\dfrac{1}{z} = \dfrac{1}{2 - 3i} = \dfrac{1}{2 - 3i}\dfrac{2 + 3i}{2 + 3i} = \dfrac{2 + 3i}{13}$.

   If $z = x + iy$, $\dfrac{1}{z} = \dfrac{1}{x + iy} \cdot \dfrac{x - iy}{x - iy} = \dfrac{x - iy}{x^2 + y^2}$.

   Note that we could also write $\dfrac{1}{z} = \dfrac{1}{z}\dfrac{\bar{z}}{\bar{z}} = \dfrac{\bar{z}}{|z|^2}$.

   [See text equation (5.1)].

23. If $z = 2 - 3i$, then

$$\frac{1 + z}{1 - z} = \frac{3 - 3i}{-1 + 3i} = \frac{3 - 3i}{-1 + 3i} \cdot \frac{-1 - 3i}{-1 - 3i} = \frac{-3 - 9i + 3i + 9i^2}{1 - 9i^2}$$

$$= \frac{-12 - 6i}{10} = \frac{-6 - 3i}{5} .$$

   If $z = x + iy$ then

$$\frac{1 + z}{1 - z} = \frac{1 + x + iy}{1 - x - iy} = \frac{(1 + x) + iy}{(1 - x) - iy} \cdot \frac{(1 - x) + iy}{(1 - x) + iy}$$

$$= \frac{(1 + x)(1 - x) + iy(1 + x + 1 - x) - y^2}{(1 - x)^2 + y^2}$$

$$= \frac{1 - x^2 - y^2 + 2iy}{(1 - x)^2 + y^2} .$$

26. From text equation (5.1), we have $|z| = \sqrt{z\bar{z}}$; then

$$\left|\frac{2i - 1}{i - 2}\right| = \sqrt{\frac{2i - 1}{i - 2} \cdot \frac{-2i - 1}{-i - 2}} = \sqrt{\frac{-4i^2 + 1}{-i^2 + 4}} = 1$$

Note carefully that here we multiply the <u>numerator</u> by the complex conjugate of the <u>numerator</u> and the <u>denominator</u> by the complex conjugate of the <u>denominator</u> to find $\sqrt{z\bar{z}} = r$. Contrast this with Problem 2 where we multiplied numerator and denominator <u>both</u> by the complex conjugate of the denominator to obtain z in another form (with no i in the denominator). Both processes are useful and correct -- they are used for different purposes.

31. Using $|z| = \sqrt{z\bar{z}}$ from text equation (5.1), we find

$$\left|\frac{5-2i}{5+2i}\right| = \sqrt{\frac{5-2i}{5+2i} \cdot \frac{5+2i}{5-2i}} = 1.$$

Note how much easier this method is than the alternative of finding $\frac{5-2i}{5+2i}$ in the $x + iy$ form (by multiplying numerator and denominator by $5 - 2i$) and then finding $\sqrt{x^2 + y^2}$. The latter method gives the same result but with <u>much</u> more computation:

$$\frac{5-2i}{5+2i} \frac{5-2i}{5-2i} = \frac{25-20i-4}{25+4} = \frac{21}{29} - \frac{20}{29}i$$

$$\sqrt{\left(\frac{21}{29}\right)^2 + \left(\frac{20}{29}\right)^2} = \sqrt{\frac{441+400}{841}} = 1.$$

By contrast, you can probably do the problem in your head using equation (5.1) as above!

32. Using $|z| = \sqrt{z\bar{z}}$, we have

$$|(2-3i)^4| = \sqrt{(2-3i)^4(2+3i)^4} = \sqrt{[(2-3i)(2+3i)]^4}$$
$$= \sqrt{(2^2+3^2)^4} = \sqrt{13^4} = 13^2 = 169.$$

35. The equation $x + iy = 3i - 4$ means that the point (x,y) is the point $(-4,3)$, that is, $x = -4$, $y = 3$. Another way of saying this is that the real terms are equal ($x = -4$) and the imaginary terms are equal ($y = 3$).

42. The right-hand side of the equation is zero; zero means the complex number $0 + 0i$. Then the real part of the left-hand side of the equation must be zero and so must the imaginary part. Thus

42. (continued)

$$x + 2y + 3 = 0$$

$$3x - y - 1 = 0$$

To find x and y we solve these equations simultaneously.

Multiply the second equation by 2 and then add the equations:

$$x + 2y + 3 = 0$$

$$6x - 2y - 2 = 0$$

$$\overline{7x \qquad + 1 = 0}$$

$$x = -\frac{1}{7}$$

From the second equation, $y = 3x - 1 = 3\left(-\frac{1}{7}\right) - 1 = -\frac{10}{7}$.

43. Given $(x + iy)^2 = 2ix$, we multiply out the left side to get

$$x^2 - y^2 + 2ixy = 2ix = 0 + 2ix.$$

Equating real terms and equating imaginary terms gives

$$x^2 - y^2 = 0 \qquad \text{and} \qquad 2xy = 2x.$$

The second equation is true if $x = 0$ or if $y = 1$. If $x = 0$, then $y = 0$ by the first equation. If $y = 1$, then $x = \pm 1$ from the first equation. Thus there are three points satisfying the given equation: $(x,y) = (0,0)$ or $(1,1)$ or $(-1,1)$.

46. Clear fractions to get

$$x + iy = -i(x - iy) = -ix - y.$$

Equating real terms gives $x = -y$, and equating imaginary terms gives $y = -x$. Thus all points on the line $x + y = 0$ are solutions of the equation.

49.   The left side of the equation (an absolute value) is a real
      number ≥ 0. Therefore the right side of the equation $(x+iy)$
      is also a real number ≥ 0. Thus $y = 0$ and $x \geq 0$. Then the
      equation gives $|1-x| = x$. Now $|1-x| = 1-x$ or $x-1$, which-
      ever is ≥ 0. Since $x-1$ cannot equal x, we find only $1-x = x$,
      $x = 1/2$. The solution is $x = 1/2$, $y = 0$.

53.   With $z = x+iy$, $|z-1| = 1$ is $|(x-1)+iy| = 1$. We want to find
      the absolute value of the complex number with real part $(x-1)$
      and imaginary part y. The absolute value of a complex number is

$$\sqrt{(\text{real part})^2 + (\text{imaginary part})^2} = \sqrt{(x-1)^2 + y^2}.$$

      This is given equal to 1. Squaring, we get

$$(x-1)^2 + y^2 = 1.$$

      This is a circle of radius 1 with center at $(1,0)$. Also see
      Problem 65 below.

57.   We find $z^2 = (x+iy)^2 = x^2 + 2ixy + i^2y^2 = x^2 - y^2 + 2ixy$ since $i^2 = -1$.
      Then the real part of z^2 is $x^2 - y^2$. We have $\text{Re}(z^2) = x^2 - y^2 = 4$;
      this is the equation of a hyperbola.

62. $|z+1| + |z-1| = 8,$

                        $|x+1+iy| + |x-1+iy| = 8,$

                  $\sqrt{(x+1)^2 + y^2} + \sqrt{(x-1)^2 + y^2} = 8.$

      This says that (x,y) is a point for which the sum of its dis-
      tances from $(-1,0)$ and from $(1,0)$ is equal to 8. You may
      recognize this as the geometrical definition of an ellipse with

62.   (continued)

foci at (-1,0) and (1,0). If not, subtract the second square
root from both sides of the equation, and then square both sides:

$$(x + 1)^2 + y^2 = 64 - 16 \sqrt{(x - 1)^2 + y^2} + (x - 1)^2 + y^2 ,$$

$$x^2 + 2x + \cancel{1} + y^2 = 64 - 16 \sqrt{(x - 1)^2 + y^2} + \cancel{x^2} - 2x + \cancel{1} + y^2 ,$$

$$16 \sqrt{(x - 1)^2 + y^2} = 64 - 4x \quad \text{or} \quad 4 \sqrt{(x - 1)^2 + y^2} = 16 - x .$$

Again square both sides of the equation:

$$16 (x^2 - 2x + 1 + y^2) = (16 - x)^2 = 256 - 32x + x^2$$

$$15x^2 + 16y^2 = 240 \quad \text{or} \quad \frac{x^2}{16} + \frac{y^2}{15} = 1.$$

This is a standard form for the equation of an ellipse with semi-
major axis $a = \sqrt{16} = 4$, semiminor axis $b = \sqrt{15}$. The foci are on the
x axis at $x = \pm c = \pm \sqrt{a^2 - b^2} = \pm \sqrt{16 - 15} = \pm 1$, that is, at (-1,0) and
(1,0).

65.

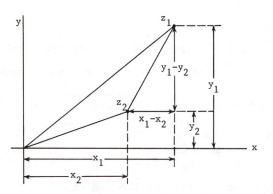

From the diagram we see that the distance between the points z_1 and z_2 is $\sqrt{(x_1 - x_2)^2 + (y_1 - y_2)^2}$. Now

$$|z_1 - z_2| = |(x_1 + iy_1) - (x_2 + iy_2)| = |(x_1 - x_2) + i(y_1 - y_2)| .$$

We want the absolute value of a complex number with real part = $x_1 - x_2$ and imaginary part = $y_1 - y_2$. This is

$$\sqrt{(\text{real part})^2 + (\text{imaginary part})^2} = \sqrt{(x_1 - x_2)^2 + (y_1 - y_2)^2}$$

Thus $|z_1 - z_2|$ is the distance between the two points z_1 and z_2.

Note the close analogy between complex numbers and two-dimensional vectors. If we draw a vector $\vec{V}_1$ from the origin to the point z_1 and a vector $\vec{V}_2$ from the origin to the point z_2 , then the vector from z_2 to z_1 is $\vec{V}_1 - \vec{V}_2$. The distance between the points is the length of this vector or $|\vec{V}_1 - \vec{V}_2|$.

Then, in Problem 53 above, $|z - 1|$ is the distance between the point z or (x,y) and the point $1 = 1 + 0 \cdot i$ or (1,0); the problem says this distance is 1. Thus z is on the circumference of a circle of radius 1 with center at (1,0) as we found before. Similarly, in Problem 60, $|z - 1 + i| = |z - (1 - i)|$ is the distance

65.   (continued)

between the point z and the point $1 - i$, that is, between the

points (x,y) and $(1,-1)$. In Problem 62, $|z+1|$ is the distance

between (x,y) and $(-1,0)$, and $|z - 1|$ is the distance between

(x,y) and $(1,0)$, as we found before.

66.   We first write z in the $x + iy$ form:

$$z = \frac{2t+i}{t-i} \frac{t+i}{t+i} = \frac{2t^2 - 1 + 3it}{t^2 + 1} = x + iy. \quad \text{Thus}$$

$$x = \frac{2t^2 - 1}{t^2 + 1} = 2 - \frac{3}{t^2 + 1} \; ; \quad y = \frac{3t}{t^2 + 1} . \quad \text{Then}$$

$$\frac{dx}{dt} = \frac{3 \cdot 2t}{(t^2 + 1)^2} = \frac{6t}{(t^2 + 1)^2} \; ; \quad \frac{dy}{dt} = \frac{3(t^2 + 1) - 3t \cdot 2t}{(t^2 + 1)^2} = \frac{3(1 - t^2)}{(t^2 + 1)^2} \; ;$$

$$v^2 = \left(\frac{dx}{dt}\right)^2 + \left(\frac{dy}{dt}\right)^2 = \frac{36t^2 + 9(1 - 2t^2 + t^4)}{(t^2 + 1)^4} = \frac{9(t^2 + 1)^2}{(t^2 + 1)^4} = \frac{9}{(t^2 + 1)^2} \; ;$$

$$v = \frac{3}{t^2 + 1} \quad \text{as in the text.}$$

$$\frac{d^2x}{dt^2} = \frac{6(t^2 + 1)^2 - 6t \cdot 2(t^2 + 1) \cdot 2t}{(t^2 + 1)^4} = \frac{6(1 - 3t^2)}{(t^2 + 1)^3} \; ;$$

$$\frac{d^2y}{dt^2} = \frac{-6t(t^2 + 1)^2 - 3(1 - t^2) \cdot 2 \cdot (t^2 + 1) \cdot 2t}{(t^2 + 1)^4} = \frac{6t(t^2 - 3)}{(t^2 + 1)^3} \; ;$$

$$a^2 = \left(\frac{d^2x}{dt^2}\right)^2 + \left(\frac{d^2y}{dt^2}\right)^2 = \frac{36(1-3t^2)^2 + 36t^2(t^2-3)^2}{(t^2 + 1)^6} = \frac{36(t^2+1)^3}{(t^2 + 1)^6} = \frac{36}{(t^2+1)^3} \; ;$$

$$a = \frac{6}{(t^2 + 1)^{3/2}} \quad \text{as in the text.}$$

Note that the text method using complex expressions is much

easier.

Section 6

1. We want to show that $\sum a_n$ and $\sum b_n$ both converge if $\sum \sqrt{a_n^2 + b_n^2}$ converges. (Remember that a_n and b_n are real.) By Chapter 1, Section 6, Test A, $\sum |a_n|$ and $\sum |b_n|$ converge because $|a_n| < \sqrt{a_n^2 + b_n^2}$ and $|b_n| < \sqrt{a_n^2 + b_n^2}$. Then $\sum a_n$ and $\sum b_n$ converge since, by Chapter 1, Problem 7.9, an absolutely convergent series converges.

2. We find $|1+i| = \sqrt{2} > 1$, so the absolute value of the nth term = $|1+i|^n = (\sqrt{2})^n$ does not tend to zero as $n \to \infty$. Thus the series diverges by the preliminary test. Alternatively, we could find, by the ratio test, $\rho = \sqrt{2} > 1$, so the series diverges.

5. $$\sum \left(\frac{1}{n^2} + \frac{i}{n} \right) = \sum \frac{1}{n^2} + i \sum \frac{1}{n} .$$

    This series will be convergent only if both $\sum \frac{1}{n^2}$ and $\sum \frac{1}{n}$ are convergent. But $\sum \frac{1}{n}$ is the harmonic series which is divergent. Therefore $\sum \left(\frac{1}{n^2} + \frac{i}{n} \right)$ is divergent.

12. By the ratio test,

    $$\rho = \lim_{n \to \infty} \left| \frac{(3+2i)^{n+1}}{(n+1)!} \div \frac{(3+2i)^n}{n!} \right| = \lim_{n \to \infty} |3+2i| \frac{n!}{(n+1)!}$$

    $$= \lim_{n \to \infty} \frac{\sqrt{13}}{n+1} = 0 < 1, \text{ so the series converges.}$$

Section 7

1. $e^z = 1 + z + \frac{z^2}{2!} + \frac{z^3}{3!} + \cdots + \frac{z^n}{n!} + \cdots$.   By ratio test

$$\rho_n = \left| \frac{z^{n+1}}{(n+1)!} \div \frac{z^n}{n!} \right| = |z| \frac{n!}{(n+1)!} = \frac{|z|}{n+1} \; ;$$

$$\rho = \lim_{n \to \infty} \rho_n = 0 \; .$$

Thus the e^z series converges for all z.

7. By ratio test,

$$\rho_n = \left| \frac{z^{2(n+1)}}{[2(n+1)]!} \div \frac{z^{2n}}{(2n)!} \right| = |z|^2 \frac{(2n)!}{(2n+2)!} = \frac{|z|^2}{(2n+2)(2n+1)} \; ;$$

$$\rho = \lim_{n \to \infty} \rho_n = 0 \; .$$

Thus the series converges for all z.   (Note that this series is

$1 - \frac{z^2}{2!} + \frac{z^4}{4!} - \cdots = \cos z$.   See text, Section 11.)

10. By ratio test

$$\rho_n = \left| \frac{(iz)^{n+1}}{(n+1)^2} \div \frac{(iz)^n}{n^2} \right| = \left| \frac{izn^2}{(n+1)^2} \right| \; ;$$

$$\rho = \lim_{n \to \infty} \left| \frac{izn^2}{(n+1)^2} \right| = |iz| = |z| \; .$$

The series converges for $|z| < 1$, that is, inside a circle of
radius 1 and center at the origin.

16. By ratio test

$$\rho_n = \left| \frac{2^{n+1}(z+i-3)^{2(n+1)}}{2^n(z+i-3)^{2n}} \right| = |2(z+i-3)^2| = \rho \; .$$

The series converges for $2|z+i-3|^2 < 1$, or $|z - (3-i)| < 1/\sqrt{2}$;
this is the interior of a circle of radius $1/\sqrt{2}$ with center at
$(3,-1)$.

Section 8

1. By multiplying the two series and tabulating the results, we
    find

$$e^{z_1} \cdot e^{z_2} = \left(1 + z_1 + \frac{z_1^2}{2!} + \frac{z_1^3}{3!} \cdots\right)\left(1 + z_2 + \frac{z_2^2}{2!} + \frac{z_2^3}{3!} \cdots\right)$$

$$= 1 + \quad z_2 \quad + \quad \frac{z_2^2}{2!} \quad + \quad \frac{z_2^3}{3!} \quad \cdots$$

$$+ \quad z_1 \quad + \quad z_1 z_2 \quad + \quad \frac{z_1 z_2^2}{2!} \quad \cdots$$

$$+ \quad \frac{z_1^2}{2!} \quad + \quad \frac{z_1^2 z_2}{2!} \quad \cdots$$

$$+ \quad \frac{z_1^3}{3!} \quad \cdots$$

$$\overline{1 + z_1 + z_2 + \frac{\left(z_1 + z_2\right)^2}{2!} + \frac{\left(z_1 + z_2\right)^3}{3!} \cdots = e^{z_1 + z_2}} \quad .$$

The sum of the third order terms may be clearer if we write them
as $\left(z_1^3 + 3z_1^2 z_2 + 3z_1 z_2^2 + z_2^3\right)\big/3!$. Similarly, the nth order terms are

$$\frac{z_1^n}{n!} + \frac{z_1^{n-1}}{(n-1)!} z_2 + \frac{z_1^{n-2}}{(n-2)!} \frac{z_2^2}{2!} + \cdots$$

$$= \left(z_1^n + n z_1^{n-1} z_2 + \frac{n(n-1)}{2!} z_1^{n-2} z_2^2 \cdots\right)\big/ n! = \frac{\left(z_1 + z_2\right)^n}{n!} \quad .$$

Section 9

In the following problems we shall sketch a diagram for each
point. However, with practice, you can often simply picture
the diagram in your mind and so do problems like these without
actually plotting the points on paper. Note that it is often
convenient to draw the diagram using a different r from the one
given and then multiply by the appropriate factor. For example,

in Problem 4.1 we found $\sqrt{2}e^{i\pi/4} = 1+i$. Then $e^{i\pi/4} = \dfrac{1+i}{\sqrt{2}}$,

$5e^{i\pi/4} = \dfrac{5(1+i)}{\sqrt{2}}$, etc. for any desired r.

3. Using the method just discussed above,
    we find $e^{3\pi i/2}$ and then multiply by 9
    to get

    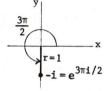

    $$9e^{3\pi i/2} = 9(-i) = -9i.$$

6. By text Figure 9.4, we see that $2^{2n\pi i} = 1$.
    We drew Figure 9.4 for positive n. However,
    the angle -2π has the same terminal side as
    $+2\pi$, and similarly any angle $\pm 2n\pi$ (n = integer)
    has the positive x axis as its terminal side.
    Thus $e^{\pm 2n\pi i} = 1$. Then

    $$e^{-2\pi i} - e^{-4\pi i} + e^{-6\pi i} = 1 - 1 + 1 = 1.$$

7. $3e^{2(1+i\pi)} = 3e^{2+2i\pi} = 3e^2 e^{2\pi i} = 3e^2 \cdot 1 = 3e^2$
    since $e^{2\pi i} = 1$ (Problem 6).

14. From text Figure 3.3, $1 + i\sqrt{3} = 2e^{i\pi/3}$. Then
    $$(1 + i\sqrt{3})^6 = (2e^{i\pi/3})^6 = 2^6 e^{2\pi i} = 64 \cdot 1 = 64.$$

17. From Problem 4.1, $1 + i = \sqrt{2}e^{i\pi/4}$. Then

    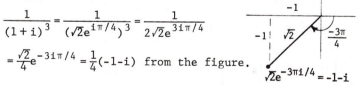

    $$\frac{1}{(1+i)^3} = \frac{1}{(\sqrt{2}e^{i\pi/4})^3} = \frac{1}{2\sqrt{2}e^{3i\pi/4}}$$

    $$= \frac{\sqrt{2}}{4}e^{-3i\pi/4} = \frac{1}{4}(-1-i) \text{ from the figure.}$$

20. $-1 + i = \sqrt{2}e^{3\pi i/4}$ so $\dfrac{\sqrt{2}}{i-1} = e^{-3\pi i/4}$.   Then

$$\left(\frac{\sqrt{2}}{i-1}\right)^{10} = \left(e^{-\frac{3\pi i}{4}}\right)^{10} = e^{-15\pi i/2} = e^{-8\pi i}e^{i\pi/2} = 1 \cdot i = i.$$

24. $1 - i\sqrt{3} = 2e^{-i\pi/3}$ from the figure.

$i - 1 = \sqrt{2}e^{3i\pi/4}$ from Problem 20.

Then

$$\frac{(1 - i\sqrt{3})^{21}}{(i-1)^{38}} = \frac{\left(2e^{-\frac{i\pi}{3}}\right)^{21}}{\left(\sqrt{2}e^{\frac{3i\pi}{4}}\right)^{38}} = \frac{2^{21}}{(\sqrt{2})^{38}}\frac{e^{-7i\pi}}{e^{3i\pi \cdot \frac{38}{4}}}$$

$$= \frac{2^{21}}{2^{19}}\frac{e^{-i\pi}}{e^{-\frac{3i\pi}{2}}} = 2^2 e^{\frac{i\pi}{2}} = 4i.$$

We have used $e^{-7i\pi} = e^{-6i\pi - i\pi} = e^{-i\pi}$ since $e^{\pm 2ni\pi} = 1$ (Problem 6)
and $3i\pi \cdot \dfrac{38}{4} = 3i\pi\left(\dfrac{40-2}{4}\right) = 30i\pi - \dfrac{3i\pi}{2}$, so

$$e^{3i\pi \cdot \frac{38}{4}} = e^{30i\pi - \frac{3i\pi}{2}} = e^{\frac{-3i\pi}{2}}.$$

27.   By text equation (5.1), $|z| = \sqrt{z\bar{z}}$. For real y, $\overline{e^{iy}} = e^{-iy}$, so
$|e^{iy}| = \sqrt{e^{iy}e^{-iy}} = \sqrt{e^0} = \sqrt{1} = 1$. Thus $|e^z| = |e^{x+iy}| = |e^x e^{iy}|$
$= e^x|e^{iy}| = e^x$. (Since e^x is real and positive, $|e^x| = e^x$.)

28. $|z_1 z_2| = \left|r_1 e^{i\theta_1} \cdot r_2 e^{i\theta_2}\right| = r_1 r_2 \left|e^{i(\theta_1 + \theta_2)}\right| = r_1 r_2$.

(Since r_1 and r_2 are real and positive, $|r_1 r_2| = r_1 r_2$.)

32. $|3e^{2+4i}| = |3e^2 e^{4i}| = 3e^2|e^{4i}| = 3e^2$.

(We have used $|e^{4i}| = 1$ by Problem 27; also $|3e^2| = 3e^2$ since $3e^2$
is real and positive.)

35.   By Problem 28, $|3e^{5i} \cdot 7e^{-2i}| = 3 \cdot 7 = 21$.

38.   From Problem 4.1, $1+i = \sqrt{2}e^{i\pi/4}$; then $\left|\dfrac{e^{i\pi}}{1+i}\right| = \left|\dfrac{e^{i\pi}}{\sqrt{2}e^{i\pi/4}}\right| = \dfrac{1}{\sqrt{2}}$

       since $|e^{i\theta}| = 1$.

Section 10

3. $1 = e^{2n\pi i}$; $\sqrt[4]{1} = e^{2n\pi i/4}$. Thus the four

      fourth roots of 1 have $r = 1$ and

      $\theta = 0$, $\dfrac{2\pi}{4}$, $\dfrac{4\pi}{4}$, $\dfrac{8\pi}{4} = 0$, $\dfrac{\pi}{2}$, π , $\dfrac{3\pi}{2}$.   Then the

      four roots are e^{0} , $e^{i\pi/2}$, $e^{i\pi}$, $e^{3\pi i/2}$ or

      1 , i , -1 , $-i$ as shown.

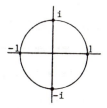

7. $16 = 16e^{2n\pi i}$; $\sqrt[8]{16} = \sqrt[8]{16}\,e^{2n\pi i/8} = \sqrt{2}e^{in\pi/4}$.

      The eighth roots of 16 have $r = \sqrt{2}$, and

      $\theta = 0$, $\dfrac{\pi}{4}$, $\dfrac{\pi}{2}$, etc., or $\theta = 0°$, $45°$, $90°,\ldots$

      We plot these numbers and read them in

      rectangular form from the figure:

      $\pm\sqrt{2}$, $\pm i\sqrt{2}$, $\pm 1 \pm i$.

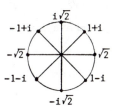

16.   Following the outline at the end of Section 10 of the text,

       we find for $\sqrt[6]{-1}$

       (a) $r = 1$

       (b) One angle of -1 is π or $180°$; one angle of

            $\sqrt[6]{-1}$ is $\dfrac{\pi}{6}$ or $30°$.

       (c) $\dfrac{2\pi}{6} = \dfrac{360°}{6} = 60°$.

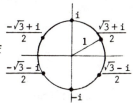

       Then the polar coordinates of the 6 sixth roots

       of -1 are $r = 1$ and $\theta = 30° + 60°n = 30°$, $90°$, $150°$, $210°$ $\cdots$ or

       $\theta = \dfrac{\pi}{6} + \dfrac{2n\pi}{6} = \dfrac{\pi}{6}$, $\dfrac{3\pi}{6}$ or $\dfrac{\pi}{2}$, $\dfrac{5\pi}{6}$, $\cdots$. We plot these numbers and

       read them in rectangular form from the figure: $\pm i$, $\dfrac{\pm\sqrt{3} \pm i}{2}$.

18. $i = e^{i(\frac{\pi}{2} + 2n\pi)}$ so $\sqrt{i} = e^{i(\frac{\pi}{4} + n\pi)} = \pm e^{i\frac{\pi}{4}}$ since $e^{in\pi} = (-1)^n = \pm 1$.

Then the two square roots of i are $\pm\dfrac{1+i}{\sqrt{2}}$.   (If needed, see

text Figure 5.1: $1+i = \sqrt{2}\, e^{i\frac{\pi}{4}}$.)

22.   Following the outline at the end of Section 10 of the text, we

find

(a) $r = \sqrt[3]{|2i - 2|} = \sqrt[3]{2\sqrt{2}} = \sqrt{2}$

(b) One angle of $-2 + 2i$ is $135°$ or $\dfrac{3\pi}{4}$ so one angle of $\sqrt[3]{2i - 2}$

is $45°$ or $\dfrac{\pi}{4}$.

(c) $2\pi/3 = \dfrac{360°}{3} = 120°$.

Then the polar coordinates of the three cube roots of $2i - 2$ are

$r = \sqrt{2}$, $\theta = 45°$, $165°$, $285°$.   We plot these

numbers and use a calculator to find them in

rectangular form as in text Example 4,

Section 10.

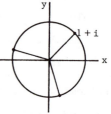

$\sqrt{2}\ \underline{/45°}\ = 1 + i$

$\sqrt{2}\ \underline{/165°}\ = 1.366 + 0.366i$

$\sqrt{2}\ \underline{/285°}\ = 0.366 - 1.366i$

25.   Following the outline at the end of Section 10 of the text,

we find

(a) $\sqrt[5]{|-1 - i|} = (\sqrt{2})^{1/5} = \sqrt[10]{2}$.

(b) One angle of $-1 - i$ is $225°$

(see figure); $\dfrac{225°}{5} = 45°$.

(c) $\dfrac{360°}{5} = 72°$.

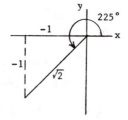

25.   (continued)

Then the polar coordinates of the 5 fifth roots of $(-1-i)$ are
$r = \sqrt[10]{2}$, $\theta = 45° + 72°n$.

We plot these and find their rectangular coordinates using a
calculator and the notation of example 4, Section 10 of the
text.

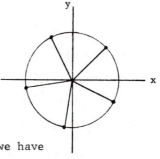

$\sqrt[10]{2} \ \underline{/45°} = 0.758(1+i)$

$\sqrt[10]{2} \ \underline{/117°} = -0.487 + 0.955i$

$\sqrt[10]{2} \ \underline{/189°} = -1.059 - 0.168i$

$\sqrt[10]{2} \ \underline{/261°} = -0.168 - 1.059i$

$\sqrt[10]{2} \ \underline{/333°} = 0.955 - 0.487i$

27.   From text equation (10.2) with $n = 2$, we have

$(e^{i\theta})^2 = (\cos \theta + i \sin \theta)^2 = \cos^2\theta + 2i \sin \theta \cos \theta - \sin^2\theta$,

$(e^{i\theta})^2 = e^{2i\theta} = \cos 2\theta + i \sin 2\theta$.

We equate the real parts and equate the imaginary parts of these
two expressions for $(e^{i\theta})^2$ to get:

$\cos 2\theta = \cos^2\theta - \sin^2\theta$,

$\sin 2\theta = 2 \sin \theta \cos \theta$.

31.   Method 1:   First consider the n nth roots of 1, namely

$1, \ e^{2\pi i/n}, \ e^{4\pi i/n}, \ e^{6\pi i/n}, \ \ldots$ or

$1, \ e^{2\pi i/n}, \ (e^{2\pi i/n})^2, \ (e^{2\pi i/n})^3, \ \ldots$.

The sum of these n roots is

$S = 1 + u + u^2 + u^3 + \cdots + u^{n-1}$ where $u = e^{2\pi i/n}$.

Multiply by u and remember that $u^n = 1$ to get

$uS = u + u^2 + u^3 + \cdots + 1 = S$.

Thus

31.   (continued)

$$uS = S, \text{ or } (u - 1)S = 0.$$

Since $u \neq 1$, we have $S = 0$.

Now the sum of the n nth roots of any complex number $z = re^{i\theta}$ is

$$S' = r^{1/n}e^{i\theta/n} + r^{1/n}e^{i(\theta+2\pi)/n} + r^{1/n}e^{i(\theta+4\pi)/n} + \cdots$$

$$= r^{1/n}e^{i\theta/n}\left[1 + e^{2\pi i/n} + e^{4\pi i/n} + \cdots e^{(n-1)2\pi i/n}\right].$$

The bracket is the sum S above; therefore $S' = 0$.

Method 2:   The n nth roots of the complex number $\underline{a}$ satisfy the equation $z^n = a$, or $z^n - a = 0$.   Call the roots z_1, z_2, z_3, $\ldots$, z_n ;   then the equation in factored form is

$$(z - z_1)(z - z_2)(z - z_3) \cdots (z - z_n) = 0.$$

If we multiply this out, we get

$$z^n - (z_1 + z_2 + z_3 + \cdots + z_n)z^{n-1} + \cdots + (-1)^n z_1 z_2 \cdots z_n = 0.$$

Since this is the equation $z^n - a = 0$, the coefficient of z^{n-1} is zero.   Note that this coefficient is just minus the sum of the roots; thus the sum of the roots is zero.

You may recall that in the quadratic equation $x^2 + \frac{b}{a}x + \frac{c}{a} = 0$, the sum of the roots is $-\frac{b}{a}$.   Now $\frac{b}{a}$ is the coefficient of x (or of x^{n-1} when $n = 2$).   Thus the sum of the roots of $x^2 - 4 = 0$ is zero since there is no x term in the equation.   Similarly for an equation of any degree n, if the term in x^{n-1} is missing, the sum of the roots is zero.

We have, incidentally, proved that the product of the n nth roots of $\underline{a}$ is always $\pm a$.   You may recall that the product of

31.   (continued)

the roots of the quadratic equation $x^2 + \frac{b}{a}x + \frac{c}{a} = 0$ is $\frac{c}{a}$.

Similarly, as we have shown above, $(-1)^n z_1 z_2 \ldots z_n = -a$; that is,

the product of the n roots of the equation $z^n - a = 0$ is $(-1)^{n-1}a$.

Section 11

3.
$$e^{-\frac{i\pi}{4} + \ell n\ 3} = e^{\ell n\ 3}e^{-\frac{i\pi}{4}} .$$

Now $e^{\ell n\ 3} = 3$ [see text equations (13.1) and (13.2)]. $e^{-\frac{i\pi}{4}} = \frac{1 - i}{\sqrt{2}}$

by text Figure 9.5. Thus

$$e^{-\frac{i\pi}{4} + \ell n\ 3} = 3 \cdot \frac{1 - i}{\sqrt{2}} = \frac{3}{\sqrt{2}}(1 - i) .$$

8.   By text equation (11.4)

$$\cos(\pi - 2i\ \ell n\ 3) = \frac{e^{i(\pi - 2i\ \ell n\ 3)} + e^{-i(\pi - 2i\ \ell n\ 3)}}{2}$$

$$= \frac{1}{2}\left(e^{i\pi}e^{2\ \ell n\ 3} + e^{-i\pi}e^{-2\ \ell n\ 3}\right) .$$

Now $e^{i\pi} = e^{-i\pi} = -1$ (see text Figure 9.2). Also $e^{2\ \ell n\ 3} = e^{\ell n\ 9} = 9$

[see text equations (13.1) and (13.2)]. Then $e^{-2\ \ell n\ 3} = \frac{1}{e^{2\ \ell n\ 3}} = \frac{1}{9}$.

Thus

$$\cos(\pi - 2i\ \ell n\ 3) = \frac{1}{2}\left(-9 - \frac{1}{9}\right) = -\frac{41}{9} .$$

11. $\int_{-\pi}^{\pi} \cos 2x \cos 3x\ dx = \frac{1}{4}\int_{-\pi}^{\pi}\left(e^{2ix} + e^{-2ix}\right)\left(e^{3ix} + e^{-3ix}\right)dx$

$$= \frac{1}{4}\int_{-\pi}^{\pi}\left(e^{5ix} + e^{ix} + e^{-ix} + e^{-5ix}\right)dx .$$

11.   (continued)

All of these integrals are of the form $\int_{-\pi}^{\pi} e^{inx} dx$ where n is a positive or negative integer (not zero).

$$\int_{-\pi}^{\pi} e^{inx} dx = \frac{e^{inx}}{in} \Big|_{-\pi}^{\pi} = \frac{e^{in\pi} - e^{-in\pi}}{in} \ .$$

You should satisfy yourself that the angle $n\pi$ and the angle $-n\pi$ have the same terminal side (see figures). Thus $e^{in\pi} = e^{-in\pi}$ and the integral is zero for any integral n. Then each of the four integrals (with n = 5, 1, -1, and -5) is zero and the original integral is zero.

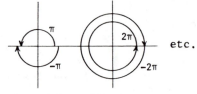

etc.

12.   The easiest way to evaluate $\int_{-\pi}^{\pi} \cos^2 3x \, dx$ is to first satisfy yourself (from the graphs) that the areas under a $\sin^2 \theta$ graph and a $\cos^2 \theta$ graph are the same over one or more quarter periods, measured from the origin.

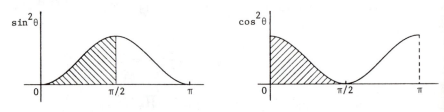

The shaded areas are equal; also all the quarter-period areas (0 to $\frac{\pi}{2}$, $\frac{\pi}{2}$ to π, etc.) are the same. For example,

12.   (continued)

$$\int_0^\pi \sin^2\theta\, d\theta = \int_0^\pi \cos^2\theta\, d\theta \, .$$

But $$\int_0^\pi (\sin^2\theta + \cos^2\theta)\, d\theta = \int_0^\pi d\theta = \pi, \ \text{so}$$

$$\int_0^\pi \sin^2\theta\, d\theta = \int_0^\pi \cos^2\theta\, d\theta = \pi/2 \, .$$

Now cos 3x has 3 complete periods between $-\pi$ and π.   Thus

$$\int_{-\pi}^\pi \cos^2 3x\, dx = \int_{-\pi}^\pi \sin^2 3x\, dx = \frac{1}{2}\int_{-\pi}^\pi (\sin^2 3x + \cos^2 3x)\, dx = \frac{1}{2}\int_{-\pi}^\pi dx = \pi.$$

You might wonder why the method used in Problem 11 could not be used here to prove (incorrectly) that the $\cos^2$ integral is zero.

$$\cos^2 3x = \left(\frac{e^{3ix} + e^{-3ix}}{2}\right)^2 = \frac{1}{4}\left(e^{6ix} + 2e^{3ix}e^{-3ix} + e^{-6ix}\right).$$

The first and last terms do integrate to zero.   But the middle term is $\frac{2}{4}e^0 = \frac{1}{2}$.   Recall that in our proof of $\int_{-\pi}^\pi e^{inx}dx = 0$ we said $n \neq 0$.   Here we have a term $\frac{1}{2}e^0$ for which $n = 0$ and our proof does not apply.   In fact, $\int_{-\pi}^\pi \frac{1}{2}\, dx = \pi$, the same result as in the area method.

Section 12

4.   Using text equations (12.2), we find

$$\cosh z = \frac{e^z + e^{-z}}{2} = \frac{e^{x+iy} + e^{-(x+iy)}}{2}$$

$$= \frac{1}{2}\left[e^x(\cos y + i \sin y) + e^{-x}(\cos y - i \sin y)\right]$$

$$= \frac{1}{2}(e^x + e^{-x})\cos y + \frac{1}{2}i(e^x - e^{-x})\sin y$$

$$= \cosh x \cos y + i \sinh x \sin y \, .$$

7.   Using text equations (12.2), we find

$$\sinh 2z = \tfrac{1}{2}(e^{2z} - e^{-2z}) = \tfrac{1}{2}(e^z - e^{-z})(e^z + e^{-z})$$

$$= 2\left(\frac{e^z - e^{-z}}{2}\right)\left(\frac{e^z + e^{-z}}{2}\right) = 2 \sinh z \cosh z \; .$$

9.   Using text equations (11.4), we find

$$\frac{d}{dz}\cos z = \frac{d}{dz}\frac{e^{iz} + e^{-iz}}{2} = \frac{ie^{iz} - ie^{-iz}}{2} = -\frac{e^{iz} - e^{-iz}}{2i} = -\sin z \; .$$

17. Using text equations (12.3), (12.2), and (11.4), we find

$$\tanh iz = \frac{\sinh iz}{\cosh iz} = \frac{e^{iz} - e^{-iz}}{2} \div \frac{e^{iz} + e^{-iz}}{2}$$

$$= i \sin z \div \cos z = i \tan z \; .$$

18. We use the results of Problems 1 and 2 (see text, page 70) to
     get:

$$\tan z = \frac{\sin z}{\cos z} = \frac{\sin x \cosh y + i \cos x \sinh y}{\cos x \cosh y - i \sin x \sinh y} \; .$$

Then divide numerator and denominator by $\cos x \cosh y$:

$$\tan z = \frac{\tan x + i \tanh y}{1 - i \tan x \tanh y} \; .$$

20. $e^{2z} = (e^z)^2 = (\cosh z + \sinh z)^2 = \cosh^2 z + 2 \cosh z \sinh z + \sinh^2 z;$

$e^{-2z} = (\cosh z - \sinh z)^2 = \cosh^2 z - 2 \cosh z \sinh z + \sinh^2 z.$

Add and subtract these two equations to find:

$$\cosh 2z = \frac{e^{2z} + e^{-2z}}{2} = \cosh^2 z + \sinh^2 z.$$

$$\sinh 2z = \frac{e^{2z} - e^{-2z}}{2} = 2 \sinh z \cosh z.$$

25.   Using the definition [text equation (11.4)] of the sine of a
complex number, we find

$$\sin(x - iy) = \frac{e^{i(x-iy)} - e^{-i(x-iy)}}{2i}$$

$$= \frac{e^{ix}e^{y} - e^{-ix}e^{-y}}{2i}$$

$$= e^{y}\frac{\cos x + i\sin x}{2i} - e^{-y}\frac{\cos x - i\sin x}{2i}$$

$$= (\cos x)\frac{e^{y} - e^{-y}}{2i} + (\sin x)\frac{e^{y} + e^{-y}}{2}$$

$$= \sin x \cosh y - i\cos x \sinh y$$

Another method is to show [from text equations (12.1) and (12.2)]
that

$$\cos(-iy) = \cosh y, \qquad \sin(-iy) = -i\sinh y.$$

Then use the addition formula for the sine

$$\sin(x - iy) = \sin x \cos(-iy) + \cos x \sin(-iy)$$

$$= \sin x \cosh y - i\cos x \sinh y,$$

as above.   Now we have:

$$\text{Re} \sin(x - iy) = \sin x \cosh y$$

$$\text{Im} \sin(x - iy) = -\cos x \sinh y$$

$$|\sin(x - iy)| = \sqrt{\sin^2 x \cosh^2 y + \cos^2 x \sinh^2 y}.$$

We can simplify the square root by using

$$\cosh^2 y = 1 + \sinh^2 y \qquad \text{and} \qquad \cos^2 x = 1 - \sin^2 x.$$

Then

$$|\sin(x - iy)| = \sqrt{(\sin^2 x)(1 + \sinh^2 y) + (1 - \sin^2 x)\sinh^2 y}$$

$$= \sqrt{\sin^2 x + \sinh^2 y}.$$

Similarly the square root can be simplified in terms of $\cos x$
and $\cosh y$.

30. Using the basic definitions, text equations (12.2) and (12.3), we find

$$\tanh(3\pi i/4) = \frac{e^{3\pi i/4} - e^{-3\pi i/4}}{e^{3\pi i/4} + e^{-3\pi i/4}} \ .$$

From text Figure 9.6, $e^{3\pi i/4} = \dfrac{-1+i}{\sqrt{2}}$, and from a similar figure $e^{-3\pi i/4} = \dfrac{-1-i}{\sqrt{2}}$. Then

$$\tanh(3\pi i/4) = \frac{\frac{1}{\sqrt{2}}[-1 + i - (-1 - i)]}{\frac{1}{\sqrt{2}}[-1 + i + (-1 - i)]} = \frac{2i}{-2} = -i.$$

Alternatively, using the result of Problem 17,

$$\tanh(3\pi i/4) = i\tan(3\pi/4) = i(-1) = -i.$$

35. Substituting $z = i + 2$ into text equation (12.2), we find

$$\cosh(i + 2) = \tfrac{1}{2}(e^{i+2} + e^{-i-2}) = \tfrac{1}{2}e^2 e^i + \tfrac{1}{2}e^{-2}e^{-i}.$$

Now e^i means $e^{i\theta}$ with $\theta = 1$ (radian), so

$$e^i = \cos 1 + i\sin 1 = 0.54 + 0.84i \ ,$$
$$e^{-i} = \cos 1 - i\sin 1 = 0.54 - 0.84i.$$

Then $\cosh(i + 2) = \tfrac{1}{2}e^2(\cos 1 + i\sin 1) + \tfrac{1}{2}e^{-2}(\cos 1 - i\sin 1)$

$$= \frac{e^2 + e^{-2}}{2}\cos 1 + i\,\frac{e^2 - e^{-2}}{2}\sin 1$$

$$= \cosh 2 \cos 1 + i\sinh 2 \sin 1$$

$$= (3.76)(0.54) + i(3.63)(0.84) = 2.03 + 3.05i.$$

Alternatively, we could have used the result of Problem 4 with $z = i + 2$, that is, $x = 2$, $y = 1$.

Section 14

5.   We first write $-\sqrt{2} - i\sqrt{2} = \sqrt{2}(-1 - i)$ in polar form; from text
     Figure 4.1, we have $-1 - i = \sqrt{2}e^{5\pi i/4}$ so $\sqrt{2}(-1 - i) = 2e^{5\pi i/4}$.
     Then, by text equation (13.5),

$$\ell n(-\sqrt{2} - i\sqrt{2}) = \text{Ln } 2 + 5\pi i/4 = 0.693 + 3.93i .$$

   (Note that we <u>must</u> use radians here because $5\pi/4$ is part of the
   final answer -- the final step is <u>not</u> to find a sine or cosine;
   read the end of Section 3 in the text.)

8.   Using text equation (14.1), we write $i^{2/3} = e^{(2/3)\,\ell n\,i}$. Now
     $(2/3)\,\ell n\,i = (2/3)\,[\text{Ln } 1 + i(\pi/2 + 2n\pi)] = (2/3)i(\pi/2 + 2n\pi)$

$$= i(\pi/3 + 4n\pi/3) = i(60° + 240°n).$$

   Then

$$i^{2/3} = e^{i(\pi/3 + 4n\pi/3)} = \cos(\pi/3 + 4n\pi/3) + i\,\sin(\pi/3 + 4n\pi/3).$$

   If we repeatedly add 240° to 60° and subtract 360° as necessary
   to find only angles $< 360°$, we find exactly 3 angles (as expected
   for a cube root): 60°, 180°, 300°, or $\pi/3$, π, $5\pi/3$. Plotting
   the three complex numbers with these angles and absolute value
   1, we can read from the sketch:

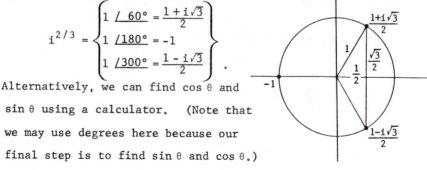

$$i^{2/3} = \begin{cases} 1 \ \underline{/\ 60°} = \dfrac{1 + i\sqrt{3}}{2} \\ 1 \ \underline{/180°} = -1 \\ 1 \ \underline{/300°} = \dfrac{1 - i\sqrt{3}}{2} \end{cases}.$$

   Alternatively, we can find $\cos\theta$ and
   $\sin\theta$ using a calculator. (Note that
   we may use degrees here because our
   final step is to find $\sin\theta$ and $\cos\theta$.)

10. By text equation (14.1), $i^{\ln i} = e^{(\ln i)(\ln i)}$. Now

$\ln i = \mathrm{Ln}\, 1 + i\pi/2 = i\pi/2$, so $(\ln i)(\ln i) = -\pi^2/4$. Then $i^{\ln i} = e^{-\pi^2/4}$.

Comment: We have found one value; there are infinitely many others, all real. Note that you can use different angles for the two $\ln i$ values. Can you find the value $i^{\ln i} = e^{4\pi^2 - (\pi^2/4)}$?

11. $2^i = e^{i\,\ln 2} = e^{i\,[\mathrm{Ln}\, 2 + 2n\pi i]} = e^{-2n\pi}e^{i\,\mathrm{Ln}\, 2}$

$$= e^{-2n\pi}(\cos \mathrm{Ln}\, 2 + i \sin \mathrm{Ln}\, 2).$$

Since n may be any positive or negative integer, we may also write $2^i = e^{\pm 2n\pi}(\cos \mathrm{Ln}\, 2 + i \sin \mathrm{Ln}\, 2)$. For $n = 0$, we find the value

$$2^i = \cos \mathrm{Ln}\, 2 + i \sin \mathrm{Ln}\, 2 = 0.769 + 0.639i.$$

20. To evaluate $\sin\!\left(i\, \ln \dfrac{1-i}{1+i}\right)$, we find

$$\frac{1-i}{1+i} = \frac{\sqrt{2}e^{-i\pi/4}}{\sqrt{2}e^{i\pi/4}} = e^{-i\pi/2},$$

$$\ln \frac{1-i}{1+i} = \mathrm{Ln}\, 1 + i\left(-\frac{\pi}{2} + 2n\pi\right) = i\left(-\frac{\pi}{2} + 2n\pi\right).$$

Multiplying this by i gives $\dfrac{\pi}{2} - 2n\pi$. Then

$$\sin\!\left(i\, \ln \frac{1-i}{1+i}\right) = \sin\!\left(\frac{\pi}{2} - 2n\pi\right) = \sin \frac{\pi}{2} = 1.$$

24. For example, compare $(i^i)^i$ and i^{-1}.

$$i^i = e^{i\,\ln i} = e^{i\left(\frac{\pi}{2} + 2n\pi\right)i} = e^{-\left(\frac{\pi}{2} + 2n\pi\right)}.$$

Now $\ln e^{-\left(\frac{\pi}{2} + 2n\pi\right)} = -\frac{\pi}{2} - 2n\pi + i(2m\pi)$, so

$$(i^i)^i = \left(e^{-\frac{\pi}{2} - 2n\pi}\right)^i = e^{i\left(-\frac{\pi}{2} - 2n\pi + 2m\pi i\right)} = e^{-2m\pi}e^{-i\pi/2} = -ie^{-2m\pi}.$$

But $i^{-1} = e^{-\ln i} = e^{-\left(\frac{\pi}{2} + 2k\pi\right)i} = e^{-i\pi/2} = -i.$

24.   (continued)

Thus i^{-1} has only one value, but $(i^i)^i$ has an infinite number of values, namely $e^{-2m\pi}$ times the value of i^{-1}.

Section 15

2.   If $z = \arctan 2i$, then

$$2i = \tan z = \frac{e^{iz} - e^{-iz}}{i(e^{iz} + e^{-iz})} = \frac{u - \frac{1}{u}}{i(u + \frac{1}{u})} \text{ where } u = e^{iz}$$

$$= \frac{u^2 - 1}{i(u^2 + 1)} \text{ by multiplying numerator and denominator by } u.$$

Clearing the equation of fractions and solving for u^2, we get

$$u^2 = \left(-\frac{1}{3}\right) = (e^{iz})^2 = e^{2iz}.$$

To solve for z, we take logarithms of both sides of this equation.

$$2iz = \ell n \left(-\frac{1}{3}\right) = \text{Ln}\left(\frac{1}{3}\right) + i(\pi + 2n\pi) = -\text{Ln } 3 + (2n+1)i\pi.$$

Multiply by $-\frac{1}{2}i$ to get

$$\arctan 2i = z = \frac{1}{2}i \text{ Ln } 3 + \left(n + \frac{1}{2}\right)\pi.$$

4.   Let $z = \sinh^{-1}(i/2)$; then

$$\frac{i}{2} = \sinh z = \frac{e^z - e^{-z}}{2} = \frac{u - u^{-1}}{2} = \frac{u^2 - 1}{2u},$$

where $u = e^z$.   Then

$$u^2 - 1 = iu \qquad \text{or} \qquad u^2 - iu - 1 = 0.$$

We solve for u and then find $z = \ell n \, u$:

4. (continued)

$$u = \frac{i \pm \sqrt{-1+4}}{2} = \frac{i \pm \sqrt{3}}{2}$$

$$z = \ell n \ \frac{i \pm \sqrt{3}}{2} = Ln \ 1 \ + \ i\begin{cases} \frac{\pi}{6} + 2n\pi \\ \frac{5\pi}{6} + 2n\pi, \end{cases}$$

$$\sinh^{-1}(i/2) = i\left(\frac{\pi}{6} + 2n\pi\right), \quad \text{or} \quad i\left(\frac{5\pi}{6} + 2n\pi\right).$$

5. If $z = arc \cos (i\sqrt{8})$, then

$$i\sqrt{8} = \cos z = \frac{e^{iz} + e^{-iz}}{2} = \frac{1}{2}\left(u + \frac{1}{u}\right) \text{ where } u = e^{iz}.$$

$$i\sqrt{8} = \frac{u^2 + 1}{2u} \ , \quad \text{or } 2i\sqrt{8}\,u = u^2 + 1 \ . \quad \text{Then}$$

$$u^2 - 2i\sqrt{8}\,u + 1 = 0$$

$$u = \frac{2i\sqrt{8} \pm \sqrt{-32-4}}{2} = i(\sqrt{8} \pm 3) = e^{iz}$$

$$iz = \begin{cases} \ell n\,[i(\sqrt{8} + 3)] \\ \ell n\,[i(\sqrt{8} - 3)] \end{cases} = \begin{cases} Ln(3 + \sqrt{8}) + i\left(\frac{\pi}{2} + 2n\pi\right) \\ Ln(3 - \sqrt{8}) + i\left(-\frac{\pi}{2} + 2n\pi\right) \end{cases}$$

We separated the $\sqrt{8} + 3$ solution and the $\sqrt{8} - 3$ solution because
the angles are different. Note that $\sqrt{8} - 3$ is negative; then
$\ell n\,[i(\sqrt{8} - 3)] = \ell n\,[-i(3 - \sqrt{8})]$, and $-i$ has angle $\frac{3\pi}{2}$ or $-\frac{\pi}{2}$ whereas
$+i$ has angle $\frac{\pi}{2}$. Solving for $z = arc \cos(i\sqrt{8})$, we get

$$arc \cos(i\sqrt{8}) = \begin{cases} \frac{\pi}{2} + 2n\pi - i\,Ln(3 + \sqrt{8}) \\ -\frac{\pi}{2} + 2n\pi - i\,Ln(3 - \sqrt{8}) \end{cases} .$$

We can write this in a more compact form. First we show that

$$Ln(3 - \sqrt{8}) = -Ln\,\frac{1}{3 - \sqrt{8}} = -Ln\,\frac{3 + \sqrt{8}}{9 - 8} = -Ln(3 + \sqrt{8}).$$

We also note that we can write $-2n\pi$ instead of $2n\pi$ since n is
any integer, positive or negative or zero. Then our two

5.   (continued)

solutions are just negatives of each other, and we have

$$\text{arc cos}(i\sqrt{8}) = \pm \left[\frac{\pi}{2} + 2n\pi - i \ Ln(3 + \sqrt{8}) \right]$$

$$= \pm \left[3.14\left(\frac{1}{2} + 2n\right) - 1.76i \right] \ .$$

We might have expected this; since $\cos(-z) = \cos z$, then if z
is a value of arc $\cos(i\sqrt{8})$ so is $-z$.

8.   If $z = \text{arc sin}(5/3)$, then

$$\frac{5}{3} = \sin z = \frac{e^{iz} - e^{-iz}}{2i} = \frac{u - u^{-1}}{2i} = \frac{u^2 - 1}{2iu}$$

where $u = e^{iz}$.   Then

$$u^2 - 1 = \frac{5}{3}(2iu) \qquad \text{or} \qquad 3u^2 - 10iu - 3 = 0.$$

$$u = \frac{10i \pm \sqrt{-100 + 36}}{6} = \frac{10i \pm 8i}{6} = \begin{cases} 3i \\ i/3 \ . \end{cases}$$

$$iz = \ell n \ u = \begin{cases} \ell n(3i) \\ \ell n(i/3) \end{cases} = \begin{cases} \ell n \ i + Ln \ 3 \\ \ell n \ i - Ln \ 3 \end{cases}$$

$$= \pm \ Ln \ 3 + \ell n \ i = \pm \ Ln \ 3 + i(\pi/2 + 2n\pi),$$

$$\text{arc sin}(5/3) = z = \frac{\pi}{2} + 2n\pi \pm i \ Ln \ 3.$$

18.   We try to solve $\tanh z = \pm 1$ by attempting to find $\tanh^{-1}(\pm 1)$.   If
$z = \tanh^{-1}(\pm 1)$, then

$$\pm 1 = \tanh z = \frac{e^z - e^{-z}}{e^z + e^{-z}} = \frac{u - u^{-1}}{u + u^{-1}} = \frac{u^2 - 1}{u^2 + 1}$$

where $u = e^z$.   Thus we need to solve the equations $u^2 - 1 = \pm(u^2 + 1)$.
Now $u^2 - 1 \neq u^2 + 1$ for any u.   The equation $u^2 - 1 = -(u^2 + 1)$ has
only the solution $u = 0$; but $u = e^z$ and $e^z \neq 0$ for any (finite) z.
Thus $\tanh z$ never takes the values $+1$ or -1.

Section 16

1. If $z = re^{i\theta}$ then

$$iz = e^{\frac{i\pi}{2}} re^{i\theta} = re^{i(\theta + \frac{\pi}{2})}.$$

Thus multiplying a complex number by i leaves its absolute value the same but increases its angle by $\frac{\pi}{2}$. We can describe this by saying that the line segment from the origin to the point z is rotated $90°$.

If $z = ae^{i\omega t}$ then

$$\frac{dz}{dt} = i\omega ae^{i\omega t} = i\omega z.$$

$$\frac{d^2 z}{dt^2} = (i\omega)^2 ae^{i\omega t} = -\omega^2 z.$$

Consider a particle at the point $z = x + iy = ae^{i\omega t}$ at time t. Since $|z| = a$, the particle travels in a circle. The speed of the particle is $v = \sqrt{\left(\frac{dx}{dt}\right)^2 + \left(\frac{dy}{dt}\right)^2} = \left|\frac{dz}{dt}\right| = \omega a$. Similarly the magnitude of its acceleration is

$\left|\frac{d^2 z}{dt^2}\right| = \omega^2 a = \frac{v^2}{a}$. Since $\frac{dz}{dt} = i\omega z$, the

direction of v is a $90°$ rotation from z (assuming $\omega > 0$), and the direction of the acceleration $(-\omega^2 z)$ is opposite z, that is, toward the center of the circle.

5. Given $z = z_1 t + z_2 (1 - t)$, we find

$$\frac{dz}{dt} = z_1 - z_2 , \qquad \frac{d^2 z}{dt^2} = 0.$$

5.   (continued)

Since the acceleration is zero, the particle moves along a straight line at constant speed. When $t = 0$, $z = z_1$, and when $t = 1$, $z = z_2$; thus the particle moves along the line through z_1 and z_2.

6.   (a) In series: $Z = Z_1 + Z_2 = 2 + 3i + 1 - 5i = 3 - 2i$.

In parallel: $\dfrac{1}{Z} = \dfrac{1}{Z_1} + \dfrac{1}{Z_2} = \dfrac{1}{2 + 3i} + \dfrac{1}{1 - 5i}$

$$= \dfrac{2 - 3i}{13} + \dfrac{1 + 5i}{26} = \dfrac{5 - i}{26}$$

$$Z = \dfrac{26}{5 - i} = \dfrac{26(5 + i)}{26} = 5 + i.$$

11.   We write

$$\cos\theta + \cos 3\theta + \cos 5\theta + \cdots + \cos(2n - 1)\theta$$

(1) $$+ i[\sin\theta + \sin 3\theta + \sin 5\theta + \cdots + \sin(2n - 1)\theta]$$

$$= e^{i\theta} + e^{3i\theta} + e^{5i\theta} + \cdots + e^{i(2n-1)\theta} = S_n.$$

Now S_n is the sum of n terms of a geometric progression with first term $a = e^{i\theta}$ and ratio $r = e^{2i\theta}$. Then by text equation (1.4) of Chapter 1, we have

$$S_n = \dfrac{e^{i\theta}\left[1 - (e^{2i\theta})^n\right]}{1 - e^{2i\theta}} = \dfrac{e^{i\theta}(1 - e^{2in\theta})}{1 - e^{2i\theta}}.$$

We write this in terms of sines and cosines as follows:

$$1 - e^{2i\theta} = e^{i\theta}(e^{-i\theta} - e^{i\theta}) = -2ie^{i\theta}\left(\dfrac{e^{i\theta} - e^{-i\theta}}{2i}\right)$$

$$= -2ie^{i\theta}\sin\theta$$

and similarly

$$1 - e^{2in\theta} = -2ie^{in\theta}\sin n\theta.$$

11.   (continued)

Then

$$S_n = \frac{e^{i\theta}(-2ie^{in\theta}\sin n\theta)}{-2ie^{i\theta}\sin\theta} = \frac{\sin n\theta}{\sin\theta}e^{in\theta}$$

(2) $= \dfrac{\sin n\theta}{\sin\theta}(\cos n\theta + i\sin n\theta)$

$= \dfrac{\sin 2n\theta}{2\sin\theta} + i\,\dfrac{\sin^2 n\theta}{\sin\theta}$,   since $2\sin n\theta \cos n\theta = \sin 2n\theta$.

We equate real parts of (1) and (2), and equate imaginary parts,
to obtain the desired results.

Section 17

2.   We write $(1+i\sqrt{3}) = 2e^{i\pi/3}$ (see text Figure 3.3) and

$\sqrt{2}+i\sqrt{2} = \sqrt{2}(1+i) = \sqrt{2}\cdot\sqrt{2}e^{i\pi/4} = 2e^{i\pi/4}$ (see text Figure 5.1).

Then

$$\left(\frac{1+i\sqrt{3}}{\sqrt{2}+i\sqrt{2}}\right)^{50} = \left(\frac{2e^{i\pi/3}}{2e^{i\pi/4}}\right)^{50} = (e^{i\pi/12})^{50}$$

$$= e^{4i\pi}\cdot e^{i\pi/6} = (\sqrt{3}+i)/2$$

(see text Figures 9.4 and 9.1).

6.   By text equation (14.1),

$$(-e)^{i\pi} = e^{i\pi\,\ell n(-e)} .$$

By text equation (13.5),

$$\ell n(-e) = \text{Ln } e + i(\pi + 2n\pi) = 1 + i\pi(2n+1).$$

Then

$$(-e)^{i\pi} = e^{i\pi[1+i\pi(2n+1)]} = e^{i\pi}e^{-\pi^2(2n+1)} = -e^{-\pi^2(2n+1)}.$$

For $n = 0$, we find the value

$$(-e)^{i\pi} = -e^{-\pi^2} = -5.17 \times 10^{-5}.$$

9. Hint: First find $\left(\dfrac{\sqrt{3}+i}{\sqrt{3}-i}\right)^{12}$.

11. Using the result of Problem 22 below, we find

$$\tanh^{-1}i = \frac{1}{2}\ell n\,\frac{1+i}{1-i} = \frac{1}{2}\ell n\,\frac{\sqrt{2}e^{i\pi/4}}{\sqrt{2}e^{-i\pi/4}} = \frac{1}{2}\ell n\,e^{i\pi/2} = \frac{i\pi}{4}\ .$$

Then $e^{2\tanh^{-1}i} = e^{i\pi/2} = i$.

15. By text equation (14.1), $|z|^{\ell n\,n} = e^{(\ell n\,n)(\ell n|z|)} = n^{\ell n|z|}$. Then
by the integral test (text Chapter 1, Section 6, Test B) or the
p-series test (Chapter 1, Problem 6.15), the series $\sum z^{\ell n\,n}$ is
absolutely convergent if $\ell n|z| < -1$, that is, if $|z| < e^{-1}$.

17. We let $w = \arcsin z$; then

$$z = \sin w = \frac{e^{iw} - e^{-iw}}{2i} = \frac{s - s^{-1}}{2i} = \frac{s^2 - 1}{2is}\ ,$$

where $s = e^{iw}$. Then

$$s^2 - 1 = 2isz \qquad \text{or} \qquad s^2 - 2izs - 1 = 0.$$

We solve the quadratic equation for s and then find w from
$s = e^{iw}$:

$$s = \frac{2iz \pm \sqrt{-4z^2 + 4}}{2} = iz \pm \sqrt{1 - z^2} = e^{iw},$$

$$iw = \ell n\,s = \ell n\left(iz \pm \sqrt{1 - z^2}\right), \qquad w = -i\,\ell n\left(iz \pm \sqrt{1 - z^2}\right).$$

22. If $w = \tanh^{-1}z$, then

$$z = \tanh w = \frac{e^w - e^{-w}}{e^w + e^{-w}} = \frac{s - s^{-1}}{s + s^{-1}} = \frac{s^2 - 1}{s^2 + 1}\ ,$$

where $s = e^w$. We clear the fractions and solve for s:

22.   (continued)

$$z(s^2 + 1) = s^2 - 1 \qquad \text{or} \qquad s^2(1 - z) = 1 + z;$$

$$s^2 = \frac{1 + z}{1 - z}, \qquad s = \sqrt{\frac{1 + z}{1 - z}}.$$

Then from $s = e^w$, we find

$$w = \ell n\, s = \ell n \sqrt{\frac{1 + z}{1 - z}} = \frac{1}{2} \ell n \frac{1 + z}{1 - z}.$$

26.   Using $|z| = \sqrt{z\bar{z}}$, we find

$$\left| \frac{2e^{i\theta} - i}{ie^{i\theta} + 2} \right| = \left(\frac{2e^{i\theta} - i}{ie^{i\theta} + 2} \cdot \frac{2e^{-i\theta} + i}{-ie^{-i\theta} + 2} \right)^{\frac{1}{2}}$$

$$= \left(\frac{4 + 1 + 2i(e^{i0} - e^{-i0})}{1 + 4 + 2i(e^{i\theta} - e^{-i\theta})} \right)^{\frac{1}{2}} = 1.$$

Here is an alternative way to do this problem without calculation.
Since $|\bar{z}| = |z|$ (verify this), and $|z_1/z_2| = |z_1|/|z_2|$ (see Problem
9.28), it does not change the absolute value of the fraction if
we take the complex conjugate of the numerator. Also, since
$|z_1 z_2| = |z_1| \cdot |z_2|$, it does not change the absolute value of the
numerator if we multiply by $e^{i\theta}$. Observe that $e^{i\theta}$ times the complex conjugate of the numerator is $e^{i\theta}(2e^{-i\theta} + i) = 2 + ie^{i\theta}$ which
is the denominator. Thus the absolute value is 1 as we found
above.

30.   Using text equations (9.3) and (8.1), and $(1 + i) = \sqrt{2}e^{i\pi/4}$, we
have

30.   (continued)

$$e^{x(1+i)} = e^x e^{ix} = e^x(\cos x + i \sin x) \qquad \text{and}$$

$$e^{x(1+i)} = \sum \frac{x^n(1+i)^n}{n!} = \sum \frac{x^n(\sqrt{2}e^{i\pi/4})^n}{n!}$$

$$= \sum \frac{x^n 2^{n/2}(\cos \frac{n\pi}{4} + i \sin \frac{n\pi}{4})}{n!} \quad .$$

We equate real parts and equate imaginary parts of the two

expressions for $e^{x(1+i)}$ to get

$$e^x \cos x = \sum \frac{x^n 2^{n/2} \cos \frac{n\pi}{4}}{n!}$$

$$e^x \sin x = \sum \frac{x^n 2^{n/2} \sin \frac{n\pi}{4}}{n!}$$

To see quickly the values of $e^{in\pi/4}$, we think of these points

plotted in the complex plane (they are the 8 eighth roots of 1).

We find for $n = 0$ to 7, the values:

1, $(1+i)/\sqrt{2}$, i, $(-1+i)/\sqrt{2}$, -1,

$(-1-i)/\sqrt{2}$, -i, $(1-i)/\sqrt{2}$, and repeat for

$n = 8$ to 15, etc. Thus we see immediately

that the $e^x \cos x$ series (<u>real</u> part of the

$e^{x(1+i)}$ series) has zero coefficients

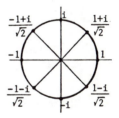

when $e^{in\pi/4} = \pm i$, that is, for $n = 2$, 6, 10,

etc., and similarly, the $e^x \sin x$ series has zero coefficients

when $e^{in\pi/4} = \pm 1$, that is, for $n = 4$, 8, 12, etc.

Chapter 3

Section 2

4. We first rewrite the equations in standard form:

$$x - 5y \quad\quad = 14$$
$$2x \quad + \quad 7z = 15$$
$$x - \quad y + 3z = \quad 9$$

Then we write and row reduce the corresponding matrix following the notation of the text example on page 85.

$$\begin{pmatrix} 1 & -5 & 0 & 14 \\ 2 & 0 & 7 & 15 \\ 1 & -1 & 3 & 9 \end{pmatrix} \xrightarrow[\substack{R2 - 2R1 \\ R3 - R1}]{} \begin{pmatrix} 1 & -5 & 0 & 14 \\ 0 & 10 & 7 & -13 \\ 0 & 4 & 3 & -5 \end{pmatrix}$$

$$\xrightarrow[R2 - 2R3]{} \begin{pmatrix} 1 & -5 & 0 & 14 \\ 0 & 2 & 1 & -3 \\ 0 & 4 & 3 & -5 \end{pmatrix} \xrightarrow[R3 - 2R2]{} \begin{pmatrix} 1 & -5 & 0 & 14 \\ 0 & 2 & 1 & -3 \\ 0 & 0 & 1 & 1 \end{pmatrix}.$$

The equations corresponding to this row reduced matrix are

$$x - 5y = 14, \quad\quad 2y + z = -3 \ , \quad\quad z = 1.$$

Now working from the last equation backward (back substitution), we find

$$z = 1$$
$$2y + 1 = -3, \quad\quad y = -2$$
$$x - 5(-2) = 14, \quad\quad x = 4.$$

The solution is $(x,y,z) = (4,-2,1)$.

10.   We write the matrix for the given set of equations with the
unknowns in the order p,q,r,s,t, and then row reduce it.

$$\begin{vmatrix} 2 & 0 & 1 & -2 & 0 & -2 \\ 0 & 5 & -1 & 4 & 0 & 7 \\ 0 & 0 & 2 & 1 & 1 & 0 \\ 3 & 1 & 0 & 0 & -4 & 1 \\ 1 & 1 & -1 & 0 & 2 & 3 \end{vmatrix} \xrightarrow{R1 \longleftrightarrow R5} \begin{vmatrix} 1 & 1 & -1 & 0 & 2 & 3 \\ 0 & 5 & -1 & 4 & 0 & 7 \\ 0 & 0 & 2 & 1 & 1 & 0 \\ 3 & 1 & 0 & 0 & -4 & 1 \\ 2 & 0 & 1 & -2 & 0 & -2 \end{vmatrix} \xrightarrow[R5 - 2R1]{R4 - 3R1}$$

$$\begin{vmatrix} 1 & 1 & -1 & 0 & 2 & 3 \\ 0 & 5 & -1 & 4 & 0 & 7 \\ 0 & 0 & 2 & 1 & 1 & 0 \\ 0 & -2 & 3 & 0 & -10 & -8 \\ 0 & -2 & 3 & -2 & -4 & -8 \end{vmatrix} \xrightarrow[5R5 + 2R2]{R4 - R5} \begin{vmatrix} 1 & 1 & -1 & 0 & 2 & 3 \\ 0 & 5 & -1 & 4 & 0 & 7 \\ 0 & 0 & 2 & 1 & 1 & 0 \\ 0 & 0 & 0 & 2 & -6 & 0 \\ 0 & 0 & 13 & -2 & -20 & -26 \end{vmatrix} \xrightarrow[R4 \div 2]{2R5-13R3}$$

$$\begin{vmatrix} 1 & 1 & -1 & 0 & 2 & 3 \\ 0 & 5 & -1 & 4 & 0 & 7 \\ 0 & 0 & 2 & 1 & 1 & 0 \\ 0 & 0 & 0 & 1 & -3 & 0 \\ 0 & 0 & 0 & -17 & -53 & -52 \end{vmatrix} \xrightarrow{R5 + 17R4} \begin{vmatrix} 1 & 1 & -1 & 0 & 2 & 3 \\ 0 & 5 & -1 & 4 & 0 & 7 \\ 0 & 0 & 2 & 1 & 1 & 0 \\ 0 & 0 & 0 & 1 & -3 & 0 \\ 0 & 0 & 0 & 0 & -104 & -52 \end{vmatrix}$$

The equations corresponding to this reduced matrix are

$$p + q - r + 2t = 3$$
$$5q - r + 4s = 7$$
$$2r + s + t = 0$$
$$s - 3t = 0$$
$$2t = 1 \quad \text{(divide last row by -52)}$$

By back substitution (work up from the last equation), we find

$$t = \frac{1}{2}$$
$$s = \frac{3}{2}$$
$$r = -\frac{1}{2}(s+t) = -1$$
$$5q = 7 + r - 4s = 7 - 1 - 6 = 0, \quad q = 0$$
$$p = 3 - q + r - 2t = 3 - 0 - 1 - 1 = 1$$

The solution is $(p,q,r,s,t) = \left(1, 0, -1, \frac{3}{2}, \frac{1}{2}\right)$.

11.   The matrix to be row reduced is:

$$\begin{pmatrix} 1-i & 1+i & 4i & 0 \\ 0 & 2+3i & 3+2i & 3+4i \\ 1 & 0 & -2 & 1-i \end{pmatrix}$$

We interchange rows 1 and 3 so as to use 1 as the pivot.

$$\begin{pmatrix} 1 & 0 & -2 & 1-i \\ 0 & 2+3i & 3+2i & 3+4i \\ 1-i & 1+i & 4i & 0 \end{pmatrix}$$

Subtract $(1 - i)$ times row 1 from row 3:

$$\begin{pmatrix} 1 & 0 & -2 & 1-i \\ 0 & 2+3i & 3+2i & 3+4i \\ 0 & 1+i & 2+2i & 2i \end{pmatrix}$$

Divide row 3 by $(1 + i)$ and interchange it with row 2:

$$\begin{pmatrix} 1 & 0 & -2 & 1-i \\ 0 & 1 & 2 & 1+i \\ 0 & 2+3i & 3+2i & 3+4i \end{pmatrix}$$

Subtract $(2 + 3i)$ times row 2 from row 3:

$$\begin{pmatrix} 1 & 0 & -2 & 1-i \\ 0 & 1 & 2 & 1+i \\ 0 & 0 & -1-4i & 4-i \end{pmatrix}$$

Then $w = \dfrac{4 - i}{-(1 + 4i)} \cdot \dfrac{1 - 4i}{1 - 4i} = \dfrac{-17i}{-17} = i,$

$v = 1 + i - 2w = 1 - i,$

$u = 1 - i + 2w = 1 + i.$

14.   We write the Kirchhoff's law equations as described in the text, pages 86-87.

$$\text{Currents:} \quad I_1 = I_2 + I_3$$
$$\text{Upper loop:} \quad 3 - I_3 + 2I_2 + 4 = 0$$
$$\text{Lower loop:} \quad -3I_1 + 2 + 3 - I_3 = 0$$

Arrange the equations in standard form and solve:

$$I_1 - I_2 - I_3 = 0$$
$$2I_2 - I_3 = -7$$
$$3I_1 \qquad + I_3 = 5$$

$$\begin{pmatrix} 1 & -1 & -1 & 0 \\ 0 & 2 & -1 & -7 \\ 3 & 0 & 1 & 5 \end{pmatrix} \xrightarrow{R3 - 3R1} \begin{pmatrix} 1 & -1 & -1 & 0 \\ 0 & 2 & -1 & -7 \\ 0 & 3 & 4 & 5 \end{pmatrix}$$

$$\xrightarrow{R2 - R3} \begin{pmatrix} 1 & -1 & -1 & 0 \\ 0 & -1 & -5 & -12 \\ 0 & 3 & 4 & 5 \end{pmatrix} \xrightarrow[R2 \times (-1)]{R3 + 3R2} \begin{pmatrix} 1 & -1 & -1 & 0 \\ 0 & 1 & 5 & 12 \\ 0 & 0 & -11 & -31 \end{pmatrix}$$

$$I_3 = 31/11, \quad I_2 = 12 - 5I_3 = -23/11, \quad I_1 = I_2 + I_3 = 8/11.$$

The problem is a little simpler if we eliminate one I, say I_1, at the beginning (see "loop currents" in a text which discusses circuits). Then we have

$$2I_2 - I_3 = -7$$
$$3I_2 + 4I_3 = 5$$

$$\begin{pmatrix} 2 & -1 & -7 \\ 3 & 4 & 5 \end{pmatrix} \xrightarrow{R2 - R1} \begin{pmatrix} 2 & -1 & -7 \\ 1 & 5 & 12 \end{pmatrix} \xrightarrow{R1 \longleftrightarrow R2} \begin{pmatrix} 1 & 5 & 12 \\ 2 & -1 & -7 \end{pmatrix}$$

$$\xrightarrow{R2 - 2R1} \begin{pmatrix} 1 & 5 & 12 \\ 0 & -11 & -31 \end{pmatrix}, \quad \text{which gives the same answers as above.}$$

Section 3

4.

$$D = \begin{vmatrix} -2 & 4 & 7 & 3 \\ 8 & 2 & -9 & 5 \\ -4 & 6 & 8 & 4 \\ 2 & -9 & 3 & 8 \end{vmatrix} \longrightarrow \left\{ \begin{array}{l} \text{Row reduce: } R2 + 4R1 \\ \qquad\qquad\qquad R3 - 2R1 \\ \qquad\qquad\qquad R4 + R1 \end{array} \right\}$$

$$= \begin{vmatrix} -2 & 4 & 7 & 3 \\ 0 & 18 & 19 & 17 \\ 0 & -2 & -6 & -2 \\ 0 & -5 & 10 & 11 \end{vmatrix} \longrightarrow \{\text{Laplace development}\}$$

$$= -2 \begin{vmatrix} 18 & 19 & 17 \\ -2 & -6 & -2 \\ -5 & 10 & 11 \end{vmatrix} \longrightarrow \left\{ \begin{array}{l} \text{Subtract column 3 from} \\ \text{column 1 and from column 2} \end{array} \right\}$$

$$= -2 \begin{vmatrix} 1 & 2 & 17 \\ 0 & -4 & -2 \\ -16 & -1 & 11 \end{vmatrix} \longrightarrow \text{Row reduce: } R3 + 16R1$$

$$= -2 \begin{vmatrix} 1 & 2 & 17 \\ 0 & -4 & -2 \\ 0 & 31 & 283 \end{vmatrix} = -2 \begin{vmatrix} -4 & -2 \\ 31 & 283 \end{vmatrix} = 2 \begin{vmatrix} 4 & 2 \\ 31 & 283 \end{vmatrix}$$

$$= 2(4 \cdot 283 - 31 \cdot 2) = 2140.$$

7.   The point of this problem is to show examples of the manipula-
tions which can be done on a determinant (using Facts 1-4, text
page 89) to put it in different form or evaluate it without
complicated algebra.

Multiply the first row by $\underline{a}$,
the 2nd row by $\underline{b}$, and the third
row by $\underline{c}$. This multiplies the
determinant by abc, so we must
divide by abc; we do this by
dividing the last column by abc.

$$\begin{vmatrix} 1 & a & bc \\ 1 & b & ac \\ 1 & c & ab \end{vmatrix}$$

$$= \frac{1}{abc} \begin{vmatrix} a & a^2 & abc \\ b & b^2 & abc \\ c & c^2 & abc \end{vmatrix}$$

$$= \begin{vmatrix} a & a^2 & 1 \\ b & b^2 & 1 \\ c & c^2 & 1 \end{vmatrix}$$

Make two column interchanges;
each multiplies the determinant
by -1.

$$= \begin{vmatrix} 1 & a & a^2 \\ 1 & b & b^2 \\ 1 & c & c^2 \end{vmatrix}$$

Subtract row 1 from rows 2 and
3; factor $b^2-a^2 = (b-a)(b+a)$ and
similarly for c^2-a^2. Factor (b-a)
from the 2nd row and (c-a) from
the 3rd row.

$$= \begin{vmatrix} 1 & a & a^2 \\ 0 & b-a & (b-a)(b+a) \\ 0 & c-a & (c-a)(c+a) \end{vmatrix}$$

$$= (b-a)(c-a) \begin{vmatrix} 1 & a & a^2 \\ 0 & 1 & b+a \\ 0 & 1 & c+a \end{vmatrix}$$

Subtract the 2nd row from
the 3rd.

$$= (b-a)(c-a) \begin{vmatrix} 1 & a & a^2 \\ 0 & 1 & b+a \\ 0 & 0 & c-b \end{vmatrix}$$

Factor (c-b) from the 3rd row,
and find that the value of the
resulting determinant is 1.

$$= (b-a)(c-a)(c-b) \begin{vmatrix} 1 & a & a^2 \\ 0 & 1 & b+a \\ 0 & 0 & 1 \end{vmatrix}$$

$$= (b-a)(c-a)(c-b).$$

7.   (continued)

If we had wanted just the final result (and not the intermediate determinants), we could have found it as follows:

$$
\begin{vmatrix}
1 & a & bc \\
1 & b & ac \\
1 & c & ab
\end{vmatrix}
$$

Subtract row 1 from rows 2 and 3. $=
\begin{vmatrix}
1 & a & bc \\
0 & b-a & c(a-b) \\
0 & c-a & b(a-c)
\end{vmatrix}$

Factor (b-a) from row 2 and
(c-a) from row 3.
$= (b-a)(c-a)
\begin{vmatrix}
1 & a & bc \\
0 & 1 & -c \\
0 & 1 & -b
\end{vmatrix}$

Expand by a Laplace development
on the 1st column and evaluate
$= (b-a)(c-a)
\begin{vmatrix}
1 & -c \\
1 & -b
\end{vmatrix}$

the resulting 2 by 2 determinant. $= (b-a)(c-a)(c-b).$

10.   If $a_{ij} = -a_{ij}$, then the elements on the main diagonal are
       zero ($a_{11} = -a_{11}$ means $a_{11} = 0$, etc.). If we interchange rows
       and columns, the off diagonal elements a_{ij} and a_{ji} are
       exchanged (for example, 2 and -2 in Problem 9). Thus every
       element in the determinant has its sign changed. This
       multiplies the determinant by $(-1)^n$ where n is the order of
       the determinant (number of rows or columns) since multiply-
       ing one row by (-1) multiplies the determinant by (-1)
       (see Fact 1, text, page 89). By Fact 4a, the determinant
       is not changed by interchanging rows and columns. Thus
       we have $D = (-1)^n D$. If n is odd, we have $D = -D$, or $D = 0$.

12.   Caution: The value of this determinant is not zero. It
       is skew-symmetric (see Problem 10), but it is of _even_ order
       (namely 4). Thus it does not satisfy the requirements of
       Problem 10 and there is no reason to expect it to be zero.
       Evaluate it as you would any determinant.

14.   Let us call the n-rowed determinant D_n .   Then for $n = 2$,

$$D_2 = \begin{vmatrix} \cos\theta & 1 \\ 1 & 2\cos\theta \end{vmatrix} = 2\cos^2\theta - 1 = \cos 2\theta.$$

Next we show that if the result is true for determinants of order $<n$, then it is true for D_n; that is, if $D_m = \cos m\theta$ for $m < n$, then $D_n = \cos n\theta$. Since we have proved $D_2 = \cos 2\theta$, the result is therefore true for each higher n in turn. (This is mathematical induction.) We expand D_n using the last row:

$$D_n = - \begin{vmatrix} \cos\theta & 1 & 0 & \cdot & \cdot & \cdot & 0 \\ 1 & 2\cos\theta & 1 & & & & \cdot \\ 0 & 1 & 2\cos\theta & & & & \cdot \\ \cdot & & & \cdot & & & \cdot \\ \cdot & & & & 2\cos\theta & 1 & 0 \\ \cdot & & & & 1 & 2\cos\theta & 0 \\ 0 & \cdot & \cdot & \cdot & 0 & 1 & 1 \end{vmatrix} + (2\cos\theta)D_{n-1}$$

Expand this determinant using elements of the last column to get 1 times D_{n-2}.   Thus

$$D_n = (2\cos\theta)D_{n-1} - D_{n-2} = 2\cos\theta\cos(n-1)\theta - \cos(n-2)\theta$$

(remember that we are assuming $D_m = \cos m\theta$ for determinants of order $m < n$). By a trigonometric addition formula

$$\cos(n-2)\theta = \cos[(n-1)\theta - \theta] = \cos(n-1)\theta\cos\theta + \sin(n-1)\theta\sin\theta.$$

Substitute this into D_n and again use an addition formula to get:

$$D_n = \cos(n-1)\theta\cos\theta - \sin(n-1)\theta\sin\theta = \cos n\theta.$$

15. To solve text equations (2.1) by Cramer's rule, we first

evaluate D, the determinant of the coefficients. We find

$$D = \begin{vmatrix} 2 & 0 & -1 \\ 6 & 5 & 3 \\ 2 & -1 & 0 \end{vmatrix} = \begin{vmatrix} 2 & 0 & -1 \\ 12 & 5 & 0 \\ 2 & -1 & 0 \end{vmatrix} = -1 \begin{vmatrix} 12 & 5 \\ 2 & -1 \end{vmatrix} = 22.$$

(Add three times the first row to the second and expand using

the last column.) To find the x numerator, we replace the

first column in D by the constants on the right-hand side

of text equation (2.1). (Evaluate the x determinant just as

we did D.) Then

$$x = \frac{1}{D}\begin{vmatrix} 2 & 0 & -1 \\ 7 & 5 & 3 \\ 4 & -1 & 0 \end{vmatrix} = \frac{1}{D}\begin{vmatrix} 2 & 0 & -1 \\ 13 & 5 & 0 \\ 4 & -1 & 0 \end{vmatrix} = -\frac{1}{D}\begin{vmatrix} 13 & 5 \\ 4 & -1 \end{vmatrix} = \frac{33}{22} = \frac{3}{2}.$$

To find the y numerator, we replace the second column in D by

the constants. Then

$$y = \frac{1}{D}\begin{vmatrix} 2 & 2 & -1 \\ 6 & 7 & 3 \\ 2 & 4 & 0 \end{vmatrix} = \frac{1}{D}\begin{vmatrix} 2 & 2 & -1 \\ 12 & 13 & 0 \\ 2 & 4 & 0 \end{vmatrix} = -\frac{1}{D}\begin{vmatrix} 12 & 13 \\ 2 & 4 \end{vmatrix} = -\frac{22}{22} = -1.$$

To find the z numerator, replace the third column of D by the

constants. Then

$$z = \frac{1}{D}\begin{vmatrix} 2 & 0 & 2 \\ 6 & 5 & 7 \\ 2 & -1 & 4 \end{vmatrix} = \frac{1}{D}\begin{vmatrix} 2 & 0 & 0 \\ 6 & 5 & 1 \\ 2 & -1 & 2 \end{vmatrix} = \frac{2}{D}\begin{vmatrix} 5 & 1 \\ -1 & 2 \end{vmatrix} = \frac{2}{22}\cdot 11 = 1.$$

(This time we subtracted the first column from the third.)

23.   Let b = bonus, t = tax.   Then

$$b = 0.20(230,000 - t)$$
$$t = 0.40(230,000 - b)$$

First we arrange the equations in standard form:

$$b + 0.2t = 0.2(230,000) = 46,000$$
$$0.4b + t = 0.4(230,000) = 92,000$$

Then by row reduction:

$$\begin{pmatrix} 1 & 0.2 & 46000 \\ 0.4 & 1 & 92000 \end{pmatrix} \xrightarrow{R2 - 0.4R1} \begin{pmatrix} 1 & 0.2 & 46000 \\ 0 & 0.92 & 73600 \end{pmatrix}$$

Then

$$\begin{array}{ccc} b + 0.2t = 46000 & & t = 80,000 \\ & \text{or} & \\ 0.94t = 73600 & & b = 30,000 \end{array}$$

Alternatively, by Cramer's rule:

$$D = \begin{vmatrix} 1 & 0.2 \\ 0.4 & 1 \end{vmatrix} = 1 - 0.08 = 0.92$$

$$b = \frac{1}{D} \begin{vmatrix} 46000 & 0.2 \\ 92000 & 1 \end{vmatrix} = \frac{46000 - 18400}{0.92} = 30,000$$

$$t = \frac{1}{D} \begin{vmatrix} 1 & 46000 \\ 0.4 & 92000 \end{vmatrix} = \frac{92000 - 18400}{0.92} = 80,000$$

25. We shall do Problems 2.4 and 2.14 (previously done by row
     reduction) by Cramer's rule.

     Problem 2.4

     The equations in standard form are

$$\begin{cases} x - 5y \quad\;\;\; = 14 \\ 2x \quad\;\;\; + 7z = 15 \\ x - y + 3z = 9 \end{cases}$$

     Then we find

$$D = \begin{vmatrix} 1 & -5 & 0 \\ 2 & 0 & 7 \\ 1 & -1 & 3 \end{vmatrix} = \begin{vmatrix} 1 & -5 & 0 \\ 0 & 10 & 7 \\ 0 & 4 & 3 \end{vmatrix} = \begin{vmatrix} 10 & 7 \\ 4 & 3 \end{vmatrix} = 2.$$

$$x = \frac{1}{D} \begin{vmatrix} 14 & -5 & 0 \\ 15 & 0 & 7 \\ 9 & -1 & 3 \end{vmatrix} = \frac{1}{D} \begin{vmatrix} -31 & 0 & -15 \\ 15 & 0 & 7 \\ 9 & -1 & 3 \end{vmatrix} = \frac{1}{D} \begin{vmatrix} -31 & -15 \\ 15 & 7 \end{vmatrix} = \frac{1}{2} \cdot 8 = 4.$$

$$\overrightarrow{R5 - 5R3}$$

$$y = \frac{1}{D} \begin{vmatrix} 1 & 14 & 0 \\ 2 & 15 & 7 \\ 1 & 9 & 3 \end{vmatrix} = \frac{1}{D} \begin{vmatrix} 1 & 14 & 0 \\ 0 & -13 & 7 \\ 0 & -5 & 3 \end{vmatrix} = \frac{1}{D} \begin{vmatrix} -13 & 7 \\ -5 & 3 \end{vmatrix} = \frac{1}{2}(-4) = -2.$$

$$z = \frac{1}{D} \begin{vmatrix} 1 & -5 & 14 \\ 2 & 0 & 15 \\ 1 & -1 & 9 \end{vmatrix} = \frac{1}{D} \begin{vmatrix} 1 & -5 & 14 \\ 0 & 10 & -13 \\ 0 & 4 & -5 \end{vmatrix} = \frac{1}{D} \begin{vmatrix} 10 & -13 \\ 4 & -5 \end{vmatrix} = \frac{1}{2} \cdot 2 = 1.$$

25.   (continued)

Problem 2.14

We write the equations in standard form and solve by Cramer's
rule:

$$\begin{cases} 2I_2 - I_3 = -7 \\ 3I_2 + 4I_3 = 5 \end{cases}$$

$$D = \begin{vmatrix} 2 & -1 \\ 3 & 4 \end{vmatrix} = 11$$

and

$$I_2 = \frac{1}{D}\begin{vmatrix} -7 & -1 \\ 5 & 4 \end{vmatrix} = -\frac{23}{11}$$

$$I_1 = I_2 + I_3$$

$$I_3 = \frac{1}{D}\begin{vmatrix} 2 & -7 \\ 3 & 5 \end{vmatrix} = \frac{31}{11}$$

$$I_1 = I_2 + I_3 = \frac{8}{11}$$

Section 4

5.

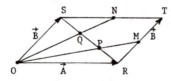

Define P and Q as the
trisection points on the
diagonal $\vec{RS} = \vec{B} - \vec{A}$. Then
the vector $\vec{OP} = \vec{OR} + \vec{RP} =$
$\vec{A} + \frac{1}{3}(\vec{B} - \vec{A}) = \frac{1}{3}(2\vec{A} + \vec{B})$.
The vector from O to M,
the midpoint of $\vec{RT}$, is

$$\vec{OM} = \vec{A} + \frac{1}{2}\vec{B} = \frac{1}{2}(2\vec{A} + \vec{B}).$$

We see that $\vec{OP}$ and $\vec{OM}$ are in the same direction; thus OPM is a
straight line passing through both the trisection point P and
the midpoint M.

Similarly show that $\vec{OQ}$ and $\vec{ON}$ are parallel.

8.

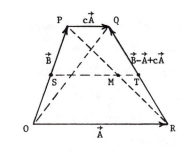

In the figure, $\vec{PQ}$ is parallel to $\vec{OR} = \vec{A}$, so $\vec{PQ} = c\vec{A}$ where $c = |\vec{PQ}| / |\vec{OR}|$. In terms of the vectors $\vec{A}$, $\vec{B}$, and $c\vec{A}$, the fourth side of the trapezoid is

$$\vec{RQ} = \vec{RO} + \vec{OP} + \vec{PQ} = -\vec{A} + \vec{B} + c\vec{A}$$

as shown in the figure.

Let ST be the median of the trapezoid (that is, S is the midpoint of OP and T is the midpoint of RQ). Let M be the midpoint of the diagonal PR. We want to show that SMT is a straight line. We find

$$\vec{ST} = \vec{SP} + \vec{PQ} + \vec{QT} = \tfrac{1}{2}\vec{B} + c\vec{A} - \tfrac{1}{2}(\vec{B} - \vec{A} + c\vec{A}) = \tfrac{1}{2}(\vec{A} + c\vec{A}).$$

$$\vec{SM} = \vec{SP} + \vec{PM} = \tfrac{1}{2}\vec{B} + \tfrac{1}{2}(\vec{A} - \vec{B}) = \tfrac{1}{2}\vec{A}.$$

Since vectors $\vec{SM}$ and $\vec{ST}$ are parallel (both are multiples of $\vec{A}$) and both go through S, SMT is a straight line which bisects the diagonal PR.

We found vector $\vec{ST} = \tfrac{1}{2}(\vec{A} + c\vec{A})$. This vector is half the sum of the vectors along the two parallel bases, that is, it is parallel to them and half as long as their sum.

14.   (c)

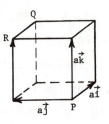

Let $\underline{a}$ = length of an edge of the cube. Then the vectors along the three edges of the cube that meet at P are $a\vec{i}$, $a\vec{j}$ and $a\vec{k}$.

The vector from P to Q is a space diagonal;

vector $\vec{PQ} = a\vec{i} + a\vec{j} + a\vec{k} = a(\vec{i} + \vec{j} + \vec{k})$.

Vector $\vec{PR}$ is a face diagonal; $\vec{PR} = a\vec{j} + a\vec{k}$. By text equation (4.1), $|\vec{PR}| = \sqrt{a^2 + a^2} = a\sqrt{2}$, and $|\vec{PQ}| = a\sqrt{3}$. We find the angle between $\vec{PQ}$ and $\vec{PR}$ by using the dot product formulas, text equations (4.2) and (4.10):

$$\vec{PQ}\cdot\vec{PR} = |\vec{PQ}|\cdot|\vec{PR}|\cos\theta = a\sqrt{3}\cdot a\sqrt{2}\cos\theta = a^2\sqrt{6}\cos\theta \ ,$$

$$\vec{PQ}\cdot\vec{PR} = (a\vec{i} + a\vec{j} + a\vec{k})\cdot(a\vec{i} + a\vec{j}) = a^2 + a^2 = 2a^2 \ .$$

Then

$$a^2\sqrt{6}\cos\theta = 2a^2, \quad \cos\theta = 2/\sqrt{6}, \quad \theta \cong 35.3° \ .$$

15.   (d) Examples are $(\vec{i} + 2\vec{j})/\sqrt{2}$, $(\vec{i} - \vec{j})/\sqrt{2}$, $(\vec{i} + 4\vec{j} + \vec{k})/(3\sqrt{2})$. Find some more examples. The idea is to find vectors whose dot product with the given vector $\vec{A}$ is zero (there are infinitely many such vectors -- any vector in a plane perpendicular to $\vec{A}$). Divide each vector by its magnitude [text equation (4.1)] to make it a unit vector.

18.   The dot product of the given vectors is

$(2\vec{i} - \vec{j} + 4\vec{k}) \cdot (5\vec{i} + 2\vec{j} - 2\vec{k}) = 10 - 2 - 8 = 0$, so they are perpendicular.

The cross product of the given vectors is perpendicular to both. Thus

$$\begin{vmatrix} \vec{i} & \vec{j} & \vec{k} \\ 2 & -1 & 4 \\ 5 & 2 & -2 \end{vmatrix} = \vec{i}(2 - 8) - \vec{j}(-4 - 20) + \vec{k}(4 + 5)$$

$$= -6\vec{i} + 24\vec{j} + 9\vec{k} \qquad \text{or} \qquad 2\vec{i} - 8\vec{j} - 3\vec{k}$$

is perpendicular to both given vectors. To check this, show that the dot products are zero.

21.   Hints: To show that two vectors are orthogonal (perpendicular), show that their dot product = 0. In finding, say, the dot product $(\vec{B}|\vec{A}|) \cdot (\vec{A}|\vec{B}|) = (|\vec{A}|)(|\vec{B}|)(\vec{B} \cdot \vec{A})$, remember that $|\vec{A}|$ and $|\vec{B}|$ are just <u>numbers</u> (scalars); you find the dot product of the <u>vectors</u> $\vec{B}$ and $\vec{A}$.

23.   Hint for $\vec{A} \cdot \vec{B} = \vec{A} \times \vec{B} = 0$: Can $\sin\theta$ and $\cos\theta$ both equal 0 for the same θ ? Alternatively, can two vectors be both parallel and perpendicular to each other?

24.   Hint: Use text equations (4.2) and (4.14).

26.   From the figure:

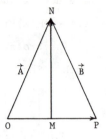

Base $\overrightarrow{OP} = \vec{A} - \vec{B}$

Median $\overrightarrow{MN} = \frac{1}{2}(\vec{A} - \vec{B}) + \vec{B} = \frac{1}{2}(\vec{A} + \vec{B})$

We show that the median is perpendicular

to the base by showing that the dot

product of $\overrightarrow{MN}$ and $\overrightarrow{OP}$ is zero.

$$\overrightarrow{MN} \cdot \overrightarrow{OP} = \frac{1}{2}(\vec{A} - \vec{B}) \cdot (\vec{A} + \vec{B}) = \frac{1}{2}(A^2 - B^2)$$

Since the triangle is isoceles, $|\vec{A}| = |\vec{B}|$, or $A^2 = B^2$.   Thus

$$\overrightarrow{MN} \cdot \overrightarrow{OP} = \frac{1}{2}(A^2 - B^2) = 0.$$

Section 5

4.   A vector along the given line is $\vec{A} = \hat{i} - 2\hat{j}$ [see text equation

(5.4) and Figure 5.2].   A vector in the (x,y) plane perpen-

dicular to $\hat{i} - 2\hat{j}$ is $2\hat{i} + \hat{j} = (2,1)$ (make the dot product equal

to zero).   Then the desired perpendicular line through $\vec{r}_0 = (1,0)$

is [see text equation (5.4)]

$$\vec{r} = (1,0) + (2,1)t.$$

In terms of components, this is

$$x = 1 + 2t, \quad y = t, \quad \text{or} \quad x = 1 + 2y.$$

6.   A vector $\vec{A}$ along the line is $\vec{A} = (2,-3,-3) - (1,-1,-5) = (1,-2,2)$

or $\vec{A} = \hat{i} - 2\hat{j} + 2\hat{k}$.   We may use any point on the line as (x_0, y_0, z_0).

If we use the point $(1,-1,-5)$, the symmetric equations, text

equation (5.6), are $\frac{x-1}{1} = \frac{y+1}{-2} = \frac{z+5}{2}$.   The parametric equations

of the line are given by text equation (5.8), where $\vec{r}_0 = (1,-1,-5)$

or $(2,-3,-3)$ or any other point on the line, and $\vec{A}$ is given above.

Thus we have

$$\vec{r} = (1,-1,-5) + (1,-2,2)t \quad \text{or} \quad \vec{r} = \hat{i} - \hat{j} - 5\hat{k} + (\hat{i} - 2\hat{j} + 2\hat{k})t.$$

9.   A vector $\vec{A}$ along the line is

$$\vec{A} = (-1,3,7) - (-1,-2,7) = (0,5,0), \quad \text{or} \quad \vec{A} = 5\vec{j}.$$

The equations of this line cannot be written in the form of
text equation (5.6) (there would be zeros in denominators!).
Consider text equations (5.7); $c = 0$ simply means that the z
coordinates of all the points on the line are all the same,
namely, $z = z_0$. In other words, the line lies in a plane perpendicular to the z axis. In this problem, both x and z have
constant values; the line is the intersection of the two
planes $x = -1$ and $z = 7$. These are the equations of the line
in the form of text equations (5.7). (Satisfy yourself that
these equations do determine the line and that y does not
appear because there is no restriction on y -- it can have any
value. The line is parallel to the y axis.) You might think
first of writing the parametric equations of the line as

$$\vec{r} = (-1,3,7) + (0,5,0)t \quad \text{or} \quad \vec{r} = -\vec{i} + 3\vec{j} + 7\vec{k} + 5\vec{j}t.$$

However, these can be written in the simpler form $\vec{r} = -\vec{i} + 7\vec{k} + \vec{j}t'$,
where $t' = 3 + 5t$ is a simpler parameter (whatever letter we use
for the parameter, it takes all values). Another way of finding
the parametric equations is to realize that the point $(-1,0,7)$
is on the line since $x = -1$, $z = 7$, and y can have any value.
Also the vector $\vec{j}$ is along the line and is a simpler choice for
$\vec{A}$ than $5\vec{j}$. Then $\vec{r}_0 = -\vec{i} + 7\vec{k}$, $\vec{A} = \vec{j}$ and text equation (5.8) gives
$\vec{r} = -\vec{i} + 7\vec{k} + \vec{j}t.$

18.   The vector $\vec{A} = (5,-2,1)$ is parallel to both lines and so parallel
       to the plane. Another vector parallel to (or in) the plane is
       the vector joining the two points $(5,-4,2)$ and $(1,-1,0)$, one on
       each line. This vector is $(5,-4,2) - (1,-1,0) = (4,-3,2)$; its
       cross product with $\vec{A}$ is perpendicular to the plane:

$$\vec{N} = \begin{vmatrix} \vec{i} & \vec{j} & \vec{k} \\ 4 & -3 & 2 \\ 5 & -2 & 1 \end{vmatrix} = \vec{i}(-3+4) - \vec{j}(4-10) + \vec{k}(-8+15) = \vec{i} + 6\vec{j} + 7\vec{k}.$$

Using the point $(1,-1,0)$ on the plane, we find that the equation
of the plane is [see text equation (5.10)]:

$$(x-1) + 6(y+1) + 7z = 0 \quad \text{or}$$
$$x + 6y + 7z + 5 = 0.$$

20.   As in text Example 1, two vectors in the plane are

$$(2,1,3) - (0,1,1) = (2,0,2) \quad \text{and} \quad (4,2,1) - (2,1,3) = (2,1,-2).$$

The cross product of these two vectors is perpendicular to the
plane.

$$\vec{N} = \begin{vmatrix} \vec{i} & \vec{j} & \vec{k} \\ 2 & 0 & 2 \\ 2 & 1 & -2 \end{vmatrix} = -2\vec{i} + 8\vec{j} + 2\vec{k}.$$

For $\vec{r}_0$, we can use any one of the three points, say $(0,1,1)$.
Then by text equation (5.10), the equation of the plane is

$$-2x + 8(y-1) + 2(z-1) = 0 \quad \text{or} \quad x - 4y - z + 5 = 0.$$

(You might like to verify that the result is the same if you
use either of the other points.)

21. Comparing the given equations with text equations (5.10), we see
that vectors normal (perpendicular) to the planes are:
$\vec{N}_1 = 2\vec{i} + 6\vec{j} - 3\vec{k}$ and $\vec{N}_2 = 5\vec{i} + 2\vec{j} - \vec{k}$. The angle θ between the
planes is the same as the angle between the normals. By text
equations (4.2) and (4.10):

$$\vec{N}_1 \cdot \vec{N}_2 = N_1 N_2 \cos\theta \text{ and } \vec{N}_1 \cdot \vec{N}_2 = (2,6,-3) \cdot (5,2,-1) = 25,$$

$$N_1 = \sqrt{2^2 + 6^2 + (-3)^2} = \sqrt{49} = 7, \qquad N_2 = \sqrt{30},$$

$$\cos\theta = (\vec{N}_1 \cdot \vec{N}_2)/N_1 N_2 = 25/(7\sqrt{30}) = 0.652,$$

$$\theta \cong 49.3° \quad \text{by calculator.}$$

24. We want (x,y,z) satisfying both equations

$$2x + 6y - 3z = 10 \quad \text{and} \quad 5x + 2y - z = 12.$$

Since there are many such points, we may be able to find one by
inspection, for example $(2,1,0)$ (also see below). A vector
along the line of intersection of the planes lies in both planes
and so is perpendicular to both normals $\vec{N}_1$ and $\vec{N}_2$ (Problem 21).
Then a vector along the line of intersection is

$$\vec{N}_1 \times \vec{N}_2 = \begin{vmatrix} \vec{i} & \vec{j} & \vec{k} \\ 2 & 6 & -3 \\ 5 & 2 & -1 \end{vmatrix} = (0,-13,-26).$$

A simpler vector along this line is $\vec{A} = (0,1,2)$. Using the point
$\vec{r}_0 = (2,1,0)$ that we found above, we write the equations of the
line of intersection of the planes in the parametric form and in
the symmetric form [see text equations (5.8) and (5.7)]:

$$\vec{r} = (2,1,0) + (0,1,2)t \quad \text{or} \quad \vec{r} = 2\vec{i} + \vec{j} + (\vec{j} + 2\vec{k})t \quad \text{(parametric)},$$

$$x = 2, \quad \frac{y-1}{1} = \frac{z}{2} \quad \text{(symmetric)}.$$

24.   (continued)

If a point on both planes cannot easily be found by inspection, as above, then we can row reduce the matrix of the equations as in Section 2 to obtain a simpler set of equations (also see Section 8). In this problem we find

$$\begin{pmatrix} 2 & 6 & -3 & 10 \\ 5 & 2 & -1 & 12 \end{pmatrix} \xrightarrow{2R2 - 5R1} \begin{pmatrix} 2 & 6 & -3 & 10 \\ 0 & -26 & 13 & -26 \end{pmatrix} .$$

Cancel -13 in the last row and write the equations:

$$2x + 6y - 3z = 10$$

$$2y - z = 2$$

Then for any arbitrary value of z, we find

$$y = 1 + z/2, \qquad x = 2.$$

The point we chose above corresponded to $z = 0$. Other simple points are $(2,2,2)$, $(2,0,-2)$, $(2,3,4)$, etc.

To find the distance from a point P to the line QR, we follow text Figure 5.7. For this problem P is the origin and Q may be the point $(2,1,0)$, so $\vec{PQ} = (2,1,0) - (0,0,0) = (2,1,0)$. We found above that a vector along the line is $\vec{A} = \vec{j} + 2\vec{k}$; then $\vec{u} = (\vec{j} + 2\vec{k})/\sqrt{5}$ is a unit vector along the line. Now

$$\vec{PQ} \times \vec{u} = \frac{1}{\sqrt{5}} \begin{vmatrix} \vec{i} & \vec{j} & \vec{k} \\ 2 & 1 & 0 \\ 0 & 1 & 2 \end{vmatrix} = \frac{1}{\sqrt{5}}(2\vec{i} - 4\vec{j} + 2\vec{k}).$$

Thus we find

$$d = |\vec{PQ} \times u| = \frac{1}{\sqrt{5}} \sqrt{2^2 + 4^2 + 2^2} = \sqrt{\frac{24}{5}} = 2\sqrt{\frac{6}{5}} .$$

31. We want a point on the plane $2x + 6y - 3z = 10$. This is easy by
     inspection. For example, take $y = z = 0$, $x = 5$. Then (using text
     Figure 5.6), $P = (-2,4,5)$, $Q = (5,0,0)$, $\overrightarrow{PQ} = (7,-4,-5)$. A vector
     $\vec{N}$ normal to the plane is $(2,6,-3)$, and a unit vector in this
     direction is $\vec{n} = \vec{N}/|\vec{N}| = \vec{N}/7$. Then (see text Figure 5.6)

$$|\overrightarrow{PR}| = |\overrightarrow{PQ} \cdot \vec{n}| = |(7,-4,-5) \cdot (2,6,-3)/7| = \frac{14 - 24 + 15}{7} = \frac{5}{7} \ .$$

34. With $P = (3,0,-5)$ and $Q = (2,1,-5)$, we find
     $\overrightarrow{PQ} = (-1,1,0)$. A unit vector parallel to the
     lines is $\vec{u} = (-3\vec{j} + \vec{k})/\sqrt{10}$. Then $d = |\overrightarrow{PQ} \times \vec{u}|$.

$$\overrightarrow{PQ} \times \vec{u} = \frac{1}{\sqrt{10}} \begin{vmatrix} \vec{i} & \vec{j} & \vec{k} \\ -1 & 1 & 0 \\ 0 & -3 & 1 \end{vmatrix}$$

$$= \frac{1}{\sqrt{10}}(\vec{i} + \vec{j} + 3\vec{k}),$$

$$d = |\overrightarrow{PQ} \times \vec{u}| = \sqrt{11/10}.$$

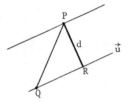

39. If the lines intersect, there are values of t_1 and t_2 for which
     the two $\vec{r}$ vectors are the same. (Note that t_1 and t_2 do not have
     to be equal; it is not necessary for two trains to collide even
     though their tracks cross. See text Figure 5.3.) If the two $\vec{r}$
     vectors are equal, their corresponding components are equal.
     Thus, if the lines intersect, we have

$$\vec{r} = (0,2,1) + (3,0,-1)t_1 = (7,0,2) + (2,-1,1)t_2 \quad \text{or}$$
$$(x,y,z) = (3t_1,2,1-t_1) = (7+2t_2,-t_2,2+t_2).$$

The y components give $t_2 = -2$, $y = 2$; then the x-components give
$t_1 = 1$, $x = 3$. These values both give $z = 0$; thus the lines intersect at $\vec{r} = (3,-2,0)$.

39.   (continued)

We find the acute angle between the lines by using text equations
(4.2) and (4.10):

$$|(3\vec{i} - \vec{k}) \cdot (2\vec{i} - \vec{j} + \vec{k})| = \sqrt{10}\,\sqrt{6}\cos\theta. \quad \text{Then}$$

$$\cos\theta = \frac{5}{\sqrt{60}} = 0.645, \qquad \theta = 49.8°.$$

43.   Refer to text Figure 5.8. Here we have

$$P = (1,-2,1/2), \qquad Q = (-2,2,1/2), \qquad \vec{PQ} = (-3,4,0).$$

A vector along the first line is $\vec{A} = (2,3,2)$; note that we
must write $\frac{2z-1}{4} = \frac{z-\frac{1}{2}}{2}$. A vector along the second line is
$\vec{B} = (-1,-2,0)$; again note that we must write $\frac{2-y}{2} = \frac{y-2}{-2}$, and
also note from text equation (5.7) that $z = 1/2$ means $c = 0$.
Then a vector normal to both lines is

$$\vec{N} = \vec{A} \times \vec{B} = \begin{vmatrix} \vec{i} & \vec{j} & \vec{k} \\ 2 & 3 & 2 \\ -1 & -2 & 0 \end{vmatrix} = 4\vec{i} - 2\vec{j} - \vec{k},$$

and a unit vector in this direction is $\vec{n} = \vec{N}/\sqrt{21}$. The distance
between the lines is

$$|\vec{PQ} \cdot \vec{n}| = |(-3,4,0) \cdot (4,-2,-1)| = |(-12-8)/\sqrt{21}| = 20/\sqrt{21} = 4.36.$$

45.   Refer to text Figure 5.7. We find $\vec{PQ} = (3,-1,1)$, $\vec{A} = (2,-2,1)$,
$\vec{u} = \vec{A}/3$. Then

$$\vec{PQ} \times \vec{u} = \frac{1}{3}\begin{vmatrix} \vec{i} & \vec{j} & \vec{k} \\ 3 & -1 & 1 \\ 2 & -2 & 1 \end{vmatrix} = (\vec{i} - \vec{j} - 4\vec{k})/3,$$

$$d = |\vec{PQ} \times \vec{u}| = \sqrt{18}/3 = \sqrt{2}.$$

45.   (continued)

The shortest distance from a point to a line is the distance along the perpendicular to the line. Thus (see text Figure 5.3), at the time of closest approach to the origin, the vector $\vec{r} = \vec{r}_0 + \vec{A}t$ is perpendicular to $\vec{A}$. Then $\vec{A} \cdot \vec{r} = 0 = \vec{A} \cdot \vec{r}_0 + \vec{A} \cdot \vec{A}t$, so $t = -\vec{A} \cdot \vec{r}_0 / A^2$. In this problem, we are given $\vec{r} = (3,-1,1) + (2,-2,1)t$, so $\vec{r}_0 = (3,-1,1)$ and $\vec{A} = (2,-2,1)$. Then we find for the time of closest approach

$$t = -\vec{A} \cdot \vec{r}_0 / A^2 = -9/9 = -1.$$

At that time $\vec{r} = (3,-1,1) + (2,-2,1)(-1) = (1,1,0)$, so $d = |\vec{r}| = \sqrt{2}$, as we found above.

Section 6

2.   Following the "row times column" rule [text equation (6.2)], we find

$$AB = \begin{pmatrix} 2 & -5 \\ -1 & 3 \end{pmatrix} \begin{pmatrix} -1 & 4 \\ 0 & 2 \end{pmatrix} = \begin{pmatrix} 2(-1)-5(0) & 2 \cdot 4 - 5 \cdot 2 \\ -1(-1)+3 \cdot 0 & -1 \cdot 4 + 3 \cdot 2 \end{pmatrix} = \begin{pmatrix} -2 & -2 \\ 1 & 2 \end{pmatrix}$$

$$BA = \begin{pmatrix} -1 & 4 \\ 0 & 2 \end{pmatrix} \begin{pmatrix} 2 & -5 \\ -1 & 3 \end{pmatrix} = \begin{pmatrix} -6 & 17 \\ -2 & 6 \end{pmatrix} \quad \text{Note that } AB \neq BA.$$

$$A + B = \begin{pmatrix} 2-1 & -5+4 \\ -1+0 & 3+2 \end{pmatrix} = \begin{pmatrix} 1 & -1 \\ -1 & 5 \end{pmatrix} = B + A$$

$$5A = 5\begin{pmatrix} 2 & -5 \\ -1 & 3 \end{pmatrix} = \begin{pmatrix} 10 & -25 \\ -5 & 15 \end{pmatrix} \quad \text{(Multiply \underline{every} \underline{element} by 5)}$$

$$3B = 3\begin{pmatrix} -1 & 4 \\ 0 & 2 \end{pmatrix} = \begin{pmatrix} -3 & 12 \\ 0 & 6 \end{pmatrix}$$

$$5A - 3B = \begin{pmatrix} 13 & -37 \\ -5 & 9 \end{pmatrix}$$

2.   (continued)

      det $A = 1$,   det $B = -2$

$$\det(AB) = \begin{vmatrix} -2 & -2 \\ 1 & 2 \end{vmatrix} = -2, \qquad\qquad \det(BA) = \begin{vmatrix} -6 & 17 \\ -2 & 6 \end{vmatrix} = -2$$

$$\det(A+B) = \begin{vmatrix} 1 & -1 \\ -1 & 5 \end{vmatrix} = 5 - 1 = 4$$

$$\det(5A) = \begin{vmatrix} 10 & -25 \\ -5 & 15 \end{vmatrix} = 25, \qquad\qquad \det(3B) = \begin{vmatrix} -3 & 12 \\ 0 & 6 \end{vmatrix} = -18$$

We see that $\det(AB) = \det(BA) = (\det A)(\det B)$ but
$\det(A+B) \neq \det A + \det B$.   Also $\det(5A) = 25 = 25\det A$, $\det(3B) = -18$
$= 9\det B$. We could have found these last results without computation. Multiplying a matrix by k means multiplying <u>every</u>
element by k, whereas multiplying <u>one</u> row (or column) by k multiplies a determinant by k. Here 5A has 2 rows multiplied by 5
which multiplies det A by 5^2 , so $\det(5A) = 25\det A = 25$.   Similarly
$\det(3B) = 3^2\det B = 9(-2) = -18$.

4.   We shall do part of this problem; the rest is done in a similar
way. We cannot find AB because a row of A has 4 elements and a
column of B has only 3. We <u>can</u> find BA because a row of B has
2 elements and a column of A has 2 elements. Multiplying "row
times column", we get

$$BA = \begin{pmatrix} 2 & 4 \\ 1 & -1 \\ 3 & -1 \end{pmatrix} \begin{pmatrix} 2 & 3 & 1 & -4 \\ 2 & 1 & 0 & 5 \end{pmatrix} = \begin{pmatrix} 12 & 10 & 2 & 12 \\ 0 & 2 & 1 & -9 \\ 4 & 8 & 3 & -17 \end{pmatrix} .$$

Satisfy yourself that we cannot find A^2 or B^2, but we <u>can</u> find
C^2. We cannot multiply BA (above) times A or B or C (we must
multiply BA times a matrix with 4 rows, that is, 4 elements in

4.   (continued)

a column), but we can multiply C times BA:

$$CBA = \begin{pmatrix} 2 & 1 & 3 \\ 4 & -1 & -2 \\ -1 & 0 & 1 \end{pmatrix} \begin{pmatrix} 12 & 10 & 2 & 12 \\ 0 & 2 & 1 & -9 \\ 4 & 8 & 3 & -17 \end{pmatrix} = \begin{pmatrix} 36 & 46 & 14 & -36 \\ 40 & 22 & 1 & 91 \\ -8 & -2 & 1 & -29 \end{pmatrix}.$$

You should verify that you get the same result if you first find

CB and then multiply the result times A (matrix multiplication

obeys the associative law).

7. $z' = x' + iy' = e^{-i\theta}(x + iy) = (\cos \theta - i \sin \theta)(x + iy)$

$$= x \cos \theta + y \sin \theta - ix \sin \theta + iy \cos \theta .$$

Then $x' = $ (real part of z') $= x \cos \theta + y \sin \theta$,

$y' = $ (imaginary part of z') $= -x \sin \theta + y \cos \theta$.

These are the rotation equations (6.3).

10.   We write the two sets of equations using matrix notation as in

text equations (6.8) ff.

$$r' = \begin{pmatrix} x' \\ y' \end{pmatrix} = \frac{1}{2} \begin{pmatrix} 1 & \sqrt{3} \\ -\sqrt{3} & 1 \end{pmatrix} \begin{pmatrix} x \\ y \end{pmatrix} = Ar,$$

$$r'' = \begin{pmatrix} x'' \\ y'' \end{pmatrix} = \frac{1}{2} \begin{pmatrix} -1 & \sqrt{3} \\ -\sqrt{3} & -1 \end{pmatrix} \begin{pmatrix} x' \\ y' \end{pmatrix} = Br'.$$

Then $r'' = B(Ar) = BAr$ so we want the matrix product BA.

$$BA = \frac{1}{4} \begin{pmatrix} -1 & \sqrt{3} \\ -\sqrt{3} & -1 \end{pmatrix} \begin{pmatrix} 1 & \sqrt{3} \\ -\sqrt{3} & 1 \end{pmatrix} = \frac{1}{4} \begin{pmatrix} -4 & 0 \\ 0 & -4 \end{pmatrix} = \begin{pmatrix} -1 & 0 \\ 0 & -1 \end{pmatrix}.$$

The equations $r'' = BAr$ are

$$r'' = \begin{pmatrix} x'' \\ y'' \end{pmatrix} = \begin{pmatrix} -1 & 0 \\ 0 & -1 \end{pmatrix} \begin{pmatrix} x \\ y \end{pmatrix} = \begin{pmatrix} -x \\ -y \end{pmatrix} \quad \text{or} \quad \begin{matrix} x'' = -x \\ y'' = -y \end{matrix} .$$

10.   (continued)

This corresponds to a rotation of axes of 180°; we see this
either from the equations (x" = -x, y" = -y) or by comparing
the matrix $\begin{pmatrix} -1 & 0 \\ 0 & -1 \end{pmatrix}$ with text equation (6.4) to see that
$\cos \theta = -1$, $\sin \theta = 0$. By comparing the matrices A and B with
text equation (6.4), we see that

$$\text{for A:} \quad \cos \theta = 1/2 , \qquad \sin \theta = \sqrt{3}/2 , \qquad \theta = 60°;$$

$$\text{for B:} \quad \cos \theta = -1/2 , \qquad \sin \theta = \sqrt{3}/2 , \qquad \theta = 120°.$$

Thus the total rotation is through $60° + 120° = 180°$ as we found
from the matrix product.

13.   (a) First we multiply

$$\begin{pmatrix} -1 & 4 \\ 2 & -1 \end{pmatrix} \begin{pmatrix} -1 \\ 2 \end{pmatrix} = \begin{pmatrix} (-1)(1) + 4 \cdot 2 \\ 2(-1) + (-1)(2) \end{pmatrix} = \begin{pmatrix} 9 \\ -4 \end{pmatrix} .$$

Then

$$(2 \quad 3) \begin{pmatrix} -1 & 4 \\ 2 & -1 \end{pmatrix} \begin{pmatrix} -1 \\ 2 \end{pmatrix} = (2 \quad 3) \begin{pmatrix} 9 \\ -4 \end{pmatrix} = (6).$$

Note that the result is a matrix of one row and one column,
that is, one element. Verify that you get the same result by
multiplying the first two matrices together and then the
result times the third matrix.

13.   (b) We can write the equation as $r^T Ar = 30$ where $r = \begin{pmatrix} x \\ y \end{pmatrix}$,

$r^T = (x \quad y)$ is the transpose of the matrix r, and

$A = \begin{pmatrix} 5 & -7 \\ 7 & 3 \end{pmatrix}$.   Then

$$Ar = \begin{pmatrix} 5 & -7 \\ 7 & 3 \end{pmatrix} \begin{pmatrix} x \\ y \end{pmatrix} = \begin{pmatrix} (5x - 7y) \\ (7x + 3y) \end{pmatrix}.$$

This is a matrix of one column; note carefully that $(5x - 7y)$ is one element [compare part (a)].   Then

$$r^T Ar = (x \quad y) \begin{pmatrix} 5x - 7y \\ 7x + 3y \end{pmatrix} = \left(x(5x - 7y) + y(7x + 3y) \right) = (5x^2 + 3y^2).$$

This is a matrix of one row and one column, that is, one element. The equation says this one element equals 30.   Thus we have

$$5x^2 + 3y^2 = 30 \qquad \text{or} \qquad \frac{x^2}{6} + \frac{y^2}{10} = 1,$$

which is the equation of an ellipse with semi-axes $\sqrt{6}$ and $\sqrt{10}$.

21.   Using the row reduction method of finding A^{-1} [see text (6.23),
       page 122], we write the following matrix and find its reduced
       echelon form:

$$\begin{pmatrix} -2 & 0 & 1 & 1 & 0 & 0 \\ 1 & -1 & 2 & 0 & 1 & 0 \\ 3 & 1 & 0 & 0 & 0 & 1 \end{pmatrix} \xrightarrow[\substack{R3-3R2 \\ R1 \longleftrightarrow R2}]{R1+2R2} \begin{pmatrix} 1 & -1 & 2 & 0 & 1 & 0 \\ 0 & -2 & 5 & 1 & 2 & 0 \\ 0 & 4 & -6 & 0 & -3 & 1 \end{pmatrix}$$

$$\xrightarrow[R2 \div (-2)]{R3+2R2} \begin{pmatrix} 1 & -1 & 2 & 0 & 1 & 0 \\ 0 & 1 & -5/2 & -1/2 & -1 & 0 \\ 0 & 0 & 4 & 2 & 1 & 1 \end{pmatrix} \xrightarrow[R3 \div 4]{R1+R2} \begin{pmatrix} 1 & 0 & -1/2 & -1/2 & 0 & 0 \\ 0 & 1 & -5/2 & -1/2 & -1 & 0 \\ 0 & 0 & 1 & 1/2 & 1/4 & 1/4 \end{pmatrix}$$

$$\xrightarrow[R2+\frac{5}{2}R3]{R1+\frac{1}{2}R3} \begin{pmatrix} 1 & 0 & 0 & -1/4 & 1/8 & 1/8 \\ 0 & 1 & 0 & 3/4 & -3/8 & 5/8 \\ 0 & 0 & 1 & 1/2 & 1/4 & 1/4 \end{pmatrix}.$$

Then

$$A^{-1} = \begin{pmatrix} -1/4 & 1/8 & 1/8 \\ 3/4 & -3/8 & 5/8 \\ 1/2 & 1/4 & 1/4 \end{pmatrix} = \frac{1}{8} \begin{pmatrix} -2 & 1 & 1 \\ 6 & -3 & 5 \\ 4 & 2 & 2 \end{pmatrix}.$$

We can also find A^{-1} by using the formula [text equation (6.24)].
We first find $\det A = 8$. Next we find the cofactors of the
elements of A:

$$\text{Row 1:} \quad \begin{vmatrix} -1 & 2 \\ 1 & 0 \end{vmatrix} = -2, \qquad -\begin{vmatrix} 1 & 2 \\ 3 & 0 \end{vmatrix} = 6, \qquad \begin{vmatrix} 1 & -1 \\ 3 & 1 \end{vmatrix} = 4$$

$$\text{Row 2:} \quad -\begin{vmatrix} 0 & 1 \\ 1 & 0 \end{vmatrix} = 1, \qquad \begin{vmatrix} -2 & 1 \\ 3 & 0 \end{vmatrix} = -3, \qquad -\begin{vmatrix} -2 & 0 \\ 3 & 1 \end{vmatrix} = 2$$

$$\text{Row 3:} \quad \begin{vmatrix} 0 & 1 \\ -1 & 2 \end{vmatrix} = 1, \qquad -\begin{vmatrix} -2 & 1 \\ 1 & 2 \end{vmatrix} = 5, \qquad \begin{vmatrix} -2 & 0 \\ 1 & -1 \end{vmatrix} = 2$$

Then, by text equation (6.24), we have

$$C = \begin{pmatrix} -2 & 6 & 4 \\ 1 & -3 & 2 \\ 1 & 5 & 2 \end{pmatrix}, \quad A^{-1} = \frac{1}{8}C^{T} = \frac{1}{8} \begin{pmatrix} -2 & 1 & 1 \\ 6 & -3 & 5 \\ 4 & 2 & 2 \end{pmatrix}.$$

23.   See text Chapter 10, equation (1.3).

27.   We write the equations in the matrix form $Ar = k$ and find $r = A^{-1}k$
       [see text equations (6.18) and (6.19)].

$$Ar = \begin{pmatrix} 1 & -1 & 1 \\ 2 & 1 & -1 \\ 3 & 2 & 2 \end{pmatrix} \begin{pmatrix} x \\ y \\ z \end{pmatrix} = \begin{pmatrix} 4 \\ -1 \\ 5 \end{pmatrix} = k.$$

We can find A^{-1} either by row reduction or by the formula (see
Problem 21). By row reduction:

$$\begin{pmatrix} 1 & -1 & 1 & 1 & 0 & 0 \\ 2 & 1 & -1 & 0 & 1 & 0 \\ 3 & 2 & 2 & 0 & 0 & 1 \end{pmatrix} \xrightarrow[R3 - 3R1]{R2 - 2R1} \begin{pmatrix} 1 & -1 & 1 & 1 & 0 & 0 \\ 0 & 3 & -3 & -2 & 1 & 0 \\ 0 & 5 & -1 & -3 & 0 & 1 \end{pmatrix}$$

$$\xrightarrow{2R2 - R3} \begin{pmatrix} 1 & -1 & 1 & 1 & 0 & 0 \\ 0 & 1 & -5 & -1 & 2 & -1 \\ 0 & 5 & -1 & -3 & 0 & 1 \end{pmatrix} \xrightarrow[R1 + R2]{R3 - 5R2} \begin{pmatrix} 1 & 0 & -4 & 0 & 2 & -1 \\ 0 & 1 & -5 & -1 & 2 & -1 \\ 0 & 0 & 24 & 2 & -10 & 6 \end{pmatrix}$$

$$\xrightarrow{R3 \div 24} \begin{pmatrix} 1 & 0 & -4 & 0 & 2 & -1 \\ 0 & 1 & -5 & -1 & 2 & -1 \\ 0 & 0 & 1 & \frac{1}{12} & -\frac{5}{12} & \frac{1}{4} \end{pmatrix} \xrightarrow[R2 + 5R3]{R1 + 4R3} \begin{pmatrix} 1 & 0 & 0 & \frac{1}{3} & \frac{1}{3} & 0 \\ 0 & 1 & 0 & -\frac{7}{12} & -\frac{1}{12} & \frac{1}{4} \\ 0 & 0 & 1 & \frac{1}{12} & -\frac{5}{12} & \frac{1}{4} \end{pmatrix}$$

$$A^{-1} = \begin{pmatrix} \frac{1}{3} & \frac{1}{3} & 0 \\ -\frac{7}{12} & -\frac{1}{12} & \frac{1}{4} \\ \frac{1}{12} & -\frac{5}{12} & \frac{1}{4} \end{pmatrix} = \frac{1}{12} \begin{pmatrix} 4 & 4 & 0 \\ -7 & -1 & 3 \\ 1 & -5 & 3 \end{pmatrix}$$

27. (continued)

To use the formula for A^{-1} [text (6.24)], we first find $\det A = 12$. The cofactors are:

Row 1: $\begin{vmatrix} 1 & -1 \\ 2 & 2 \end{vmatrix} = 4,$ $-\begin{vmatrix} 2 & -1 \\ 3 & 2 \end{vmatrix} = -7,$ $\begin{vmatrix} 2 & 1 \\ 3 & 2 \end{vmatrix} = 1$

Row 2: $-\begin{vmatrix} -1 & 1 \\ 2 & 2 \end{vmatrix} = 4,$ $\begin{vmatrix} 1 & 1 \\ 3 & 2 \end{vmatrix} = -1,$ $-\begin{vmatrix} 1 & -1 \\ 3 & 2 \end{vmatrix} = -5$

Row 3: $\begin{vmatrix} -1 & 1 \\ 1 & -1 \end{vmatrix} = 0,$ $-\begin{vmatrix} 1 & 1 \\ 2 & -1 \end{vmatrix} = 3,$ $\begin{vmatrix} 1 & -1 \\ 2 & 1 \end{vmatrix} = 3$

Then

$$A^{-1} = \frac{1}{\det A} C^T = \frac{1}{12} \begin{pmatrix} 4 & 4 & 0 \\ -7 & -1 & 3 \\ 1 & -5 & 3 \end{pmatrix} \quad \text{as before.}$$

Now we find x, y, z from the matrix equation $r = A^{-1}k$:

$$r = \begin{pmatrix} x \\ y \\ z \end{pmatrix} = \frac{1}{12} \begin{pmatrix} 4 & 4 & 0 \\ -7 & -1 & 3 \\ 1 & -5 & 3 \end{pmatrix} \begin{pmatrix} 4 \\ -1 \\ 5 \end{pmatrix} = \frac{1}{12} \begin{pmatrix} 12 \\ -12 \\ 24 \end{pmatrix} = \begin{pmatrix} 1 \\ -1 \\ 2 \end{pmatrix}$$

so

$$\begin{cases} x = 1 \\ y = -1 \\ z = 2 \end{cases}.$$

28. Let us write out the product AC^T for a 3 by 3 matrix; the result is explained below.

$$\begin{pmatrix} a_{11} & a_{12} & a_{13} \\ a_{21} & a_{22} & a_{23} \\ a_{31} & a_{32} & a_{33} \end{pmatrix} \begin{pmatrix} C_{11} & C_{21} & C_{31} \\ C_{12} & C_{22} & C_{32} \\ C_{13} & C_{23} & C_{33} \end{pmatrix} = \begin{pmatrix} \det A & 0 & 0 \\ 0 & \det A & 0 \\ 0 & 0 & \det A \end{pmatrix}.$$

Row 1 times column 1 is the sum of elements of row 1 times their cofactors; this is $\det A$. Row 1 times column 2 is the sum of

28.   (continued)

elements of row 1 times the cofactors of row 2; this is zero by

Problem 3.8. Similarly all the row times column results give

either zero or $\det A$ as shown. If we divide both C^T and the

matrix on the right by $\det A$, we have a unit matrix on the right,

which says that $\frac{1}{\det A} C^T$ is the inverse of A.

30.   Combining text equations (6.19) and (6.24), we have

$$\begin{pmatrix} x \\ y \\ z \end{pmatrix} = r = A^{-1}k = \frac{1}{\det A}C^T k = \frac{1}{\det A}\begin{pmatrix} C_{11} & C_{21} & C_{31} \\ C_{12} & C_{22} & C_{32} \\ C_{13} & C_{23} & C_{33} \end{pmatrix}\begin{pmatrix} k_1 \\ k_2 \\ k_3 \end{pmatrix}$$

$$= \frac{1}{\det A}\begin{pmatrix} k_1 C_{11} + k_2 C_{21} + k_3 C_{31} \\ k_1 C_{12} + k_2 C_{22} + k_3 C_{32} \\ k_1 C_{13} + k_2 C_{23} + k_3 C_{33} \end{pmatrix}$$

or

$$\begin{cases} x = \left(k_1 C_{11} + k_2 C_{21} + k_3 C_{31}\right)/\det A, \\ y = \left(k_1 C_{12} + k_2 C_{22} + k_3 C_{32}\right)/\det A, \\ z = \left(k_1 C_{13} + k_2 C_{23} + k_3 C_{33}\right)/\det A. \end{cases}$$

We note that the denominators of x, y, z are all $\det A$ as Cramer's

rule says. To see that the numerators give the Cramer's rule

result, think of evaluating each numerator determinant in Cramer's

rule using the k column; we find the value of the determinant by

multiplying each k times the cofactor of the element it replaces

and adding the results. For example,

$$y = \frac{1}{\det A}\begin{vmatrix} a_{11} & k_1 & a_{13} \\ a_{21} & k_2 & a_{23} \\ a_{31} & k_3 & a_{33} \end{vmatrix} = k_1 C_{12} + k_2 C_{22} + k_3 C_{32}$$

as above, and similarly for x and z.

Section 7

1. $f(\vec{r}_1 + \vec{r}_2) = \vec{A} \cdot (\vec{r}_1 + \vec{r}_2) + 3;$

            $f(\vec{r}_1) + f(\vec{r}_2) = \vec{A} \cdot \vec{r}_1 + 3 + \vec{A} \cdot \vec{r}_2 + 3.$

Since $f(\vec{r}_1 + \vec{r}_2) \neq f(\vec{r}_1) + f(\vec{r}_2)$, $f(\vec{r})$ is not linear. Also
$f(a\vec{R}) = \vec{A} \cdot (a\vec{r}) + 3 = a\vec{A} \cdot \vec{r} + 3 \neq a(\vec{A} \cdot \vec{r} + 3)$, so the given $f(\vec{r})$ does
not satisfy either of the requirements for linearity.

4.   The vector function $\vec{F}(\vec{r}) = \vec{r} - \vec{i}x$ gives the vector projection of $\vec{r}$
      into the yz plane. Since the projection of the sum of two vectors
      is the sum of the projections and the projection of a times a
      vector is a times the projection, this function is linear.
      Formally

            $\vec{F}(\vec{r}_1 + \vec{r}_2) = \vec{r}_1 + \vec{r}_2 - \vec{i}(x_1 + x_2) = \vec{F}(\vec{r}_1) + \vec{F}(\vec{r}_2),$

            $\vec{F}(a\vec{r}) = a\vec{r} - \vec{i}(ax) = a(\vec{r} - \vec{i}x) = a\vec{F}(\vec{r}).$

8.   We must compare $\ln(a+b)$ and $\ln a + \ln b$. Now $\ln a + \ln b = \ln(ab)$
      which is not in general equal to $\ln(a+b)$; thus $\ln x$ is not a
      linear function. Also $\ln(kx) = \ln k + \ln x \neq k \ln x$ in general, so
      $\ln x$ does not satisfy either of the requirements for a linear
      function. Or we could think of finding the logarithm as an
      operation, and say that $\ln$ is not a linear operator.

11. Recall from text, page 47, that $|z| = |x + iy| = r = \sqrt{x^2 + y^2}$. Then

$|z_1 + z_2| = |x_1 + x_2 + i(y_1 + y_2)| = \sqrt{(x_1 + x_2)^2 + (y_1 + y_2)^2}$. This is

not, in general, equal to $|z_1| + |z_2| = \sqrt{x_1^2 + y_1^2} + \sqrt{x_2^2 + y_2^2}$, so

finding the absolute value is not a linear operation. (Compare

the corresponding discussion of finding the length of a vector --

see text Figure 7.1.)

14. We want to compare the maximum value of $f(x) + g(x)$ with

$\max f(x) + \max g(x)$. For example, let $f(x) = \sin x$, $g(x) = \cos x$.

Think of the graphs of $\sin x$, $\cos x$, and $\sin x + \cos x$. The maximum

of $\sin x$ is 1 (at, say, $x = \pi/2$); the maximum of $\cos x$ is 1 (at,

say, $x = 0$). Then the sum of the maximum values is $1 + 1 = 2$. The

maximum of $\sin x + \cos x$ occurs at, for example, $x = \pi/4$ (set the

derivative equal to zero), so the maximum value is

$\sin \frac{\pi}{4} + \cos \frac{\pi}{4} = 2 \cdot \frac{1}{\sqrt{2}} = \sqrt{2}$. We see that the sum of the maximum

values is not equal to the maximum of the sum of the functions.

Thus finding the maximum is not a linear operation. You can see

the reason for this from our example. In general, two different

functions take their maximum values at different points and the

sum of the functions is a maximum at still another point. Thus

there is no general relation between the maximum of the sum and

the sum of the maxima.

Section 8

4.   We write and row reduce the augmented matrix.

$$\begin{pmatrix} -1 & 1 & -1 & 4 \\ 1 & -1 & 2 & 3 \\ 2 & -2 & 4 & 6 \end{pmatrix} \xrightarrow[\text{R3 + 2R1}]{\text{R2 + R1}} \begin{pmatrix} -1 & 1 & -1 & 4 \\ 0 & 0 & 1 & 7 \\ 0 & 0 & 2 & 14 \end{pmatrix}$$

$$\xrightarrow{\text{R3 - 2R2}} \begin{pmatrix} -1 & 1 & -1 & 4 \\ 0 & 0 & 1 & 7 \\ 0 & 0 & 0 & 0 \end{pmatrix}$$

The equations corresponding to the reduced matrix are

$$\begin{cases} -x + y - z = 4 \\ \qquad z = 7 \end{cases} \quad \text{or} \quad \begin{cases} x = y - 11 \\ z = 7 \end{cases}$$

Thus y is arbitrary, but if we choose a value for y, then x and z are determined. This problem is an example of text equation (8.12c) with $R = 2$ (both the matrix of the coefficients and the augmented matrix have rank 2) and $n = 3$ (number of unknowns). We solve for $R = 2$ unknowns (x and z), corresponding to the non-zero pivots, in terms of $n - r = 1$ unknown (y).

The solution set consists of all points on the line of intersection of the two planes $x = y - 11$ and $z = 7$. We can write the solution in the vector form [see text equation (8.16)]:

$$\vec{r} = (x, y, z) = (y - 11, y, 7) = (-11, 0, 7) + (1, 1, 0)y,$$

where y is arbitrary. This is a straight line through the point $(-11, 0, 7)$. The parallel straight line through the origin gives the solution of the corresponding homogeneous equations [see text equations (8.17) and (8.18)].

5.   We row reduce the augmented matrix as follows.

$$\begin{pmatrix} 1 & -1 & 2 & 5 \\ 2 & 3 & -1 & 4 \\ 2 & -2 & 4 & 6 \end{pmatrix} \xrightarrow[R3 - 2R1]{R2 - 2R1} \begin{pmatrix} 1 & -1 & 2 & 5 \\ 0 & 5 & -5 & -6 \\ 0 & 0 & 0 & -4 \end{pmatrix}$$

The "equation" corresponding to the last row would be
$0 \cdot z = -4$, which is not possible. Thus the equations are incon-
sistent. This problem is an example of text equation (8.12a):
the rank of the coefficient matrix is 2 (2 non-zero rows in
the reduced matrix) but the rank of the augmented matrix is
3; these ranks are not equal and there is no solution.

11.   Note that the equations in this problem are the homogeneous
equations corresponding to Problem 12. As in text equations
(8.13) to (8.18), the solution (below) of Problem 12 includes
the solution of Problem 11; we simply replace the last column
by zeros.

12. Interchange the first two equations and then row reduce the augmented matrix.

$$
\begin{pmatrix}
1 & 2 & -1 & 5 \\
2 & -3 & 5 & 3 \\
1 & -5 & 6 & -2 \\
4 & 1 & 3 & 13
\end{pmatrix}
\begin{array}{l} R2 - 2R1 \\ R3 - R1 \\ R4 - 4R1 \end{array}
\begin{pmatrix}
1 & 2 & -1 & 5 \\
0 & -7 & 7 & -7 \\
0 & -7 & 7 & -7 \\
0 & -7 & 7 & -7
\end{pmatrix}
\begin{array}{l} R3 - R2 \\ R4 - R2 \\ R2 \div (-7) \end{array}
$$

$$
\begin{pmatrix}
1 & 2 & -1 & 5 \\
0 & 1 & -1 & 1 \\
0 & 0 & 0 & 0 \\
0 & 0 & 0 & 0
\end{pmatrix}
\begin{array}{l} R1 - 2R2 \end{array}
\begin{pmatrix}
1 & 0 & 1 & 3 \\
0 & 1 & -1 & 1 \\
0 & 0 & 0 & 0 \\
0 & 0 & 0 & 0
\end{pmatrix}
$$

The corresponding equations are

$$
\begin{cases} x + z = 3 \\ y - z = 1 \end{cases} \quad \text{or} \quad \begin{cases} x = 3 - z, \\ y = 1 + z, \end{cases} \quad \text{with } z \text{ arbitrary.}
$$

In vector form

$$
\vec{r} = (x, y, z) = (3 - z, 1 + z, z) = (3, 1, 0) + (-1, 1, 1)z.
$$

This is a straight line through the point $(3,1,0)$. The parallel line through the origin gives the solution of Problem 11:

$$
(x, y, z) = (-1, 1, 1)z \quad \text{or} \quad \begin{cases} x = -z, \\ y = z, \end{cases} \quad z \text{ arbitrary.}
$$

19.   We row reduce the augmented matrix.

$$\begin{pmatrix} 1 & -3 & -2 & 0 & 3 & 1 \\ 0 & 0 & 1 & 3 & 5 & -1 \\ 1 & -3 & 1 & 1 & 2 & 6 \\ 0 & 0 & 1 & 1 & 1 & 1 \end{pmatrix} \xrightarrow{R3\ -\ R1} \begin{pmatrix} 1 & -3 & -2 & 0 & 3 & 1 \\ 0 & 0 & 1 & 3 & 5 & -1 \\ 0 & 0 & 3 & 1 & -1 & 5 \\ 0 & 0 & 1 & 1 & 1 & 1 \end{pmatrix} \xrightarrow[R4\ -\ R2]{R3\ -\ 3R2}$$

$$\begin{pmatrix} 1 & -3 & -2 & 0 & 3 & 1 \\ 0 & 0 & 1 & 3 & 5 & -1 \\ 0 & 0 & 0 & -8 & -16 & 8 \\ 0 & 0 & 0 & -2 & -4 & 2 \end{pmatrix} \xrightarrow[R3\ \div\ (-8)]{R4\ -\ \frac{1}{4}R3} \begin{pmatrix} 1 & -3 & -2 & 0 & 3 & 1 \\ 0 & 0 & 1 & 3 & 5 & -1 \\ 0 & 0 & 0 & 1 & 2 & -1 \\ 0 & 0 & 0 & 0 & 0 & 0 \end{pmatrix}$$

$$\xrightarrow{R2\ -\ 3R3} \begin{pmatrix} 1 & -3 & -2 & 0 & 3 & 1 \\ 0 & 0 & 1 & 0 & -1 & 2 \\ 0 & 0 & 0 & 1 & 2 & -1 \\ 0 & 0 & 0 & 0 & 0 & 0 \end{pmatrix} \xrightarrow{R1\ +\ 2R2} \begin{pmatrix} 1 & -3 & 0 & 0 & 1 & 5 \\ 0 & 0 & 1 & 0 & -1 & 2 \\ 0 & 0 & 0 & 1 & 2 & -1 \\ 0 & 0 & 0 & 0 & 0 & 0 \end{pmatrix}$$

(We have used the pivot 1's to obtain zeros above the pivots --
this is an alternative to back substitution.)   The equations
corresponding to the reduced matrix are

$$\begin{cases} x - 3y + v = 5 \\ \qquad z - v = 2 \\ \qquad u + 2v = -1 \end{cases} \quad \text{or} \quad \begin{cases} x = 5 + 3y - v \\ z = 2 + v \\ u = -1 - 2v \end{cases}$$

Here y and v are arbitrary.

23. By text (8.3), page 130, the rank of a matrix is the number of nonzero rows when the matrix is in echelon form. Thus, we row reduce the given matrix.

$$\begin{pmatrix} 1 & 1 & 4 & 3 \\ 3 & 1 & 10 & 7 \\ 4 & 2 & 14 & 10 \\ 2 & 0 & 6 & 4 \end{pmatrix} \begin{matrix} \\ R2-3R1 \\ R3-4R1 \\ R4-2R1 \end{matrix} \begin{pmatrix} 1 & 1 & 4 & 3 \\ 0 & -2 & -2 & -2 \\ 0 & -2 & -2 & -2 \\ 0 & -2 & -2 & -2 \end{pmatrix} \begin{matrix} \\ \\ R3-R2 \\ R4-R2 \end{matrix} \begin{pmatrix} 1 & 1 & 4 & 3 \\ 0 & -2 & -2 & -2 \\ 0 & 0 & 0 & 0 \\ 0 & 0 & 0 & 0 \end{pmatrix}$$

Then by text (8.3), the rank is 2 since there are two non-zero rows.

26. We want numbers a and b so that

$$(1,4,-5) \equiv (9,0,7)a + (0,-9,13)b.$$

Taking components, we get

$$1 = 9a, \qquad 4 = -9b, \qquad -5 = 7a + 13b.$$

The first two equations give $a = \frac{1}{9}$, $b = -\frac{4}{9}$; the third equation,

$-5 = \frac{7 - 4 \cdot 13}{9} = -\frac{45}{9} = -5$, checks our work. Thus

$$(1,4,-5) = \frac{1}{9}(9,0,7) - \frac{4}{9}(0,-9,13).$$

More simply, if we use as basis vectors $u = \frac{1}{9}(9,0,7) = \left(1,0,\frac{7}{9}\right)$ and $v = \frac{1}{9}(0,-9,13) = \left(0,-1,\frac{13}{9}\right)$, then u gives the right x component and -4v gives the right y component; thus we have $(1,4,5) = u - 4v$ as before. Similarly, for the second given vector, $(5,2,1) = 5u - 2v$ gives the x and y components; for z:

$$1 = 5 \cdot \frac{7}{9} - 2 \cdot \frac{13}{9} = \frac{35 - 26}{9} = 1, \text{ which checks.}$$

In the same way, we can write the other two vectors in terms of the basis vectors.

30. By text (8.11), page 133, the set of three equations in three unknowns has nontrivial solutions if and only if the determinant of the coefficients is zero. Thus we want

$$\begin{vmatrix} -(1+\lambda) & 1 & 3 \\ 1 & 2-\lambda & 0 \\ 3 & 0 & 2-\lambda \end{vmatrix} = 0.$$

We expand the determinant using the last column:

$$3 \begin{vmatrix} 1 & 2-\lambda \\ 3 & 0 \end{vmatrix} + (2-\lambda) \begin{vmatrix} -(1+\lambda) & 1 \\ 1 & 2-\lambda \end{vmatrix}$$

$$= -9(2-\lambda) + (2-\lambda)[-(1+\lambda)(2-\lambda)-1]$$

$$= (2-\lambda)(-9-2-\lambda+\lambda^2-1) = (2-\lambda)(\lambda^2-\lambda-12)$$

$$= (2-\lambda)(\lambda-4)(\lambda+3) = 0.$$

Thus the determinant of the coefficients is zero if $\lambda = 2$ or 4 or -3. For each of these values of λ we solve the set of equations for (x,y,z).

$\lambda = 2$: $\left.\begin{array}{l} -3x+y+3z=0 \\ x=0 \\ 3x=0 \end{array}\right\}$ $x=0, \qquad y=-3z, \qquad$ or
$(x,y,z) = (0,-3,1)z$

$\lambda = 4$: $\left.\begin{array}{l} -5x+y+3z=0 \\ x-2y=0 \\ 3x-2z=0 \end{array}\right\}$ $x=2y, \qquad z=3y, \qquad$ or
$(x,y,z) = (2,1,3)y$

$\lambda = -3$: $\left.\begin{array}{l} 2x+y+3z=0 \\ x+5y=0 \\ 3x+5y=0 \end{array}\right\}$ $x=-5y, \qquad z=3y, \qquad$ or
$(x,y,z) = (-5,1,3)y$

You might note that the three vectors $(0,-3,1)$, $(2,1,3)$ and $(-5,1,3)$ are mutually orthogonal (find the dot products); this is not an accident; see Chapter 10, Section 4 of the text.

31. If the four points (x_1, y_1, z_1), (x_2, y_2, z_2), etc., all satisfy the
equation $ax + by + cz = d$ of a plane, then we have

$$\begin{cases} ax_1 + by_1 + cz_1 - d = 0 \\ ax_2 + by_2 + cz_2 - d = 0 \\ ax_3 + by_3 + cz_3 - d = 0 \\ ax_4 + by_4 + cz_4 - d = 0. \end{cases}$$

By text equation (8.11), we can solve for (a, b, c, d) -- and so
find the equation of the plane -- if and only if the determinant
of the coefficients is zero. Thus the requirement is

$$\begin{vmatrix} x_1 & y_1 & z_1 & -1 \\ x_2 & y_2 & z_2 & -1 \\ x_3 & y_3 & z_3 & -1 \\ x_4 & y_4 & z_4 & -1 \end{vmatrix} = 0$$

If the four points satisfy this condition, then they lie in
a plane.

36. Using text equation (8.5), we find the Wronskian of the given
functions.

$$W = \begin{vmatrix} x & e^x & xe^x \\ 1 & e^x & xe^x + e^x \\ 0 & e^x & xe^x + 2e^x \end{vmatrix} \xrightarrow[R2 - R3]{R1 - R3} \begin{vmatrix} x & 0 & -2e^x \\ 1 & 0 & -e^x \\ 0 & e^x & e^x(x+2) \end{vmatrix}$$

$$= -e^x \begin{vmatrix} x & -2e^x \\ 1 & -e^x \end{vmatrix} = -e^x(-xe^x + 2e^x) = e^{2x}(x - 2).$$

Since $W \neq 0$, the functions are linearly independent.

40. We want $\underline{a}$ and $\underline{b}$ so that

$$\vec{V} = a\vec{A} + b\vec{B}$$

or, taking components,

$$aA_x + bB_x = V_x$$
$$aA_y + bB_y = V_y$$

The Cramer's rule solution for $\underline{a}$ and $\underline{b}$ is:

$$a = \frac{\begin{vmatrix} V_x & B_x \\ V_y & B_y \end{vmatrix}}{\begin{vmatrix} A_x & B_x \\ A_y & B_y \end{vmatrix}} = \frac{V_x B_y - V_y B_x}{A_x B_y - A_y B_x} = \frac{B_x V_y - B_y V_x}{B_x A_y - B_y A_x} = \frac{(\vec{B} \times \vec{V}) \cdot \vec{k}}{(\vec{B} \times \vec{A}) \cdot \vec{k}}$$

$$b = \frac{\begin{vmatrix} A_x & V_x \\ A_y & V_y \end{vmatrix}}{\begin{vmatrix} A_x & B_x \\ A_y & B_y \end{vmatrix}} = \frac{A_x V_y - A_y V_x}{A_x B_y - A_y B_x} = \frac{(\vec{A} \times \vec{V}) \cdot \vec{k}}{(\vec{A} \times \vec{B}) \cdot \vec{k}} ,$$

as in Problem 39. For the given vectors, we find

$$\vec{A} \times \vec{V} = \begin{vmatrix} \vec{i} & \vec{j} & \vec{k} \\ 2 & 1 & 0 \\ 3 & 5 & 0 \end{vmatrix} = \vec{k} \begin{vmatrix} 2 & 1 \\ 3 & 5 \end{vmatrix} = 7\vec{k},$$

and similarly

$$\vec{B} \times \vec{A} = \vec{k} \begin{vmatrix} 3 & -2 \\ 2 & 1 \end{vmatrix} = 7\vec{k} = -\vec{A} \times \vec{B},$$

$$\vec{B} \times \vec{V} = \vec{k} \begin{vmatrix} 3 & -2 \\ 3 & 5 \end{vmatrix} = 21\vec{k}.$$

Then $a = \dfrac{21}{7} = 3,$ $b = \dfrac{7}{-7} = -1.$

42. For Problem 11, the matrix of the coefficients and its row
     reduced form are (see the first three columns in the solution
     of Problem 12):

$$
\begin{pmatrix}
1 & 2 & -1 \\
2 & -3 & 5 \\
1 & -5 & 6 \\
4 & 1 & 3
\end{pmatrix}
\longrightarrow
\begin{pmatrix}
1 & 0 & 1 \\
0 & 1 & -1 \\
0 & 0 & 0 \\
0 & 0 & 0
\end{pmatrix}
$$

Thus the vectors $\vec{u} = (1,0,1)$ and $\vec{v} = (0,1,-1)$ are a basis for the
row space. Each of the original vectors can be written as a
linear combination of $\vec{u}$ and $\vec{v}$; for example, $(1,2,-1) = \vec{u} + 2\vec{v}$.
All these vectors (with tails at the origin) lie in a plane
passing through the origin; this plane is the row space. The
solution set of Problem 11 (see end of solution of Problem 12)
is

$$\vec{r} = (-1,1,1)t = \vec{A}t,$$

where t is arbitrary. This is a straight line through the origin
in the direction $\vec{A}$. We can verify that it is orthogonal to the
row space plane by showing that $\vec{r}$ is perpendicular to both $\vec{u}$ and
$\vec{v}$. This follows either from $\vec{r} \cdot \vec{u} = 0$, $\vec{r} \cdot \vec{v} = 0$ (verify this) or from
$\vec{u} \times \vec{v} = \vec{A}$ (verify this).

Section 9

4. We write the given matrix A, and [using the text definitions
    (9.6)] the transpose A^T, the conjugate $\bar{A}$, and the transpose
    conjugate $A^\dagger$:

$$A = \begin{pmatrix} 0 & 2i & -1 \\ -i & 2 & 0 \\ 3 & 0 & 0 \end{pmatrix}, \qquad A^T = \begin{pmatrix} 0 & -i & 3 \\ 2i & 2 & 0 \\ -1 & 0 & 0 \end{pmatrix},$$

$$\bar{A} = \begin{pmatrix} 0 & -2i & -1 \\ i & 2 & 0 \\ 3 & 0 & 0 \end{pmatrix}, \qquad A^\dagger = \begin{pmatrix} 0 & i & 3 \\ -2i & 2 & 0 \\ -1 & 0 & 0 \end{pmatrix}.$$

The adjoint of A [see text (9.6)] is the same as C^T in text
equation (6.24), that is, the transpose of the matrix of
cofactors. The cofactors are:

Row 1: $\quad \begin{vmatrix} 2 & 0 \\ 0 & 0 \end{vmatrix} = 0, \qquad -\begin{vmatrix} -i & 0 \\ 3 & 0 \end{vmatrix} = 0, \qquad \begin{vmatrix} -i & 2 \\ 3 & 0 \end{vmatrix} = -6$

Row 2: $\quad -\begin{vmatrix} 2i & -1 \\ 0 & 0 \end{vmatrix} = 0, \qquad \begin{vmatrix} 0 & -1 \\ 3 & 0 \end{vmatrix} = 3, \qquad -\begin{vmatrix} 0 & 2i \\ 3 & 0 \end{vmatrix} = 6i$

Row 3: $\quad \begin{vmatrix} 2i & -1 \\ 2 & 0 \end{vmatrix} = 2, \qquad -\begin{vmatrix} 0 & -1 \\ -i & 0 \end{vmatrix} = i, \qquad \begin{vmatrix} 0 & 2i \\ -i & 2 \end{vmatrix} = -2$

Then we have

$$C = \begin{pmatrix} 0 & 0 & -6 \\ 0 & 3 & 6i \\ 2 & i & -2 \end{pmatrix}, \qquad \text{adj } A = C^T = \begin{pmatrix} 0 & 0 & 2 \\ 0 & 3 & i \\ -6 & 6i & -2 \end{pmatrix}.$$

Using the last column of A, we find $\det A = 6$. Then by text
equation (6.4), $A^{-1} = C^T/\det A = \frac{1}{6}C^T$.

   We can check our work by showing that $AA^{-1} =$ the unit matrix U
and also $A^{-1}A = U$.

$$A^{-1}A = \frac{1}{6}\begin{pmatrix} 0 & 0 & 2 \\ 0 & 3 & i \\ -6 & 6i & -2 \end{pmatrix}\begin{pmatrix} 0 & 2i & -1 \\ -i & 2 & 0 \\ 3 & 0 & 0 \end{pmatrix} = \frac{1}{6}\begin{pmatrix} 6 & 0 & 0 \\ 0 & 6 & 0 \\ 0 & 0 & 6 \end{pmatrix} = \begin{pmatrix} 1 & 0 & 0 \\ 0 & 1 & 0 \\ 0 & 0 & 1 \end{pmatrix}.$$

Similarly, $AA^{-1} = U$.

5.   Hint: See text Chapter 10, Section 1.

8.   Hint: Multiply AA^T; see text equations (9.7).

9.   By text equations (9.7), if A is orthogonal, then $AA^T = U$, so
      $det(AA^T) = 1$.
      By text equation (6.2c), $det\ AA^T = (det\ A)(det\ A^T)$, so
      $(det\ A)(det\ A^T) = 1$.
      By Fact 4a (text page 89), $det\ A^T = det\ A$.   Thus $(det\ A)^2 = 1$,
      $det\ A = \pm 1$.

10.   By definition, if A is orthogonal, then $A^{-1} = A^T$.   By text
      equation (6.24), if $det\ A = 1$, then $A^{-1} = C^T$.   Thus $A^T = C^T$ or
      $A = C$, that is, A and the matrix C of cofactors are identical.

13.   Hint: Multiply $AA^\dagger$; you should get U [see text equations (9.6)
      and (9.7)].

16.   The Laplace development says that

$$det\ A = \sum_j a_{ij} C_{ij} \quad \text{(elements of row i times their cofactors),}$$

or

$$det\ A = \sum_i a_{ij} C_{ij} \quad \text{(elements of column j times their cofactors)}.$$

Problem 3.8 says that if we multiply the elements of row i by the
cofactors of another row, say row k, we get zero:

$$\sum_j a_{ij} C_{kj} = 0, \qquad k \neq i.$$

Thus

$$\sum_j a_{ij} C_{kj} = \begin{cases} 0, & k \neq i \\ det\ A, & k = i \end{cases} = \delta_{ik} \cdot det\ A$$

[see text equation (9.2)].   Similarly, if we multiply the
elements of column j by the cofactors of another column, say

16.   (continued)

column k, we get zero.   Thus

$$\sum_i a_{ij} C_{ik} = \begin{cases} 0, & k \neq j \\ \det A, & k = j \end{cases} = \delta_{jk} \cdot \det A.$$

Section 10

11.   Hint: Expand each determinant using elements of the second
column.   It is not necessary to expand the 2 by 2 cofactor
determinants.   For example, it should be clear that

$$(a_{12} + b_{12}) \begin{vmatrix} a_{21} & a_{23} \\ a_{31} & a_{33} \end{vmatrix} = a_{12} \begin{vmatrix} a_{21} & a_{23} \\ a_{31} & a_{33} \end{vmatrix} + b_{12} \begin{vmatrix} a_{21} & a_{23} \\ a_{31} & a_{33} \end{vmatrix}$$

without expanding the 2 by 2 determinants.

12.   Perform only one operation at a time!   Once you have added the
first row to the second row, you have a new second row.   If you
now add this new second row to the first row, the rows are not
identical.

16.   As in text Section 5, Figure 5.6, we find a point Q, say
(x_1, y_1, z_1), in the plane; then $ax_1 + by_1 + cz_1 = d$.   A unit normal
to the plane is $\vec{n} = (\vec{i}a + \vec{j}b + \vec{k}c)/\sqrt{a^2 + b^2 + c^2}$.   The vector from
Q to P is $\vec{QP} = (x_0, y_0, z_0) - (x_1, y_1, z_1)$, and the distance from P
to the plane is

$$|PR| = |\vec{n} \cdot \vec{QP}| = |a(x_0 - x_1) + b(y_0 - y_1) + c(z_0 - z_1)|/\sqrt{a^2 + b^2 + c^2}$$

$$= |ax_0 + by_0 + cz_0 - d|/\sqrt{a^2 + b^2 + c^2}.$$

23.   Hints: On the x axis, $y = z = 0$; then
       the intersection point of the plane
       $2x + 3y + 6z = 6$ with the x axis is the
       point $P = (3,0,0)$. Similarly find
       Q and R (see first figure). Now
       for any triangle PQR (see second
       figure), if $|PQ|$ is the base, the
       altitude is $|PR| \sin \theta$, where θ is
       the angle between $\vec{PQ}$ and $\vec{PR}$. Thus

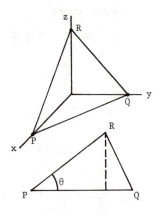

$$\text{area} = \tfrac{1}{2} |PQ| |PR| \sin \theta = \tfrac{1}{2} |\vec{PQ} \times \vec{PR}|.$$

Use this result to find the area of the given triangle.

Chapter 4

Section 1

4. Given $w = x^3 - y^3 - 2xy + 6$, we find

$$\frac{\partial w}{\partial x} = 3x^2 - 2y, \qquad\qquad \frac{\partial^2 w}{\partial x^2} = 6x,$$

$$\frac{\partial w}{\partial y} = -3y^2 - 2x, \qquad\qquad \frac{\partial^2 w}{\partial y^2} = -6y.$$

When $\frac{\partial w}{\partial x} = \frac{\partial w}{\partial y} = 0$, we find

$$0 = 3x^2 - 2y \qquad \text{or} \qquad y = \frac{3}{2}x^2,$$

$$0 = 3y^2 + 2x \qquad \text{or} \qquad x = -\frac{3}{2}y^2 = -\frac{3}{2}\left(\frac{3}{2}x^2\right)^2 = -\left(\frac{3}{2}\right)^3 x^4.$$

Thus $x = 0$ or $x^3 = -\left(\frac{2}{3}\right)^3$, $x = -\frac{2}{3}$. If $x = 0$, then $y = 0$;

if $x = -2/3$, then $y = 2/3$.

At $(0,0)$, $\qquad\qquad \frac{\partial^2 w}{\partial x^2} = 0, \quad \frac{\partial^2 w}{\partial y^2} = 0.$

At $(-2/3, 2/3)$, $\qquad\quad \frac{\partial^2 w}{\partial x^2} = -4, \quad \frac{\partial^2 w}{\partial y^2} = -4.$

9. To find $\left(\frac{\partial z}{\partial x}\right)_\theta$, we first write z in terms of x and θ. From the figure, we have $y = x \tan \theta$. Thus

$$z = x^2 + 2y^2 = x^2 + 2(x \tan \theta)^2$$
$$= x^2(1 + 2\tan^2 \theta).$$

$$\left(\frac{\partial z}{\partial x}\right)_\theta = 2x(1 + 2\tan^2 \theta).$$

$$\frac{y}{x} = \tan \theta$$

13.   From Problem 9 above, we have $z = x^2(1 + 2\tan^2\theta)$.   Thus

$$\left(\frac{\partial z}{\partial \theta}\right)_x = x^2(4\tan\theta\ \sec^2\theta).$$

We can write this in simpler form.
From the figure, $r = x\sec\theta$.   Thus

$\frac{r}{x} = \sec\theta$

$$\left(\frac{\partial z}{\partial \theta}\right)_x = 4r^2\tan\theta.$$

14.   We need z in terms of θ and y.   From
the figure, $x = y\cot\theta$.   Then

$\frac{x}{y} = \cot\theta$

$\frac{r}{y} = \csc\theta$

$$z = x^2 + 2y^2 = y^2\cot^2\theta + 2y^2,$$

$$\left(\frac{\partial z}{\partial \theta}\right)_y = y^2(2\cot\theta)(-\csc^2\theta) = -2y^2\cot\theta\ \csc^2\theta.$$

From the figure $y\csc\theta = r$, so a simpler form for the answer is

$$\left(\frac{\partial z}{\partial \theta}\right)_y = -2r^2\cot\theta.$$

20.   From either Problem 9 or Problem 13, we can find

$$\frac{\partial^2 z}{\partial x \partial \theta} = \frac{\partial^2 z}{\partial \theta \partial x}\ .$$

From Problem 9:

$$\frac{\partial^2 z}{\partial \theta \partial x} = \frac{\partial}{\partial \theta}\frac{\partial z}{\partial x} = \frac{\partial}{\partial \theta}(2x + 4x\tan^2\theta) = 4x(2\tan\theta)(\sec^2\theta)$$

$$= 8x\tan\theta\ \sec^2\theta.$$

Since $x\tan\theta = y$ (see figure, Problem 9), a simpler answer is

$$\frac{\partial^2 z}{\partial \theta \partial x} = 8y\sec^2\theta.$$

Alternatively, from the answer to Problem 13 in terms of x and θ:

$$\frac{\partial^2 z}{\partial x \partial \theta} = \frac{\partial}{\partial x}(4x^2\tan\theta\ \sec^2\theta) = 8x\tan\theta\ \sec^2\theta = 8y\sec^2\theta$$

as before.

21.   From Problem 14, we have

$$\left(\frac{\partial z}{\partial \theta}\right)_y = -2y^2 \cot \theta \csc^2 \theta.$$

Then

$$\frac{\partial^2 z}{\partial y \partial \theta} = -4y \cot \theta \csc^2 \theta.$$

Since $y \cot \theta = x$ (see figure, Problem 14), a simpler result is

$$\frac{\partial^2 z}{\partial y \partial \theta} = -4x \csc^2 \theta.$$

8'.   We first write z in terms of x and r. From the figure,

$$\tan^2 \theta = \frac{y^2}{x^2} = \frac{r^2 - x^2}{x^2} \quad .$$

Thus

$$z = r^2 \tan^2 \theta = \frac{r^2(r^2 - x^2)}{x^2} = \frac{r^4}{x^2} - r^2 .$$

$$\tan \theta = \frac{y}{x}$$

$$y^2 = r^2 - x^2$$

$$\left(\frac{\partial z}{\partial x}\right)_r = r^4 (-2x^{-3}) = -2r^4/x^3 .$$

17'.   From Problem 8', we have $z = \frac{r^4}{x^2} - r^2.$ Then

$$\left(\frac{\partial z}{\partial r}\right)_x = \frac{4r^3}{x^2} - 2r.$$

22'.   From either Problem 8' or Problem 17' we find

$$\frac{\partial^2 z}{\partial r \partial x} = \frac{\partial}{\partial r}\frac{\partial z}{\partial x} = \frac{-8r^3}{x^3} ,$$

$$\frac{\partial^2 z}{\partial x \partial r} = \frac{\partial}{\partial x}\frac{\partial z}{\partial r} = (4r^3)(-2x^{-3}) = -\frac{8r^3}{x^3} .$$

Since $\frac{r}{x} = \sec \theta$, we could write this in the simpler form

$$\frac{\partial^2 z}{\partial x \partial r} = -8 \sec^3 \theta.$$

Section 2

5. We use the binomial series [text, page 24, equation (13.5)] for $(1+u)^{1/2}$ and then replace u by xy.

$$(1+xy)^{1/2} = 1 + \frac{1}{2}xy + \frac{\left(\frac{1}{2}\right)\left(-\frac{1}{2}\right)}{2!}(xy)^2 + \frac{\left(\frac{1}{2}\right)\left(-\frac{1}{2}\right)\left(-\frac{3}{2}\right)}{3!}(xy)^3$$
$$+ \frac{\left(\frac{1}{2}\right)\left(-\frac{1}{2}\right)\left(-\frac{3}{2}\right)\left(-\frac{5}{2}\right)}{4!}(xy)^4 \cdots$$
$$= 1 + \frac{1}{2}xy - \frac{1}{8}x^2y^2 + \frac{1}{16}x^3y^3 - \frac{5}{128}x^4y^4 \cdots .$$

8. By text, page 59, equation (8.1),

$$e^z = 1 + z + \frac{z^2}{2!} + \frac{z^3}{3!} + \cdots .$$

With z = x + iy, we have

$$e^{x+iy} = 1 + x + iy + \frac{(x+iy)^2}{2!} + \frac{(x+iy)^3}{3!} + \cdots$$
$$= 1 + x + iy + \frac{1}{2}(x^2 + 2ixy - y^2) + \frac{1}{6}(x^3 + 3ix^2y - 3xy^2 - iy^3) + \cdots .$$

Now $e^{x+iy} = e^x e^{iy} = e^x(\cos y + i \sin y)$. Thus the real terms in the e^z series give the series for $e^x \cos y$ and the imaginary part of the e^z series is the series for $e^x \sin y$. Thus:

$$e^x \cos y = 1 + x + \frac{1}{2}(x^2 - y^2) + \frac{1}{6}(x^3 - 3xy^2) + \cdots ,$$

$$e^x \sin y = y + xy + \frac{1}{6}(3x^2y - y^3) + \cdots .$$

Of course these series may also be obtained separately by multiplying the e^x series times the cos y or sin y series. For example, the $e^x \cos y$ series to third order terms is:

$$e^x \cos y = \left(1 + x + \frac{x^2}{2!} + \frac{x^3}{3!} - \cdots\right)\left(1 - \frac{y^2}{2!} + \cdots\right)$$
$$= 1 + x + \frac{x^2}{2!} + \frac{x^3}{3!} - \frac{y^2}{2!} - \frac{xy^2}{2!} + \cdots$$
$$= 1 + x + \frac{1}{2}(x^2 - y^2) + \frac{1}{6}(x^3 - 3xy^2) + \cdots$$

as above.

8.   (continued)

We can also obtain these series by using the general formulas for the coefficients [text equations (2.2), (2.3) or (2.7)]. For $f(x,y) = e^x\cos y$, we find the values of f and its derivatives at $(0,0)$:

$$f = e^x\cos y = 1,$$

$$f_x = e^x\cos y = 1,$$

$$f_y = -e^x\sin y = 0,$$

$$f_{xx} = e^x\cos y = 1,$$

$$f_{xy} = -e^x\sin y = 0,$$

$$f_{yy} = -e^x\cos y = -1,\ \text{etc.}$$

Then

$$e^x\cos y = 1 + x + \frac{1}{2!}(x^2 - y^2) + \cdots$$

as before.

Section 4

4.   We take differentials of the equation $\frac{1}{i} + \frac{1}{o} = \frac{1}{f}$ to get

$$-\frac{di}{i^2} - \frac{do}{o^2} = 0.\quad (f = \text{focal length of lens} = \text{const.})$$

Then if $i = 12$, $o = 18$, and o changes to 17.5, that is, $do = -0.5$, we find

$$di = -\frac{i^2}{o^2}\,do = -\left(\frac{12}{18}\right)^2(-0.5) = \left(\frac{2}{3}\right)^2\left(\frac{1}{2}\right) = \frac{2}{9} = 0.2,$$

$$i + di = 12.2.$$

6.   Given $g = \dfrac{4\pi^2 \ell}{T^2}$, we have

$$dg = 4\pi^2 \left(\frac{d\ell}{T^2} - \frac{2\ell dT}{T^3} \right) ,$$

(Relative error in g) $= \dfrac{dg}{g} = \dfrac{d\ell}{\ell} - \dfrac{2dT}{T}$.

In the worst case, $d\ell$ and dT have opposite signs. Then if $\left| \dfrac{d\ell}{\ell} \right| = 5\%$ and $\left| \dfrac{dT}{T} \right| = 2\%$,

$$\left| \frac{dg}{g} \right| = 5\% + 2\,(2\%) = 9\%.$$

11.   We consider $f(x,y) = \sqrt{x^2 - y^2}$. When $x = 5$ and $y = 3$, then $f = 4$. We estimate the change in f if x changes to 4.98 and y changes to 3.03, that is, if $dx = -0.02$ and $dy = 0.03$.

$$df = \frac{1}{2}(x^2 - y^2)^{-1/2}(2x\,dx - 2y\,dy) = \frac{1}{f}(x\,dx - y\,dy) .$$

Then for the values $x = 5$, $y = 3$, $f = 4$, $dx = -0.02$, $dy = 0.03$, we have

$$df = \frac{1}{4}[5(-0.02) - 3(0.03)] = -\frac{0.19}{4} \cong -0.05 ,$$

$$f + df = 4 - 0.05 = 3.95 .$$

15.   Hint: The derivative of an integral with respect to the upper limit is the integrand evaluated at the upper limit. In symbols:

$$\frac{d}{dv} \int_0^v f(x)\,dx = f(v),$$

or, using differentials:

$$d \int_0^v f(x)\,dx = f(v)\,dv.$$

Section 5

3. We find the differentials of $r = e^{-p^2 - q^2}$, $p = e^s$, $q = e^{-s}$:

$$dr = e^{-p^2 - q^2}(-2p\ dp - 2q\ dq) = -2r(p\ dp + q\ dq),$$

$$dp = e^s ds = p\ ds,$$

$$dq = e^{-s}(-ds) = -q\ ds.$$

Substitute dp and dq into dr:

$$dr = -2r[p^2 ds + q(-q\ ds)] = 2r(q^2 - p^2)ds.$$

Divide by ds to find:

$$\frac{dr}{ds} = 2r(q^2 - p^2).$$

5. Hint: In text equation (5.1), we have $z(x,y)$, $x(t)$ and $y(t)$.
    Suppose $x \equiv t$.

Section 6

2. We differentiate $ye^{xy} = \sin x$ implicitly with respect to x to get:

(1) $y\frac{d}{dx}(e^{xy}) + \frac{dy}{dx} e^{xy} = \cos x$.

Again differentiate with respect to x, remembering that y and $\frac{dy}{dx}$ are functions of x:

(2) $y\frac{d^2}{dx^2}(e^{xy}) + 2\frac{dy}{dx}\frac{d}{dx}(e^{xy}) + e^{xy}\frac{d^2 y}{dx^2} = -\sin x.$

Now at (0,0), we find from equation (1)

$$0 + \frac{dy}{dx} \cdot 1 = 1 \qquad \text{or} \qquad \frac{dy}{dx} = 1.$$

In equation (2) we need the value of $\frac{d}{dx}(e^{xy})$ at (0,0):

$$\frac{d}{dx}(e^{xy}) = e^{xy}\left(x\frac{dy}{dx} + y\right) = 0 \text{ at } (0,0).$$

Thus from equation (2) at (0,0):

$$0 + 0 + \frac{d^2 y}{dx^2} = 0 \qquad \text{or} \qquad \frac{d^2 y}{dx^2} = 0.$$

2.   (continued)

Note that we must differentiate <u>before</u> we substitute the numerical values for x and y. However, it may save algebra to substitute the values of x and y before we solve for y' and y". Here
the general formulas are quite messy but the numerical values
are simple.

5.   We differentiate implicitly the equation $xy^3 - yx^3 = 6$ to get:
$$y^3 + 3xy^2y' - x^3y' - 3x^2y = 0.$$
At the point (1,2), we have
$$8 + 12y' - y' - 6 = 0 \qquad \text{or} \qquad y' = -\tfrac{2}{11}.$$
Thus the slope of the tangent line at (1,2) is -2/11 and the
equation of the tangent line is
$$\frac{y-2}{x-1} = -\tfrac{2}{11} \qquad \text{or} \qquad 2x + 11y = 24.$$

11.   Caution: Differentiate again the equation you had in Problem 10
<u>before</u> you substituted x = y = 0. Then as soon as you have differ-
entiated the second time, let x = y = 0 and solve for y".

Section 7

1.   Given x = yz and y = 2sin(y + z), we take differentials of both
equations:

(1) $dx = y\,dz + z\,dy$,

(2) $dy = 2\cos(y+z)(dy + dz)$.

Since we want $\frac{dx}{dy}$, we eliminate dz. From equation (1):

$$dz = \frac{dx - z\,dy}{y}.$$

Substitute this into equation (2):

1.   (continued)
$$dy = 2\cos(y + z)\left(dy + \frac{dx - z \, dy}{y}\right).$$

Clear fractions and solve for dx:

$$y \, dy = 2\cos(y + z)(y \, dy + dx - z \, dy),$$

$$dx = \frac{y \, dy}{2\cos(y + z)} - y \, dy + z \, dy.$$

Divide by dy and use the given equation $y = 2\sin(y + z)$:

(3) $\frac{dx}{dy} = \frac{y}{2\cos(y + z)} - y + z = z - y + \tan(y + z).$

Note that we can also find $\frac{dz}{dy}$ [from equation (2) above]:

$$dz = \frac{dy}{2\cos(y + z)} - dy = \frac{1}{2}\sec(y + z)\,dy - dy,$$

(4) $\frac{dz}{dy} = \frac{1}{2}\sec(y + z) - 1.$

To find $\frac{d^2x}{dy^2}$, we differentiate equation (3), remembering that z is a function of y, and use equation (4). We must <u>not</u> use differentials in finding second derivatives.

$$\frac{d^2x}{dy^2} = \frac{dz}{dy} - 1 + \sec^2(y + z)\left(1 + \frac{dz}{dy}\right)$$

$$= \frac{1}{2}\sec(y + z) - 2 + \frac{1}{2}\sec^3(y + z).$$

4.   Using differentials, we find

$$dw = e^{-r^2 - s^2}(-2r \, dr - 2s \, ds) = -2w(r \, dr + s \, ds),$$

$$dr = u \, dv + v \, du,$$

$$ds = du + 2 \, dv .$$

Substitute dr and ds into dw:

$$dw = -2w(ru \, dv + rv \, du + s \, du + 2s \, dv)$$

$$= -2w(ru + 2s)\,dv - 2w(rv + s)\,du.$$

4.   (continued)

We can now read the required partial derivatives from dw because the coefficient of dv is $\frac{\partial w}{\partial v}$ and the coefficient of du is $\frac{\partial w}{\partial u}$:

$$\frac{\partial w}{\partial u} = -2w(rv + s),$$

$$\frac{\partial w}{\partial v} = -2w(ru + 2s).$$

Alternatively, we can use the chain rule. For example,

$$\frac{\partial w}{\partial u} = \frac{\partial w}{\partial r}\frac{\partial r}{\partial u} + \frac{\partial w}{\partial s}\frac{\partial s}{\partial u}.$$

The needed four derivatives are:

$$\frac{\partial w}{\partial r} = \left(e^{-r^2-s^2}\right)(-2r) = -2rw, \qquad \frac{\partial w}{\partial s} = -2sw,$$

$$\frac{\partial r}{\partial u} = v, \qquad \frac{\partial s}{\partial u} = 1.$$

Thus

$$\frac{\partial w}{\partial u} = -2rwv - 2sw,$$

as above. Similarly, find $\frac{\partial w}{\partial v}$ by the chain rule.

10.   We take differentials of the given equations:

$$2x\, dx + 2y\, dy = 2t\, ds + 2s\, dt,$$

$$2y\, dx + 2x\, dy = 2s\, ds - 2t\, dt.$$

Cancel the 2's and substitute $(x, y, s, t) = (4, 2, 5, 3)$:

$$4\, dx + 2\, dy = 3\, ds + 5\, dt,$$

$$2\, dx + 4\, dy = 5\, ds - 3\, dt.$$

Solve these equations simultaneously for dx and dy:

$$dx = \frac{1}{6}(ds + 13\, dt),$$

$$dy = \frac{1}{6}(7\, ds - 11\, dt).$$

Then the coefficients of dt and ds are the required partial derivatives:

$$\frac{\partial x}{\partial s} = \frac{1}{6}, \quad \frac{\partial x}{\partial t} = \frac{13}{6}, \quad \frac{\partial y}{\partial s} = \frac{7}{6}, \quad \frac{\partial y}{\partial t} = \frac{-11}{6}.$$

15. We take differentials of both equations:

$$2xu\,dx + x^2\,du - 2yv\,dy - y^2\,dv = 0,$$

$$dx + dy = udv + vdu.$$

To find $\left(\frac{\partial x}{\partial u}\right)_v$, we set $dv = 0$ and eliminate dy between the two equations. From the second equation with $dv = 0$, we find

$$dy = vdu - dx.$$

Substitute this into the first equation (with $dv = 0$) and solve for dx:

$$2xu\,dx + x^2\,du = 2yv(v\,du - dx) = 0,$$

$$(2xu + 2yv)\,dx = (2yv^2 - x^2)\,du.$$

Then

$$\left(\frac{\partial x}{\partial u}\right)_v = \frac{2yv^2 - x^2}{2xu + 2yv}.$$

Similarly, to find $\left(\frac{\partial x}{\partial u}\right)_y$, eliminate dv with $dy = 0$.

23. Let's write $u = f(x - ct) + g(x + ct)$ as

$$u = f(y) + g(z) \qquad \text{where} \qquad \begin{cases} y = x - ct, \\ z = x + xt. \end{cases}$$

We see that $\frac{\partial y}{\partial x} = \frac{\partial z}{\partial x} = 1$ and $\frac{\partial y}{\partial t} = -c$, $\frac{\partial z}{\partial t} = c$.

Then by the chain rule

$$\frac{\partial u}{\partial x} = \frac{df}{dy}\frac{\partial y}{\partial x} + \frac{dg}{dz}\frac{\partial z}{\partial x} = \frac{df}{dy} + \frac{dg}{dz} = f' + g',$$

$$\frac{\partial u}{\partial t} = \frac{df}{dy}\frac{\partial y}{\partial t} + \frac{dg}{dz}\frac{\partial z}{\partial t} = \frac{df}{dy}(-c) + \frac{dg}{dz}(c) = c(g' - f').$$

Now $f'(y)$ and $g'(z)$ can be differentiated again just as f and g were, so we find

23.   (continued)

$$\frac{\partial^2 u}{\partial x^2} = \frac{\partial}{\partial x} \frac{\partial u}{\partial x} = \frac{\partial}{\partial x}(f' + g') = f'' \cdot \frac{\partial y}{\partial x} + g'' \cdot \frac{\partial z}{\partial x} = f'' + g'' ,$$

$$\frac{\partial^2 u}{\partial t^2} = \frac{\partial}{\partial t} \frac{\partial u}{\partial t} = c\frac{\partial}{\partial t}(g' - f') = cg'' \cdot \frac{\partial z}{\partial t} - cf'' \cdot \frac{\partial y}{\partial t}$$

$$= c^2 g'' + (-c)^2 f'' = c^2 (f'' + g'') .$$

Thus

$$\frac{\partial^2 u}{\partial x^2} = \frac{1}{c^2} \frac{\partial^2 u}{\partial t^2} .$$

25.   Hints: Call $\frac{\partial f}{\partial x} = f_1$, $\frac{\partial f}{\partial y} = f_2$, $\frac{\partial f}{\partial z} = f_3$ (see text, Section 1).

Then $df = f_1 dx + f_2 dy + f_3 dz = 0$.

To find $\left(\frac{\partial y}{\partial x}\right)_z$, find $\frac{dy}{dx}$ with $dz = 0$:

$$\left(\frac{\partial y}{\partial x}\right)_z = -\frac{f_1}{f_2} .$$

Similarly, find the other required partial derivatives and use them to verify the stated formulas.

26.   Hint: Write $df = 0$ and $dg = 0$ as in Problem 25 above and eliminate dz.

28.   Hint: Write ds in terms of dv and dT.   Write dv in terms of dp and dT and substitute this expression for dv into ds.   Now you have ds in terms of dp and dT; the coefficient of dT is equal to $\left(\frac{\partial s}{\partial T}\right)_p$.

Section 8

4.   We use the results of Problem 2. For the function
$$f(x,y) = x^2 - y^2 + 2x - 4y + 10$$
we find

$$\left. \begin{array}{l} f_x = 2x + 2 = 0 \\ f_y = -2y - 4 = 0 \end{array} \right\} \quad (x,y) = (-1,-2),$$

$$f_{xx} = 2, \qquad f_{xy} = 0, \qquad f_{yy} = -2.$$

Since $f_{xx}f_{yy} < f_{xy}^2$, the point $(-1,-2)$ is neither a maximum point
nor a minimum point. It is actually a saddle point since in the
x direction, f is a minimum ($f_{xx} > 0$) and in the y direction, f
is a maximum ($f_{yy} < 0$).

8.   We want to maximize the area of the cross section of the gutter.
The area of a trapezoid is

$$A = \frac{1}{2}h(b_1 + b_2)$$

$$= \frac{1}{2}x \sin \theta (48 - 4x + 2x \cos \theta)$$

$$= x \sin \theta (24 - 2x + x \cos \theta)$$

$$= 24x \sin \theta + x^2 (\sin \theta \cos \theta - 2\sin \theta).$$

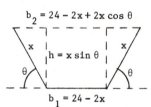

To maximize the area, we find $\frac{\partial A}{\partial x}$ and $\frac{\partial A}{\partial \theta}$ and set them equal to
zero:

$$\frac{\partial A}{\partial x} = 24\sin \theta + 2x(\sin \theta \cos \theta - 2\sin \theta) = 0,$$

$$\frac{\partial A}{\partial \theta} = 24x \cos \theta + x^2 (\cos^2 \theta - \sin^2 \theta - 2\cos \theta) = 0.$$

It is clear from the geometry that neither $x = 0$ nor $\sin \theta = 0$
gives maximum area, so we cancel these factors. Solve the first
equation for x and substitute into the second equation to get

8.   (continued)

$$x = \frac{12}{2 - \cos \theta} \, ,$$

$$24\cos \theta + \frac{12}{2 - \cos \theta}(\cos^2\theta - \sin^2\theta - 2\cos \theta) = 0 \, .$$

Cancel 12, clear the fractions, and put $\sin^2\theta = 1 - \cos^2\theta$:

$$2(\cos \theta)(2 - \cos \theta) + 2\cos^2\theta - 1 - 2\cos \theta = 0, \quad \text{or}$$

$$2\cos \theta = 1, \qquad \theta = \pi/3 \, .$$

Then

$$x = \frac{12}{2 - \frac{1}{2}} = 8 \, .$$

Thus we should bend up 8 cm on each side at an angle $\theta = \pi/3$.
It is clear from the geometry that there must be a maximum area;
therefore this must be the desired solution.

12.   The distance from the origin to the point (x,y,z) on the surface
$z = xy + 5$ is

$$d = \sqrt{x^2 + y^2 + z^2} = \sqrt{x^2 + y^2 + (xy + 5)^2} \, .$$

To minimize d, we can instead minimize d^2.

$$f(x,y) = d^2 = x^2 + y^2 + (xy + 5)^2 \, .$$

We set the partial derivatives of f equal to zero and solve for
x and y.

$$\frac{\partial f}{\partial x} = 2x + 2(xy + 5)y = 0 \, ,$$

$$\frac{\partial f}{\partial y} = 2y + 2(xy + 5)x = 0 \, .$$

If $x = 0$ or $y = 0$, then $x = y = 0$, $z = 5$, $d = 5$.

If $x \neq 0$, $y \neq 0$, then

12.   (continued)

$$xy + 5 = -\frac{x}{y} = -\frac{y}{x}, \qquad x^2 = y^2, \qquad y = \pm x.$$

$$y = x \Longrightarrow x^2 + 5 = -1, \quad \text{which is not possible.}$$

$$y = -x \Longrightarrow -x^2 + 5 = 1, \qquad x^2 = 4, \qquad x = \pm 2.$$

If $x = 2$, then $y = -2$, $z = -4 + 5 = 1$, $d = \sqrt{2^2 + (-2)^2 + 1} = 3$.

If $x = -2$, then $y = 2$, $z = 1$, $d = 3$.

Thus the minimum distance equals 3.

16.   The points $(-1, y_1)$, $(0, y_2)$, $(1, y_3)$ are on the line $y = mx + b$.
Thus

$$y_1 = m(-1) + b = -m + b,$$
$$y_2 = m(0) + b = b,$$
$$y_3 = m(1) + b = m + b.$$

We want to find m and b to minimize

$$f(m,b) = (-2 - y_1)^2 + (0 - y_2)^2 + (3 - y_3)^2$$
$$= (-2 + m - b)^2 + b^2 + (3 - m - b)^2.$$

We set the partial derivatives of f equal to zero and solve for
m and b.

$$\frac{\partial f}{\partial m} = 2(-2 + m - b) + 2(3 - m - b)(-1) = 0 \qquad \text{or} \quad 2m - 5 = 0,$$

$$\frac{\partial f}{\partial b} = 2(-2 + m - b)(-1) + 2b + 2(3 - m - b)(-1) = 0 \quad \text{or} \quad 3b - 1 = 0.$$

Thus $m = 5/2$, $b = 1/3$, and the equation of the straight line is

$$y = \frac{5}{2} x + \frac{1}{3} \qquad \text{or} \qquad 15x - 6y + 2 = 0.$$

Section 9

2.   We recall the formulas for the volume and surface area of a cone:

$$\text{volume} = \frac{1}{3}(\text{area of base}) \times \text{height},$$

$$\text{area} = \frac{1}{2}(\text{circumference of base}) \times \text{slant height}.$$

Then the total volume and total surface area of the projectile (as shown in the text figure) are:

$$V = \pi r^2 \ell + \frac{1}{3}\pi r^2 \sqrt{s^2 - r^2},$$

$$A = \pi r^2 + 2\pi r\ell + \frac{1}{2} \cdot 2\pi r \cdot s = \pi(r^2 + 2r\ell + rs).$$

We want to maximize V subject to the condition A = const. We use the method of Lagrange multipliers [text equation (9.14)] and write:

$$F = V + \lambda A = \pi \left[r^2 \ell + \frac{1}{3}r^2 \sqrt{s^2 - r^2} + \lambda(r^2 + 2r\ell + rs) \right].$$

We set the three partial derivatives of F equal to zero:

$$\frac{\partial F}{\partial r} = \pi \left[2r\ell + \frac{2}{3}r \sqrt{s^2 - r^2} - \frac{1}{3}r^2 \frac{r}{\sqrt{s^2 - r^2}} + \lambda(2r + 2\ell + s) \right] = 0,$$

$$\frac{\partial F}{\partial \ell} = \pi [r^2 + 2r\lambda] = 0,$$

$$\frac{\partial F}{\partial s} = \pi \left[\frac{1}{3}r^2 \frac{s}{\sqrt{s^2 - r^2}} + \lambda r \right] = 0.$$

From the last two equations we get

$$\lambda = -\frac{r}{2} = -\frac{rs}{3\sqrt{s^2 - r^2}}, \qquad \text{or} \qquad \sqrt{s^2 - r^2} = \frac{2s}{3}, \quad \frac{r}{s} = \frac{\sqrt{5}}{3}.$$

(Note that if the area A is not zero, then $r \neq 0$, so we may cancel it.) Divide the $\frac{\partial F}{\partial r}$ equation by r, substitute the values just found, and solve for ℓ/s:

2.   (continued)

$$2\ell + \frac{2}{3} \cdot \frac{2}{3}s - \frac{1}{3}r^2 \cdot \frac{3}{2s} - \frac{1}{2}(2r + 2\ell + s) = 0,$$

$$\frac{\ell}{s} = -\frac{4}{9} + \frac{1}{2} \frac{r^2}{s^2} + \frac{r}{s} + \frac{1}{2} = -\frac{4}{9} + \frac{1}{2} \cdot \frac{5}{9} + \frac{\sqrt{5}}{3} + \frac{1}{2} = \frac{1 + \sqrt{5}}{3} \,.$$

Thus we have

$$\frac{r}{s} = \frac{\sqrt{5}}{3}, \quad \frac{\ell}{s} = \frac{1 + \sqrt{5}}{3}, \qquad \text{or} \qquad r:\ell:s = \sqrt{5} : 1 + \sqrt{5} : 3.$$

These are the proportions to maximize the volume. If we were given a numerical value of the area A, we could find the actual dimensions by substituting for r and ℓ in terms of s into the area formula to find s, and so find r and ℓ.

3.   Hint: If the box has edges x, y, z as shown, the length is y and the girth (distance around the box in a plane perpendicular to the length) is $2x + 2z$.

6.   We want to maximize

$$V = xyz$$

subject to the condition

$$\phi = 2x + 3y + 4z = 6.$$

By the method of Lagrange multipliers [text equation (9.7)], we write

$$F = xyz + \lambda(2x + 3y + 4z)$$

and set the three partial derivatives of F equal to zero:

$$\frac{\partial F}{\partial x} = yz + 2\lambda = 0,$$

$$\frac{\partial F}{\partial y} = xz + 3\lambda = 0,$$

$$\frac{\partial F}{\partial z} = xy + 4\lambda = 0.$$

6.   (continued)

We solve these equations simultaneously with the ϕ equation

$2x + 3y + 4z = 6$.

From the first three equations we find

$$\lambda = -\frac{yz}{2} = -\frac{xz}{3} = -\frac{xy}{4} \quad \text{or} \quad y = \frac{2}{3}x, \quad z = \frac{1}{2}x,$$

(assuming $x,y,z \neq 0$ since this would give zero volume). From the ϕ equation we get $x = 1$; thus we have

$$x = 1, \quad y = 2/3, \quad z = 1/2, \quad V = xyz = 1/3.$$

Since it is clear from the geometry that there is a maximum volume, this must be it.

11.   Hint: There are two conditions so you need two λ's as in text Example 4. You might like to check your answer by vector methods (see text Chapter 3, Section 5).

<u>Section 10</u>

4.   We want to minimize the distance d from the origin to a point on the surface $\phi = 3x^2 + y^2 - 4xz = 4$. We can instead minimize $f = d^2$.

First method: Lagrange multipliers. We write $F = f + \lambda\phi$ and set the partial derivatives of F equal to zero.

$$F = x^2 + y^2 + z^2 + \lambda(3x^2 + y^2 - 4xz),$$

(1) $\frac{\partial F}{\partial x} = 2x + \lambda(6x - 4z) = 0$,

(2) $\frac{\partial F}{\partial y} = 2y + 2\lambda y = 0$,

(3) $\frac{\partial F}{\partial z} = 2z - 4\lambda x = 0$.

From equation (2) we get either (a) $\lambda = -1$ or (b) $y = 0$.

4.   (continued)

(a) If $\lambda = -1$, then equations (1) and (3) give $z = x$ and $z = -2x$
    which are satisfied only by $x = z = 0$. Then from the ϕ
    equation, $y = \pm 2$, so $d = 2$.

(b) If $y = 0$, then we see from the ϕ equation that $x \neq 0$.
    Substitute $z = 2\lambda x$ from equation (3) into equation (1) and
    cancel a factor $2x$ to get
    $$1 + 3\lambda - 4\lambda^2 = 0, \qquad \lambda = 1 \quad \text{or} \quad \lambda = -\frac{1}{4}.$$
    If $\lambda = 1$, then $z = 2x$, and the ϕ equation (with $y = 0$) gives
    $3x^2 - 8x^2 = 4$, which is not possible.
    If $\lambda = -\frac{1}{4}$, then $z = -\frac{1}{2}x$, and the ϕ equation (with $y = 0$) gives
    $$3x^2 + 2x^2 = 4, \qquad x^2 = \frac{4}{5}, \qquad x = \pm\frac{2}{\sqrt{5}}.$$

Then
$$z = -\frac{1}{2}x = \mp\frac{1}{\sqrt{5}} \qquad \text{and} \qquad d = \sqrt{\frac{4}{5} + \frac{1}{5}} = 1.$$

This distance is smaller than $d = 2$ which we found in (a). Thus
the points nearest the origin on the given surface are
$(-2/\sqrt{5}, 0, 1/\sqrt{5})$ and $(2/\sqrt{5}, 0, -1/\sqrt{5})$, both at distance $d = 1$ from
the origin.

Second Method. We can also do this problem by eliminating y^2
from f using the ϕ equation:
$$f = x^2 + y^2 + z^2 = x^2 + z^2 + 4 + 4xz - 3x^2 = z^2 + 4xz - 2x^2 + 4.$$
We set the partial derivatives of f equal to zero and solve for
x and z.
$$\frac{\partial f}{\partial x} = 4z - 4x = 0,$$
$$\frac{\partial f}{\partial z} = 2z + 4x = 0, \qquad x = z = 0 \text{ so } y = \pm 2, \qquad d = 2.$$

4.   (continued)

Now we observe that y^2 cannot take all values since $y^2 \geqslant 0$; thus we must consider the boundary $y^2 = 0$. From the ϕ equation this is

$$y^2 = 4 + 4xz - 3x^2 = 0 \qquad \text{or} \qquad z = \frac{3x^2 - 4}{4x} \ .$$

Then

$$f = x^2 + y^2 + z^2 = x^2 + 0^2 + \left(\frac{3x^2 - 4}{4x}\right)^2 = \frac{25}{16}x^2 - \frac{3}{2} + \frac{1}{x^2} \ .$$

To minimize f, we set $\frac{df}{dx} = 0$.

$$\frac{df}{dx} = \frac{25}{16}(2x) - \frac{2}{x^3} = 0, \qquad x^4 = \frac{16}{25}, \qquad x = \pm \frac{2}{\sqrt{5}} \ .$$

Then on this boundary

$$z = \frac{3x^2 - 4}{4x} = \mp \frac{1}{\sqrt{5}}, \qquad d = \sqrt{\frac{4}{5} + \frac{1}{5}} = 1 \text{ as before.}$$

10.   We want the maximum and minimum values of $T = y^2 + xz$.

    (a) if $y = 0$, $x^2 + z^2 = 1$,

    (b) if $x^2 + y^2 + z^2 = 1$,

    (c) if $x^2 + y^2 + z^2 \leqslant 1$.

(a) By the Lagrange multiplier method we write

$$F = xz + \lambda(x^2 + z^2) ,$$

$$\frac{\partial F}{\partial x} = z + 2\lambda x = 0 ,$$

$$\frac{\partial F}{\partial z} = x + 2\lambda z = 0 .$$

Note that these equations imply that if either x or z is zero then the other is zero too. But we cannot have $x = z = 0$ since we are given $x^2 + z^2 = 1$, so neither x nor z is zero. Then we can divide the equations by x or by z to write

$$\lambda = -\frac{z}{2x} = -\frac{x}{2z} , \qquad \text{so} \quad x^2 = z^2 .$$

10.   (continued)

Since $x^2 + z^2 = 1$, we have

$$x^2 = z^2 = \frac{1}{2}, \qquad x = \pm\frac{1}{\sqrt{2}}, \qquad z = \pm\frac{1}{\sqrt{2}}, \qquad T = xz = \pm\frac{1}{2}.$$

Thus the maximum temperature is 1/2 and the minimum temperature is -1/2.

(b) If we eliminate y, we want the maximum and minimum values of a function of x and z, namely

$$T = 1 - x^2 - z^2 + xz$$

in the region $x^2 + z^2 = 1 - y^2 \leqslant 1$. To find interior maximum or minimum points, we set the partial derivatives of T equal to zero:

$$\frac{\partial T}{\partial x} = -2x + z = 0, \qquad \frac{\partial T}{\partial z} = -2z + x = 0.$$

Thus $x = 0$, $z = 0$, $y^2 = 1$, so $T = 1$. Now we must consider possible maximum or minimum points on the boundary of the (x,z) region, that is, on $x^2 + z^2 = 1$. But this is just what we did in part (a). Thus the maximum temperature on the surface $x^2 + y^2 + z^2 = 1$ is $T = 1$ at $(0, \pm 1, 0)$ and the minimum temperature is $T = -1/2$ at $(\pm 1/\sqrt{2}, 0, \mp 1/\sqrt{2})$.

(c) We find maximum and minimum values in the interior of the sphere $x^2 + y^2 + z^2 < 1$ by setting the three partial derivatives of $T = y^2 + xz$ equal to zero. This gives $x = y = 0$, $T = 0$. Then the maximum and minimum temperatures in the sphere $x^2 + y^2 + z^2 \leqslant 1$ occur on the boundary (surface) as found in part (b).

13. Suppose the line L passes through the
origin and the point (a,b,c). Let
γ = the angle the line L makes with the
(x,y) plane. Then (see figure)

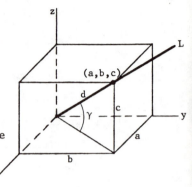

$$\sin \gamma = \frac{c}{d}, \text{ where } d^2 = a^2 + b^2 + c^2.$$

Similarly, if α and β are the angles
the line L makes with the (y,z) and the
(x,z) planes, then

$$\sin \alpha = \frac{a}{d}, \qquad \sin \beta = \frac{b}{d}.$$

Thus we have

$$\sin^2\alpha + \sin^2\beta + \sin^2\gamma = \frac{a^2 + b^2 + c^2}{d^2} = 1.$$

Now we want to find maximum and minimum values of $\alpha + \beta + \gamma$ subject
to the condition $\sin^2\alpha + \sin^2\beta + \sin^2\gamma = 1$. We use the Lagrange
multiplier method:

$$F = \alpha + \beta + \gamma + \lambda(\sin^2\alpha + \sin^2\beta + \sin^2\gamma - 1),$$

$$\frac{\partial F}{\partial \alpha} = 1 + 2\lambda\sin \alpha \cos \alpha = 0,$$

$$\frac{\partial F}{\partial \beta} = 1 + 2\lambda\sin \beta \cos \beta = 0,$$

$$\frac{\partial F}{\partial \gamma} = 1 + 2\lambda\sin \gamma \cos \gamma = 0.$$

Recall that $2 \sin \theta \cos \theta = \sin 2\theta$. Then the three equations give

$$\sin 2\alpha = \sin 2\beta = \sin 2\gamma.$$

Now if $\sin 2\alpha = \sin 2\beta$, then $2\beta = 2\alpha$ or $2\beta = 180° - 2\alpha$, that is,
$\beta = \alpha$ or $\beta = 90° - \alpha$. Similarly, $\gamma = \alpha$ or $\gamma = 90° - \alpha$. The net result
is that either (a) the three angles are equal or (b) two of them
are equal and the third is equal to 90° minus either of the
others.

13.   (continued)

(a) $\alpha = \beta = \gamma$,

$\sin^2\alpha + \sin^2\beta + \sin^2\gamma = 3\sin^2\alpha = 1$,

$\alpha = \beta = \gamma = \arcsin(1/\sqrt{3})$,

$\alpha + \beta + \gamma = 3\arcsin(1/\sqrt{3}) = 105.8°$.

(b) $\beta = \alpha$, $\gamma = 90° - \alpha$, $\sin\beta = \sin\alpha$, $\sin\gamma = \cos\alpha$.

$\sin^2\alpha + \sin^2\beta + \sin^2\gamma = 1 = \sin^2\alpha + \sin^2\alpha + \cos^2\alpha = \sin^2\alpha + 1$.

$\sin\alpha = 0$, $\alpha = 0$, $\beta = 0$, $\gamma = 90°$.

$\alpha + \beta + \gamma = 90°$.

Now α, β, γ are acute angles, so they are required to be between 0° and 90°. Therefore we must consider end point values. If, say, $\alpha = 90°$, then $\sin^2\alpha = 1$, so $\sin^2\beta + \sin^2\gamma = 0$, $\beta = \gamma = 0$. Then

$\alpha + \beta + \gamma = 90°$.

If, say, $\alpha = 0$, then $\sin^2\beta + \sin^2\gamma = 1$ or $\sin^2\beta = 1 - \sin^2\gamma = \cos^2\gamma$. Since β and γ are first quadrant angles, we have $\sin\beta = \cos\gamma$, so $\gamma = 90° - \beta$ and $\alpha + \beta + \gamma = 90°$. Note that our endpoint investigation has uncovered a new situation although not a new smallest sum of angles. Previously we had two angles of 0° and one of 90° [say a line along the x axis which makes 0° angles with the (x,y) and (x,z) planes and 90° with the (y,z) plane]. Here we have one angle of 0° and the others are any two angles adding to 90°. [For example, any line parallel to the (x,y) plane makes a 0° angle with the (x,y) plane and angles adding to 90° with the (y,z) and (x,z) planes.]

The net result of our work is that the smallest sum of the three angles is 90° and the largest sum is $3\arcsin(1/\sqrt{3}) \cong 105.8°$.

Section 11

1. Given the change of variables

$$\begin{cases} s = y + 2x, \\ t = y + 3x, \end{cases}$$

we find by the chain rule

$$\frac{\partial f}{\partial x} = \frac{\partial f}{\partial s}\frac{\partial s}{\partial x} + \frac{\partial f}{\partial t}\frac{\partial t}{\partial x} = 2\frac{\partial f}{\partial s} + 3\frac{\partial f}{\partial t},$$

$$\frac{\partial f}{\partial y} = \frac{\partial f}{\partial s}\frac{\partial s}{\partial y} + \frac{\partial f}{\partial t}\frac{\partial t}{\partial y} = \frac{\partial f}{\partial s} + \frac{\partial f}{\partial t}.$$

It is convenient to write this in operator notation:

$$\frac{\partial}{\partial x} = 2\frac{\partial}{\partial s} + 3\frac{\partial}{\partial t},$$

$$\frac{\partial}{\partial y} = \frac{\partial}{\partial s} + \frac{\partial}{\partial t}.$$

Then

$$\frac{\partial^2 z}{\partial x^2} = \frac{\partial}{\partial x}\frac{\partial z}{\partial x} = \left(2\frac{\partial}{\partial s} + 3\frac{\partial}{\partial t}\right)\left(2\frac{\partial z}{\partial s} + 3\frac{\partial z}{\partial t}\right)$$

$$= 4\frac{\partial^2 z}{\partial s^2} + 12\frac{\partial^2 z}{\partial s \partial t} + 9\frac{\partial^2 z}{\partial t^2},$$

$$\frac{\partial^2 z}{\partial x \partial y} = \left(2\frac{\partial}{\partial s} + 3\frac{\partial}{\partial t}\right)\left(\frac{\partial z}{\partial s} + \frac{\partial z}{\partial t}\right) = 2\frac{\partial^2 z}{\partial s^2} + 5\frac{\partial^2 z}{\partial s \partial t} + 3\frac{\partial^2 z}{\partial t^2},$$

$$\frac{\partial^2 z}{\partial y^2} = \left(\frac{\partial}{\partial s} + \frac{\partial}{\partial t}\right)\left(\frac{\partial z}{\partial s} + \frac{\partial z}{\partial t}\right) = \frac{\partial^2 z}{\partial s^2} + 2\frac{\partial^2 z}{\partial s \partial t} + \frac{\partial^2 z}{\partial t^2}.$$

We substitute these results into the given differential equation and collect coefficients of the s and t derivatives.

$$\frac{\partial^2 z}{\partial x^2} - 5\frac{\partial^2 z}{\partial x \partial y} + 6\frac{\partial^2 z}{\partial y^2} = 0$$

$$= (4 - 10 + 6)\frac{\partial^2 z}{\partial s^2} + (12 - 25 + 12)\frac{\partial^2 z}{\partial s \partial t} + (9 - 15 + 6)\frac{\partial^2 z}{\partial t^2}$$

$$= -\frac{\partial^2 z}{\partial s \partial t} = 0.$$

1.   (continued)

To solve this equation for z(s,t) we write

$$\frac{\partial^2 z}{\partial s\,\partial t} = \frac{\partial}{\partial s}\left(\frac{\partial z}{\partial t}\right) = 0.$$

Then $\frac{\partial z}{\partial t}$ is independent of s, that is it is some function of t.
We integrate it with respect to t to get

$$z = f(t) + \text{"const."}$$

The "constant" may be a function of s; all that is required is
that its partial derivative with respect to t must be zero.   Thus
the solution of the differential equation is

$$z = f(t) + g(s).$$

We substitute the values of t and s in terms of x and y to find
the solution of the original differential equation:

$$z = f(y + 3x) + g(y + 2x).$$

It is interesting to see why this worked and how we could find
the change of variables.   Verify that the original differential
equation can be "factored" as follows:

$$\left(\frac{\partial}{\partial x} - 3\frac{\partial}{\partial y}\right)\left(\frac{\partial}{\partial x} - 2\frac{\partial}{\partial y}\right)z = 0 .$$

Then this equation would become $\frac{\partial^2 z}{\partial s\,\partial t} = 0$ if

$$\frac{\partial}{\partial s} = \frac{\partial}{\partial x} - 3\frac{\partial}{\partial y} \qquad \text{and} \qquad \frac{\partial}{\partial t} = \frac{\partial}{\partial x} - 2\frac{\partial}{\partial y} .$$

This would be true if $x = s + t$, $y = -3s - 2t$.   (Verify this by the
chain rule.)   Solving for s and t, we find $s = -2x - y$, $t = 3x + y$
which (except for the sign of s) is the change of variables we
started with.

6.   If $x = e^z$, then $dx = e^z dz$ or $\dfrac{dz}{dx} = e^{-z}$.

Then by the chain rule

$$\frac{df}{dx} = \frac{df}{dz}\frac{dz}{dx} = e^{-z}\frac{df}{dz}$$

or, in operator notation

$$\frac{d}{dx} = e^{-z}\frac{d}{dz} \ .$$

Then

(1) $\dfrac{dy}{dx} = e^{-z}\dfrac{dy}{dz}$,

(2) $\dfrac{d^2y}{dx^2} = e^{-z}\dfrac{d}{dz}\left(e^{-z}\dfrac{dy}{dz}\right) = e^{-z}\left(e^{-z}\dfrac{d^2y}{dz^2} - e^{-z}\dfrac{dy}{dz}\right)$,

and so on. Now multiply (1) by e^z and (2) by e^{2z} and use
$e^z = x$, $e^{2z} = x^2$:

(3) $x\dfrac{dy}{dx} = \dfrac{dy}{dz}$,

(4) $x^2\dfrac{d^2y}{dx^2} = \dfrac{d^2y}{dz^2} - \dfrac{dy}{dz} \ .$

(Can you find $x^3\dfrac{d^3y}{dx^3}$?) Now we substitute (3) and (4) into the

given differential equation:

$$x^2\frac{d^2y}{dx^2} + 2x\frac{dy}{dx} - 5y = \frac{d^2y}{dz^2} - \frac{dy}{dz} + 2\frac{dy}{dz} - 5y = 0 \quad \text{or}$$

$$\frac{d^2y}{dz^2} + \frac{dy}{dz} - 5y = 0 .$$

This is the desired differential equation with constant coefficients. See text, Chapter 8, Sections 5 and 7(d).

9.   Hint: Follow the method and notation of text Example 2,
       page 188 ff.

Section 12

1. By Leibniz' rule [text (12.11)], we find

$$\frac{dy}{dx} = \frac{d}{dx} \int_0^{\sqrt{x}} \sin t^2 \, dt = \sin\left[(\sqrt{x})^2\right]\frac{d}{dx}\sqrt{x} = (\sin x)\left(\tfrac{1}{2}x^{-1/2}\right)$$

$$= \frac{\sin x}{2\sqrt{x}} \; .$$

6. Using Leibniz' rule, we write dw:

$$dw = d \int_{xy}^{2x-3y} \frac{du}{\ln u} = \frac{d(2x - 3y)}{\ln(2x - 3y)} - \frac{d(xy)}{\ln(xy)}$$

$$= \frac{2dx - 3dy}{\ln(2x - 3y)} - \frac{x \, dy + y \, dx}{\ln(xy)} \; .$$

At $x = 3$, $y = 1$:

$$dw = \frac{2dx - 3dy}{\ln 3} - \frac{3dy + dx}{\ln 3} = \frac{1}{\ln 3}(dx - 6dy) \; ,$$

$$\frac{\partial w}{\partial x} = \frac{1}{\ln 3} \; , \qquad \frac{\partial w}{\partial y} = -\frac{6}{\ln 3} \; , \qquad \left(\frac{\partial y}{\partial x}\right)_w = \frac{1}{6} \; .$$

12. By Leibniz' rule:

$$\frac{d}{dx} \int_x^{x^2} \frac{du}{\ln(x+u)} = \frac{1}{\ln(x+x^2)} \frac{d}{dx}(x^2) - \frac{1}{\ln(x+x)} \frac{d}{dx}(x) + \int_x^{x^2} \frac{\partial}{\partial x}\left(\frac{1}{\ln(x+u)}\right) du$$

$$= \frac{2x}{\ln(x+x^2)} - \frac{1}{\ln(2x)} + \int_x^{x^2} \frac{-1}{(x+u)[\ln(x+u)]^2} \, du.$$

To evaluate the integral, let $y = \ln(x + u)$. Remember that u
(or y) is the integration variable here; x is a constant during
the u (or y) integration. Then

$$dy = \frac{du}{x + u}$$

and the integral is

12.   (continued)

$$\int \frac{dy}{y^2} = -\frac{1}{y} = -\frac{1}{\ell n(x+u)} \quad \text{or}$$

$$\int_x^{x^2} \frac{-1}{(x+u)[\ell n(x+u)]^2} = \frac{1}{\ell n(x+u)} \Big|_x^{x^2} = \frac{1}{\ell n(x+x^2)} - \frac{1}{\ell n(2x)}.$$

We substitute this into the result above to get:

$$\frac{d}{dx} \int_x^{x^2} \frac{du}{\ell n(x+u)} = \frac{2x+1}{\ell n(x+x^2)} - \frac{2}{\ell n(2x)}.$$

14.   Given

$$\int_0^\infty \frac{dx}{y^2+x^2} = \frac{\pi}{2y},$$

we differentiate both sides of the equation with respect to y:

$$\frac{d}{dy}\int_0^\infty \frac{dx}{y^2+x^2} = \int_0^\infty -\frac{2y}{(y^2+x^2)^2}\,dx = \frac{d}{dy}\frac{\pi}{2y} = -\frac{\pi}{2y^2}.$$

Divide by -2y to get:

$$\int_0^\infty \frac{dx}{(y^2+x^2)^2} = \frac{\pi}{4y^3}.$$

19.   Given $y = \int_0^x f(u)\sin(x-u)\,du$, we find y' and y".

$$y' = \frac{d}{dx}\int_0^x f(u)\sin(x-u)\,du = f(x)\sin(x-x) + \int_0^x \frac{\partial}{\partial x}[f(u)\sin(x-u)]\,du$$

$$= 0 + \int_0^x f(u)\cos(x-u)\,du,$$

$$y" = \frac{d}{dx}y' = \frac{d}{dx}\int_0^x f(u)\cos(x-u)\,du$$

$$= f(x)\cos 0 + \int_0^x \frac{\partial}{\partial x}[f(u)\cos(x-u)]\,du$$

$$= f(x) - \int_0^x f(u)\sin(x-u)\,du.$$

We add y and y" to get $y" + y = f(x)$.

Section 13

1. By definition, $f(x,y,z)$ is homogeneous of degree n if

$$f(tx,ty,tz) = t^n f(x,y,z).$$

Let $u = tx$, $v = ty$, $w = tz$; then $\dfrac{du}{dt} = x$, $\dfrac{dv}{dt} = y$, $\dfrac{dw}{dt} = z$, and

$$f(u,v,w) = t^n f(x,y,z).$$

Differentiate with respect to t:

$$\frac{\partial f}{\partial u}\frac{du}{dt} + \frac{\partial f}{\partial v}\frac{dv}{dt} + \frac{\partial f}{\partial w}\frac{dw}{dt} = x\frac{\partial f}{\partial u} + y\frac{\partial f}{\partial v} + z\frac{\partial f}{\partial w} = nt^{n-1}f.$$

Now let $t = 1$ (which means $u = x$, $v = y$, $w = z$) to get:

$$x\frac{\partial f}{\partial x} + y\frac{\partial f}{\partial y} + z\frac{\partial f}{\partial z} = nf.$$

2. (c) We want to minimize the distance D from (x_0, y_0) to (x,y) subject to the condition $ax + by = c$. It is simpler to minimize D^2 so we write

$$f = D^2 + \lambda(ax + by) = (x - x_0)^2 + (y - y_0)^2 + \lambda(ax + by)$$

and set the partial derivatives of f equal to zero.

(1) $\dfrac{\partial f}{\partial x} = 2(x - x_0) + \lambda a = 0$,

(2) $\dfrac{\partial f}{\partial y} = 2(y - y_0) + \lambda b = 0$.

Multiply equation (1) by a, equation (2) by b, add the equations using $ax + by = c$, and solve for λ:

$$2a(x - x_0) + 2b(y - y_0) + \lambda(a^2 + b^2) = 0,$$

$$2c - 2ax_0 - 2by_0 + \lambda(a^2 + b^2) = 0,$$

(3) $\lambda = \left| \dfrac{2(ax_0 + by_0 - c)}{a^2 + b^2} \right|$.

Now we find from equations (1), (2) and (3):

2.   (continued)

$$D^2 = (x - x_0)^2 + (y - y_0)^2 = \frac{\lambda^2 a^2}{4} + \frac{\lambda^2 b^2}{4} = \frac{4(ax_0 + by_0 - c)^2(a^2 + b^2)}{4(a^2 + b^2)^2},$$

$$D = \left| \frac{ax_0 + by_0 - c}{\sqrt{a^2 + b^2}} \right|.$$

Also see the solution of Chapter 3 Problem 10.16.

8.   Hint: Use differentials to approximate Δr. Further questions:
      What would be the answer to the same problem on the Moon? On a
      spherical asteroid of radius 100 miles? What is the relation
      between dr and Δr in each case?

9.   Given the equations

      (1) $z = xy$,

      (2) $2x^3 + 2y^3 = 3t^2$,

      (3) $3x^2 + 3y^2 = 6t$ or $x^2 + y^2 = 2t$,

      we take differentials of the equations to get:

      (4) $dz = x\,dy + y\,dx$,

      (5) $6x^2 dx + 6y^2 dy = 6t\,dt$,

      (6) $6x\,dx + 6y\,dy = 6\,dt$.

      Multiply equation (6) by y and subtract from equation (5) to
      find $\frac{dx}{dt}$:

$$6(x^2 - xy)dx = 6(t - y)dt \qquad \text{or} \qquad \frac{dx}{dt} = \frac{t - y}{x(x - y)}.$$

      Similarly find

$$\frac{dy}{dt} = \frac{t - x}{y(y - x)} = \frac{x - t}{y(x - y)}.$$

      Then from equation (4)

9.   (continued)

$$\frac{dz}{dt} = x\frac{dy}{dt} + y\frac{dx}{dt} = \frac{x(x-t)}{y(x-y)} + \frac{y(t-y)}{x(x-y)}$$

$$= \frac{x^2(x-t) + y^2(t-y)}{xy(x-y)} = \frac{x^3 - y^3 - t(x^2 - y^2)}{xy(x-y)}$$

$$= \frac{(x-y)(x^2 + xy + y^2) - t(x-y)(x+y)}{xy(x-y)}$$

$$= \frac{x^2 + xy + y^2 - t(x+y)}{xy} .$$

We can simplify this using the given equations (1) and (3):

$$\frac{dz}{dt} = \frac{2t + z - t(x+y)}{z} = 1 + \frac{t}{z}(2 - x - y), \qquad z \ne 0.$$

Further comment: Note that $t \geqslant 0$ (because $2t = x^2 + y^2$). Also $t \leqslant 32/9$ because the two (x,y) graphs do not intersect if t is larger. For $16/9 < t < 32/9$, the given equations define two different functions $z(t)$ but the formula for $\frac{dz}{dt}$ correctly gives the slope for both (just as $y' = -x/y$ gives the slope of the tangent to either the upper or lower half of a circle with center at the origin). At $t = 32/9$, $z = 0$ and the derivative is infinite (vertical tangent). However, at $t = z = 0$, we can show that there is a derivative (right-hand, since $t \geqslant 0$), $\left(\frac{dz}{dt}\right)_{t=0} = -1$. From equation (2), we have

$$\frac{x^3 + y^3}{t^{3/2}} = \frac{3}{2}\sqrt{t} \to 0 \text{ as } t \to 0.$$

Factor:

(7) $$\frac{x^3 + y^3}{t^{3/2}} = \frac{(x+y)}{\sqrt{t}} \frac{(x^2 - xy + y^2)}{t} .$$

Now

9. (continued)

$$x^2 + y^2 - xy \over t = \frac{\frac{1}{2}(x^2+y^2)+\frac{1}{2}(x-y)^2}{t} = \frac{t+\frac{1}{2}(x-y)^2}{t} = 1+\frac{1}{2t}(x-y)^2 \geq 1.$$

Thus the second factor in equation (7) cannot approach zero so the first factor must approach zero. Squaring it, we get

$$\left(\frac{x+y}{\sqrt{t}}\right)^2 = \frac{x^2+2xy+y^2}{t} \to 0$$

or using equations (1) and (3)

$$\frac{2t+2z}{t} \to 0, \qquad \frac{z}{t} \to -1.$$

Thus

$$\left(\frac{dz}{dt}\right)_{t=0} = -1.$$

14. If $w = f(x,s,t)$, $s = 2x + y$, $t = 2x - y$, then w can be expressed as a function of x and y, so we can find $\left(\frac{\partial w}{\partial x}\right)_y$. We use the notation

$$f_1 = \left(\frac{\partial f}{\partial x}\right)_{s,t}, \qquad f_2 = \left(\frac{\partial f}{\partial s}\right)_{x,t}, \qquad f_3 = \left(\frac{\partial f}{\partial t}\right)_{x,s}$$

(see text, page 146). This notation avoids confusion about the meaning of $\frac{\partial f}{\partial x}$; note that $\left(\frac{\partial f}{\partial x}\right)_{s,t}$ is not the same as $\left(\frac{\partial f}{\partial x}\right)_y$. Then

$$\left(\frac{\partial w}{\partial x}\right)_y = f_1(x,s,t) + f_2(x,s,t)\frac{\partial s}{\partial x} + f_3(x,s,t)\frac{\partial t}{\partial x}$$

$$= f_1 + 2f_2 + 2f_3.$$

Now can you show that $\left(\frac{\partial w}{\partial y}\right)_x = f_2 - f_3$?

16. Hints: (a) Let $\frac{y}{x} = v$; then $\frac{df}{dv} = f'$. (b) As another method, can you use the theorem in Problem 1?

20. Hint: Don't forget that $|\cos\theta| \leq 1$.

24.   This problem is straightforward by Leibniz' rule. However, there
       is an interesting alternative method: Make the change of vari-
       ables $v = xt$. Here t and v are variables; during the integration,
       x is constant. Then

$$v = xt, \qquad \frac{dv}{v} = \frac{dt}{t} \; .$$

For the limits, we have

$$t = \frac{1}{x} \Longrightarrow v = 1, \qquad \text{and} \qquad t = \frac{2}{x} \Longrightarrow v = 2.$$

The integral is

$$\int_{t=1/x}^{t=2/x} \frac{\cosh xt}{t} \, dt = \int_{1}^{2} \frac{\cosh v}{v} \, dv.$$

Thus the integral is a constant so its derivative with respect to
x is zero.

28.   The expression which is called F in the text, Section 9(c), is
       in this problem

$$\ln F + \lambda \phi = \ln f(x) + \ln f(y) + \ln f(z) + \lambda (x^2 + y^2 + z^2).$$

We set the three partial derivatives of this expression equal to
zero to get:

$$\frac{f'(x)}{f(x)} + 2\lambda x = 0,$$

$$\frac{f'(y)}{f(y)} + 2\lambda y = 0,$$

$$\frac{f'(z)}{f(z)} + 2\lambda z = 0.$$

We integrate each of these equations to get

$$\ln f(x) + \lambda x^2 = \text{const.} \qquad \text{or} \qquad f(x) = Ce^{-\lambda x^2}$$

and similar y and z equations. Then

$$F = f(x) f(y) f(z) = Ae^{-\lambda (x^2 + y^2 + z^2)}.$$

Chapter 5

<u>Section 2</u>

1. Since the limits are constant [integral over a rectangle -- see text equation (2.8)], we may evaluate the double integral as a product of two single integrals:

$$\left(\int_0^1 3x\,dx\right)\left(\int_2^4 dy\right) = \left(\frac{3x^2}{2}\bigg|_0^1\right)\left(y\bigg|_2^4\right) = \frac{3}{2}(4-2) = 3.$$

5. We first evaluate the y integral:

$$\int_{y=x}^{e^x} y\,dy = \frac{y^2}{2}\bigg|_x^{e^x} = \frac{1}{2}(e^{2x} - x^2).$$

Then we integrate this result with respect to x:

$$\frac{1}{2}\int_0^1 (e^{2x} - x^2)\,dx = \frac{1}{2}\left(\frac{e^{2x}}{2} - \frac{x^3}{3}\right)\bigg|_0^1 = \frac{1}{4}(e^2 - e^0) - \frac{1}{6}(1-0)$$

$$= \frac{e^2}{4} - \frac{5}{12}.$$

7. Compare the given area with text Figure 2.7; we may integrate in either order. The equation of the slanted line is $y = x/2$ or $x = 2y$. Thus we find

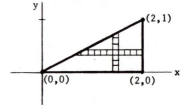

$$\int_{x=0}^2 \int_{y=0}^{x/2} (2x - 3y)\,dy\,dx \qquad \text{or} \qquad \int_{y=0}^1 \int_{x=2y}^2 (2x - 3y)\,dx\,dy$$

$$= \int_{x=0}^2 \left(2xy - \frac{3y^2}{2}\bigg|_{y=0}^{x/2}\right)dx \qquad = \int_{y=0}^1 \left(x^2 - 3xy\bigg|_{x=2y}^2\right)dy$$

$$= \int_0^2 \left(\frac{5x^2}{8}\right)dx = \frac{5x^3}{24}\bigg|_0^2 = \frac{5}{3} \qquad = \int_0^1 (4 - 6y + 2y^2)\,dy = \frac{5}{3}$$

11.   We first find the points of

intersection of the line

$y = 2x + 8$ and the parabola $y = x^2$.

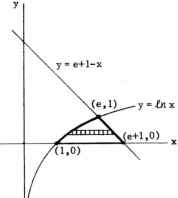

$$x^2 = 2x + 8$$

$$x^2 - 2x - 8 = (x - 4)(x + 2) = 0$$

$$x = -2, \qquad x = 4.$$

Compare the given area with text

Figure 2.5. We first integrate with respect to y from the

parabola to the line:

$$\int_{y=x^2}^{2x+8} dy = 2x + 8 - x^2$$

Then

$$\int_A \int x \, dy \, dx = \int_{-2}^{4} x(2x + 8 - x^2) \, dx = \frac{2x^3}{3} + \frac{8x^2}{2} - \frac{x^4}{4} \bigg|_{-2}^{4} = 36.$$

15.   First verify the points of

intersection as shown on the figure.

Then compare the given area with

text Figure 2.6. We integrate with

respect to x from the ℓn curve to

the straight line. If $y = \ell n \, x$,

then $x = e^y$, and on the line

$x = e + 1 - y$, so the x limits are

from e^y to $e + 1 - y$. Then we sum

the horizontal strips from $y = 0$ to $y = 1$.

$$\int_{y=0}^{1} \int_{x=e^y}^{e+1-y} dx \, dy = \int_{0}^{1} (e + 1 - y) \, dy = e + 1 - \frac{1}{2} - (e - 1) = \frac{3}{2} \, .$$

21. We evaluate $\iint z\, dx\, dy$ over the given
     triangular area. The equation of the
     line through $(0,0)$ and $(2,1)$ is $y = x/2$.
     If we integrate with respect to y first,
     we have

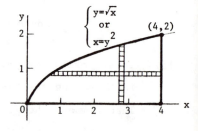

$$\int_{x=0}^{2}\int_{y=0}^{x/2}(24 - x^2 - y^2)\,dy\,dx = \int_{0}^{2}\left[24\cdot\frac{x}{2} - x^2\cdot\frac{x}{2} - \frac{1}{3}\left(\frac{x}{2}\right)^3\right]dx$$

$$= 12\cdot\frac{x^2}{2} - \frac{1}{2}\cdot\frac{x^4}{4} - \frac{1}{24}\cdot\frac{x^4}{4}\Bigg|_{0}^{2} = 24 - 2 - \frac{1}{6} = \frac{131}{6}\ .$$

We can also evaluate this by integrating first with respect to x
from 2y to 2, and then with respect to y from 0 to 1. Try it.

27. The area of integration is shown in
     the sketch. As the iterated integral
     is given in the text, we integrate
     first with respect to y from $y = 0$ to
     the curve $y = \sqrt{x}$ (for any fixed value
     of x); this gives a vertical strip as
     shown in the figure. Then we sum these
     vertical strips from $x = 0$ to $x = 4$. Now consider doing the problem
     by integrating in the opposite order. If we integrate with re-
     spect to x first, then x goes from the curve $x = y^2$ to the line
     $x = 4$ (horizontal strip in the figure). Then we sum these horizon-
     tal strips from $y = 0$ to $y = 2$. We evaluate the double integral
     both ways:

27.   (continued)

$$\int_{x=0}^{4}\int_{y=0}^{\sqrt{x}} y\sqrt{x}\ dy\ dx = \int_{x=0}^{4}\sqrt{x}\ dx\left(\frac{y^2}{2}\Big|_{0}^{\sqrt{x}}\right) = \frac{1}{2}\int_{0}^{4}\sqrt{x}\cdot x\ dx = \frac{1}{2}\cdot\frac{2}{5}x^{5/2}\Big|_{0}^{4} = \frac{32}{5}\ .$$

$$\int_{y=0}^{2}\int_{x=y^2}^{4} y\sqrt{x}\ dx\ dy = \int_{y=0}^{2} y\ dy\left(\frac{2}{3}x^{3/2}\Big|_{y^2}^{4}\right) = \frac{2}{3}\int_{0}^{2} y\ dy(8 - y^3)$$

$$= \frac{2}{3}\int_{0}^{2}(8y - y^4)\,dy = \frac{2}{3}\left(8\cdot\frac{y^2}{2} - \frac{y^5}{5}\ \Big|_{0}^{2}\right) = \frac{32}{5}\ .$$

31.   The indefinite integral $\int\frac{dy}{\ell n\ y}$ cannot be evaluated in terms of
elementary functions. Let us try to
evaluate the double integral by
integrating first with respect to x.
We sketch the area of integration.
If we integrate first with respect to
x, we see that x goes from $x = 0$ to
the curve $x = 2\,\ell n\ y$. Then we sum the
horizontal strips from $y = 1$ to $y = 4$.
Thus

$$\int_{x=0}^{\ell n\ 16}\int_{y=e^{x/2}}^{4}\frac{dy\ dx}{\ell n\ y} = \int_{y=1}^{4}\int_{x=0}^{2\,\ell n\ y}\frac{dx\ dy}{\ell n\ y} = \int_{y=1}^{4}\frac{dy}{\ell n\ y}\left(x\Big|_{x=0}^{2\,\ell n\ y}\right)$$

$$= \int_{1}^{4}\frac{dy}{\ell n\ y}\cdot 2\,\ell n\ y = 2\int_{1}^{4}dy = 6\ .$$

36. Let I_0 be the uniform intensity of the light (energy per unit
     area per second). Then the total light (energy per second)
     incident on the whole mirror is $4I_0$ (intensity times area). The
     light incident on the area element dx dy is I_0 dx dy, and the light
     reflected from dx dy is $I_0 \frac{(x-y)^2}{4}$ dx dy. Then the total light
     reflected from the whole mirror is

$$\frac{I_0}{4} \int_{x=-1}^{1} \int_{y=-1}^{1} (x-y)^2 \, dx \, dy = \frac{I_0}{4} \int_{x=-1}^{1} \left[\frac{(x-y)^3}{3} \Big|_{x=-1}^{1} \right] dy$$

$$= \frac{I_0}{12} \int_{-1}^{1} \left[(1-y)^3 - (-1-y)^3 \right] dy = \frac{I_0}{12} \left[-\frac{(1-y)^4}{4} + \frac{(1+y)^4}{4} \right]_{-1}^{1} = \frac{2}{3} I_0 \; .$$

The fraction reflected is

$$\frac{\text{total reflected}}{\text{total incident}} = \frac{2I_0/3}{4I_0} = \frac{1}{6} \; .$$

39.
$$\int_{y=-2}^{3} \int_{z=1}^{2} \int_{x=y+z}^{2y+z} 6y \, dx \, dz \, dy = \int_{y=-2}^{3} \int_{z=1}^{2} 6y(2y+z-y-z) \, dz \, dy$$

$$= \int_{y=-2}^{3} \int_{z=1}^{2} 6y^2 \, dz \, dy = \left(\int_{-2}^{3} 6y^2 \, dy \right) \left(\int_{1}^{2} dz \right) = \left(6 \cdot \frac{y^3}{3} \Big|_{-2}^{3} \right)(2-1) = 2\left[3^3 - (-2)^3 \right] = 70 \; .$$

43. First we must find out whether the surfaces intersect either each
     other or the triangle of integration. We see that both
     $z_1 = 2x^2 + y^2 + 12$ and $z_2 = x^2 + y^2 + 8$ are positive; also $z_1 > z_2$, so
     we integrate from z_2 to z_1. The x and y limits are found from
     the triangle as in Problem 21. Thus the volume is

$$V = \int_{x=0}^{1} \int_{y=0}^{2x} \int_{z=x^2+y^2+8}^{2x^2+y^2+12} dz \, dy \, dx = \int_{x=0}^{1} \int_{y=0}^{2x} (x^2+4) \, dy \, dx$$

$$= \int_{0}^{1} (x^2+4) \cdot 2x \, dx = 9/2 \; .$$

45. The element of mass is $dM = \rho dV$ where ρ is the density and
    $dV = dx\,dy\,dz$ is the volume element. Here $\rho = kx$ (k is the propor-
    tionality constant), so $dM = kx\,dx\,dy\,dz$. In Problem 43, we found
    the volume by evaluating the triple integral of the volume
    element dV. The same triple integral (that is, with the same
    limits) of the mass element dM gives the total mass. The z and
    y integrals are found as in Problem 43.

$$M = k \int_{x=0}^{1} \int_{y=0}^{2x} \int_{z=x^2+y^2+8}^{2x^2+y^2+12} x\,dz\,dy\,dx = k \int_{0}^{1} x(x^2+4)\cdot 2x\,dx = \frac{46k}{15}.$$

Section 3

1. We want to compare the moment of inertia I of a body about a
   given axis A with the moment of inertia I_m of the body about a
   parallel axis A' through the center of mass. Let us choose our
   coordinate system so that A is the
   z axis; then A' is a line parallel
   to the z axis. Now, by definition
   (see text Example 1f) the moment
   of inertia of a body about the z
   axis is

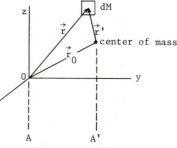

$$I = \int \int \int (x^2 + y^2)\,dM$$

where (see figure) the vector $\vec{r} = \hat{i}x + \hat{j}y + \hat{k}z$ is the vector from
the origin to dM. Similarly, the moment of inertia about the
parallel axis through the center of mass is

$$I_m = \int \int \int (x'^2 + y'^2)\,dM$$

1.   (continued)

where (see figure) $\vec{r}' = \vec{i}x' + \vec{j}y' + \vec{k}z'$ is the vector from the center of mass to dM.

Now (see figure) $\vec{r} = \vec{r}_0 + \vec{r}'$, where $\vec{r}_0 = \vec{i}a + \vec{j}b + \vec{k}c$ is the vector from the origin to the center of mass. Then

$$x = x' + a, \qquad y = y' + b, \qquad \text{so}$$

$$I = \iiint (x^2 + y^2)\, dM = \iiint \left[(x' + a)^2 + (y' + b)^2 \right] dM$$

$$= \iiint (x'^2 + y'^2)\, dM + (a^2 + b^2) \iiint dM + 2a \iiint x'\, dM + 2b \iiint y'\, d$$

$$= I_m + (a^2 + b^2) M.$$

We have dropped the last two integrals because they are zero; let us see why. The x coordinate of the center of mass (see text Example 1d) is $\frac{1}{M} \iiint x\, dM$. Then $\frac{1}{M} \iiint x'\, dM$ and $\frac{1}{M} \iiint y'\, dM$ give the x' and y' coordinates of the center of mass; these are zero since the center of mass is the origin of the x',y' system. Now $d^2 = a^2 + b^2$ is the square of the distance between the two axes (projection of $\vec{r}_0$ on the (x,y) plane). Thus

$$I = I_m + Md^2.$$

3.   Place the rod along the x axis with the light end at the origin. Then the density is a linear function of x. We have $\rho = a + bx$ where $\rho = 4$ when $x = 0$ and $\rho = 24$ when $x = 10$; thus $\rho = 4 + 2x$.

(a) $M = \int dM = \int \rho\, dx = \int_0^{10} (4 + 2x)\, dx = 140$

(b) $M\bar{x} = \int x\, dM = \int x\, \rho\, dx = \int_0^{10} x(4 + 2x)\, dx = \dfrac{2600}{3}$, $\qquad \bar{x} = \dfrac{2600}{3 \cdot 140} = \dfrac{13}{2}$

3.   (continued)

(c) $I_m = \int (x - \bar{x})^2 dM = \int_0^{10} \left(x - \frac{130}{21}\right)^2 (4 + 2x) dx = \frac{61}{63} \times 10^3$,

$I_m/M = \frac{61}{63} \times 10^3 \div 140 = 6.92$, $I_m = 6.92M$.

(d) $I = \int_0^{10} (10 - x)^2 (4 + 2x) dx = 3000 = \frac{3000}{140} M = \frac{150}{7} M$.

Since the integral in (c) involves somewhat messier arithmetic
than (d), we could use the result (d) and the parallel axis
theorem (Problem 1) to find I_m. The distance between the heavy
end at $x = 10$ and the center of mass $\bar{x}$ is $d = \left(10 - \frac{130}{21}\right) = \frac{80}{21}$. By
Problem 1, $I = I_m + Md^2$ or $I_m = I - Md^2$. Thus

$$I_m = \left[\frac{150}{7} - \left(\frac{80}{21}\right)^2\right] M = 6.92M.$$

7.   (a) $M = \iint_A xy \; dx \; dy = \int_{y=0}^{2} \int_{x=0}^{3} xy \; dx \; dy = \left(\int_0^3 x \; dx\right)\left(\int_0^2 y \; dy\right) = \frac{9}{2} \cdot 2 = 9$.

(b) $\bar{x}M = \iint_A x(xy \; dx \; dy) = \left(\int_0^3 x^2 dx\right)\left(\int_0^2 y \; dy\right) = 9 \cdot 2 = 18$, $\bar{x} = 2$,

$\bar{y}M = \iint_A y(xy \; dx \; dy) = 12$, $\bar{y} = \frac{4}{3}$.

(c) $I_x = \iint_A y^2 (xy \; dx \; dy) = 18 = 2M$,

$I_y = \iint_A x^2 (xy \; dx \; dy) = \frac{81}{2} = \frac{9}{2} M$.

(d)   By the perpendicular axis theorem (text Example 1f),
$I_z = I_x + I_y = \frac{13}{2} M$.

Then by the parallel axis theorem (Problem 1 above), I about an
axis parallel to the z axis through the center of mass is

7.   (continued)

$$I_{zm} = I_z - Md^2 \qquad \text{where} \qquad d^2 = \bar{x}^2 + \bar{y}^2 = \frac{52}{9} \ .$$

Thus

$$I_{zm} = \frac{13}{2}M - \frac{52}{9}M = \frac{13}{18}M.$$

You might check this (a) by direct integration; (b) by using the parallel axis theorem first to find I_{xm} and I_{ym} and then the perpendicular axis theorem to find I_{zm}.

13.   The surface area generated when an arc revolves around the x axis is given by $A = 2\pi \int y \, ds$. [See text equation (3.11).] The y co-ordinate of the centroid of the arc is given by $\ell \bar{y} = \int y \, ds$ where $\ell = \int ds$ is the length of the arc [see text equation (3.6)]. Thus $A = 2\pi \bar{y} \ell$. Since $2\pi \bar{y}$ is the circumference of the circle traced by the centroid, the surface area is equal to this circumference times the arc length. This is one of the theorems of Pappus.

16.   If the curve $y = f(x)$ is revolved about the x axis, each point on it travels in a circle of radius $f(x)$ in a plane parallel to the y,z plane. The equation of this circle is

$$y^2 + z^2 = (\text{radius})^2 = [f(x)]^2.$$

Since this gives, for each x, a cross section of the desired sur-face, it is the equation of the surface. For $f(x) = x^2$, we have, as in text equation (3.9), $y^2 + z^2 = x^4$.

21.   With $y = \sqrt{x}$, we find

$$y' = \frac{1}{2} x^{-1/2}, \qquad ds = \sqrt{1 + y'^2}\, dx = \sqrt{1 + \frac{1}{4x}}\, dx.$$

Then, by text equation (3.6) and integral tables, we find

$$s\bar{x} = \int \bar{x}\, ds = \int x\, ds = \int_0^2 x\, \sqrt{1 + \frac{1}{4x}}\, dx = \frac{1}{2} \int_0^2 \sqrt{4x^2 + x}\, dx$$

$$= \frac{8x + 1}{32} \sqrt{4x^2 + x} - \frac{1}{128} \ell n \left(\sqrt{4x^2 + x} + 2x + \frac{1}{4} \right) \Big|_0^2$$

$$= \frac{51}{32} \sqrt{2} - \frac{1}{128} \ell n \left(12\sqrt{2} + 17 \right) = \frac{51\sqrt{2} - \ell n (1 + \sqrt{2})}{32} = 2.23.$$

From Problem 18,

$$s = \int ds = \frac{1}{2} \left[3\sqrt{2} + \ell n (1 + \sqrt{2}) \right] = 2.56.$$

Thus

$$\bar{x} = \frac{2.23}{2.56} = 0.87.$$

Similarly

$$s\bar{y} = \int \bar{y}\, ds = \int y\, ds = \int_0^2 \sqrt{x} \sqrt{1 + \frac{1}{4x}}\, dx = \int_0^2 \sqrt{x + \frac{1}{4}}\, dx$$

$$= \frac{2}{3} \left(x + \frac{1}{4} \right)^{3/2} \Big|_0^2 = \frac{13}{6}$$

$$\bar{y} = \frac{13}{6s} = 0.85.$$

23. The surface area of a surface of revolution is given by [text
     equation (3.11)]

     $$A = \int 2\pi y \, ds.$$

     The y and z coordinates of the centroid are zero by symmetry.
     The x coordinate is given by [text equation (3.4)]

     $$A\bar{x} = \int \bar{x} \, dA = \int x \, dA = 2\pi \int xy \, ds.$$

     Here we have $y = \sqrt{x}$, $ds = \sqrt{1 + \frac{1}{4x}} \, dx$ as in Problem 21. Then, using
     integral tables, we find

     $$A\bar{x} = 2\pi \int_0^2 x\sqrt{x}\sqrt{1 + \frac{1}{4x}} \, dx = \pi \int_0^2 x\sqrt{4x + 1} \, dx$$

     $$= \frac{\pi}{60}(6x - 1)(4x + 1)^{3/2}\Big|_0^2 = \frac{149\pi}{30}.$$

     Using $A = 13\pi/3$ from Problem 20, we find $\bar{x} = 149/130$. Thus the
     coordinates of the centroid are $(\bar{x},\bar{y},\bar{z}) = \left(\frac{149}{130}, 0, 0\right).$

26. The moment of inertia of a solid about the x axis is given by

     $$I_x = \int\int\int (y^2 + z^2) \, dM.$$

     We find the limits of integration as in text Example 2, Figure
     3.5. The equation of the surface of the solid is $y^2 + z^2 = (\sqrt{x})^2 = x$
     [compare text equation (3.9) and see Problem 16 above]. Thus we
     have

     $$I_x = \rho \int_{x=0}^2 \int_{z=-\sqrt{x}}^{\sqrt{x}} \int_{y=-\sqrt{x-z^2}}^{\sqrt{x-z^2}} (y^2 + z^2) \, dy \, dz \, dx.$$

     By symmetry, the y,z plane integral (see text Figure 3.5) is just
     4 times the corresponding integral over the first quadrant. Then

26.   (continued)

$$I_x = 4\rho \int_{x=0}^{2} \int_{z=0}^{\sqrt{x}} \int_{y=0}^{\sqrt{x-z^2}} (y^2 + z^2)\, dy\, dz\, dx$$

$$= 4\rho \int_{0}^{2} dx \int_{0}^{\sqrt{x}} dz \left[\frac{1}{3}(x - z^2)^{3/2} + z^2(x - z^2)^{1/2} \right]$$

$$= 4\rho \int_{0}^{2} dx \left(\frac{\pi x^2}{8} \right) \qquad \text{(by integral tables)}$$

$$= 4\pi\rho/3 = 2M/3$$

since $M = \rho V = 2\pi\rho$ by Problem 19.

29.   The limits are the same as in Problem 26. Since $|xyz|$ is the same for negative y and z as for positive y and z, we can again (as in Problem 26) integrate over positive y and z and multiply the result by 4. For $y \geqslant 0$, $z \geqslant 0$, $0 \leqslant x \leqslant 2$, we do not need the absolute value sign. Thus

$$M = 4 \int_{x=0}^{2} \int_{y=0}^{\sqrt{x}} \int_{z=0}^{\sqrt{x-y^2}} xyz\, dx\, dy\, dz$$

$$= 4 \int_{0}^{2} dx \int_{0}^{\sqrt{x}} dy \left[xy \cdot \frac{1}{2}(x - y^2) \right] = 2 \int_{0}^{2} dx \left(x^2 \cdot \frac{x}{2} - x \cdot \frac{x^2}{4} \right) = 2.$$

31.   (a) For $y = x^{-1}$, we have [see text equations (3.1), (3.8), and (3.11)]

$$y' = -x^{-2}, \qquad ds = \sqrt{1 + y'^2} = \sqrt{1 + x^{-4}}$$

$$V = \int_{1}^{\infty} \pi y^2\, dx = \pi \int_{1}^{\infty} \frac{dx}{x^2} = \pi$$

$$A = \int_{1}^{\infty} 2\pi y\, ds = 2\pi \int_{1}^{\infty} x^{-1}\sqrt{1 + x^{-4}}\, dx$$

31.   (continued)

Since $\sqrt{1 + x^{-4}} > 1$,

$$A > 2\pi \int_1^\infty x^{-1} dx = 2\pi \ln x \Big|_1^\infty = \infty.$$

Thus the volume is finite but the surface area is infinite.

(b) Hint: See the solution of Chapter 1, Problem 15.31c. Here consider a monomolecular layer of paint of thickness, say, 10^{-7}cm. For what x will the layer more than fill the volume?

Section 4

1.   (a) $A = \int_{r=0}^a \int_{\theta=0}^{2\pi} r\, dr\, d\theta = \int_0^a r\, dr \int_0^{2\pi} d\theta = \pi a^2$

(b) $\iint \bar{x} r\, dr\, d\theta = \iint x r\, dr\, d\theta = \int_{r=0}^a \int_{\theta=0}^{\pi/2} r \cos\theta\, r dr\, d\theta = a^3/3$

Since the area of one quadrant $= \dfrac{\pi a^2}{4}$, $\bar{x} = \dfrac{a^3}{3} \div \dfrac{\pi a^2}{4} = \dfrac{4a}{3\pi}$.

In a similar way, or by symmetry, $\bar{y} = \dfrac{4a}{3\pi}$.

(c) $I_x = \rho \iint y^2 r\, dr\, d\theta = \rho \int_{r=0}^a \int_{\theta=0}^{2\pi} r^2 \sin^2\theta\, r\, dr\, d\theta = \dfrac{\pi a^4 \rho}{4} = Ma^2/4.$

(See solution of Chapter 2 Problem 11.12 for an easy way to find the integral of $\sin^2\theta$.)

(d) $C = \int_0^{2\pi} a\, d\theta = 2\pi a$

(e) $\int \bar{x} a\, d\theta = \int xa\, d\theta = \int_0^{\pi/2} a \cos\theta\, a\, d\theta = a^2.$

Since the quarter circle arc length $= \dfrac{\pi a}{2}$, $\bar{x} = a^2 \div \dfrac{\pi a}{2} = \dfrac{2a}{\pi}$.

Similarly, or by symmetry, $\bar{y} = 2a/\pi$.

3.   (a) Let the disk occupy the region $x^2 + y^2 \leqslant a^2$, $z = 0$. Then in cylindrical coordinates the distance from the z axis to the mass element $dM = \rho r \, dr \, d\theta$ is r, so

$$I_z = \int_0^{2\pi} \int_0^a r^2 \cdot \rho r \, dr \, d\theta = 2\pi\rho a^4/4.$$

The total mass is $M = \pi a^2 \rho$ so

$$I_z = \frac{1}{2} M a^2.$$

5.   (a) In spherical coordinates (see text Figure 4.5), the equation of the cone $z^2 = x^2 + y^2$ becomes $r^2\cos^2\theta = r^2\sin^2\theta$ or $\theta = \pi/4$. This is a cone of half angle $\pi/4$; the sketch shows a vertical cross section of it. For any given θ, we see from the sketch that r goes from its value when $z = r\cos\theta = 1$, namely $r = \frac{1}{\cos\theta}$, to its value when $z = 2$, namely $r = \frac{2}{\cos\theta}$. Then θ goes from 0 to $\pi/4$, and ϕ from 0 to 2π (rotate the cross section shown about the z axis). Using the spherical coordinate volume element [text equation (4.5)], we have

$$V = \int_{\phi=0}^{2\pi} \int_{\theta=0}^{\pi/4} \int_{r=\frac{1}{\cos\theta}}^{\frac{2}{\cos\theta}} r^2 \sin\theta \, dr \, d\theta \, d\phi$$

$$= 2\pi \int_{\theta=0}^{\pi/4} \sin\theta \, d\theta \left(\frac{r^3}{3} \Big|_{\frac{1}{\cos\theta}}^{\frac{2}{\cos\theta}} \right) = \frac{2\pi}{3} \int_0^{\pi/4} \sin\theta \, d\theta \left(\frac{2^3 - 1}{\cos^3\theta} \right)$$

$$= \frac{14\pi}{3} \int_0^{\pi/4} \frac{\sin\theta \, d\theta}{\cos^3\theta} = \frac{14\pi}{3} \left(\frac{1}{2\cos^2\theta} \right) \Big|_0^{\pi/4} = \frac{14\pi}{3} \left(1 - \frac{1}{2} \right) = \frac{7\pi}{3}.$$

11.   In cylindrical coordinates (see text Figure 4.4), the equation of

the cylinder $x^2 + y^2 = 4$ is $r^2 = 4$ or $r = 2$. The cylinder is circu-

lar with radius 2 and axis along the z axis. It is cut off at

the bottom by the (x,y) plane ($z = 0$) and at the top by the sur-

face (paraboloid)

$z = 2x^2 + y^2 = 2r^2\cos^2\theta + r^2\sin^2\theta = r^2(\sin^2\theta + 2\cos^2\theta) = r^2(1 + \cos^2\theta).$

Thus z goes from 0 to $r^2(1 + \cos^2\theta)$; then the r limits are 0 to 2,

and the θ limits are 0 to 2π. The cylindrical coordinate volume

element is $dV = r\, dr\, d\theta\, dz$.   Then

$$V = \int_{r=0}^{2} \int_{\theta=0}^{2\pi} \int_{z=0}^{r^2(1+\cos^2\theta)} r\, dr\, d\theta\, dz$$

$$= \int_{r=0}^{2} \int_{\theta=0}^{2\pi} r\, dr\, d\theta \cdot r^2(1 + \cos^2\theta) = \int_{0}^{2} r^3 dr \int_{0}^{2\pi} (1 + \cos^2\theta)\, d\theta$$

$$= \frac{2^4}{4}(2\pi + \pi) = 12\pi$$

(Recall, Chapter 2, Problem 11.12, $\int_{0}^{2\pi} \cos^2\theta\, d\theta = \pi$.)

14.   To write $I = \int_{x=0}^{1} \int_{y=0}^{\sqrt{1-x^2}} e^{-x^2-y^2} dx\, dy$ in polar coordinates, first

consider the area of integration. Here y goes from 0 (that is,

the x axis) to $y = \sqrt{1 - x^2}$ (that is, the circle $x^2 + y^2 = 1$);   then x

goes from 0 to 1. Thus the area of integration is the first quad-

rant part of the area inside the circle of radius 1 and center at

the origin. To cover this same area in polar coordinates, we

must use the limits $r = 0$ to 1 and $\theta = 0$ to $\pi/2$. We write the inte-

grand in polar coordinates $\left(e^{-x^2-y^2} = e^{-r^2} \right)$ and use the polar co-

14.   (continued)

ordinate area element [text equation (4.2)].   Then

$$I = \int_0^{\pi/2} d\theta \int_0^1 e^{-r^2} r \, dr = \frac{\pi}{2} \left(-\frac{1}{2} e^{-r^2} \right) \Big|_0^1 = \frac{\pi}{4} (1 - e^{-1}).$$

15.   Using text equations (4.12) and (4.4), we find

$$\frac{\partial(x,y,z)}{\partial(r,\theta,z)} = \begin{vmatrix} \cos\theta & -r\sin\theta & 0 \\ \sin\theta & r\cos\theta & 0 \\ 0 & 0 & 1 \end{vmatrix} = r(\cos^2\theta + \sin^2\theta) = r.$$

Thus the cylindrical coordinate volume element is $dV = r \, dr \, d\theta \, dz$
as in text equation (4.6).

18.

$$\frac{\partial(x,y)}{\partial(u,v)} \frac{\partial(u,v)}{\partial(s,t)} = \begin{vmatrix} \dfrac{\partial x}{\partial u} & \dfrac{\partial x}{\partial v} \\[2mm] \dfrac{\partial y}{\partial u} & \dfrac{\partial y}{\partial v} \end{vmatrix} \begin{vmatrix} \dfrac{\partial u}{\partial s} & \dfrac{\partial u}{\partial t} \\[2mm] \dfrac{\partial v}{\partial s} & \dfrac{\partial v}{\partial t} \end{vmatrix}.$$

If we multiply, for example, the first row times the first column
(see text, Chapter 3, Section 6), we get

$$\frac{\partial x}{\partial u} \frac{\partial u}{\partial s} + \frac{\partial x}{\partial v} \frac{\partial v}{\partial s} = \frac{\partial x}{\partial s}$$

(chain rule -- see text, Chapter 4, Section 7).   This is the
element in the first row and first column of the determinant
$\frac{\partial(x,y)}{\partial(s,t)}$.   Similarly you can find the other three elements and so
obtain the result $\frac{\partial(x,y)}{\partial(s,t)}$.

Now if $s = x$ and $t = y$ then we have

$$\frac{\partial(x,y)}{\partial(u,v)} \frac{\partial(u,v)}{\partial(x,y)} = \frac{\partial(x,y)}{\partial(x,y)} = \begin{vmatrix} 1 & 0 \\ 0 & 1 \end{vmatrix} = 1.$$

19. Using the theorem stated in text equations (4.11) to (4.13) we
     write

$$\int\int f(u,v)\, du\, dv = \int\int f \cdot \frac{\partial(u,v)}{\partial(x,y)}\, dx\, dy,$$

where f and the Jacobian in the (x,y) integral must be expressed
in terms of x and y, and the limits must be properly adjusted.
Given $u = x^2 - y^2$, $v = 2xy$, we find

$$\frac{\partial(u,v)}{\partial(x,y)} = \begin{vmatrix} 2x & -2y \\ 2y & 2x \end{vmatrix} = 4(x^2 + y^2).$$

Then we see that

$$I = \int_0^\infty \int_0^\infty \frac{e^{-2xy}}{1 + (x^2 - y^2)^2}(x^2 + y^2)\, dx\, dy = \frac{1}{4}\int\int \frac{e^{-v}}{1 + u^2}\, du\, dv$$

where we must determine the limits for u and v. Now x and y both
go from 0 to ∞; then v = 2xy also takes all values from 0 to ∞.
But $u = x^2 - y^2$ takes both positive values (when x > y) and negative
values (when x < y), so the u limits are -∞ to ∞.   Thus

$$I = \frac{1}{4}\int_{u=-\infty}^{\infty} \int_{v=0}^{\infty} \frac{e^{-v}}{1 + u^2}\, du\, dv = \frac{1}{4}\int_{-\infty}^{\infty} \frac{du}{1 + u^2}\int_0^\infty e^{-v}\, dv =$$

$$\frac{1}{4}\left(\arctan u \Big|_{-\infty}^{\infty}\right)\left(-e^{-v}\Big|_0^\infty\right) = \frac{1}{4}\left[\frac{\pi}{2} - \left(-\frac{\pi}{2}\right)\right]\cdot 1 = \frac{\pi}{4}.$$

21.   See Chapter 10, Problem 8.1.

Section 5

3.   We want the area of part of the paraboloid $z = x^2 + y^2$ so we find
     $\sec \gamma$ from this equation.   Using text equation (5.6), we get

$$\sec \gamma = \sqrt{\left(\frac{\partial z}{\partial x}\right)^2 + \left(\frac{\partial z}{\partial y}\right)^2 + 1} = \sqrt{4x^2 + 4y^2 + 1}.$$

The part of the paraboloid inside the cylinder $x^2 + y^2 = 9$ lies
directly above the circle $x^2 + y^2 \leqslant 9$ in the (x,y) plane.   Thus we
integrate text equation (5.2) over this area.   This is most
easily done in polar coordinates.

$$A = \iint \sec \gamma \; dx \; dy = \iint \sqrt{4(x^2 + y^2) + 1} \; dx \; dy$$

$$= \int_{r=0}^{3} \int_{\theta=0}^{2\pi} \sqrt{4r^2 + 1} \; r \; dr \; d\theta = 2\pi \cdot \frac{1}{12}(4r^2 + 1)^{3/2} \Big|_0^3$$

$$= \frac{\pi}{6}(37^{3/2} - 1)$$

5.   Caution: Where do the sphere and cone intersect?

6.   Since the cylinder is vertical, its projection into the (x,y)
     plane is the circumference of a circle, not an area.   We can
     either project into the (x,z) [or (y,z)] plane, or, more simply,
     we can interchange x and z in the equation (that is, relabel the
     axes).   Interchanging x and z leaves the equation of the sphere
     unchanged; the cylinder equation becomes

$$z^2 + y^2 - y = 0.$$

Since we want the area of part of the _cylinder_, we find $\sec \gamma$
from the cylinder equation: Using text equation (5.5) with
$\phi = z^2 + y^2 - y$, we get

6.   (continued)

$$\sec \gamma = \frac{\sqrt{\left(\frac{\partial \phi}{\partial x}\right)^2 + \left(\frac{\partial \phi}{\partial y}\right)^2 + \left(\frac{\partial \phi}{\partial z}\right)^2}}{\left|\frac{\partial \phi}{\partial z}\right|} = \frac{\sqrt{(2y - 1)^2 + 4z^2}}{|2z|} .$$

Using the cylinder equation, we eliminate z:

$$\sec \gamma = \frac{\sqrt{4y^2 - 4y + 1 + 4(y - y^2)}}{2\sqrt{y - y^2}} = \frac{1}{2\sqrt{y(1 - y)}} .$$

We find the limits of integration of text equation (5.2) by con-
sidering the intersection of the cylinder $y^2 + z^2 - y = 0$ and the
sphere $x^2 + y^2 + z^2 = 1$. Eliminating z,
we have $y = 1 - x^2$. On the cylinder,
$y = y^2 + z^2 \geqslant 0$; thus the area of integration is bounded by the x axis and
the parabola $y = 1 - x^2$. Since the
cylinder lies both above and below the (x,y) plane, we must
double our result, and if we integrate only over the first quadrant, we must double it again. Thus

$$A = 4 \int_{y=0}^{1} \int_{x=0}^{\sqrt{1-y}} \frac{dx \; dy}{2\sqrt{y(1 - y)}} = 2 \int_{y=0}^{1} \frac{\sqrt{1 - y} \; dy}{\sqrt{y(1 - y)}}$$

$$= 2 \int_{0}^{1} \frac{dy}{\sqrt{y}} = 4\sqrt{y} \Big|_{0}^{1} = 4 .$$

9.   We want the area of part of the cone $x^2 + y^2 = z^2$; thus we use text
equation (5.5) with $\phi = x^2 + y^2 - z^2$ to find $\sec \gamma$ and then eliminate
z using the equation of the cone.

$$\sec \gamma = \frac{\sqrt{4x^2 + 4y^2 + 4z^2}}{2|z|} = \frac{\sqrt{8z^2}}{|2z|} = \sqrt{2} .$$

9.   (continued)

Now by text equation (5.2), $A = \iint \sqrt{2} \, dx \, dy$ over the area of the circle $(x - 1)^2 + y^2 \leqslant 1$. This is just $\sqrt{2}$ times the area of the circle (radius = 1) so the desired area of the cone is

$$A = \sqrt{2} \, \pi \cdot 1^2 = \pi\sqrt{2}.$$

14.   We find $\sec \gamma$ and the limit of integration as in text Example 1. The density $|x|$ takes the same values in the second quadrant as in the first quadrant. Thus we can integrate over the first quadrant semicircle and double the result as in text equation (5.7). Then we have

$$M = 2 \int_{y=0}^{1} \int_{x=0}^{\sqrt{y-y^2}} \frac{x \, dx \, dy}{\sqrt{1 - x^2 - y^2}} = 2 \int_{y=0}^{1} dy \left[-\sqrt{1 - x^2 - y^2} \right]_{x=0}^{\sqrt{y-y^2}}$$

$$= 2 \int_{0}^{1} dy \left(\sqrt{1 - y^2} - \sqrt{1 - y} \right)$$

$$= 2 \left[\tfrac{1}{2} \left(y\sqrt{1 - y^2} + \sin^{-1} y \right) + \tfrac{2}{3}(1 - y)^{3/2} \right] \qquad \text{(by tables)}$$

$$= 2 \left(0 + \tfrac{1}{2} \, \tfrac{\pi}{2} - \tfrac{2}{3} \right) = \tfrac{\pi}{2} - \tfrac{4}{3}.$$

(You can check this by doing the integration in polar coordinates as in the text Example 1.)

Section 6

4.   The easiest way to do this problem is by a combination of the
     parallel and perpendicular axis theorems. If the hoop is in the
     (x,y) plane with center at the origin, then I_z is just MR^2 (by
     definition, since all points of M are at distance R from the
     axis).   Then

     (a) $I_x = I_y = \frac{1}{2} I_z = \frac{1}{2} MR^2$ (See text Section 3, Example 1f).

     (b) I about a tangent line $= \frac{1}{2} MR^2 + MR^2 = \frac{3}{2} MR^2$ (See Problem 3.1).

     However, in a similar problem with less symmetry we might need to
     write integrals for these moments of inertia, so let us do this
     problem by integration.

     (a) We want to evaluate, say, $I_x = \rho \int y^2 ds$. With the center of
     the hoop at the origin, we have

$$x^2 + y^2 = R^2$$

$$2x\,dx + 2y\,dy = 0, \qquad y' = -\frac{x}{y}, \qquad ds = \sqrt{1 + \frac{x^2}{y^2}}\ dx$$

$$y^2 ds = y\sqrt{y^2 + x^2}\ dx = yR = R\sqrt{R^2 - x^2}$$

$$I_x = \rho R \int_0^R \sqrt{R^2 - x^2}\ dx \cdot 4 \qquad\qquad \text{(4 quadrants)}$$

$$= 4\rho R \cdot \frac{1}{2}\left[x\sqrt{R^2 - x^2} + R^2 \arcsin\frac{x}{R} \right]_0^R \qquad \text{(from tables)}$$

$$= \rho R^3 \pi = \frac{1}{2} MR^2 \qquad\qquad \text{since } M = 2\pi R\rho.$$

Alternatively, we could use polar coordinates (a good choice
when, as here, the equation of the curve is simple in polar coordinates). We have

4.   (continued)

$ds = R\,d\theta,\qquad y = R\sin\theta$ on the circle;

$$I_x = \rho\int_0^{2\pi} R^2\sin^2\theta\; R\,d\theta = \rho R^3\pi = \frac{1}{2}MR^2\quad\text{as before.}$$

   (b) To do this part of the problem directly (that is, not using the parallel axis theorem) we sketch the hoop as in Problem 4.2 (see text). Then

$r = 2R\cos\theta,\qquad dr = -2R\sin\theta\,d\theta,$

$ds^2 = dr^2 + r^2 d\theta^2 = (4R^2\sin^2\theta + 4R^2\cos^2\theta)\,d\theta^2,\qquad ds = 2R\,d\theta,$

$x = r\cos\theta = 2R\cos^2\theta$ on the circle;

$$I_y = \rho\int x^2 ds = 8R^3\rho\int_{-\pi/2}^{\pi/2}\cos^4\theta\,d\theta$$

$$= 8R^3\rho\left[\frac{3\theta}{8} + \frac{\sin 2\theta}{4} + \frac{\sin 4\theta}{32}\right]_{-\pi/2}^{\pi/2}\qquad\text{from tables}$$

$$= 3\pi R^3\rho = \frac{3}{2}MR^2\quad\text{as before.}$$

6.   (a) From the sketch of the area to be found (inside the heavy lines) we see that, over this area, θ varies from $\pi/6$ to $\pi/2$, and (for each θ) r varies from its value on the line $y = 1$ (that is, $r = \dfrac{1}{\sin\theta} = \csc\theta$) to its value on the circle ($r = 2$). Then the area is

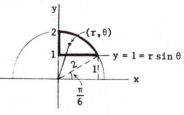

$$A = \int_{\theta=\pi/6}^{\pi/2}\int_{r=\csc\theta}^{2} r\,dr\,d\theta = \frac{1}{2}\int_{\pi/6}^{\pi/2}(4 - \csc^2\theta)\,d\theta$$

$$= \frac{1}{2}(4\theta + \cot\theta)\Big|_{\pi/6}^{\pi/2} = \frac{2\pi}{3} - \frac{\sqrt{3}}{2}\,.$$

10.   (a) $A = \int_0^\pi \sin x \, dx = -\cos x \Big|_0^\pi = 2$

$A\bar{y} = \int_{x=0}^\pi \int_{y=0}^{\sin x} y \, dx \, dy = \int_0^\pi \frac{1}{2} \sin^2 x \, dx = \frac{1}{2} \frac{\pi}{2}, \qquad \bar{y} = \frac{\pi}{8}$

By symmetry we can see that $\bar{x} = \frac{\pi}{2}$;   alternatively,

$A\bar{x} = \int_{x=0}^\pi \int_{y=0}^{\sin x} x \, dx \, dy = \int_0^\pi x \sin x \, dx$

$\qquad = \sin x - x \cos x \Big|_0^\pi = \pi, \qquad \bar{x} = \frac{\pi}{2}$.

14.   We want the volume inside $x^2 + y^2 = a^2$ (cylinder with axis along
the z axis) cut off at the top and bottom by $z = \pm \sqrt{a^2 - x^2}$
(cylinder along the y axis). This is $V = \iint 2z \, dx \, dy$ where
$2z = 2\sqrt{a^2 - x^2}$ is the distance (for each x) from the bottom surface to the top and the double integral is over the (x,y) plane
area inside the circle $x^2 + y^2 = a^2$. By symmetry, we can integrate
over the first quadrant and multiply by 4.   Thus

$$V = 8 \int_{x=0}^a \int_{y=0}^{\sqrt{a^2-x^2}} \sqrt{a^2 - x^2} \, dy \, dx = 8 \int_0^a (a^2 - x^2) \, dx = \frac{16}{3} a^3 .$$

19.   The cone $r^2 = z^2$, $0 < z < h$, is shown in text Figure 4.6. To find
I_x, say, in cylindrical coordinates, we evaluate
$\iiint (y^2 + z^2) \rho \, r \, dr \, d\theta$, where $y^2 + z^2 = r^2 \sin^2\theta + z^2$ is the square of
the distance from the mass element to the x axis, and $\rho = r^2$. The
integration limits are as discussed in text Section 4, Example 2.
Thus

19.   (continued)

$$I_x = \int_{\theta=0}^{2\pi} \int_{z=0}^{h} \int_{r=0}^{z} (r^2 \sin^2\theta + z^2) r^2 \, r \, dr \, d\theta \, dz$$

$$= \int_{\theta=0}^{2\pi} \int_{z=0}^{h} \left(\frac{1}{6} z^6 \sin^2\theta + \frac{1}{4} z^6 \right) d\theta \, dz$$

$$= \frac{1}{6} \frac{h^7}{7} \pi + \frac{1}{4} \frac{h^7}{7} 2\pi = \frac{h^7}{7} \frac{2}{3} \pi.$$

Using the result from Problem 18, $M = \frac{\pi h^5}{10}$, we have

$$\frac{I_x}{M} = \frac{20}{21} h^2 \qquad \text{or} \qquad I_x = \frac{20}{21} Mh^2.$$

I_y and I_z are evaluated in a similar way. The moment of inertia I_{mx} about a line through the center of mass parallel to the x axis is most easily obtained using the parallel axis theorem (see Problem 3.1). From Problem 18, the center of mass is the point $(0,0,5h/6)$; then the distance from the center of mass to the x axis is $5h/6$. Thus

$$I_{xm} = I_x - M\left(\frac{5}{6} h\right)^2 = \left(\frac{20}{21} - \frac{25}{36} \right) Mh^2 = \frac{65}{252} Mh^2.$$

With more effort, this result can be obtained by evaluating $\iiint \ell^2 r^2 r \, dr \, d\theta \, dz$ where ℓ is the distance from (x,y,z) to the line $z = \frac{5}{6}$, $y = 0$, that is,

$$\ell^2 = y^2 + \left(z - \frac{5}{6} h \right)^2 = r^2 \sin^2\theta + \left(z - \frac{5}{6} h \right)^2.$$

21.   This is the same cone as in Problems 18 and 19.   There we used
       cylindrical coordinates because the density was $x^2 + y^2 =$
       (cylindrical coordinate r)^2.   Here we shall use spherical co-
       ordinates because the gravitational force depends on
       $r = \sqrt{x^2 + y^2 + z^2}$, that is, the spherical coordinate r.   The magni-
       tude of the force on a unit mass at the origin due to the element
       of mass $dM = \rho dV = \rho r^2 \sin\theta\, dr\, d\theta\, d\phi$ at the point (r,θ,ϕ) is (see
       Problem 4.24) $G\, dM/r^2$.   This force is a vector acting in the
       radial direction.   By symmetry, the horizontal components of the
       force due to the whole mass are zero.   We find the z component of
       the force by integrating $(G\, dM/r^2)\cos\theta$ over the volume of the
       cone.   The limits on r (see solution of Problem 4.6) are from
       $r = 0$ to $r = \dfrac{h}{\cos\theta}$ (on the plane $z = r\cos\theta = h$); then θ goes from 0
       to $\pi/4$ (the half angle of the cone) and ϕ from 0 to 2π.   Thus

$$F_z = G\rho \int_{\phi=0}^{2\pi} \int_{\theta=0}^{\pi/4} \int_{r=0}^{\frac{h}{\cos\theta}} \left(\frac{1}{r^2} r^2 \sin\theta\, dr\, d\theta\, d\phi\right) \cos\theta$$

$$= 2\pi G\rho \int_0^{\pi/4} (\sin\theta\, d\theta)\cos\theta \cdot \frac{h}{\cos\theta} = 2\pi G\rho h \int_0^{\pi/4} \sin\theta\, d\theta$$

$$= 2\pi G\rho h (-\cos\theta)\Big|_0^{\pi/4} = 2\pi G\rho h\left(1 - \frac{1}{\sqrt{2}}\right) = \pi G\rho h(2 - \sqrt{2}).$$

24.   We want to evaluate $V = \iiint dV$ and $V\bar{z} = \iiint z \, dV$. Following the
method of text Example 4, Section 4, we make the change of vari-
ables $x = ax'$, $y = by'$, $z = cz'$; then the equation of the paraboloid
becomes

$$z' = x'^2 + y'^2$$

and $dV = dx \, dy \, dz$ becomes $abc \, dx' \, dy' \, dz'$. We change to cylindrical
coordinates r', θ', z' in the primed system. The equation of the
paraboloid is then $z' = r'^2$, so the integration limits for r' are
0 to $\sqrt{z'}$. Then (since z goes from 0 to c), z' goes from 0 to 1;
θ' goes from 0 to 2π. Thus we have

$$V = \iiint dx \, dy \, dz = abc \int_{\theta'=0}^{2\pi} \int_{z'=0}^{1} \int_{r'=0}^{\sqrt{z'}} r' \, dr' \, d\theta' \, dz'$$

$$= abc \cdot 2\pi \int_{0}^{1} dz' \cdot \frac{1}{2} z' = \pi abc/2.$$

$$V\bar{z} = \iiint z \, dx \, dy \, dz = abc \int_{\theta'=0}^{2\pi} \int_{z'=0}^{1} \int_{r'=0}^{\sqrt{z'}} (cz')r' \, dr' \, d\theta' \, dz'$$

$$= abc^2 \cdot 2\pi \int_{0}^{1} dz' \cdot \frac{1}{2} z'^2 = \pi abc^2/3$$

$$\bar{z} = \frac{2}{3} c.$$

By symmetry, $\bar{x} = \bar{y} = 0$.

27. We use the theorem of text equations (4.11) to (4.13). For
u = y/x, v = x + y, we find

$$J = \frac{\partial(u,v)}{\partial(x,y)} = \begin{vmatrix} -\dfrac{y}{x^2} & \dfrac{1}{x} \\ 1 & 1 \end{vmatrix} = -\left(\frac{y}{x^2} + \frac{1}{x}\right) = -\frac{y+x}{x^2}$$

$$du\ dv \longleftrightarrow |J|\ dx\ dy = \frac{x+y}{x^2}\ dx\ dy$$

Then

$$I = \int_{x=0}^{1}\int_{y=0}^{x} e^{x+y}\ \frac{x+y}{x^2}\ dx\ dy = \iint e^{v} du\ dv$$

where the (u,v) limits must be found. To do this we sketch the
areas of integration in the (x,y) plane and in the (u,v) plane.
In the (x,y) integral, for
each x value, y goes from 0
to its value on the line
y = x; then x goes from 0 to
1. Thus the area of integration is the triangle shown
in the (x,y) plane. Now the
corresponding boundaries in the (u,v) plane are:

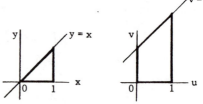

y = x	u = y/x = 1
y = 0, x ≠ 0	u = 0
x = 1	v = 1 + u
x = y = 0	v = 0

We sketch these boundaries in the (u,v) plane; then the (u,v)
limits are

27.   (continued)

        v: 0 to 1 + u

        u: 0 to 1

Thus we find

$$I = \int_{u=0}^{1} \int_{v=0}^{1+u} e^v \, du \, dv = \int_{0}^{1} du(e^{1+u} - 1)$$

$$= (e^{1+u} - u) \Big|_{0}^{1} = e^2 - e - 1.$$

Chapter 6

<u>Section 3</u>

6. The diagram is just schematic, that is, the given $\vec{A}$, $\vec{B}$, $\vec{C}$ are not used in drawing the diagram. Think of the whole figure as rotating about $\vec{A}$ at 2 rad/sec. Then the velocity of the head of $\vec{B}$ is $\vec{v} = \vec{\omega} \times \vec{r}$, where $\vec{\omega} = \dfrac{2\vec{A}}{|\vec{A}|}$ (that is, a vector of length 2 in direction $\vec{A}$) and $\vec{r} = \vec{A} + \vec{B}$ (that is, a vector from 0 to the head of $\vec{B}$). Then

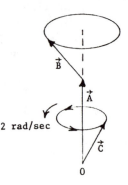

2 rad/sec

$$\vec{v} = \frac{2}{|A|} \vec{A} \times (\vec{A} + \vec{B}) = \frac{2}{|A|} (\vec{A} \times \vec{B}) \qquad \text{since } \vec{A} \times \vec{A} = 0.$$

Using the given vectors $\vec{A}$ and $\vec{B}$, we find $|\vec{A}| = \sqrt{1+1+4}$,

$$\vec{v} = \frac{2}{\sqrt{6}} \begin{vmatrix} \vec{i} & \vec{j} & \vec{k} \\ 1 & 1 & -2 \\ 2 & -1 & 3 \end{vmatrix} = \frac{2}{\sqrt{6}} (\vec{i} - 7\vec{j} - 3\vec{k}).$$

The torque of $\vec{B}$ about the head of $\vec{C}$ is $\vec{r} \times \vec{F}$ where $\vec{F} = \vec{B}$ and $\vec{r} = \vec{A} - \vec{C}$, that is, the vector from the head of $\vec{C}$ (point the torque is about) to the head of $\vec{A}$ (point of application of the force). Thus the vector torque $= (\vec{A} - \vec{C}) \times \vec{B}$.

The scalar torque about line $\vec{C}$ is $\vec{n} \cdot (\vec{r} \times \vec{F})$ where $\vec{n} = \dfrac{\vec{C}}{|\vec{C}|}$; thus for the scalar torque we find

$$\frac{\vec{C}}{|C|} \cdot (\vec{A} - \vec{C}) \times \vec{B} = \frac{1}{\sqrt{26}} \begin{vmatrix} 0 & 1 & -5 \\ 1 & 0 & 3 \\ 2 & -1 & 3 \end{vmatrix} = \frac{8}{\sqrt{26}}$$

7.   (a) By text equation (3.10), the torque of $\vec{F}$ about the origin is
$\vec{r} \times \vec{F}$ where $\vec{r}$ is the vector <u>from</u> the origin <u>to</u> the point (1,5,2)
at which the force acts.

$$\vec{r} \times \vec{F} = \begin{vmatrix} \vec{i} & \vec{j} & \vec{k} \\ 1 & 5 & 2 \\ 2 & -3 & 1 \end{vmatrix} = 11\vec{i} + 3\vec{j} - 13\vec{k}.$$

(b) The torque about the y axis is $\vec{j} \cdot (\vec{r} \times \vec{F}) = 3.$

(c) The torque about the given line through the origin is
$\vec{n} \cdot (\vec{r} \times \vec{F})$ where $\vec{n}$ is a unit vector along the line:
$\vec{n} = (2\vec{i} + \vec{j} - 2\vec{k})/3.$ Thus the torque is

$$\vec{n} \cdot (\vec{r} \times \vec{F}) = \frac{1}{3}(2\vec{i} + \vec{j} - 2\vec{k}) \cdot (11\vec{i} + 3\vec{j} - 13\vec{k}) = \frac{51}{3} = 17.$$

11. For example, consider $\vec{A} \cdot (\vec{B} \times \vec{C})$, $(\vec{A} \times \vec{B}) \cdot \vec{C}$, $\vec{C} \cdot (\vec{A} \times \vec{B})$, $(\vec{B} \times \vec{A}) \cdot \vec{C}$;
these are 4 of the 12 triple scalar products. As in text
equation (3.2), we can write

$$\vec{C} \cdot (\vec{A} \times \vec{B}) = \begin{vmatrix} C_x & C_y & C_z \\ A_x & A_y & A_z \\ B_x & B_y & B_z \end{vmatrix}$$

Recall from text page 89, that interchanging two rows of a determinant changes its sign. If we make the two interchanges $C \longleftrightarrow A$,
and then $B \longleftrightarrow C$, in the determinant above, we have the determi-
nant in text equation (3.2). Thus $\vec{A} \cdot (\vec{B} \times \vec{C}) = \vec{C} \cdot (\vec{A} \times \vec{B})$; note the
cyclic permutation of the vectors. Now $(\vec{A} \times \vec{B}) \cdot \vec{C} = \vec{C} \cdot (\vec{A} \times \vec{B})$ since
scalar multiplication is commutative [see text page 100, equation
(4.3)]. Also $\vec{B} \times \vec{A} = -(\vec{A} \times \vec{B})$, so $(\vec{B} \times \vec{A}) \cdot \vec{C} = -(\vec{A} \times \vec{B}) \cdot \vec{C}$. Thus we have
$$\vec{A} \cdot (\vec{B} \times \vec{C}) = (\vec{A} \times \vec{B}) \cdot \vec{C} = \vec{C} \cdot (\vec{A} \times \vec{B}) = -(\vec{B} \times \vec{A}) \cdot \vec{C}.$$

13.   We want to evaluate

(1) $[(\vec{A} \times \vec{B}) \times (\vec{B} \times \vec{C})] \cdot (\vec{C} \times \vec{A})$.

First let $\vec{B} \times \vec{C} = \vec{D}$; then by the text rule (3.9)

(2) $(\vec{A} \times \vec{B}) \times \vec{D} = (\vec{A} \cdot \vec{D})\vec{B} - (\vec{B} \cdot \vec{D})\vec{A}$.

Since $\vec{D} = \vec{B} \times \vec{C}$, we find

(3)
$\vec{A} \cdot \vec{D} = \vec{A} \cdot (\vec{B} \times \vec{C})$

$\vec{B} \cdot \vec{D} = \vec{B} \cdot (\vec{B} \times \vec{C}) = 0$ since $(\vec{B} \times \vec{C}) \perp \vec{B}$.

Substitute (3) and $\vec{D} = \vec{B} \times \vec{C}$ into (2) to get

(4) $(\vec{A} \times \vec{B}) \times (\vec{B} \times \vec{C}) = [\vec{A} \cdot (\vec{B} \times \vec{C})]\vec{B}$.

Substitute (4) into (1) to get:

$[(\vec{A} \times \vec{B}) \times (\vec{B} \times \vec{C})] \cdot (\vec{C} \times \vec{A}) = [\vec{A} \cdot (\vec{B} \times \vec{C})]\vec{B} \cdot (\vec{C} \times \vec{A})$.

By Problem 11, $\vec{B} \cdot (\vec{C} \times \vec{A}) = \vec{A} \cdot (\vec{B} \times \vec{C})$. Thus the result is the square of the triple scalar product $\vec{A} \cdot (\vec{B} \times \vec{C})$.

17.   Using the text rule (3.9), we find

$$\vec{a} = \vec{\omega} \times (\vec{\omega} \times \vec{r}) = (\vec{\omega} \cdot \vec{r})\vec{\omega} - (\vec{\omega} \cdot \vec{\omega})\vec{r} = (\vec{\omega} \cdot r)\vec{\omega} - \omega^2 \vec{r}.$$

If $\vec{r} \perp \vec{\omega}$, then $\vec{r} \cdot \vec{\omega} = 0$, so $\vec{a} = -\omega^2 \vec{r}$. By text Figure 2.6, $\vec{v} = \vec{\omega} \times \vec{r}$. If $\vec{r} \perp \vec{\omega}$, $|\vec{v}| = \omega r \sin 90° = \omega r$; then $|\vec{a}| = \omega^2 r = \dfrac{(\omega r)^2}{r} = \dfrac{v^2}{r}$.

19.   (a) By text equation (3.10), $\vec{r} \times \vec{F}$ gives the vector torque about a point if $\vec{r}$ goes <u>from</u> the point we are taking torques about <u>to</u> the point of action of the force. Thus we find

$$\vec{r} = (1,1,1) - (2,-1,5) = (-1,2,-4),$$

$$\vec{r} \times \vec{F} = \begin{vmatrix} \vec{i} & \vec{j} & \vec{k} \\ -1 & 2 & -4 \\ 1 & 3 & 2 \end{vmatrix} = 16\vec{i} - 2\vec{j} - 5\vec{k}.$$

19.   (continued)

(b) We first verify that the line goes through the point $(2,-1,5)$
which we took the torque about in part (a). Then the torque
about the line is $\vec{n} \cdot (\vec{r} \times \vec{F})$ where $\vec{n}$ is a unit vector along the
line, and $\vec{r} \times \vec{F}$ is the vector torque in part (a). We find

$$\vec{n} = \frac{1}{\sqrt{6}}(\vec{i} - \vec{j} + 2\vec{k})$$

$$\vec{n} \cdot (\vec{r} \times \vec{F}) = \frac{1}{\sqrt{6}}(\vec{i} - \vec{j} + 2\vec{k}) \cdot (16\vec{i} - 2\vec{j} - 5\vec{k}) = 8/\sqrt{6}.$$

Section 4

2.   (a) We verify that

$$\vec{r} = t^2\vec{i} - 2t\vec{j} + (t^2 + 2t)\vec{k} = 4\vec{i} - 4\vec{j} + 8\vec{k}$$

when $t = 2$.

(b) The velocity is

$$\vec{v} = \frac{d\vec{r}}{dt} = 2t\vec{i} - 2\vec{j} + (2t + 2)\vec{k}.$$

The speed is the magnitude of the velocity; we find

$$|\vec{v}| = [4t^2 + 4 + (2t + 2)^2]^{1/2} = (8t^2 + 8t + 8)^{1/2}.$$

At $t = 2$,

$$\vec{v} = 4\vec{i} - 2\vec{j} + 6\vec{k},$$

$$|\vec{v}| = 2\sqrt{14}.$$

(c) The velocity vector is along the tangent line to the curve.
The vector equation of the tangent line at $(4,-4,8)$ is then

$$\vec{r} = (4,-4,8) + \vec{v}t = (4,-4,8) + (4,-2,6)t.$$

The equation of the plane normal to the path at $(4,-4,8)$ is

$$4(x - 4) - 2(y + 4) + 6(z - 8) = 0, \quad \text{or}$$

$$2x - y + 3z = 36.$$

8.   For $\vec{r} = r\vec{e}_r$, we find
$$\vec{v} = \frac{d\vec{r}}{dt} = \frac{dr}{dt}\vec{e}_r + r\frac{d}{dt}\vec{e}_r .$$

Using text equation (4.13), we get
$$\vec{v} = \frac{dr}{dt}\vec{e}_r + r\frac{d\theta}{dt}\vec{e}_\theta .$$

To find $\vec{a} = \frac{d\vec{v}}{dt}$, differentiate $\vec{v}$ and again use (4.13).

10.   Hint: Look at Problem 9 and remember that an indefinite integral is an antiderivative.

<u>Section 6</u>

2. $\phi = x^2 - y^2 + 2xy.$

By text equation (6.3), $\nabla\phi = \vec{i}(2x + 2y) + \vec{j}(-2y + 2x)$. At $(1,1)$,
$\nabla\phi = 4\vec{i}$. The direction of most rapid <u>decrease</u> in ϕ is $-\nabla\phi = -4\vec{i}$,
that is, the negative x direction.

7.   The equation of the surface $x^2y + y^2z + z^2x + 1 = 0$ may be written
$\phi = \text{const.} = -1$ where $\phi = x^2y + y^2z + z^2x$. We find
$$\nabla\phi = (2xy + z^2)\vec{i} + (x^2 + 2yz)\vec{j} + (y^2 + 2zx)\vec{k}.$$

Then the normal to the surface at $(1,2,-1)$ is
$$\nabla\phi\big|_{(1,2,-1)} = 5\vec{i} - 3\vec{j} + 2\vec{k}.$$

By text page 108, the equation of the plane tangent to the
surface at $(1,2,-1)$ is
$$5(x - 1) - 3(y - 2) + 2(z + 1) = 0 \qquad \text{or} \qquad 5x - 3y + 2z + 3 = 0.$$

The vector equation of the normal line is (see text, page 107)
$$\vec{r} = (1,2,-1) + (5,-3,2)t.$$

10. $T = xy - x = x(y - 1)$ is the temperature. We sketch some isothermals
       (curves of constant temperature). The direction in which the
       temperature changes most rapidly at the point (x,y) is given by
       the gradient

$$\nabla T = (y - 1)\vec{i} + x\vec{j}.$$

Heat flows in the direction $-\nabla T$. We sketch some curves along
which heat flows; these are perpendicular to the isothermals.

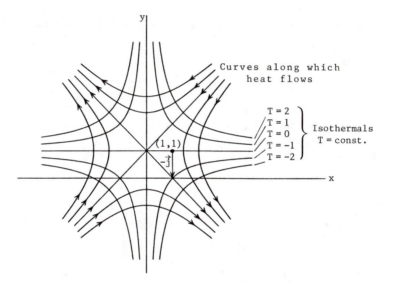

At the point (1,1), $\nabla T = \vec{j}$; thus the temperature is increasing
most rapidly in the direction $\vec{j}$ and heat flows in the direction
$-\vec{j}$. The maximum value of $\frac{dT}{ds}$ at (1,1) is $|\nabla T| = 1$. The direc-
tional derivative of T in the direction $3\vec{i} - 4\vec{j}$ at (1,1) is given
by text equation (6.4) where $\nabla T = \vec{j}$ and $\vec{u} = \dfrac{3\vec{i} - 4\vec{j}}{5}$ is a unit vector.
Then

$$\frac{dT}{ds} = \nabla T \cdot \vec{u} = \vec{j} \cdot \frac{3\vec{i} - 4\vec{j}}{5} = -\frac{4}{5}.$$

14.   (b) We find ∇z at the point $(3,2)$, where $z = 32 - x^2 - 4y^2$.

$$\nabla z = -2x\vec{i} - 8y\vec{j}\Big|_{(3,2)} = -6\vec{i} - 16\vec{j}.$$

In the direction $\vec{i} + \vec{j}$, $\dfrac{dz}{ds} = (\nabla z) \cdot \dfrac{\vec{i} + \vec{j}}{\sqrt{2}} = \dfrac{-6 - 16}{\sqrt{2}} = -11\sqrt{2}$.

Thus z is decreasing rapidly; we are going downhill.

17.   Using text equation (6.7), we find

$$\nabla r = \vec{e}_r\left(\frac{\partial r}{\partial r}\right) + \vec{e}_\theta\left(\frac{1}{r}\frac{\partial r}{\partial \theta}\right) = \vec{e}_r$$

$\left(\dfrac{\partial r}{\partial r} = 1 \text{ and } \dfrac{\partial r}{\partial \theta} = 0 \text{ since } r \text{ and } \theta \text{ are independent variables}\right)$. Using
text equation (6.3),

$$\nabla r = \left(\vec{i}\,\frac{\partial}{\partial x} + \vec{j}\,\frac{\partial}{\partial y}\right)\sqrt{x^2 + y^2} = \frac{\vec{i}x}{\sqrt{x^2 + y^2}} + \frac{\vec{j}y}{\sqrt{x^2 + y^2}}$$

$$= \frac{\vec{i}x + \vec{i}y}{r} = \vec{i}\cos\theta + \vec{j}\sin\theta = \vec{e}_r$$

by text equation (4.11).

Section 7

1.   We use text equations (7.2) and (7.3).

$$\nabla \cdot \vec{r} = \nabla \cdot (x\vec{i} + y\vec{j} + z\vec{k}) = \frac{\partial x}{\partial x} + \frac{\partial y}{\partial y} + \frac{\partial z}{\partial z} = 3.$$

$$\nabla \times \vec{r} = \begin{vmatrix} \vec{i} & \vec{j} & \vec{k} \\ \dfrac{\partial}{\partial x} & \dfrac{\partial}{\partial y} & \dfrac{\partial}{\partial z} \\ x & y & z \end{vmatrix} = 0.$$

4. $$\nabla \cdot \vec{V} = \nabla \cdot (y\vec{i} + z\vec{j} + x\vec{k}) = \frac{\partial y}{\partial x} + \frac{\partial z}{\partial y} + \frac{\partial x}{\partial z} = 0.$$

$$\nabla \times \vec{V} = \begin{vmatrix} \vec{i} & \vec{j} & \vec{k} \\ \dfrac{\partial}{\partial x} & \dfrac{\partial}{\partial y} & \dfrac{\partial}{\partial z} \\ y & z & x \end{vmatrix} = -\vec{i} - \vec{j} - \vec{k}.$$

10.   Using text equation (7.4), we find for $\phi = \ell n(x^2 + y^2)$:

$$\frac{\partial \phi}{\partial x} = \frac{2x}{x^2 + y^2} \; , \qquad \frac{\partial^2 \phi}{\partial x^2} = \frac{(x^2 + y^2) \cdot 2 - (2x)^2}{(x^2 + y^2)^2} = \frac{2(y^2 - x^2)}{(x^2 + y^2)^2} \; ,$$

$$\frac{\partial \phi}{\partial y} = \frac{2y}{x^2 + y^2} \; , \qquad \frac{\partial^2 \phi}{\partial y^2} = \frac{2(x^2 - y^2)}{(x^2 + y^2)^2} \; ,$$

$$\nabla^2 \phi = \frac{\partial^2 \phi}{\partial x^2} + \frac{\partial^2 \phi}{\partial y^2} = 0 \, .$$

13.   For $\phi = xy(x^2 + y^2 - 5z^2) = x^3 y + xy^3 - 5xyz^2$, we find

$$\nabla^2 \phi = \frac{\partial^2 \phi}{\partial x^2} + \frac{\partial^2 \phi}{\partial y^2} + \frac{\partial^2 \phi}{\partial z^2} = 6xy + 6xy - 10xy = 2xy.$$

18.
$$\vec{k} \times \vec{r} = \begin{vmatrix} \vec{i} & \vec{j} & \vec{k} \\ 0 & 0 & 1 \\ x & y & z \end{vmatrix} = -y\vec{i} + x\vec{j} \, .$$

Then

$$\nabla \times (\vec{k} \times \vec{r}) = \begin{vmatrix} \vec{i} & \vec{j} & \vec{k} \\ \frac{\partial}{\partial x} & \frac{\partial}{\partial y} & \frac{\partial}{\partial z} \\ -y & x & 0 \end{vmatrix} = 0\vec{i} + 0\vec{j} + \vec{k}\left(\frac{\partial x}{\partial x} + \frac{\partial y}{\partial y}\right) = 2\vec{k}.$$

A simpler method of doing this problem is to use the formula for
a triple vector product [text (3.8) or (3.9)], but we must be
careful since ∇ is both a vector and a differential operator. By
text (3.9) we write

$$\nabla \times (\vec{k} \times \vec{r}) = (\nabla \cdot \vec{r})\vec{k} - (\vec{k} \cdot \nabla)\vec{r} = 3\vec{k} - \left(\frac{\partial}{\partial z}\right)\vec{r} = 3\vec{k} - \vec{k} = 2\vec{k}.$$

In the second term we have written $\vec{k} \cdot \nabla$ rather than $\nabla \cdot \vec{k}$ to
emphasize that the partial derivatives in ∇ must operate on $\vec{r}$.
[For another example similar to this, see text, page 282,
equation (11.1) ff.]

Section 8

1.   We find $\int (x^2 - y^2)\, dx - 2xy\, dy$ along several paths from $(0,0)$ to $(1,2)$.

(a) Along $y = 2x^2$, let x be the integration variable; the x limits are 0 to 1. We substitute $y = 2x^2$, $dy = 4x\, dx$ into the given integral:

$$\int (x^2 - y^2)\, dx - 2xy\, dy = \int_0^1 (x^2 - 4x^4)\, dx - 2x\cdot 2x^2 \cdot 4x\, dx$$

$$= \int_0^1 (x^2 - 20x^4)\, dx = \frac{x^3}{3} - 4x^5 \Big|_0^1 = \frac{1}{3} - 4 = -\frac{11}{3}\, .$$

(b) Along the path $x = t^2$, $y = 2t$, let t be the integration variable; at $(0,0)$, $t = 0$, and at $(1,2)$, $t = 1$. We find $dx = 2t\, dt$, $dy = 2dt$. Then, along this path,

$$\int (x^2 - y^2)\, dx - 2xy\, dy = \int_0^1 (t^4 - 4t^2)\cdot 2t\, dt - 2t^2 \cdot 2t \cdot 2dt$$

$$= \int_0^1 (2t^5 - 16t^3)\, dt = \frac{t^6}{3} - 4t^4 \Big|_0^1 = \frac{1}{3} - 4 = -\frac{11}{3}\, .$$

(c) Along $y = 0$ from $x = 0$ to $x = 2$, we have $dy = 0$, so

$$\int (x^2 - y^2)\, dx - 2xy\, dy = \int_0^2 x^2\, dx = \frac{x^3}{3} \Big|_0^2 = \frac{8}{3}\, .$$

Along the straight line from $(2,0)$ to $(1,2)$, we have $y = 4 - 2x$ (equation of the straight line through two points), $dy = -2dx$. Then along this part of the path

$$\int (x^2 - y^2)\, dx - 2xy\, dy = \int_2^1 [x^2 - (4 - 2x)^2]\, dx - 2x(4 - 2x)(-2dx)$$

$$= \int_2^1 (-11x^2 + 32x - 16)\, dx = -\frac{11x^3}{3} + 16x^2 - 16x \Big|_2^1 = -\frac{19}{3}\, .$$

1.   (continued)

Combining the two results, we get $\frac{8}{3} - \frac{19}{3} = -\frac{11}{3}$.

You may note that we have obtained the same results for paths

(a), (b) and (c). This is not an accident; we easily verify that

we are integrating an exact differential since

$$\frac{\partial}{\partial y}(x^2 - y^2) = -2y = \frac{\partial}{\partial x}(-2xy).$$

Another way to say this is that, if we write the integral as

$\int dW = \int \vec{F} \cdot d\vec{r}$, then curl $\vec{F} = 0$. We find

$$W(x,y) = \int_{(0,0)}^{(x,y)} dW = \int_{(0,0)}^{(x,0)} dW + \int_{(x,0)}^{(x,y)} dW.$$

From $(0,0)$ to $(x,0)$, we have $y = 0$, $dy = 0$, so the integral is

$\int x^2 dx = x^3/3$. From $(x,0)$ to (x,y), we have $x = $ const., $dx = 0$, so

the integral is $\int -2xy \, dy = -xy^2$.   Thus

$$W(x,y) = \frac{x^3}{3} - xy^2.$$

Then $\int dW$ along <u>any</u> path from $(0,0)$ to $(1,2)$ is

$$W(1,2) - W(0,0) = \frac{1}{3} - 4 = -\frac{11}{3} .$$

4.   We evaluate $\int y^2 dx + 2x \, dy + dz$ along two different paths from

$(0,0,0)$ to $(1,1,1)$.

(a) Along the x axis from $(0,0,0)$ to $(1,0,0)$, we have $y = 0$, $z = 0$,

$dy = 0$, $dz = 0$, so the integral is 0. Along a straight line from

$(1,0,0)$ to $(1,0,1)$, we have $x = 1$, $dx = 0$, $y = 0$, $dy = 0$, and z goes

from 0 to 1. The integral is

$$\int_0^1 dz = 1.$$

4.   (continued)

Along a straight line from $(1,0,1)$ to $(1,1,1)$, $x = z = 1$,
$dx = dz = 0$, and y goes from 0 to 1. The integral is

$$\int_0^1 2 \cdot 1 \cdot dy = 2.$$

Thus the integral from $(0,0,0)$ to $(1,1,1)$ along the broken line
is $1 + 2 = 3$.

(b) Along the circle $x^2 + y^2 - 2y = 0$ from $(0,0,0)$ to $(1,1,0)$, we
have $z = 0$, $dz = 0$. We write the equation of the circle as
$x^2 + (y - 1)^2 = 1$, and (as in text Example 2) use the parameter θ as
the integration variable where

$$x = \cos \theta, \qquad\qquad dx = -\sin \theta \, d\theta,$$
$$y - 1 = \sin \theta, \qquad\qquad dy = \cos \theta \, d\theta.$$

At $(0,0)$, $\theta = -\pi/2$ and at $(1,1)$, $\theta = 0$. The integral is

$$\int_{-\pi/2}^0 (1 + \sin \theta)^2 (-\sin \theta \, d\theta) + 2\cos^2 \theta \, d\theta$$

$$= \int_{-\pi/2}^0 (-\sin \theta - 2\sin^2 \theta - \sin^3 \theta + 2\cos^2 \theta) \, d\theta$$

$$= -\int_{-\pi/2}^0 (\sin \theta + \sin^3 \theta) \, d\theta$$

since the integrals over a quadrant of $\sin^2 \theta$ and $\cos^2 \theta$ are the
same (see Problem 12 on page 56 of these Solutions). Now

$$\sin \theta + \sin^3 \theta = (1 + \sin^2 \theta) \sin \theta = (2 - \cos^2 \theta) \sin \theta$$

so the integral becomes

$$-\int_{-\pi/2}^0 (2 - \cos^2 \theta) \sin \theta \, d\theta = 2\cos \theta - \frac{\cos^3 \theta}{3} \Big|_{-\pi/2}^0 = 2 - \frac{1}{3} = \frac{5}{3}.$$

4.   (continued)

We add to this the integral along the straight line from $(1,1,0)$

to $(1,1,1)$; this is $\int_0^1 dz = 1$. Thus the total integral is

$5/3 + 1 = 8/3$.

We note that the answers to (a) and (b) are different and that

the given differential is not exact.

9.   We find

$$\nabla \times \vec{F} = \begin{vmatrix} \vec{i} & \vec{j} & \vec{k} \\ \dfrac{\partial}{\partial x} & \dfrac{\partial}{\partial y} & \dfrac{\partial}{\partial z} \\ 3x^2yz - 3y & x^3z - 3x & x^3y + 2z \end{vmatrix} = 0,$$

so $\vec{F}$ is conservative. To find a potential function, we choose

the origin as reference point (zero potential), and evaluate

$\phi = -\int \vec{F} \cdot d\vec{r}$ from the origin to the

point (x,y,z) along a broken line

path as indicated in the figure.

From $(0,0,0)$ to $(x,0,0)$, we have

$y = z = 0$, $dy = dz = 0$, so the integral

is zero. From $(x,0,0)$ to $(x,y,0)$,

we have $x = $ const., $dx = 0$, $z = dz = 0$;

the integral is

$$-\int_0^y -3x\, dy = +3xy.$$

From $(x,y,0)$ to (x,y,z), we have $x = $ const., $y = $ const., $dx = dy = 0$,

so the integral is

$$-\int_0^z (x^3y + 2z)\, dz = -x^3yz - z^2.$$

Thus the scalar potential is $\phi = 3xy - x^3yz - z^2$.

11. As in Problem 9 above, and text Example 3, we integrate along a
     broken line. From $(0,0)$ to $(x,0)$, we have $y = dy = 0$, so the
     integral is zero. From $(x,0)$ to (x,y), $x = $ const., $dx = 0$, and
     we get

$$\phi = -\int_0^y \sin^2 x \, dy = -y \sin^2 x.$$

18. We want to evaluate $\int \vec{F} \cdot d\vec{r} = \int -y \, dx + x \, dy + z \, dz$ along paths from
     $(1,0,0)$ to $(-1,0,\pi)$.

     (a) Along the helix, we have

$$x = \cos t, \qquad\qquad dx = -\sin t \, dt,$$
$$y = \sin t, \qquad\qquad dy = \cos t \, dt$$
$$z = t, \qquad\qquad dz = dt,$$

     where t goes from 0 to π. Then

$$\int \vec{F} \cdot d\vec{r} = \int_0^\pi (-\sin t)^2 dt + \cos^2 t \, dt + t \, dt$$

$$= \int_0^\pi (1 + t) \, dt = \pi + \frac{\pi^2}{2} .$$

     (b) The straight line joining $(1,0,0)$ and $(-1,0,\pi)$ has the para-
     metric equations $\vec{r} = \vec{i} + (-2\vec{i} + \pi\vec{k})t$ or $x = 1 - 2t$, $y = 0$, $z = \pi t$ (see
     text, page 107). At $(1,0,0)$, $t = 0$, and at $(-1,0,\pi)$, $t = 1$. Then

$$\int \vec{F} \cdot d\vec{r} = \int_0^1 \pi t \cdot \pi \, dt = \pi^2/2.$$

     The answers are different. We find curl $\vec{F} = 2\vec{k} \neq 0$; thus $\vec{F}$ is not
     conservative, so in general we expect different results for dif-
     ferent paths.

21. For m outside a solid sphere, the potential is $\phi = -\frac{CM}{r}$ where M is
     the total mass of the sphere. Then $\vec{F} = -\nabla\phi = -\frac{CM}{r^2}\vec{e}_r$ [see text
     equation (6.7)]. If the solid sphere is the earth, then
     $\vec{F} = -mg\vec{e}_r$ at r = R; we find $CM = mgR^2$, so

$$\vec{F} = -\frac{mgR^2}{r^2}\vec{e}_r , \qquad \phi = -\frac{mgR^2}{r} , \qquad m \text{ outside earth.}$$

For m inside a solid sphere at r < R, think of the solid sphere as
made up of two parts: a smaller solid sphere of radius r plus a
thick shell extending from r to R. There is no force on m due
to the shell since the potential inside any spherical shell is
constant. There is a force due to the small sphere; the corresponding potential is $-\frac{CM'}{r}$ where M' is the mass of the small
sphere of radius r. This mass is $M' = \frac{4}{3}\pi r^3 \rho$ (where ρ = density),
so the potential is

$$-\frac{CM'}{r} = -\frac{4C\pi r^3 \rho}{3r} = Ar^2$$

where A is a constant. We can add to this a constant B without
affecting the force. Thus

$$\phi = Ar^2 + B, \quad \text{and} \quad \vec{F} = -2Ar\vec{e}_r, \quad m \text{ inside a solid sphere.}$$

For the earth, $\vec{F} = -mg\vec{e}_r$ at r = R, so $A = \frac{mg}{2R}$. By the first part
of the problem, $\phi = -mgR$ at r = R so

$$AR^2 + B = -mgR, \qquad B = -mgR - \frac{1}{2}mgR = -\frac{3}{2}mgR.$$

Thus

$$\vec{F} = -\frac{mgr}{R}\vec{e}_r , \qquad \phi = \frac{mg}{2R}(r^2 - 3R^2), \ m \text{ inside earth.}$$

Section 9

4.   Using text equation (9.7), we have

$$P = e^x \cos y, \qquad\qquad Q = -e^x \sin y,$$

$$\frac{\partial Q}{\partial x} - \frac{\partial P}{\partial y} = 0.$$

Thus the line integral around the closed curve ADBA is zero.   The
integral along ADB which we want is then the negative of the
integral from B to A.   Along BA, $y = dy = 0$, so we have

$$\int_{ADB} = -\int_B^A = -\int_{-\ell n 2}^{\ell n 2} e^x \cos 0 \; dx = -e^x \Big|_{-\ell n 2}^{\ell n 2} = -\left(2 - \frac{1}{2}\right) = -\frac{3}{2} \; .$$

6.   By text equation (9.7), with $P = -y$ and $Q = x$:

$$\frac{1}{2}\oint x \; dy - y \; dx = \frac{1}{2}\iint_A 2 \; dx \; dy = \iint_A dx \; dy = A.$$

8.   We write the equation of the curve $x^{2/3} + y^{2/3} = 4$ in the
parametric form

$$x = 8\cos^3 \theta, \qquad\qquad y = 8\sin^3 \theta.$$

Using Problem 6 above, we evaluate $\frac{1}{2}\oint x \; dy - y \; dx$ around the given
curve.   Then θ goes from 0 to 2π.   We find

$$A = \frac{1}{2}\int_0^{2\pi} 8\cos^3 \theta \cdot 8 \cdot 3\sin^2 \theta \cos \theta \; d\theta - 8\sin^3 \theta \cdot 8 \cdot 3\cos^2 \theta (-\sin \theta) d\theta$$

$$= \frac{1}{2} \cdot 64 \cdot 3 \int_0^{2\pi} \sin^2 \theta \cos^2 \theta (\sin^2 \theta + \cos^2 \theta) d\theta = 96 \int_0^{2\pi} \frac{1}{4} \sin^2 2\theta \; d\theta$$

$$= \frac{96}{4} \cdot \frac{1}{2} \cdot 2\pi = 24\pi$$

(See page 56 of these Solutions, Problem 12.)

10.   By Green's theorem with $P = 2y$, $Q = -3x$,

$$(2y\, dx - 3x\, dy) = \int_{y=1}^{3} \int_{x=3}^{5} (-3 - 2)\, dx\, dy = -5 \cdot 2 \cdot 2 = -20.$$

Section 10

Comment on notation:

In Sections 10 to 12, we use $d\tau$ rather than dV to mean the volume element (to avoid confusion with the vector $\vec{V}$ in $\nabla \cdot \vec{V}$) and we use $d\sigma$ for the area element. In various books, you may find the following notations:

For area element: dA, da, dS, $d\sigma$ [Any of these mean $dx\, dy$ or $r\, dr\, d\theta$ in a plane parallel to the (x,y) plane, $dy\, dz$ in a plane parallel to the (y,z) plane, $a^2\sin\theta\, d\theta\, d\phi$ on the surface of a sphere, etc.]

For volume element: dV, d^3x, d^3r, $d\tau$ [Any of these mean $dx\, dy\, dz$ or $r\, dr\, d\theta\, dz$ or $r^2\sin\theta\, dr\, d\theta\, d\phi$, etc.]

2.   On the top surface of the cube, the unit vector $\vec{n}$ pointing out of the cube is $\vec{k}$. Similarly, on the bottom surface $\vec{n} = -\vec{k}$, and on the other surfaces $\vec{n} = \vec{i}$, $-\vec{i}$, $\vec{j}$, $-\vec{j}$. The area $\int d\sigma$ for each surface of the cube is 1. Thus we find:

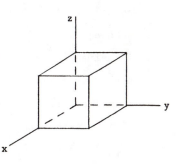

2.   (continued)

Top: $\vec{V} \cdot \vec{n} = \vec{V} \cdot \vec{k}$ $= z^2 \big|_{z=1}$ $= 1,$ $\int 1 \cdot d\sigma = 1$

Bottom: $\vec{V} \cdot \vec{n} = \vec{V} \cdot (-\vec{k})$ $= -z^2 \big|_{z=0}$ $= 0,$ $\int 0 \cdot d\sigma = 0$

Front: $\vec{V} \cdot \vec{n} = \vec{V} \cdot \vec{i}$ $= x^2 \big|_{x=1}$ $= 1,$ $\int 1 \cdot d\sigma = 1$

Back: $\vec{V} \cdot \vec{n} = \vec{V} \cdot (-\vec{i})$ $= -x^2 \big|_{x=0}$ $= 0,$ $\int 0 \cdot d\sigma = 0$

Right: $\vec{V} \cdot \vec{n} = \vec{V} \cdot \vec{j}$ $= y^2 \big|_{y=1}$ $= 1,$ $\int 1 \cdot d\sigma = 1$

Left: $\vec{V} \cdot \vec{n} = \vec{V} \cdot (-\vec{j})$ $= -y^2 \big|_{y=0}$ $= 0,$ $\int 0 \cdot d\sigma = 0$

Then the integral over the whole surface $= 1 + 1 + 1 = 3$. We find
the same result using the divergence theorem [text equation
(10.17)]:

$$\nabla \cdot \vec{V} = 2x + 2y + 2z$$

$$\nabla \cdot \vec{V} d\tau = \int_0^1 \int_0^1 \int_0^1 (2x + 2y + 2z) \, dx \, dy \, dz = 1 + 1 + 1 = 3.$$

5.   In order to use the divergence theorem, we need the unit vector $\vec{n}$
normal to the sphere. The vector $\vec{r} = \vec{i}x + \vec{j}y + \vec{k}z$ is normal to the
sphere; if we did not see this from the geometry, we could find
it using the gradient: $\nabla(x^2 + y^2 + z^2) = 2\vec{r}$. Then the desired unit
vector is $\vec{n} = \vec{r}/|\vec{r}| = (\vec{i}x + \vec{j}y + \vec{k}z)/5$ since $|\vec{r}| = r = (x^2 + y^2 + z^2)^{1/2} =$
5 on the surface of the sphere. We also want $\vec{F} \cdot \vec{n}$ when $r = 5$ since
$\int \vec{F} \cdot \vec{n} d\sigma$ is an integral over the surface of the sphere. We find

$$\vec{F} \cdot \vec{n} = (x^2 + y^2 + z^2)(\vec{r} \cdot \vec{r}/5) = r^4/5 = 5^4/5 = 5^3.$$

Then, by the divergence theorem,

$$\int \nabla \cdot \vec{F} d\tau = \int \vec{F} \cdot \vec{n} d\sigma = 5^3 \cdot (\text{area of sphere}) = 4\pi \cdot 5^5 = 12500\pi.$$

9. For the surface integral $\int \vec{F} \cdot \vec{n} \, d\sigma$ over the circle $x^2 + y^2 \leqslant 4$ in the (x,y) plane, the outward normal $\vec{n}$ is $-\vec{k}$. Then $\vec{F} \cdot \vec{n} = (x\vec{i} + y\vec{j}) \cdot (-\vec{k}) = 0$ so this part of the surface integral is zero. Thus the desired integral over the surface above the (x,y) plane is equal to the volume integral given by the divergence theorem. We find

$$\nabla \cdot \vec{F} = \nabla \cdot (x\vec{i} + y\vec{j}) = 2,$$

$$\int \vec{F} \cdot \vec{n} \, d\sigma = \int \nabla \cdot \vec{F} \, d\tau = 2 \int d\tau = 2 \cdot \text{volume} = 2 \int\int z \, dx \, dy$$

$$= 2 \int\int (4 - x^2 - y^2) \, dx \, dy = 2 \int_0^{2\pi} \int_0^2 (4 - r^2) r \, dr \, d\theta = 16\pi.$$

15. By the chain rule (text, Chapter 4, Section 5)

$$\frac{d}{dt} \rho(x,y,z,t) = \frac{\partial \rho}{\partial x} \frac{dx}{dt} + \frac{\partial \rho}{\partial y} \frac{dy}{dt} + \frac{\partial \rho}{\partial z} \frac{dz}{dt} + \frac{\partial \rho}{\partial t} \frac{dt}{dt}$$

$$= (\nabla \rho) \cdot \vec{v} + \frac{\partial \rho}{\partial t} \ .$$

From text equation (10.9) with $\vec{V} = \rho \vec{v}$, we find

$$\frac{\partial \rho}{\partial t} = -\nabla \cdot (\rho \vec{v}) = -\nabla \rho \cdot \vec{v} - \rho \nabla \cdot \vec{v}.$$

Combining the two results gives

$$\frac{d\rho}{dt} = -\rho \nabla \cdot \vec{v} \qquad \text{or} \qquad \rho \nabla \cdot \vec{v} + \frac{d\rho}{dt} = 0.$$

If $\frac{d\rho}{dt} = 0$ (incompressible fluid), then $\nabla \cdot \vec{v} = 0$.

16.   In text equation (10.17), let $\vec{V} = \phi\nabla\psi$. Then [see (f) in Table
       of Vector Identities, text, page 296]

$$\nabla \cdot \vec{V} = \nabla \cdot (\phi\nabla\psi) = \phi\nabla^2\psi + \nabla\phi \cdot \nabla\psi,$$

(1) $\int \nabla \cdot \vec{V} d\tau = \int (\phi\nabla^2\psi + \nabla\phi \cdot \nabla\psi) d\tau = \oint \vec{V} \cdot \vec{n} \, d\sigma = \oint (\phi\nabla\psi) \cdot \vec{n} \, d\sigma.$

Now write (1) with ϕ and ψ interchanged and subtract the result
from (1) to get

(2) $\int (\phi\nabla^2\psi - \psi\nabla^2\phi) d\tau = \oint (\phi\nabla\psi - \psi\nabla\phi) \cdot \vec{n} \, d\sigma.$

Section 11

3.   The easiest way to do this problem (and many other similar
      problems) is to use Stokes' theorem twice as in text Example
      1(c). First, Stokes' theorem tells us that the surface integral
      is equal to a line integral around the bounding curve. A second
      application of Stokes' theorem then says that the line integral
      is equal to the surface integral over any surface with the same
      bounding curve. The simplest surface to use is the area in the
      (x,y) plane bounded by the curve. Think of collapsing the sur-
      face into the (x,y) plane; then $\vec{n} = \vec{k}$ if we integrate in the
      counter-clockwise direction around the bounding curve. [See text
      discussion after equations (11.6) and (11.9).] Here we evaluate
      the given double integral over the circular area $x^2 + y^2 \leqslant 4$ in the
      (x,y) plane. We find

$$\nabla \times \vec{V} = \nabla \times (x^2\vec{i} + z^2\vec{j} - y^2\vec{k}) = \vec{i}(-2y - 2z),$$

$$(\nabla \times \vec{V}) \cdot \vec{n} = (\nabla \times \vec{V}) \cdot \vec{k} = (-2y - 2z)\vec{i} \cdot \vec{k} = 0.$$

Thus the given surface integral is zero.

4. As in Problem 3 above, we can integrate over the area of the
    triangle in the (x,y) plane. The plane $2x + 3y + 4z = 12$ inter-
    cepts the x axis at $x = 6$, and the y axis at $y = 4$, so the (x,y)
    plane triangle is a right triangle with legs 6 and 4. We find

    $$\nabla \times \vec{V} = \nabla \times (y\vec{i} + 2\vec{j}) = -\vec{k}, \quad \text{so} \quad (\nabla \times \vec{V}) \cdot \vec{n} = -\vec{k} \cdot \vec{k} = -1,$$

    $$\int (\nabla \times \vec{V}) \cdot n \, d\sigma = -(\text{area of triangle}) = -\frac{1}{2} \cdot 6 \cdot 4 = -12.$$

5. As in Problem 10.9, we see that $\vec{r} \cdot \vec{n} = 0$ on the (x,y) plane
    $[\vec{n} = -\vec{k}, \ \vec{r} \cdot \vec{n} = -z = 0$ on the (x,y) plane]. Thus, by the diver-
    gence theorem, $\int \vec{r} \cdot \vec{n} \, d\sigma$ over the rest of the surface equals
    $\int \nabla \cdot \vec{r} \, d\tau$ over the volume. We find

    $$\nabla \cdot \vec{r} = 3, \qquad \int \nabla \cdot \vec{r} \, d\tau = 3 \cdot \text{volume} = 3 \cdot \frac{1}{3} \cdot 12 \cdot 3 = 36.$$

    (The volume of a tetrahedron is $\frac{1}{3} \cdot$(area of base)$\cdot$height, or you
    can integrate $\int\int z \, dx \, dy$ as in Chapter 5 to find the volume.)

7. As discussed in Problem 3 above, we can replace any surface
    whose bounding curve is in the (x,y) plane by the area in the
    (x,y) plane with the same bounding curve. Then $\vec{n} = \vec{k}$, so here
    we have

    $$\nabla \times \vec{V} = \vec{i}(x^2 - 3yz^2) - \vec{j}(2xy - z + x^2), \qquad (\nabla \times \vec{V}) \cdot \vec{k} = 0.$$

    Thus the given integral is zero.

15. We use Stokes' theorem and integrate over the area of the plane
     $x + y = 2$ bounded by the curve C. This curve is the circle of
     intersection of the plane $x + y = 2$ with the sphere $x^2 + y^2 + z^2 =$
     $2(x+y)$ or $(x-1)^2 + (y-1)^2 = 2$. We see that the plane passes
     through the center $(1,1)$ of the sphere, so the radius of C is
     the same as the radius of the sphere, namely $\sqrt{2}$. We find
     $\nabla \times \vec{V} = \nabla \times (\vec{i}y + \vec{j}z + \vec{k}x) = -\vec{i} - \vec{j} - \vec{k}$. The vector $\vec{n}$ is a unit vector
     normal to the plane $x + y = 2$. Thus

$$\vec{n} = \frac{\vec{i} + \vec{j}}{\sqrt{2}} \ , \qquad\qquad (\nabla \times \vec{V}) \cdot \vec{n} = \frac{-1 - 1}{\sqrt{2}} = -\sqrt{2}.$$

Then the desired integral is $-\sqrt{2}$ times the area of the circle
or $-\sqrt{2}\,\pi(\sqrt{2})^2 = -2\pi\,\sqrt{2}$.

20. As in text Example 2, there are many $\vec{A}$'s for which the given $\vec{V}$
     satisfies $\vec{V} = \nabla \times \vec{A}$. In simple cases we may be able to find one
     by inspection; let us try this. We want A_x, A_y, A_z so that

$$\frac{\partial A_z}{\partial y} - \frac{\partial A_y}{\partial z} = ze^{zy} + x \sin zx,$$

$$\frac{\partial A_x}{\partial z} - \frac{\partial A_z}{\partial x} = x \cos xz,$$

$$\frac{\partial A_y}{\partial x} - \frac{\partial A_x}{\partial y} = -z \sin zx.$$

We can satisfy the last two equations by taking $\dfrac{\partial A_z}{\partial x} = 0$, $\dfrac{\partial A_x}{\partial y} = 0$,
and

$$A_x = \sin xz,$$

$$A_y = \cos xz.$$

Then the first equation is $\dfrac{\partial A_z}{\partial y} + x \sin xz = ze^{zy} + x \sin zx$; this is
satisfied if

20.   (continued)

$$A_z = e^{zy}.$$

These are the components of one $\vec{A}$; to these we may add the components of ∇u where u is any function $u(x,y,z)$. Thus we find

$$\vec{A} = \vec{i} \sin xz + \vec{j} \cos xz + \vec{k} e^{zy} + \nabla u.$$

You might like to find an $\vec{A}$ with, say, $A_z = 0$, and then verify
that the difference between your $\vec{A}$ and the $\vec{A}$ found here is equal
to the gradient of a function $u(x,y,z)$.

Section 12

3.   Hint: See Problem 10.15.

4.   Hint: See Table of Vector Identities on text page 296. Also
       see Problem 7.1.

9.   First we must find the torque of $\vec{F}$ about a point on the line,
       say the point $(0,3,4)$. This torque is $\vec{r} \times \vec{F}$ where $\vec{r}$ is the
       vector from $(0,3,4)$ to the point $(2,1,0)$ at which $\vec{F}$ acts. We
       find

$$\vec{r} = (2,1,0) - (0,3,4) = (2,-2,-4),$$

$$\vec{r} \times \vec{F} = \begin{vmatrix} \vec{i} & \vec{j} & \vec{k} \\ 2 & -2 & -4 \\ 1 & -5 & 2 \end{vmatrix} = (-24,-8,-8).$$

Then the scalar torque about the line is $\vec{n} \cdot (\vec{r} \times \vec{F})$ where $\vec{n}$ is a
unit vector along the line. Here a vector along the line is
$-2\vec{i}$, so $\vec{n} = -\vec{i}$. Thus

$$\vec{n} \cdot (\vec{r} \times \vec{F}) = 24.$$

(Note: This is the torque which tends to rotate the object about
the $\vec{n} = -\vec{i}$ direction according to the right-hand rule. Alterna-

9.   (continued)

tively we could say that the torque is -24 tending to rotate the
object about the $\vec{\imath}$ direction. More simply we could state the
vector torque about the line; this is $\vec{n} \cdot (\vec{r} \times \vec{F})$ in the $\vec{n}$ direction, or here, $-24\vec{\imath}$.)

13.   For $\phi = x^2 - yz$ at the point $P(3,4,1)$, we find:

(a) By text equation (6.3),

$$(\nabla\phi)_P = 2x\vec{\imath} - z\vec{\jmath} - y\vec{k}\big|_P = 6\vec{\imath} - \vec{\jmath} - 4\vec{k}.$$

(b) We first note that $\phi = 5$ at P. By text equation (6.6), the
vector $(\nabla\phi)_P$ is normal to the surface. A unit normal vector is
$(\nabla\phi)/|\nabla\phi|$ at P, namely $(6\vec{\imath} - \vec{\jmath} - 4\vec{k})/\sqrt{36+1+16} = (6\vec{\imath} - \vec{\jmath} - 4\vec{k})/\sqrt{53}$.

(c) The direction of most rapid increase in ϕ at P is $(\nabla\phi)_P$ as
in (a). See text equation (6.5) and the discussion after it.

(d) $|(\nabla\phi)_P| = \sqrt{53}$ as in (b).

(e) We use text equation (6.4) with $\vec{u}$ a unit vector along the
line and $(\nabla\phi)_P$ as in (a). Thus

$$\vec{u} = (6\vec{\imath} - \vec{\jmath} - 4\vec{k})/\sqrt{53},$$

$$\frac{d\phi}{ds} = (\nabla\phi)_P \cdot \vec{u} = (6\vec{\imath} - \vec{\jmath} - 4\vec{k}) \cdot (6\vec{\imath} - \vec{\jmath} - 4\vec{k})/\sqrt{53} = \sqrt{53}.$$

17.   In polar coordinates, the equation of the circle is $r = \sqrt{2}$. Then

$$x = \sqrt{2} \cos\theta, \qquad\qquad dx = -\sqrt{2} \sin\theta \, d\theta,$$

$$y = \sqrt{2} \sin\theta, \qquad\qquad dy = \sqrt{2} \cos\theta \, d\theta,$$

and from $(1,1)$ to $(1,-1)$, θ goes from $\pi/4$ to $-\pi/4$. We find

17.   (continued)

$$\int \vec{F} \cdot d\vec{r} = \int (2x - 3y) dx - (3x - 2y) dy$$

$$= \int_{\pi/4}^{-\pi/4} (2\sqrt{2} \cos \theta - 3\sqrt{2} \sin \theta)(-\sqrt{2} \sin \theta \, d\theta)$$

$$- (3\sqrt{2} \cos \theta - 2\sqrt{2} \sin \theta)(\sqrt{2} \cos \theta \, d\theta)$$

$$= 6 \int_{\pi/4}^{-\pi/4} (\sin^2 \theta - \cos^2 \theta) d\theta = -6 \int_{\pi/4}^{-\pi/4} \cos 2\theta \, d\theta$$

$$= -3 \sin 2\theta \Big|_{\pi/4}^{-\pi/4} = -3(-1 - 1) = 6.$$

A simpler way to do this problem is to observe that the line integral is independent of the path of integration (since curl $\vec{F} = 0$). If we integrate along the straight line from $(1,1)$ to $(1,-1)$, we have

$$x = 1, \qquad dx = 0,$$

$$\int \vec{F} \cdot d\vec{r} = \int_{1}^{-1} -(3 - 2y) dy = (-3)(-2) + (-1)^2 - 1^2 = 6.$$

21.   As in Example 1c, text page 285, and Problem 11.3, we use Stokes' theorem twice to say that the surface integral over the four slanting faces is the same as the surface integral over the square base. Then $\vec{n} = \vec{k}$ and

$$(\nabla \times \vec{V}) \cdot \vec{k} = \begin{vmatrix} \vec{i} & \vec{j} & \vec{k} \\ \dfrac{\partial}{\partial x} & \dfrac{\partial}{\partial y} & \dfrac{\partial}{\partial z} \\ x^2 z - 2 & x + y - z & -xyz \end{vmatrix} \cdot \vec{k} = 1 - 0 = 1.$$

$$\iint (1) \, d\sigma = \text{area of base} = 2^2 = 4.$$

23.   We use the divergence theorem, text equation (10.17).

$$\text{div } \vec{F} = -2x + (2x - 1) + 3 = 2$$

$$\iint \vec{F} \cdot \vec{n} \, d\sigma = \iiint \text{div } \vec{F} \, d\tau = 2 \iiint d\tau$$

$$= 2(\text{volume of cylinder}) = 2\pi \cdot 4^2 \cdot 6 = 192\pi.$$

29.   The line integral is equal to a surface integral over the area
       of the square by either Stokes' theorem [text equation (11.9)]
       or Green's theorem in the plane [text equation (9.7)] -- these
       theorems are the same for a surface in the (x,y) plane. Using
       Stokes' theorem, we have $\vec{n} = \vec{k}$,

$$(\text{curl } \vec{V}) \cdot \vec{k} = \begin{vmatrix} \frac{\partial}{\partial x} & \frac{\partial}{\partial y} \\ x^2 & 5x \end{vmatrix} = 5$$

$$\int \vec{V} \cdot d\vec{r} = \iint (\text{curl } \vec{V}) \cdot \vec{k} \, d\sigma = 5(\text{area of square})$$

$$= 5(\sqrt{2})^2 = 10.$$

(Plot the four given points to see that the side of the square
is $\sqrt{2}$.

       Alternatively, by Green's theorem, we have $P = x^2$, $Q = 5x$, so

$$\frac{\partial Q}{\partial x} - \frac{\partial P}{\partial y} = 5, \qquad \text{and}$$

$$\iint 5 \, dx \, dy \text{ over the square} = 5(\text{area of square})$$

as above.

Chapter 7

Section 2

2. For $s = 2 \sin(4t - 1)$, the amplitude is 2; the period is $2\pi/4 = \pi/2$; the frequency is $2/\pi$ (that is, the reciprocal of the period); the velocity is $ds/dt = 8 \cos(4t - 1)$, so the velocity amplitude is 8.

6. Hint: First use the trigonometric formula:
$$\sin \theta + \sin \phi = 2 \sin \tfrac{1}{2}(\theta + \phi) \cos \tfrac{1}{2}(\theta - \phi).$$

10. $z = -4e^{i(2t+3\pi)} = 4e^{2it} = 4(\cos 2t + i \sin 2t)$ by text page 47. Thus $x = 4 \cos 2t$, $y = 4 \sin 2t$. Both of these equations describe simple harmonic motion of amplitude 4, period $2\pi/2 = \pi$, frequency $1/\pi$, and velocity amplitude $= 8$. Note that we can also find the amplitude as $|z| = 4$ and the velocity amplitude as $|dz/dt| = 8$.

13. The acceleration of the mass m in the direction perpendicular to the string is $\ell\, d^2\theta/dt^2$. The component of the force perpendicular to the string is $mg \sin \theta$ acting in the direction of decreasing θ. Thus
$$m\ell \frac{d^2\theta}{dt^2} = -mg \sin \theta.$$
For small θ, $\sin \theta \cong \theta$, so the differential equation is approximately
$$\frac{d^2\theta}{dt^2} = -\frac{g}{\ell}\, \theta.$$
You can easily verify that $\theta = A \sin \omega t$ is a solution of this differential equation if $\omega^2 = g/\ell$. (Another solution is $\theta = A \cos \omega t$.) We see that the motion is simple harmonic; $\omega = \sqrt{g/\ell}$ is the angular frequency, and A is the maximum value of θ. Since $x = \ell \sin \theta \cong \ell\theta$ for small θ, we have $x = \ell A \sin \omega t$.

205

14.   The period of $x = 4 \sin(\pi t/3)$ is $T = 2\pi \div (\pi/3) = 6$, and similarly
the period of $x = 3 \sin(\pi t/4)$ is $T = 8$. From the sketch, we see
that the first crossing point when $x = 0$ is at $t = 12$.

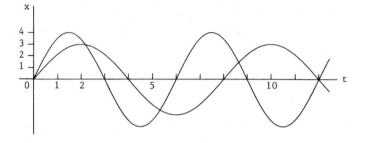

20.   By comparing $y = 3 \sin \pi \left(x - \frac{1}{2} t \right)$ with the formulas in Problem 17,
we find $A = 3$, $T = 4$, $f = 1/4$, $v = 1/2$, $\lambda = 2$. We first sketch y as
a function of x when $t = 0$. Then the $t = 1$ graph is just the $t = 0$
graph moved a distance 1/2 to the right (since now $y = 0$ when
$x = 1/2$). Similarly the $t = 2$ graph is the $t = 0$ graph moved a
distance 1 to the right.

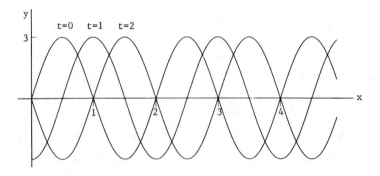

20.   (continued)

The graphs of y as a function of t are constructed similarly.
When $x = 0$, we have $y = 3 \sin \pi \left(-\frac{1}{2} t \right) = -3 \sin \left(\frac{1}{2} \pi t \right)$.   When $x = 1$,
$y = -3 \sin \frac{\pi}{2} (t - 2)$, so the $x = 1$ graph is the $x = 0$ graph moved 2
to the right.   The $x = 2$ graph coincides with the $x = 0$ graph
(move the $x = 0$ graph 4 to the right and note that the period
is 4).

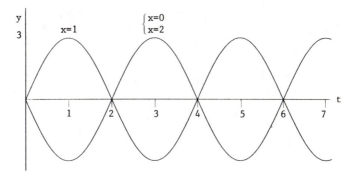

25.   We are given $A = 10$, $f = 600 \cdot 10^3 = 6 \cdot 10^5 \text{ sec}^{-1}$, $v = 3 \cdot 10^8 \text{ m/sec}$.   Then
using the formula $y = A \sin \omega \left(\frac{x}{v} - t \right)$ with $\omega = 2\pi f$ from Problem 17,
we have

$$y = 10 \sin 12\pi \cdot 10^5 \left(\frac{x}{3 \cdot 10^8} - t \right) = 10 \sin \frac{\pi}{250} \left(x - 3 \cdot 10^8 t \right).$$

Alternatively, we can find $T = 1/f = (1/6) \cdot 10^{-5}$ sec and $\lambda = v/f = 500$;
then we can use any of the forms given in Problem 17.

Section 3

3.   We sketch each of the sine curves all on the same set of axes
     and then add them.

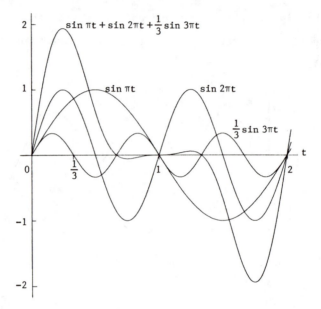

6.   See Problem 2.6.

9.   We use the trigonometry formula

$$\sin \theta \sin \phi = \frac{1}{2}[\cos(\theta - \phi) - \cos(\theta + \phi)]$$

to write

$$y = (A + B \sin 2\pi ft)\sin 2\pi f_c \left(t - \frac{x}{v}\right) = A \sin 2\pi f_c \left(t - \frac{x}{v}\right)$$

$$+ \frac{1}{2}B\left\{\cos 2\pi\left[(f - f_c)t + f_c x/v\right] - \cos 2\pi\left[(f + f_c)t - f_c x/v\right]\right\} .$$

Using the values given for A, B, f, and f_c, we sketch, at $x = 0$,
the graph of

$$y = (3 + \sin 2\pi t)\sin 40\pi t.$$

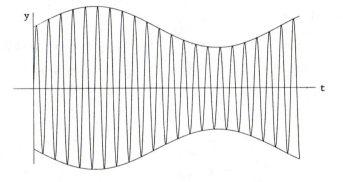

Section 4

5.   Since $\cos^2(x/2)$ has period 4π, we see that 0 to $\pi/2$ is less than
a quarter period so we cannot use the text discussion following
Figure 4.2. By text equation (4.3) and integral tables, we find

$$\text{Average of } \cos^2 \frac{x}{2} \text{ on } (0, \pi/2) \quad \text{is} \quad \frac{1}{\pi/2}\int_0^{\pi/2}\cos^2 \frac{x}{2} \, dx$$

$$= \frac{2}{\pi}\left(\frac{1}{2}x + \frac{1}{2}\sin x\right)\Big|_0^{\pi/2} = \frac{2}{\pi}\left(\frac{\pi}{4} + \frac{1}{2}\right) = \frac{1}{2} + \frac{1}{\pi} .$$

11.   Since 2π is a period of $\sin x$, the average of $\sin x$ on $(0, 2\pi)$
       is 0. Also the average of $\sin^2 x$ is 1/2 by text equation (4.8).
       Thus the average of $\sin x + \sin^2 x$ is 1/2 without calculation.

13.   Comments: If, as in text Figure 4.2, ka and kb are both
       multiples of $\pi/2$, then the result stated is valid even if
       $k(b - a)$ is an odd multiple of $\pi/2$ and so not a multiple of π.
       However, if ka and kb are not multiples of $\pi/2$, the result is
       valid as stated if $k(b - a)$ is a multiple of π. Sketch $\sin^2 x$
       and $\cos^2 x$ over any interval of length π to see this (for
       example, from $\pi/4$ to $5\pi/4$).

15.   (a) We find $k(b - a) = \pi[11/4 - (-1/4)] = 3\pi$; since this is a
       multiple of π, then by Problem 13, the integral is
       $$\frac{1}{2}(b - a) = \frac{1}{2}\left(\frac{11}{4} + \frac{1}{4}\right) = \frac{3}{2} \ .$$

Section 5

2.   We are given $f(x) = \begin{cases} 0, & -\pi < x < 0 \quad \text{and} \quad \pi/2 < x < \pi, \\ 1, & 0 < x < \pi/2. \end{cases}$

We first sketch several periods of the corresponding periodic function of period 2π:

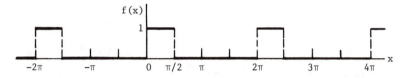

We use text equations (5.9) and (5.10) to find the Fourier coefficients.

$$a_n = \frac{1}{\pi}\int_{-\pi}^{\pi} f(x)\cos nx\,dx = \frac{1}{\pi}\int_0^{\pi/2}\cos nx\,dx = \frac{1}{n\pi}\sin nx\Big|_0^{\pi/2}$$

$$= \frac{1}{n\pi}\sin\frac{n\pi}{2} = \begin{cases} 0 &, \text{ even } n \neq 0, \\ \dfrac{1}{n\pi} &, n = 1,5,9,\cdots, \\ -\dfrac{1}{n\pi} &, n = 3,7,11,\cdots. \end{cases}$$

$$a_0 = \frac{1}{\pi}\int_0^{\pi/2} dx = \frac{1}{2}\,, \qquad \text{so} \qquad \frac{1}{2}a_0 = \frac{1}{4}\,.$$

$$b_n = \frac{1}{\pi}\int_0^{\pi/2}\sin nx\,dx = -\frac{1}{n\pi}\cos nx\Big|_0^{\pi/2} = \begin{cases} \dfrac{1}{n\pi} &, \text{ odd } n, \\ \dfrac{2}{n\pi} &, n = 2,6,10,\cdots, \\ 0 &, n = 4,8,12,\cdots. \end{cases}$$

Then the Fourier series for $f(x)$ is

$$f(x) = \frac{1}{4} + \frac{1}{\pi}\left(\cos x - \frac{\cos 3x}{3} + \frac{\cos 5x}{5}\cdots\right.$$

$$\left. + \frac{\sin x}{1} + \frac{2\sin 2x}{2} + \frac{\sin 3x}{3} + \frac{\sin 5x}{5}\cdots\right).$$

7. $f(x) = \begin{cases} 0, & -\pi < x < 0, \\ x, & 0 < x < \pi. \end{cases}$

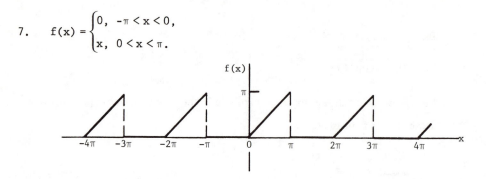

To evaluate the Fourier coefficients, use integral tables or see Appendix A of these Solutions, page 603.

$$a_n = \frac{1}{\pi}\int_0^\pi x \cos nx \, dx = \frac{1}{\pi}\left(\frac{\cos nx + nx \sin nx}{n^2}\right)\Bigg|_0^\pi$$

$$= \frac{\cos n\pi - 1}{\pi n^2} = \begin{cases} 0, & \text{even } n \neq 0, \\ -\frac{2}{\pi n^2}, & \text{odd } n. \end{cases}$$

$$a_0 = \frac{1}{\pi}\int_0^\pi x \, dx = \frac{\pi}{2} \quad \text{so} \quad \frac{1}{2}a_0 = \frac{\pi}{4}.$$

$$b_n = \frac{1}{\pi}\int_0^\pi x \sin nx \, dx = \frac{1}{\pi}\left(\frac{\sin nx - nx \cos nx}{n^2}\right)\Bigg|_0^\pi$$

$$= -\frac{\cos n\pi}{n} = \frac{(-1)^{n+1}}{n}.$$

$$f(x) = \frac{\pi}{4} - \frac{2}{\pi}\left(\cos x + \frac{\cos 3x}{3^2} + \frac{\cos 5x}{5^2} \cdots\right) + \left(\sin x - \frac{\sin 2x}{2} + \frac{\sin 3x}{3} \cdots\right).$$

Section 6

2. We use the sketch of f(x) in Problem 5.2. At $x = \pm\pi$, and at
    $x = -\pi/2$, we see that f(x) is continuous; thus, at these points,
    the Fourier series converges to the value of f(x), namely 0.
    At $x = \pm 2\pi$, $x = 0$, and $x = \pi/2$, there is a jump in f(x) from 0 to
    1; at these points, the Fourier series converges to the mid-
    point of the jump, namely 1/2.

7. Look at the sketch in Problem 5.7. Note that f(x) is continuous
    at $x = \pm 2\pi$, $\pm\pi/2$, and 0. Thus, at these points, the Fourier
    series converges to the value of f(x). At $x = \pm 2\pi$, $-\pi/2$, and 0,
    $f(x) = 0$; at $x = \pi/2$, $f(x) = \pi/2$.

Section 7

2. We expand the function sketched in Problem 5.2 in a complex
    exponential series. By text equation (7.6), we find:

$$c_0 = \frac{1}{2\pi} \int_{-\pi}^{\pi} f(x)\ dx = \frac{1}{2\pi} \int_{0}^{\pi/2} dx = \frac{1}{4}\ ,$$

$$c_n = \frac{1}{2\pi} \int_{-\pi}^{\pi} f(x)e^{-inx}\ dx = \frac{1}{2\pi} \int_{0}^{\pi/2} e^{-inx}\ dx = \frac{1}{-2\pi in}\ e^{-inx} \Big|_{0}^{\pi/2}$$

$$= \frac{e^{-in\pi/2} - 1}{-2\pi in}\ .$$

We can tabulate the values of $e^{-in\pi/2}$
rapidly by visualizing the points z
in the complex plane (see text,
pages 60-61) with $r = 1$, and $\theta = -\pi/2$,
$-\pi$, $-3\pi/2$, -2π, $\cdots$. Thus for $n = 1,2,3,4$, we find
$e^{-in\pi/2} = -i$, -1, i, 1, and these four values repeat in sequence
for other n values. For negative n, we use $c_{-n} = $ complex conjugate of c_n since $f(x)$ is real (see Problem 12 below). Thus

$$c_n = \frac{1}{-2\pi in} \left\{ -i - 1,\ -1 - 1,\ i - 1,\ 1 - 1,\ \text{and repeat} \right\}$$

$$= \frac{1}{2\pi n} \left\{ 1 - i,\ -2i,\ -1 - i,\ 0,\ \text{and repeat} \right\},\quad n > 0,$$

$$c_{-n} = \overline{c_n}.$$

Then the Fourier series is

$$f(x) = \frac{1}{4} + \frac{1}{2\pi} \left[(1 - i)e^{ix} + (1 + i)e^{-ix} - \frac{2i}{2} e^{2ix} - \frac{2i}{-2} e^{-2ix} \right.$$

$$\left. - \frac{(1 + i)}{3} e^{3ix} - \frac{(1 - i)}{3} e^{-3ix} + \frac{(1 - i)}{5} e^{5ix} + \frac{1 + i}{5} e^{-5ix} \cdots \right.$$

2.   (continued)

We can write this series in the sine-cosine form by combining

terms as follows:

$$e^{ix} + e^{-ix} = 2 \cos x,$$
$$-i(e^{ix} - e^{-ix}) = 2 \sin x,$$

(see text, page 67)

and so on for other n values.   Thus we find

$$f(x) = \frac{1}{4} + \frac{1}{\pi}\left(\cos x - \frac{\cos 3x}{3} + \frac{\cos 5x}{5} \cdots \right)$$
$$+ \frac{1}{\pi}\left(\sin x + \sin 2x + \frac{\sin 3x}{3} \cdots \right)$$

as in Problem 5.2.

7.   We expand the function sketched in Problem 5.7 in a complex

exponential series.

$$c_0 = \frac{1}{2\pi}\int_0^\pi x\, dx = \frac{1}{2\pi}\left(\frac{\pi^2}{2}\right) = \frac{\pi}{4} \; ,$$

$$c_n = \frac{1}{2\pi}\int_0^\pi x\, e^{-inx}\, dx = \frac{1}{2\pi}\frac{e^{-inx}}{(-in)^2}(-inx - 1) \Big|_0^\pi \quad \text{(by integral tables or Appendix A)}$$

$$= \frac{1}{-2\pi n^2}\left[e^{-in\pi}(-in\pi - 1) + 1\right] = \begin{cases} \dfrac{i}{2n} \,, & \text{even } n \neq 0; \\[2mm] -\left(\dfrac{1}{\pi n^2} + \dfrac{i}{2n}\right) & \text{odd } n; \end{cases}$$

$$f(x) = \frac{\pi}{4} - \left(\frac{1}{\pi} + \frac{i}{2}\right)e^{ix} - \left(\frac{1}{\pi} - \frac{i}{2}\right)e^{-ix}$$
$$+ \frac{i}{4}\left(e^{2ix} - e^{-2ix}\right) - \left(\frac{1}{9\pi} + \frac{i}{6}\right)e^{3ix} - \left(\frac{1}{9\pi} - \frac{i}{6}\right)e^{-3ix} \cdots .$$

(Note carefully that $\frac{i}{2n}$ is negative for negative n, but $\frac{1}{\pi n^2}$ is

positive.   In other words, $c_{-n} = \overline{c_n}$;   see Problem 12.)   We can

write this series in the sine-cosine form by combining terms:

7.   (continued)

$$\frac{i}{2n}(e^{inx} - e^{-inx}) = -\frac{1}{n}\frac{e^{inx} - e^{-inx}}{2i} = -\frac{1}{n}\sin nx,$$

$$-\frac{1}{\pi n^2}(e^{inx} + e^{-inx}) = -\frac{2}{\pi n^2}\cos nx$$

(see text, page 67). Thus we find

$$f(x) = \frac{\pi}{4} - \frac{2}{\pi}\sum_{\text{odd } n}\frac{1}{n^2}\cos nx - \sum_{\text{even } n \neq 0}\frac{1}{n}\sin nx + \sum_{\text{odd } n}\frac{1}{n}\sin nx$$

$$= \frac{\pi}{4} - \frac{2}{\pi}\sum_{\text{odd } n}\frac{1}{n^2}\cos nx - \sum_{1}^{\infty}\frac{(-1)^n}{n}\sin nx$$

as in Problem 5.7.

12.   Comment: The result stated in this problem is very useful in
      computing the coefficients c_n when n is negative since we are
      usually expanding real functions in Fourier series. Hint for
      proof: Using text equation (7.6), find c_{-n} by replacing n by
      -n. Find $\overline{c_n}$ by taking the complex conjugate of text equation
      (7.6). Now compare your two results assuming that f(x) is real.

Section 8

Problems 1 to 8, hint: If you put $\ell = \pi$ in your answers, you should
get the answers to the corresponding problems given in Section 5.

1.   We first sketch several periods of the function to be expanded:

$$f(x) = \begin{cases} 1, & -\ell < x < 0, \\ 0, & 0 < x < \ell. \end{cases}$$

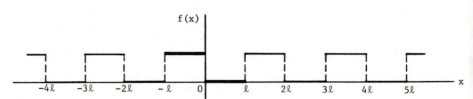

1.   (continued)

Now compare the sketch with Figure 8.3 of the text; we see that
they are identical. Although the basic interval used to define
f(x) is different in the two cases, the periodic functions are
the same. Since the average value of a periodic function over
a period is the same no matter which period we use, the values
of c_n (= average of $f(x)e^{-inx}$) will be the same for this problem
and the text example, and so the Fourier series are the same.
In the text, c_n was found as an integral from ℓ to 2ℓ. Here we
would naturally find c_n as an integral from $-\ell$ to 0. By direct
evaluation, we find

$$c_n = \frac{1}{2\ell}\int_{-\ell}^{0} e^{-in\pi x/\ell}\, dx = \frac{1}{2\ell(-in\pi/\ell)}\, e^{-in\pi x/\ell}\Big|_{-\ell}^{0} = \frac{1 - e^{in\pi}}{-2\pi in}$$

as in the text. Similarly the a_n's and b_n's are the same as for
the text example.

7.   The sketch is the same as for Problem 5.7 if we replace the
π's by ℓ's. We find

$$a_0 = \frac{1}{\ell}\int_{0}^{\ell} x\, dx = \frac{1}{\ell}\left(\frac{\ell^2}{2}\right) = \frac{\ell}{2}, \qquad \text{so} \qquad \frac{1}{2}a_0 = \frac{\ell}{4};$$

$$a_n = \frac{1}{\ell}\int_{0}^{\ell} x\cos\frac{n\pi x}{\ell}\, dx = \frac{1}{\ell}\left(\frac{\ell}{n\pi}\right)^2 \left[\cos\frac{n\pi x}{\ell} + \frac{n\pi x}{\ell}\sin\frac{n\pi x}{\ell}\right]_{0}^{\ell}$$

$$= \frac{\ell}{n^2\pi^2}(\cos n\pi - 1) = \begin{cases} 0, & \text{even } n \neq 0, \\[2mm] \dfrac{-2\ell}{n^2\pi^2}, & \text{odd } n. \end{cases}$$

(For the integrals, see Appendix A, Solutions page 603.)

7.   (continued)

$$b_n = \frac{1}{\ell} \int_0^\ell x \sin\frac{n\pi x}{\ell}\,dx = \frac{1}{\ell}\left(\frac{\ell}{n\pi}\right)^2 \left[\sin\frac{n\pi x}{\ell} - \frac{n\pi x}{\ell}\cos\frac{n\pi x}{\ell}\right]_0^\ell$$

$$= \frac{\ell}{n^2\pi^2}(\sin n\pi - n\pi\cos n\pi) = -\frac{\ell}{n\pi}(-1)^n.$$

$$c_0 = \frac{1}{2\ell}\int_0^\ell x\,dx = \frac{1}{2\ell}\frac{\ell^2}{2} = \frac{\ell}{4}\ ;$$

$$c_n = \frac{1}{2\ell}\int_0^\ell xe^{-in\pi x/\ell}\,dx = \frac{1}{2\ell}\frac{\ell^2}{-n^2\pi^2}e^{-in\pi x/\ell}\left(\frac{-in\pi x}{\ell} - 1\right)\Big|_0^\ell$$

$$= -\frac{\ell}{2n^2\pi^2}\left[e^{-in\pi}(-in\pi - 1) + 1\right] = \begin{cases} \dfrac{i\ell}{2n\pi}\,, & \text{even } n \neq 0; \\[2mm] -\dfrac{i\ell}{2n\pi} - \dfrac{\ell}{n^2\pi^2}\,, & \text{odd } n. \end{cases}$$

Then the sine-cosine Fourier series is

$$f(x) = \frac{\ell}{4} - \frac{2\ell}{\pi^2}\sum_{\substack{1\\ \text{odd } n}}^\infty \frac{1}{n^2}\cos\frac{n\pi x}{\ell} - \frac{\ell}{\pi}\sum_{n=1}^\infty \frac{(-1)^n}{n}\sin\frac{n\pi x}{\ell}$$

and the complex exponential Fourier series is

$$f(x) = \frac{\ell}{4} - \frac{\ell}{\pi^2}\sum_{\substack{-\infty\\ \text{odd } n}}^\infty \frac{1}{n^2}e^{in\pi x/\ell} + \frac{i\ell}{2\pi}\sum_{\substack{-\infty\\ n\neq 0}}^\infty \frac{(-1)^n}{n}e^{in\pi x/\ell}\,.$$

To see that these two series are equivalent, we can use the
formulas for sine and cosine in terms of complex exponentials
(see text, page 67). Then

$$\frac{-2\ell}{\pi^2}\sum_{\substack{1\\ \text{odd } n}}^\infty \frac{1}{n^2}\cos\frac{n\pi x}{\ell} = -\frac{2\ell}{\pi^2}\sum_{\substack{1\\ \text{odd } n}}^\infty \frac{1}{n^2}\frac{e^{in\pi x/\ell} + e^{-in\pi x/\ell}}{2}$$

$$= -\frac{\ell}{\pi^2}\sum_{\substack{-\infty\\ \text{odd } n}}^\infty \frac{1}{n^2}e^{in\pi x/\ell}\,.$$

7.   (continued)

(Note that the sine-cosine series sums are over positive n but
the exponential series sums are over positive and negative n.)
Similarly

$$-\frac{\ell}{\pi} \sum_1^\infty \frac{(-1)^n}{n} \sin\frac{n\pi x}{\ell} = -\frac{\ell}{\pi} \sum_1^\infty \frac{(-1)^n}{n} \frac{e^{in\pi x/\ell} - e^{-in\pi x/\ell}}{2i}$$

$$= \frac{i\ell}{2\pi} \sum_1^\infty \frac{(-1)^n}{n}(e^{in\pi x/\ell} - e^{-in\pi x/\ell}) = \frac{i\ell}{2\pi} \sum_{\substack{-\infty \\ n\neq 0}}^\infty \frac{(-1)^n}{n} e^{in\pi x/\ell} .$$

(Note, in the last step, that $1/n$ is negative when n is negative.)

11.   The sketches are in the text: (a) is Figure 8.1 and (b) is
Figure 8.2. Using text equation (8.3) with $\ell = \pi$ for part (a)
and text equation (8.1) for part (b), we find the Fourier coefficients. To evaluate the integrals, see Appendix A, page 603.

(a) $a_0 = \frac{1}{\pi}\int_{-\pi}^\pi x^2 dx = \frac{1}{\pi}\left(\frac{\pi^3}{3} + \frac{\pi^3}{3}\right) = \frac{2\pi^2}{3}$, $\frac{1}{2}a_0 = \frac{\pi^2}{3}$.

$$a_n = \frac{1}{\pi}\int_{-\pi}^\pi x^2\cos nx\ dx = \frac{1}{\pi}\frac{1}{n^3}\left[2nx\cos nx + (n^2x^2 - 2)\sin nx\right]_{-\pi}^\pi$$

$$= \frac{4n\pi\cos n\pi}{\pi n^3} = \frac{4}{n^2}(-1)^n .$$

$$b_n = \frac{1}{\pi}\int_{-\pi}^\pi x^2\sin nx\ dx = \frac{1}{\pi}\frac{1}{n^3}\left[2nx\sin nx - (n^2x^2 - 2)\cos nx\right]_{-\pi}^\pi = 0 .$$

$$c_0 = \frac{1}{2\pi}\int_{-\pi}^\pi x^2 dx = \frac{\pi^2}{3} .$$

$$c_n = \frac{1}{2\pi}\int_{-\pi}^\pi x^2 e^{-inx}\ dx = \frac{1}{2\pi}\frac{1}{(-in)^3}\left[(-in)^2x^2 + 2inx + 2\right]e^{-inx}\Big|_{-\pi}^\pi$$

$$= \frac{4in\pi(-1)^n}{2\pi in^3} = \frac{2}{n^2}(-1)^n .$$

11.   (continued)

$$f(x) = \frac{\pi^2}{3} + 4 \sum_1^\infty \frac{(-1)^n}{n^2} \cos nx \qquad \text{or}$$

$$f(x) = \frac{\pi^2}{3} + 2 \sum_{-\infty}^\infty \frac{(-1)^n}{n^2} e^{inx} .$$

(b) $\displaystyle a_0 = \frac{1}{\pi} \int_0^{2\pi} x^2 dx = \frac{1}{\pi} \left. \frac{x^3}{3} \right|_0^{2\pi} = \frac{8\pi^2}{3}$, $\displaystyle \frac{1}{2} a_0 = \frac{4\pi^2}{3}$.

$$a_n = \frac{1}{\pi} \int_0^{2\pi} x^2 \cos nx \, dx = \frac{1}{\pi} \frac{1}{n^3} \left[2nx \cos nx + (n^2 x^2 - 2) \sin nx \right]_0^{2\pi}$$

$$= \frac{1}{\pi} \frac{1}{n^3} (4n\pi) = \frac{4}{n^2} .$$

$$b_n = \frac{1}{\pi} \int_0^{2\pi} x^2 \sin nx \, dx = \frac{1}{\pi} \frac{1}{n^3} \left[2nx \sin nx - (n^2 x^2 - 2) \cos nx \right]_0^{2\pi}$$

$$= \frac{1}{\pi} \frac{-(n^2 \cdot 4\pi^2 - 2) - 2}{n^3} = -\frac{4\pi}{n} .$$

$$c_0 = \frac{1}{2\pi} \int_0^{2\pi} x^2 dx = \frac{1}{2\pi} \frac{(2\pi)^3}{3} = \frac{4\pi^2}{3} .$$

$$c_n = \frac{1}{2\pi} \int_0^{2\pi} x^2 e^{-inx} dx = \frac{1}{2\pi} \frac{1}{(-in)^3} \left[(-in)^2 x^2 + 2inx + 2 \right] e^{-inx} \Big|_0^{2\pi}$$

$$= \frac{-n^2 \cdot 4\pi^2 + 4in\pi}{2\pi in^3} = 2\left(\frac{i\pi}{n} + \frac{1}{n^2} \right) .$$

$$f(x) = \frac{4\pi^2}{3} + 4 \sum_1^\infty \frac{1}{n^2} \cos nx - 4\pi \sum_1^\infty \frac{1}{n} \sin nx \qquad \text{or}$$

$$f(x) = \frac{4\pi^2}{3} + 2 \sum_{-\infty}^\infty \left(\frac{i\pi}{n} + \frac{1}{n^2} \right) e^{inx} .$$

19.   Given $f(x) = \begin{cases} 0, & -\frac{1}{2} < x < 0, \\ x, & 0 < x < \frac{1}{2}, \end{cases}$ we first sketch several periods of the

function of period 1 which is equal to $f(x)$ on $\left(-\frac{1}{2}, \frac{1}{2}\right)$.

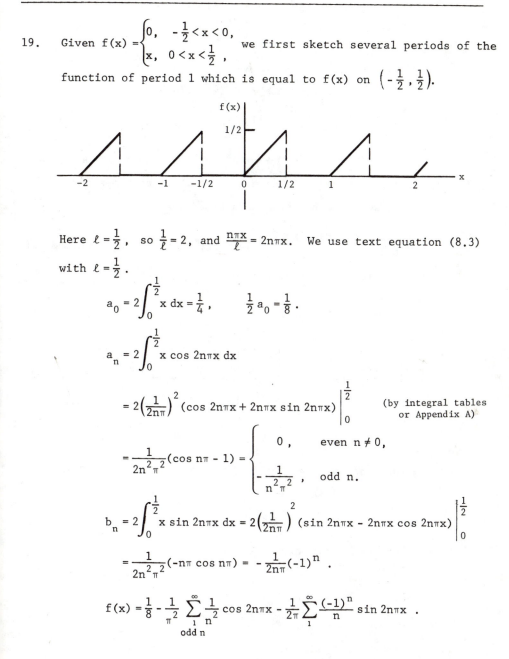

Here $\ell = \frac{1}{2}$, so $\frac{1}{\ell} = 2$, and $\frac{n\pi x}{\ell} = 2n\pi x$. We use text equation (8.3)

with $\ell = \frac{1}{2}$.

$$a_0 = 2\int_0^{\frac{1}{2}} x \, dx = \frac{1}{4}, \qquad \frac{1}{2} a_0 = \frac{1}{8}.$$

$$a_n = 2\int_0^{\frac{1}{2}} x \cos 2n\pi x \, dx$$

$$= 2\left(\frac{1}{2n\pi}\right)^2 (\cos 2n\pi x + 2n\pi x \sin 2n\pi x)\Big|_0^{\frac{1}{2}} \qquad \text{(by integral tables or Appendix A)}$$

$$= \frac{1}{2n^2\pi^2}(\cos n\pi - 1) = \begin{cases} 0, & \text{even } n \neq 0, \\ -\frac{1}{n^2\pi^2}, & \text{odd } n. \end{cases}$$

$$b_n = 2\int_0^{\frac{1}{2}} x \sin 2n\pi x \, dx = 2\left(\frac{1}{2n\pi}\right)^2 (\sin 2n\pi x - 2n\pi x \cos 2n\pi x)\Big|_0^{\frac{1}{2}}$$

$$= \frac{1}{2n^2\pi^2}(-n\pi \cos n\pi) = -\frac{1}{2n\pi}(-1)^n.$$

$$f(x) = \frac{1}{8} - \frac{1}{\pi^2}\sum_{\substack{1 \\ \text{odd } n}}^{\infty} \frac{1}{n^2} \cos 2n\pi x - \frac{1}{2\pi}\sum_{1}^{\infty} \frac{(-1)^n}{n} \sin 2n\pi x.$$

Section 9

1.   (b) Given $f(x) = xe^x$, we write

$$f(x) = \frac{1}{2}[f(x) + f(-x)] + \frac{1}{2}[f(x) - f(-x)]$$

$$= \frac{1}{2}\left[xe^x + (-x)e^{-x}\right] + \frac{1}{2}\left[xe^x - (-x)e^{-x}\right]$$

$$= \frac{1}{2}x(e^x - e^{-x}) + \frac{1}{2}x(e^x + e^{-x}) = x\sinh x + x\cosh x.$$

Observe that $x\sinh x$ is an even function (odd times odd) and $x\cosh x$ is an odd function (odd times even).

5.   Given $f(x) = \begin{cases} -1, & -\pi < x < 0, \\ 1, & 0 < x < \pi, \end{cases}$ we first sketch several periods of the periodic function of period 2π.

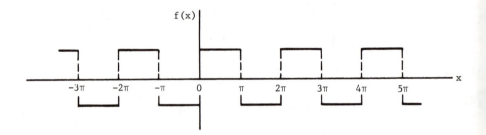

From the graph we see that $f(x)$ is an odd function.   Then by text equation (9.4):

$$a_n = 0.$$

$$b_n = \frac{2}{\pi}\int_0^\pi 1 \cdot \sin nx\, dx = \frac{2}{\pi}\left(-\frac{\cos nx}{n}\right)\Big|_0^\pi$$

$$= -\frac{2}{n\pi}(\cos n\pi - 1) = \begin{cases} 0, & \text{even } n \neq 0, \\ \dfrac{4}{n\pi}, & \text{odd } n. \end{cases}$$

$$f(x) = \frac{4}{\pi}\sum_{\substack{1 \\ \text{odd } n}}^{\infty} \frac{1}{n}\sin nx.$$

11. Given $f(x) = \cosh x$ on $(-\pi, \pi)$, we observe that $f(x)$ is an even
     function. Then by text equation (9.5):

$$b_n = 0.$$

$$a_0 = \frac{2}{\pi} \int_0^\pi \cosh x \, dx = \frac{2}{\pi} \sinh x \Big|_0^\pi = \frac{2}{\pi} \sinh \pi, \quad \text{so} \quad \frac{1}{2} a_0 = \frac{1}{\pi} \sinh \pi.$$

$$a_n = \frac{2}{\pi} \int_0^\pi \cosh x \cos nx \, dx$$

$$= \frac{2}{\pi} \frac{1}{n^2 + 1} (\cos nx \sinh x + n \sin nx \cosh x) \Big|_0^\pi \qquad \text{(see hints below)}$$

$$= \frac{2(-1)^n \sinh \pi}{\pi (n^2 + 1)},$$

since $\cos n\pi = (-1)^n$, $\sin n\pi = \sin 0 = \sinh 0 = 0$. Then

$$f(x) = \frac{1}{\pi} \sinh \pi + \frac{2}{\pi} \sinh \pi \sum_1^\infty \frac{(-1)^n}{n^2 + 1} \cos nx$$

$$= \frac{2 \sinh \pi}{\pi} \left(\frac{1}{2} - \frac{1}{2} \cos x + \frac{1}{5} \cos 2x - \frac{1}{10} \cos 3x \, \cdots \right).$$

Hints for evaluating the integral of $\cosh x \cos nx$: Write
$$\cosh x \cos nx = \frac{1}{2}(e^x + e^{-x}) \cos nx = \frac{1}{2}(e^x \cos nx + e^{-x} \cos nx)$$
and use Problem 17, text page 69 with $b = n$, $a = 1$ and $a = -1$.
Then combine terms using equation (12.2), text page 69.

14. (b) Let $f(x)$ be an even function; then [text equation (9.1)]

$$f(x) = f(-x).$$

Let $g(x) = \frac{d}{dx} f(x)$. Then

$$g(-x) = \frac{d}{d(-x)} f(-x) = -\frac{d}{dx} f(-x) = -\frac{d}{dx} f(x) = -g(x)$$

so $g(x)$ is odd. Similarly prove that the derivative of an odd
function is even.

18.   We first sketch the three required functions.

   (a) Even function f_c (cosine series). Period $= 6$, $\ell = 3$.

   (b) Odd function f_s (sine series). Period $= 6$, $\ell = 3$.

   (c) Function f_p. Period $= 3$, $\ell = 3/2$.

   (a) Using text equation (9.5) with $\ell = 3$, we find (see sketch of f_c):

$$b_n = 0.$$

$$a_0 = \frac{2}{3}\int_0^3 f(x)\,dx = \frac{2}{3}\int_0^1 dx = \frac{2}{3}, \qquad \frac{1}{2}a_0 = \frac{1}{3}.$$

$$a_n = \frac{2}{3}\int_0^1 \cos\frac{n\pi x}{3}\,dx = \frac{2}{3}\,\frac{3}{n\pi}\sin\frac{n\pi x}{3}\Big|_0^1 = \frac{2}{n\pi}\sin\frac{n\pi}{3}.$$

$$f_c(x) = \frac{1}{3} + \frac{2}{\pi}\sum_{n=1}^{\infty}\frac{1}{n}\sin\frac{n\pi}{3}\cos\frac{n\pi x}{3}.$$

18.   (continued)

For n = 1 to 6, $\sin \frac{n\pi}{3} = \left\{\frac{1}{2}\sqrt{3}, \frac{1}{2}\sqrt{3}, 0, -\frac{1}{2}\sqrt{3}, -\frac{1}{2}\sqrt{3}, 0\right\}$ and

for larger n, these six numbers repeat. Thus

$$f_c(x) = \frac{1}{3} + \frac{\sqrt{3}}{\pi}\left(\cos\frac{\pi x}{3} + \frac{1}{2}\cos\frac{2\pi x}{3} - \frac{1}{4}\cos\frac{4\pi x}{3} - \frac{1}{5}\cos\frac{5\pi x}{3} + \frac{1}{7}\cos\frac{7\pi x}{3} \cdots\right).$$

(b) Using text equation (9.4) with $\ell = 3$, we find (see sketch of

f_s):

$a_n = 0.$

$$b_n = \frac{2}{3}\int_0^1 \sin\frac{n\pi x}{3}\, dx = -\frac{2}{3}\frac{3}{n\pi}\cos\frac{n\pi x}{3}\Big|_0^1 = \frac{2}{n\pi}\left(1 - \cos\frac{n\pi}{3}\right).$$

For n = 1 to 6, $\cos\frac{n\pi}{3} = \left\{\frac{1}{2}, -\frac{1}{2}, -1, -\frac{1}{2}, \frac{1}{2}, 1\right\}$ so

$$b_n = \frac{1}{n\pi}\{1, 3, 4, 3, 1, 0, \text{ and repeat}\}.$$

$$f_s(x) = \frac{1}{\pi}\left(\sin\frac{\pi x}{3} + \frac{3}{2}\sin\frac{2\pi x}{3} + \frac{4}{3}\sin\frac{3\pi x}{3}\right.$$

$$\left. + \frac{3}{4}\sin\frac{4\pi x}{3} + \frac{1}{5}\sin\frac{5\pi x}{3} + \frac{1}{7}\sin\frac{7\pi x}{3} \cdots\right).$$

(c) Here we can integrate over any period, say from 0 to 3, or

from -3/2 to 3/2. Using text equation (8.3) with $\ell = 3/2$, we

find (see sketch of f_p):

$$a_0 = \frac{1}{\ell}\int_{-\ell}^{\ell} f(x)\, dx = \frac{2}{3}\int_0^1 dx = \frac{2}{3}, \qquad \frac{1}{2}a_0 = \frac{1}{3}.$$

$$a_n = \frac{2}{3}\int_0^1 \cos\frac{2n\pi x}{3}\, dx = \frac{2}{3}\frac{3}{2n\pi}\sin\frac{2n\pi x}{3}\Big|_0^1$$

$$= \frac{1}{n\pi}\sin\frac{2n\pi}{3} = \frac{1}{n\pi}\left\{\frac{1}{2}\sqrt{3}, -\frac{1}{2}\sqrt{3}, 0, \text{ and repeat}\right\}.$$

$$b_n = \frac{2}{3}\int_0^1 \sin\frac{2n\pi x}{3}\, dx = -\frac{2}{3}\frac{3}{2n\pi}\cos\frac{2n\pi x}{3}\Big|_0^1$$

$$= \frac{1}{n\pi}\left(1 - \cos\frac{2n\pi}{3}\right) = \frac{1}{n\pi}\left\{\frac{3}{2}, \frac{3}{2}, 0, \text{ and repeat}\right\}.$$

$$f(x) = \frac{1}{3} + \frac{\sqrt{3}}{2\pi}\left(\cos\frac{2\pi x}{3} - \frac{1}{2}\cos\frac{4\pi x}{3} + \frac{1}{4}\cos\frac{8\pi x}{3} - \frac{1}{5}\cos\frac{10\pi x}{3} \cdots\right)$$

$$+ \frac{3}{2\pi}\left(\sin\frac{2\pi x}{3} + \frac{1}{2}\sin\frac{4\pi x}{3} + \frac{1}{4}\sin\frac{8\pi x}{3} + \frac{1}{5}\sin\frac{10\pi x}{3} \cdots\right).$$

23. We want a Fourier sine series to represent a function given on $(0,\ell)$. We must then extend the function to be odd on $(-\ell,\ell)$ and continue it periodically with period 2ℓ. Note, however, from text equation (9.4) that we actually use only the values of the function on $(0,\ell)$ in computing the coefficients. From the text figure, we have

$$f(x,0) = \begin{cases} \dfrac{2h}{\ell}\,x, & 0 < x < \dfrac{\ell}{2}, \\[2mm] \dfrac{2h}{\ell}(\ell - x), & \dfrac{\ell}{2} < x < \ell. \end{cases}$$

Then by text equation (9.4):

$$b_n = \frac{2}{\ell}\left[\int_0^{\ell/2} \frac{2h}{\ell}\,x \sin\frac{n\pi x}{\ell}\,dx + \int_{\ell/2}^{\ell} \frac{2h}{\ell}(\ell - x)\sin\frac{n\pi x}{\ell}\,dx\right]$$

$$= \frac{4h}{\ell^2}\frac{\ell^2}{n^2\pi^2}\left(\sin\frac{n\pi x}{\ell} - \frac{n\pi x}{\ell}\cos\frac{n\pi x}{\ell}\right)\Bigg|_0^{\ell/2} - \frac{4h}{\ell}\frac{\ell}{n\pi}\cos\frac{n\pi x}{\ell}\Bigg|_{\ell/2}^{\ell}$$

$$- \frac{4h}{\ell^2}\frac{\ell^2}{n^2\pi^2}\left(\sin\frac{n\pi x}{\ell} - \frac{n\pi x}{\ell}\cos\frac{n\pi x}{\ell}\right)\Bigg|_{\ell/2}^{\ell}$$

$$= \frac{4h}{n^2\pi^2}\left(\sin\frac{n\pi}{2} - \frac{n\pi}{2}\cos\frac{n\pi}{2}\right) - \frac{4h}{n\pi}\left(\cos n\pi - \cos\frac{n\pi}{2}\right)$$

$$- \frac{4h}{n^2\pi^2}\left(\sin n\pi - n\pi\cos n\pi - \sin\frac{n\pi}{2} + \frac{n\pi}{2}\cos\frac{n\pi}{2}\right)$$

$$= \frac{8h}{n^2\pi^2}\sin\frac{n\pi}{2} = \frac{8h}{n^2\pi^2}\begin{cases} 1, & n = 1 + 4k, \\ 0, & n \text{ even}, \\ -1, & n = 3 + 4k. \end{cases}$$

$$f(x) = \frac{8h}{\pi^2}\left(\sin\frac{\pi x}{\ell} - \frac{1}{9}\sin\frac{3\pi x}{\ell} + \frac{1}{25}\sin\frac{5\pi x}{\ell} \cdots\right).$$

Section 10

1. We observe that p(t) is an even function so we expand it in a
cosine series. From the text figure, $\ell = 1/220$. Then by text
equation (9.5) and the text figure:

$$a_0 = \frac{2}{\ell}\int_0^\ell f(t)\,dt = 440\left[\int_0^{1/660} dt - \int_{1/330}^{1/220} dt\right] = 0.$$

$$a_n = 440\left[\int_0^{1/660} \cos 220n\pi t\,dt - \int_{1/330}^{1/220} \cos 220n\pi t\,dt\right]$$

$$= 440 \cdot \frac{1}{220n\pi}\left(\sin\frac{n\pi}{3} - \sin n\pi + \sin\frac{2n\pi}{3}\right)$$

$$= \frac{2}{n\pi}\left(\sin\frac{n\pi}{3} + \sin\frac{2n\pi}{3}\right).$$

$$p(t) = \frac{2}{\pi}\sum_1^\infty \frac{1}{n}\left(\sin\frac{n\pi}{3} + \sin\frac{2n\pi}{3}\right)\cos 220n\pi t.$$

To find the relative intensities, we need to evaluate the coefficients. Note that except for the 1/n factor the coefficients
repeat in blocks of six.

n	1	2	3	4	5	6	7	8	9	$\cdots$
$\sin\frac{n\pi}{3}$	$\frac{1}{2}\sqrt{3}$	$\frac{1}{2}\sqrt{3}$	0	$-\frac{1}{2}\sqrt{3}$	$-\frac{1}{2}\sqrt{3}$	0	$\frac{1}{2}\sqrt{3}$	$\frac{1}{2}\sqrt{3}$	0	$\cdots$
$\sin\frac{2n\pi}{3}$	$\frac{1}{2}\sqrt{3}$	$-\frac{1}{2}\sqrt{3}$	0	$\frac{1}{2}\sqrt{3}$	$-\frac{1}{2}\sqrt{3}$	0	$\frac{1}{2}\sqrt{3}$	$-\frac{1}{2}\sqrt{3}$	0	$\cdots$
a_n	$\frac{2\sqrt{3}}{\pi}$	0	0	0	$\frac{-2\sqrt{3}}{5\pi}$	0	$\frac{2\sqrt{3}}{7\pi}$	0	0	$\cdots$

The intensities are proportional to the squares of the coefficients. If we divide out the common factor $2\sqrt{3}/\pi$ and then square,
we find for the relative intensities

$$1 : 0 : 0 : 0 : \frac{1}{25} : 0 : \frac{1}{49} : 0 : 0 : 0 : \frac{1}{121} : 0 \cdots$$

1. (continued)

    Thus the first harmonic is most important. Also the only har-

    monics present are those for

    $$n = 6k \pm 1 = 1, \; 5, \; 7, \; 11, \; 13, \; 17, \; 19, \; \cdots$$

    and the intensity of each of these is proportional to $1/n^2$.

4. Hint: To integrate the absolute value of a function $f(x)$, we

    integrate $f(x)$ where $f(x) > 0$, and we integrate $-f(x)$ where

    $f(x) < 0$. So in this problem we <u>could</u> write

    $$a_0 = \frac{1}{\ell} \int_0^{2\ell} V(t) \, dt$$
    $$= 120 \left[\int_0^{1/120} 100 \sin 120\pi t \, dt + \int_{1/120}^{1/60} (-100 \sin 120\pi t) \, dt \right]$$

    and similar formulas for the other coefficients. However, it is

    much easier to observe that $V(t)$ is an even function (sketch it

    from $t = -1/120$ to 0) and use text equation (9.5). Then $b_n = 0$,

    and

    $$a_0 = \frac{2}{\ell} \int_0^{\ell} V(t) \, dt = 240 \int_0^{1/120} 100 \sin 120\pi t \, dt,$$

    and similarly the a_n's are integrals from 0 to 1/120 where the

    sine is positive.

5. Since $I(t)$ is neither even nor odd (see sketch), we

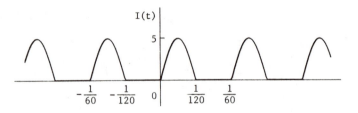

5.   (continued)

use text equation (8.3) with $\ell = 1/120$. We can integrate either
from $-\ell$ to ℓ or from 0 to 2ℓ. In either case, since $I(t) = 0$
from $-1/120$ to 0 (or from $1/120$ to $1/60$), we find

$$a_0 = 120 \int_0^{1/120} 5 \sin 120\pi t \, dt = 120 \cdot \frac{-5}{120\pi} \cos 120\pi t \Big|_0^{1/120}$$

$$= \frac{10}{\pi}, \qquad \frac{1}{2} a_0 = \frac{5}{\pi}.$$

$$a_n = 120 \int_0^{1/120} 5 \sin 120\pi t \cos 120 n\pi t \, dt$$

$$= 600 \left(\frac{\cos 120(n-1)\pi t}{2 \cdot 120\pi (n-1)} - \frac{\cos 120(n+1)\pi t}{2 \cdot 120\pi (n+1)} \right) \Big|_0^{1/120}$$

$$= \frac{5}{2\pi} \left(\frac{\cos(n-1)\pi - 1}{n-1} - \frac{\cos(n+1)\pi - 1}{n+1} \right) = \begin{cases} 0, & \text{odd } n \neq 1, \\ \dfrac{5}{\pi} \dfrac{-2}{n^2 - 1}, & \text{even } n \neq 0. \end{cases}$$

Because of the $n-1$ factor in the denominator, we must find a_1
separately:

$$a_1 = 120 \int_0^{1/120} 5 \sin 120\pi t \cos 120\pi t \, dt$$

$$= 600 \cdot \frac{1}{120\pi} \frac{\sin^2 120\pi t}{2} \Big|_0^{1/120} = 0.$$

For b_n we have

(A) $b_n = 120 \int_0^{1/120} 5 \sin 120\pi t \sin 120 n\pi t \, dt .$

We could evaluate this using tables as we did for a_n. However,
an easier way is to observe that a product of two sines is an
even function so we can evaluate the integral in (A) by integrating from $-1/120$ to $1/120$ and dividing by 2:

5.   (continued)

(B) $b_n = 60 \int_{-1/120}^{1/120} 5 \sin 120\pi t \sin 120 n\pi t \, dt.$

This is now an average (over a period) of a product of two sines, so by text equation (5.2)

$$b_n = \begin{cases} 5 \cdot \dfrac{1}{2} = 5/2, & n = 1, \\[2mm] 5 \cdot 0 = 0, & n \neq 1. \end{cases}$$

[Note carefully that (A) is an integral of $I(t) \sin 120 n\pi t$ from $-1/120$ to $1/120$ since $I(t) = 0$ from $-1/120$ to 0. Now (B) is not an integral of $I(t) \sin 120 n\pi t$, but is just an integral which is mathematically equivalent to (A) and is easy to evaluate.]

Thus we have

$$I(t) = \frac{5}{\pi}\left(1 - 2\sum_{n=2}^{\infty} \frac{1}{n^2 - 1} \cos 120 n\pi t\right) + \frac{5}{2} \sin 120\pi t.$$

The intensities corresponding to the various frequencies are proportional to the squares of the coefficients. (The constant term is called the d-c component; it does not have a frequency. Here we are concerned just with the a-c terms.) The frequencies and their relative intensities are:

n	frequency	relative intensity
1	60	$\left(\dfrac{5}{2}\right)^2 = 6.25$
2	120	$\left(\dfrac{10}{\pi(2^2 - 1)}\right)^2 = 1.13$
3	180	0
4	240	$\left(\dfrac{10}{15\pi}\right)^2 = 0.045$ and so on.

8.   Comment: In this problem, you find both sine and cosine terms in the Fourier series. To find the relative intensity of a harmonic, add the squares of the corresponding sine and cosine terms. That is, the intensity of harmonic n is proportional to $a_n^2 + b_n^2$. See text, Section 11, Example 1, and replace equation (11.7) by equation (11.4).

Section 11

3.   Given that

$$f(x) = \sum_{-\infty}^{\infty} c_n e^{in\pi x/\ell}$$

then the complex conjugate of $f(x)$ is

$$\overline{f(x)} = \sum_{-\infty}^{\infty} \overline{c_m} e^{-im\pi x/\ell}.$$

We want the average of $|f(x)|^2 = f(x)\,\overline{f(x)}$. We must multiply every term of $f(x)$ times every term of $\overline{f(x)}$; this gives products of the forms:

$$c_n e^{in\pi x/\ell}\,\overline{c_n}\,e^{-in\pi x/\ell} = |c_n|^2,$$

$$c_n e^{in\pi x/\ell}\,\overline{c_m}\,e^{-im\pi x/\ell} = c_n\overline{c_m}\,e^{i(n-m)\pi x/\ell}, \qquad n \neq m.$$

(The second form includes the cases $m = -n$, for example $c_2 e^{2i\pi x/\ell}\cdot\overline{c_{-2}}e^{2i\pi x/\ell}$. The first form includes $c_0\overline{c_0} = |c_0|^2$.)
Now the average value of $e^{i(n-m)\pi x/\ell}$ on $(-\ell,\ell)$ is (for $n \neq m$)

$$\frac{1}{2\ell}\int_{-\ell}^{\ell} e^{i(n-m)\pi x/\ell}\,dx = \frac{e^{i(n-m)\pi x/\ell}}{(n-m)\pi/\ell}\Bigg|_{-\ell}^{\ell} = \frac{e^{i(n-m)\pi} - e^{-i(n-m)\pi}}{(n-m)\pi/\ell} = 0$$

since $e^{ik\pi} = e^{-ik\pi}$. Also the average value of $|c_n|^2$ is $|c_n|^2$.
Thus

$$\text{Average of } |f(x)|^2 \text{ on } (-\ell,\ell) \text{ is } \sum_{-\infty}^{\infty} |c_n|^2.$$

7.   From Problem 5.8, we find

$$\text{Average of } |f(x)|^2 = \frac{1}{2\pi}\int_{-\pi}^{\pi}(1+x)^2 dx = \frac{1}{2\pi}\int_{-\pi}^{\pi}(1+2x+x^2)\,dx$$

$$= \frac{1}{2\pi}\left(2\pi + 0 + \frac{2\pi^3}{3}\right) = 1 + \frac{\pi^2}{3}\ .$$

The series in Problem 5.8 is

$$f(x) = 1 + \sum_{1}^{\infty}\frac{2}{n}(-1)^{n+1}\sin nx.$$

Then Parseval's theorem [text equation (11.4)] gives

$$1 + \frac{\pi^2}{3} = 1 + \frac{1}{2}\sum_{1}^{\infty}\frac{4}{n^2} \qquad \text{so} \qquad \sum_{1}^{\infty}\frac{1}{n^2} = \frac{\pi^2}{6}\ .$$

10.   Hint: Follow the text discussion of equations (11.1) to (11.4).
      As in text equation (11.3) and the discussion following it,
      write the possible product terms and find their average values.

Section 12

2.   If $x = 2.37$, then $x - [x] = 2.37 - 2 = 0.37$; if $x = -4.8$, then
      $x - [x] = -4.8 - (-5) = 0.2$. Thus we see that $x - [x]$ is just x for
      $0 < x < 1$, and the rest of the graph of $x - [x]$ is a repeat of the
      $(0,1)$ graph. If we then drop this graph down a distance $1/2$,
      we have the graph of

$$f(x) = x - [x] - \frac{1}{2}:$$

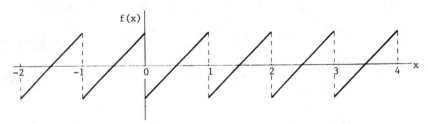

2.   (continued)

The period of $f(x)$ is 1, say from 0 to 1; then $2\ell = 1$, $\ell = \frac{1}{2}$.

We find

$$c_0 = \frac{1}{1}\int_0^1 \left(x - \frac{1}{2}\right)dx = \frac{x^2}{2} - \frac{1}{2}x\Big|_0^1 = 0.$$

$$c_n = \frac{1}{1}\int_0^1 \left(x - \frac{1}{2}\right)e^{-2in\pi x}\,dx = \frac{e^{-2in\pi x}}{(-2in\pi)^2}(-2in\pi x - 1) - \frac{1}{2}\frac{e^{-2in\pi x}}{-2in\pi}\Big|_0^1$$

$$= \frac{e^{-2in\pi}(-2in\pi - 1) - e^0(0 - 1)}{(-2in\pi)^2} + \frac{1}{4in\pi}(e^{-2in\pi} - 1) = \frac{1}{-2in\pi}.$$

$$f(x) = \sum_{-\infty}^{\infty} c_n e^{2inx} = \frac{i}{2\pi}\sum_{-\infty}^{\infty}\frac{e^{2in\pi x}}{n}.$$

3.   See the solution of Problem 9.5 above. We found

$$f(x) = \frac{4}{\pi}\sum_{\substack{1\\ \text{odd } n}}^{\infty}\frac{1}{n}\sin nx = \frac{4}{\pi}\left(\sin x + \frac{1}{3}\sin 3x + \frac{1}{5}\sin 5x + \cdots\right).$$

Note from the sketch that $f(x)$ is a discontinuous function. Thus
we know from Chapter 1 of the text that $f(x)$ cannot be repre-
sented by a power series. Let us see what happens if we try. We
write the power series for $\sin nx$ [text, page 24, equation (13.1)]

$$\sin nx = nx - \frac{n^3 x^3}{3!} + \frac{n^5 x^5}{5!}\cdots$$

and try to substitute it into the Fourier series for $f(x)$. There
is an x term in each $\sin nx$ and the sum of these is:

$$x + \frac{1}{3}\cdot 3x + \frac{1}{5}\cdot 5x + \cdots + \frac{1}{n}nx + \cdots = x(1 + 1 + 1 + 1 + \cdots)$$

which is a divergent series. For the x^3 terms, we find

$$-\frac{x^3}{3!} + \frac{1}{3}\left(-\frac{3^3 x^3}{3!}\right) + \frac{1}{5}\left(-\frac{5^3 x^3}{3!}\right) + \cdots = -\frac{x^3}{3!}\left(1 + 3^2 + 5^2 + \cdots\right),$$

3.   (continued)

      another divergent series. Similarly, as you can verify, the
      coefficient of each (odd) power of x is a divergent series.
      Thus, as we expected, our attempt to find a power series for
      a discontinuous function has failed.

7.   We first sketch $f(x) = |x|$ on $(-\pi, \pi)$, and extend it to be
      periodic with period 2π.

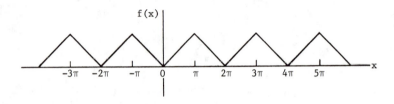

      Since $f(x)$ is an even function, we expand it in a cosine series.
      By text equation (9.5) with $\ell = \pi$:

$$a_n = \frac{2}{\pi}\int_0^\pi x \cos nx\, dx = \frac{2}{\pi n^2}(\cos nx + nx \sin nx)\Big|_0^\pi = \frac{2}{\pi}\frac{\cos n\pi - 1}{n^2}$$

$$= \begin{cases} -\dfrac{4}{\pi n^2}, & \text{odd } n \\[2mm] 0, & \text{even } n. \end{cases}$$

$$a_0 = \frac{2}{\pi}\int_0^\pi x\, dx = \pi, \qquad \frac{1}{2}a_0 = \frac{\pi}{2}.$$

$$f(x) = \frac{\pi}{2} - \frac{4}{\pi}\sum_{\text{odd } n}\frac{1}{n^2}\cos nx.$$

12.   Hint: What value of n makes the denominator smallest?

Chapter 8

Section 1

1.   Hint: See text, Chapter 2, Section 12.

4.   Integrate the differential equation

$$\frac{d^2x}{dt^2} = g\, e^{-kt}$$

with respect to t to get

$$\frac{dx}{dt} = -\frac{g}{k}\, e^{-kt} + C.$$

Since the object starts from rest, $\frac{dx}{dt} = 0$ when $t = 0$:

$$0 = -\frac{g}{k} + C, \qquad C = \frac{g}{k},$$

(A) $$\frac{dx}{dt} = \frac{g}{k}(1 - e^{-kt}).$$

Integrate again with respect to t:

$$x = \frac{g}{k}\left(t + \frac{1}{k}e^{-kt}\right) + C'.$$

Since x is the distance the object moves in time t, we must have $x = 0$ when $t = 0$.   Then $C' = -g/k^2$, so

(B) $$x = \frac{g}{k^2}(e^{-kt} + kt - 1).$$

For small values of t, we expand e^{-kt} in a Maclaurin series (see text, page 24):

$$e^{-kt} = 1 - kt + \frac{1}{2}k^2t^2 \cdots .$$

Then equation (B) gives

$$x = \frac{g}{k^2}(1 - kt + \frac{1}{2}k^2t^2 \cdots + kt - 1) \cong \frac{1}{2}gt^2$$

as in text equation (1.10).   For large values of t, e^{-kt} is negligible compared to 1 and we have from equation (A):

$$\frac{dx}{dt} \cong \frac{g}{k}.$$

6. Hint: Write the volume and the surface area of a sphere in
    terms of the radius r and find dr/dt.

Section 2

2. We separate the variables by dividing the equation by

$$\sqrt{1 - y^2} \sqrt{1 - x^2} \text{ (assuming } x^2 \neq 1, \; y^2 \neq 1; \text{ see Problem 13 below):}$$

$$\frac{x \, dx}{\sqrt{1 - x^2}} + \frac{y \, dy}{\sqrt{1 - y^2}} = 0.$$

Integrate both terms to get:

$$\int \frac{x \, dx}{\sqrt{1 - x^2}} + \int \frac{y \, dy}{\sqrt{1 - y^2}} = \text{const.} = K.$$

If $u = 1 - x^2$, then $du = -2x \, dx$, so

$$\int \frac{x \, dx}{\sqrt{1 - x^2}} = \int \frac{-\frac{1}{2} du}{\sqrt{u}} = -\frac{1}{2} \int u^{-1/2} \, du = -\frac{1}{2} \cdot 2u^{1/2} = -\sqrt{1 - x^2}$$

and a similar result for the y integral. Then

$$-\sqrt{1 - x^2} - \sqrt{1 - y^2} = K.$$

If $y = 1/2$ when $x = 1/2$, then $K = -2\sqrt{3/4} = -\sqrt{3}$, so

$$\sqrt{1 - x^2} + \sqrt{1 - y^2} = \sqrt{3}.$$

6.   Factor the numerator and denominator of the fraction to get:

$$y' = \frac{dy}{dx} = \frac{x(2y^2+1)}{y(x^2-1)} \ .$$

Separate variables by multiplying by $\dfrac{y}{2y^2+1}$ and by dx:

$$\frac{y\,dy}{2y^2+1} = \frac{x\,dx}{x^2-1} \ .$$

Integrate both sides:

$$\tfrac{1}{4}\ell n(2y^2+1) = \tfrac{1}{2}\ell n|x^2-1| + C.$$

Multiply by 4 and combine the ℓn terms:

$$\ell n(2y^2+1) - 2\,\ell n|x^2-1| = 4C = \ell n\,\frac{2y^2+1}{(x^2-1)^2} \qquad \text{so}$$

$$\frac{2y^2+1}{(x^2-1)^2} = K \quad \text{or} \quad 2y^2+1 = K(x^2-1)^2.$$

If $y = 0$ when $x = \sqrt{2}$, then $K = 1$.

12.   We separate variables in $y' = \sqrt{1-y^2}$ and integrate to get

$$\int \frac{dy}{\sqrt{1-y^2}} = \int dx,$$

arc $\sin y = x + C$,

(A) $y = \sin(x+C).$

We also see that $y = 1$ (or $y = -1$) is a solution of $y' = \sqrt{1-y^2}$
since, if $y = 1$, then $y' = 0$ and $\sqrt{1-y^2} = 0$. But we can't get
the solution $y = 1$ from equation (A) since $y = \sin(x+C)$ depends
on x rather than being identically equal to 1, no matter how
we choose the constant C. The solutions $y = 1$ and $y = -1$ are

12.   (continued)

singular solutions not contained in the "general solution" (A).
(Read the problem in the text.) We sketch the solutions (A)
for several values of C. Observe that any sine curve of amplitude 1 is one of the solutions (A) and all of them are tangent

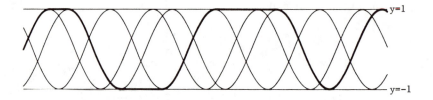

to the singular solutions $y = \pm 1$. Thus there are two solutions
passing through any point on $y = 1$ (the sine curve and the
straight line $y = 1$) and similarly for $y = -1$. Any continuous
curve made up of pieces of $y = 1$, $y = -1$, and the sine curves
satisfies the given differential equation. An example of such
a solution is indicated by the heavy curve in the sketch.

13.   Recall from Problem 2 above that we assumed $x^2 \neq 1$, $y^2 \neq 1$ in
order to separate variables. Look at the original differential
equation; if $y = 1$, then $dy = 0$ so both terms are zero and the
differential equation is satisfied. Similarly, the equation is
satisfied by $x = 1$, or $y = -1$, or $x = -1$. These are singular
solutions since they cannot be obtained from the "general
solution" in Problem 2 for any choice of K.

22.   Let $N(t)$ = number of bacteria at time t,

$\qquad R$ = constant removal rate,

$\qquad KN$ = rate of increase.

Then

(A) $\dfrac{dN}{dt} = KN - R.$

We separate variables and integrate:

$$\frac{dN}{KN - R} = dt,$$

$$K^{-1} \, \ell n \, |KN - R| = t + C.$$

If $N = N_0$ at $t = 0$, then $C = K^{-1} \, \ell n \, |KN_0 - R|$, so

(B) $\ell n \left| \dfrac{KN - R}{KN_0 - R} \right| = Kt.$

Now if $KN_0 - R > 0$, then N increases with time [see equation (A), above], so $KN - R > 0$. If $KN_0 - R < 0$ then N decreases with time and $KN - R < 0$. In either case the quantity $(KN - R)/(KN_0 - R)$ in equation (B) is positive. Thus we find

$$KN - R = (KN_0 - R) e^{Kt},$$

(C) $N = \dfrac{R}{K} + \left(N_0 - \dfrac{R}{K}\right) e^{Kt}.$

We see that if $KN_0 - R > 0$, the number of bacteria steadily increases. However, if $KN_0 - R < 0$, then N decreases and becomes zero when

$$R + (KN_0 - R) e^{Kt} = 0, \qquad t = K^{-1} \, \ell n \, \frac{R}{R - KN_0} \, .$$

(After this time the differential equation no longer applies;

22.   (continued)

since there are no bacteria left, neither growth nor removal

can occur.)

Further comment: If $KN_0 - R = 0$, then $N = N_0 = R/K$ satisfies

the differential equation as you can verify directly.   In this

case our method of solution is not valid (we divided by $KN - R = 0$)

but, since the equation is linear, the solution (C) includes

this case.

25.   Consider a block of ice of area A

and thickness x.   The rate of

formation of ice below a given

area A is $\frac{d}{dt}(Ax) = A\frac{dx}{dt}$.   This is

proportional to the rate at which heat flows through the area A

from the water to the air.   By text equation (1.1) this rate is

$$\frac{dQ}{dt} = kA\frac{dT}{dx} \qquad \text{where} \qquad \frac{dT}{dx} = \frac{20}{x}.$$

Thus

$$A\frac{dx}{dt} \propto \frac{dQ}{dt} = kA \cdot \frac{20}{x}$$

so

$$\frac{dx}{dt} = \frac{C}{x}$$

where C is a constant.   Integrate to get

$$\int x\,dx = C\int dt,$$

$$\frac{1}{2}x^2 = Ct + C'.$$

If $x = 0$ at $t = 0$, then $C' = 0$, and we have

$$x = \sqrt{2C}\,\sqrt{t} \quad \text{or} \quad x \propto \sqrt{t}.$$

27. Let $T(t)$ be the temperature of the coffee at time t. Then in a
     room at temperature $70°$,

$$\frac{dT}{dt} = k(T - 70).$$

Separate variables and integrate to get

$$\frac{dT}{T - 70} = k \, dt,$$

$$\ln(T - 70) = kt + C.$$

Use the given data to find k and C:

At $t = 0$, $T = 200$: $\ln(200 - 70) = C$.

At $t = 10$, $T = 100$: $\ln(100 - 70) = 10k + C$.

$C = \ln 130$, $k = \frac{1}{10}(\ln 30 - \ln 130) = \frac{1}{10} \ln \frac{3}{13}$.

We want the value of t when $T = 120$:

$$\ln(120 - 70) = kt + C = \frac{t}{10} \ln \frac{3}{13} + \ln 130,$$

$$\frac{t}{10} \ln \frac{3}{13} = \ln 50 - \ln 130 = \ln \frac{5}{13},$$

$$t = 10 \left(\ln \frac{5}{13} \right) \Big/ \left(\ln \frac{3}{13} \right) = 6.6 \text{ min.}$$

33. We are given the family of curves $y = kx^n$ (n fixed, k different
     for different curves of the family). Then

$$\frac{dy}{dx} = knx^{n-1} = n \frac{kx^n}{x} = n \frac{y}{x} .$$

Note that we have eliminated k so that we have a formula for
dy/dx which is correct for all curves of the family. Now at
each point the slope of the orthogonal trajectory curve is given
by

33.   (continued)

$$\frac{dy}{dx} = -\frac{x}{ny} \quad \text{or} \quad x\,dx + ny\,dy = 0.$$

We integrate this equation to find

$$\frac{1}{2}x^2 + n \cdot \frac{1}{2}y^2 = \text{const.},$$

$$x^2 + ny^2 = C.$$

Thus the orthogonal trajectories are ellipses if $n > 0$, and
hyperbolas if $n < 0$. If $n = 0$, the original curves are $y = k$
(that is, straight lines parallel to the x axis) so the orthogonal trajectories are $x = \text{const.}$ (that is, straight lines
parallel to the y axis).

Section 3

3.   Write the differential equation in the form of text equation
(3.1):

$$y' + 2xy = x\,e^{-x^2}.$$

Then $P = 2x$, $Q = x\,e^{-x^2}$. Following text equation (3.9), we find:

$$I = \int 2x\,dx = x^2, \qquad e^I = e^{x^2},$$

$$ye^{x^2} = \int x\,e^{-x^2}\,e^{x^2}\,dx = \int x\,dx = \frac{1}{2}x^2 + C,$$

$$y = \left(\frac{1}{2}x^2 + C\right)e^{-x^2}.$$

8. Divide the differential equation by $x \ln x$ to put it in the form
    of text equation (3.1):

    $$y' + \frac{1}{x \ln x} y = \frac{1}{x} .$$

    Then

    $$P = \frac{1}{x \ln x} , \qquad Q = \frac{1}{x} .$$

    By text equation (3.9):

    $$I = \int \frac{dx}{x \ln x} = \ln(\ln x),$$

    $$e^I = e^{\ln(\ln x)} = \ln x,$$

    $$y \ln x = \int \frac{1}{x} \ln x \, dx = \frac{1}{2}(\ln x)^2 + C,$$

    $$y = \frac{1}{2} \ln x + \frac{C}{\ln x} .$$

    To evaluate the integrals above, let $u = \ln x$.   Then $du = \frac{dx}{x}$,

    $$\int \frac{dx}{x \ln x} = \int \frac{du}{u} = \ln u = \ln(\ln x) \qquad \text{and}$$

    $$\int \frac{\ln x}{x} \, dx = \int u \, du = \frac{1}{2} u^2 = \frac{1}{2}(\ln x)^2 .$$

13. Write the differential equation as

    (A) $\frac{dx}{dy} + x = e^y$

    and compare it with the equation

    $$\frac{dx}{dy} + P(y)x = Q(y)$$

    which is text equation (3.1) with x and y interchanged. Thus (A)
    is a first-order linear differential equation for x as a function
    of y. We find

13.   (continued)

$$P(y) = 1, \quad Q(y) = e^y, \quad I = \int P \, dy = y, \quad e^I = e^y,$$

$$xe^y = \int e^y \, Q \, dy = \int e^{2y} \, dy = \frac{1}{2} e^{2y} + C,$$

$$x = \frac{1}{2} e^y + Ce^{-y}.$$

15.   First we find the volume of the lake at time t. At $t = 0$, the volume is 10^9 gal. Water is flowing in at 4×10^5 gal/hr and out at 10^5 gal/hr. Thus at time t hours,

$$V = (10^9 + 3 \times 10^5 t) \, \mathrm{gal} = 10^5 (10^4 + 3t) \, \mathrm{gal}.$$

Let $S = $ number of pounds of salt in lake at time t. Then

$$\frac{dS}{dt} = (\text{rate at which salt comes in}) - (\text{rate out})$$

$$= \left(\frac{5}{1000} \frac{\mathrm{lbs}}{\mathrm{gal}} \right) \left(4 \times 10^5 \frac{\mathrm{gal}}{\mathrm{hr}} \right) - \left(\frac{S}{10^5 (10^4 + 3t)} \frac{\mathrm{lbs}}{\mathrm{gal}} \right) \left(10^5 \frac{\mathrm{gal}}{\mathrm{hr}} \right)$$

$$= \left(2 \times 10^3 - \frac{S}{10^4 + 3t} \right) \frac{\mathrm{lbs}}{\mathrm{hr}},$$

$$\frac{dS}{dT} + \frac{S}{10^4 + 3t} = 2 \times 10^3.$$

This is a first-order linear equation. We use text equation (3.9) with y replaced by S and x replaced by t. Then

$$I = \int \frac{dt}{10^4 + 3t} = \frac{1}{3} \ell n (10^4 + 3t),$$

$$e^I = (10^4 + 3t)^{1/3},$$

$$Se^I = \int 2 \times 10^3 (10^4 + 3t)^{1/3} \, dt = \frac{2 \times 10^3}{3} (10^4 + 3t)^{4/3} \left(\frac{3}{4} \right) + C$$

$$= \frac{1}{2} \times 10^3 (10^4 + 3t)^{4/3} + C.$$

15.   (continued)

At $t = 0$, $S = 10^7$, so

$$C = 10^7(10^4)^{1/3} - \frac{1}{2} \times 10^3(10^4)^{4/3} = \frac{1}{2} \times 10^7 \times 10^{4/3}.$$

Thus

$$S = \frac{1}{2} \times 10^3(10^4 + 3t) + \frac{1}{2} \times 10^7 \times 10^{4/3}(10^4 + 3t)^{-1/3}$$

$$= \frac{1}{2} \times 10^7 \left[\left(1 + \frac{3t}{10^4}\right) + \left(1 + \frac{3t}{10^4}\right)^{-1/3} \right].$$

18.   For an RC circuit with $V = V_0 e^{i\omega t}$, text equation (1.3) is

$$R\frac{dI}{dt} + \frac{I}{C} = \frac{dV}{dt} = i\omega V_0 e^{i\omega t} \qquad \text{or}$$

$$\frac{dI}{dt} + \frac{I}{RC} = \frac{i\omega V_0}{R} e^{i\omega t}.$$

This is a linear first-order equation; we use text equation (3.9) with $y(x)$ replaced by $I(t)$. [Caution: Don't confuse $I(t)$ in this problem with the integral I in text equation (3.9) which becomes in this problem $\int \frac{dt}{RC} = t/(RC)$.] Then

$$I(t) e^{t/(RC)} = \frac{i\omega V_0}{R} \int e^{t/(RC)} e^{i\omega t} \, dt$$

$$= \frac{i\omega V_0}{R} \int e^{t[(RC)^{-1} + i\omega]} \, dt$$

$$= \frac{i\omega V_0}{R[(RC)^{-1} + i\omega]} e^{t[(RC)^{-1} + i\omega]} + A,$$

$$I(t) = Ae^{-t/(RC)} + i\omega V_0 C(1 + i\omega RC)^{-1} e^{i\omega t}.$$

We can now find the solution of Problem 17 by taking the real part of $I(t)$. For the second term in $I(t)$ we have

18.   (continued)

$$i(1 + i\omega RC)^{-1}e^{i\omega t} = \frac{i(1 - i\omega RC)(\cos \omega t + i \sin \omega t)}{1 + \omega^2 R^2 C^2}$$

and the real part of the numerator is

$\omega RC \cos \omega t - \sin \omega t.$

Then

$$Re \; I(t) = Ae^{-t/(RC)} + \frac{\omega V_0 C(\omega RC \cos \omega t - \sin \omega t)}{1 + \omega^2 R^2 C^2} \; .$$

21.   Hints and comments: We want to solve the set of differential equations:

$$\frac{dN_1}{dt} = -\lambda_1 N_1 \; ,$$

$$\frac{dN_2}{dt} = \lambda_1 N_1 - \lambda_2 N_2 \; ,$$

$$\frac{dN_3}{dt} = \lambda_2 N_2 - \lambda_3 N_3 \; ,$$

$$\cdots$$

$$\frac{dN_n}{dt} = \lambda_{n-1} N_{n-1} - \lambda_n N_n \; .$$

Assume that no two λ's are equal and that, at $t = 0$, $N_1 = N_0$, $N_2 = N_3 = \cdots = N_n = 0$. It is easy to show that each N is a linear combination of exponentials (see answer in text, page 761). Verifying the coefficients can be done by mathematical induction but the algebra is messy. A simpler method is to solve the equations using Laplace transforms and partial fractions -- see text Chapter 15, Section 3.

Section 4

1. $y' + y = xy^{2/3}$ is a Bernoulli equation. We follow text equations
     (4.1) and (4.2); here $n = 2/3$, $1 - n = 1/3$, so

     (A) $z = y^{1/3}$, $z' = \frac{1}{3} y^{-2/3} y'$.

     We multiply the differential equation by $\frac{1}{3} y^{-2/3}$ and use equations
     (A) to get

     $$\frac{1}{3} y^{-2/3} y' + \frac{1}{3} y^{1/3} = \frac{1}{3} x \quad \text{or} \quad z' + z/3 = x/3.$$

     This is a linear first order equation. By text equation (3.9):

     $$I = \int \frac{1}{3} dx = x/3,$$

     $$ze^{x/3} = \int \frac{x}{3} e^{x/3} dx = e^{x/3}(x - 3) + C \qquad \text{(by tables)},$$

     $$z = x - 3 + Ce^{-x/3},$$

     $$y^{1/3} = x - 3 + Ce^{-x/3}.$$

4. $(2xe^{3y} + e^x) dx + (3x^2 e^{3y} - y^2) dy = 0$ is an equation of the form
     $P\, dx + Q\, dy = 0$. We find

     $$\frac{\partial P}{\partial y} = 6x\, e^{3y}, \qquad \frac{\partial Q}{\partial x} = 6x\, e^{3y}.$$

     Thus the given equation is exact; that is, there is a function
     $F(x,y)$ such that $dF = P\, dx + Q\, dy$, or

     $$\frac{\partial F}{\partial x} = P = 2xe^{3y} + e^x,$$

     $$\frac{\partial F}{\partial y} = Q = 3x^2 e^{3y} - y^2.$$

     To find F we integrate $\frac{\partial F}{\partial x}$ partially with respect to x and deter-
     mine the integration "constant" (which may be a function of y)
     using $\frac{\partial F}{\partial y}$:

4.   (continued)

$$F = \int \frac{\partial F}{\partial x}\, dx + C(y) = \int (2xe^{3y} + e^{x})\, dx + C(y)$$

$$= x^2 e^{3y} + e^x + C(y),$$

$$\frac{\partial F}{\partial y} = 3x^2 e^{3y} + C'(y) = 3x^2 e^{3y} - y^2,$$

$$C(y) = -y^3 / 3.$$

Since the given differential equation is $dF = 0$, the solution is $F = \text{const.}$:

$$F = x^2 e^{3y} + e^x - \frac{y^3}{3} = \text{const.}$$

You might like to verify that you get the same solution by finding $\int \frac{\partial F}{\partial y}\, dy + K(x)$.

9. $xy\, dx + (y^2 - x^2)\, dy = 0$ is a homogeneous equation.   By text equation (4.12) we make the change of variables

$$y = vx, \qquad dy = v\, dx + x\, dv.$$

Then the differential equation becomes

$$xvx\, dx + (v^2 x^2 - x^2)(v\, dx + x\, dv) = 0 \qquad \text{or}$$

$$x^2 (v + v^3 - v)\, dx + x^2 (v^2 - 1)x\, dv = 0,$$

$$v^3 dx + (v^2 - 1)x\, dv = 0.$$

This is a separable equation:

$$\frac{dx}{x} + \frac{(v^2 - 1)\, dv}{v^3} = 0,$$

$$\int \frac{dx}{x} + \int \frac{dv}{v} - \int \frac{dv}{v^3} = \ln|x| + \ln|v| - \frac{v^{-2}}{-2} = C,$$

$$\ln|xv| = C - \frac{1}{2v^2}.$$

9.   (continued)

Put back $y = vx$, $v = y/x$, and multiply by 2:

$$2 \ln |y| = 2C - \frac{x^2}{y^2},$$

$$y^2 = e^{2C} e^{-x^2/y^2} \qquad \text{or} \qquad y^2 = Ae^{-x^2/y^2}.$$

Another way of solving this equation is to write it in the form

$$\frac{dx}{dy} + \frac{y^2 - x^2}{xy} = 0 \qquad \text{or} \qquad \frac{dx}{dy} - \frac{x}{y} = \frac{-y}{x} = -yx^{-1}$$

which is a Bernoulli equation for x as a function of y. Let

$$z = x^{1-n} = x^2, \qquad z' = 2xx'.$$

Multiply the differential equation by 2x and solve for z as a function of y:

$$2xx' - \frac{2x^2}{y} = -2y \qquad \text{or} \qquad z' - \frac{2z}{y} = -2y,$$

$$I = \int \frac{-2}{y} \, dy = -2 \ln |y| = -\ln y^2, \qquad e^I = \frac{1}{y^2},$$

$$\frac{z}{y^2} = -\int \frac{1}{y^2} 2y \, dy = -\int \frac{2 \, dy}{y} = -2 \ln |y| = -\ln y^2 + C,$$

$$\frac{x^2}{y^2} = -\ln y^2 + C \qquad \text{or} \qquad \ln y^2 = C - \frac{x^2}{y^2},$$

$$y^2 = Ae^{-x^2/y^2}.$$

11.   Given $y' = \cos(x + y)$, we let $u = x + y$, $u' = 1 + y'$; then

$$u' - 1 = y' = \cos u \qquad \text{or} \qquad \frac{du}{dx} = 1 + \cos u.$$

This is a separable equation:

$$\int \frac{du}{1 + \cos u} = \int dx + C,$$

$$\tan \frac{u}{2} = x + C \qquad \text{(by integral tables)},$$

$$\tan \frac{x + y}{2} = x + C.$$

18.   Square $(x - h)$ and simplify the given equation:

$$(x - h)^2 + y^2 = x^2 - 2xh + h^2 + y^2 = h^2 \qquad \text{or}$$

(A) $x^2 - 2xh + y^2 = 0.$

Differentiate equation (A) implicitly with respect to x:

(B) $2x - 2h + 2yy' = 0.$

Now (B) gives the slope y' of the tangent to <u>one</u> circle of the
given family at a point (x,y) on that circle. (See text, page
343.) We want a formula for the slope, at any point (x,y), of
the tangent to the circle which passes through that point. Thus
we eliminate h between (A) and (B). Multiply (B) by x and
subtract (A):

$$x^2 + 2xyy' - y^2 = 0,$$

$$y' = (y^2 - x^2)/(2xy) \qquad \text{(given circles)}.$$

This gives the slopes of curves of the given family of circles.
The slopes of the orthogonal trajectories are the negative
reciprocals of these slopes:

$$y' = -2xy/(y^2 - x^2) \qquad \text{or}$$

(C) $2xy\,dx + (y^2 - x^2)dy = 0 \qquad \text{(orthogonal trajectories)}.$

We can solve equation (C) in several ways:

(1) It is homogeneous (text, page 351); let $y = vx$.

(2) Write it as

$$x' - \frac{1}{2y}x = -\frac{y}{2}x^{-1}$$

to see that it is a Bernoulli equation for x as a function
of y (text, page 350); let $z = x^2$ and multiply the
equation by 2x.

18.   (continued)

(3) $\dfrac{1}{y^2}$ is an integrating factor (text, page 350), that is,

$$\frac{2xy \, dx}{y^2} + \frac{y^2 - x^2}{y^2} \, dy = \frac{2x}{y} \, dx + \left(1 - \frac{x^2}{y^2}\right) dy = 0$$

is an exact equation. Integrating, we find

$$\frac{x^2}{y} + y = \text{const.} \qquad \text{or} \qquad x^2 + y^2 = Cy.$$

If we put $C = 2k$, we can write this as

$$x^2 + (y - k)^2 = k^2.$$

Thus we see that the orthogonal trajectories are circles with centers on the y axis just as the original family consisted of circles with centers on the x axis. All circles of both families pass through the origin. The sketch shows a few circles of each family. Note that the circles always cross at right angles.

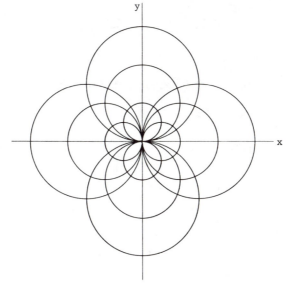

18.   (continued)

Although (3) is probably the easiest way to solve the equation, finding integrating factors is a matter of luck plus experience, so let's look briefly at methods (1) and (2).

(1) Let $y = vx$ in equation (C):

$$2xvx\, dx + (v^2 x^2 - x^2)(v\, dx + x\, dv) = 0,$$

$$(v + v^3)\, dx + (v^2 - 1) x\, dv = 0,$$

$$\frac{dx}{x} + \frac{(v^2 - 1)\, dv}{v(v^2 + 1)} = 0,$$

Write

$$\frac{v^2 - 1}{v(v^2 + 1)} = \frac{2v^2 - (v^2 + 1)}{v(v^2 + 1)} = \frac{2v}{v^2 + 1} - \frac{1}{v};$$

then

$$\int \frac{dx}{x} + \int \frac{2v\, dv}{v^2 + 1} - \int \frac{dv}{v} = \ell n\, |x| + \ell n\, |v^2 + 1| - \ell n\, |v| = \text{const.}$$

Put back $v = y/x$:

$$\ell n\, \left| \frac{x(v^2 + 1)}{v} \right| = \ell n\, \frac{y^2 + x^2}{|y|} + \text{const.},$$

$$x^2 + y^2 = Cy \qquad \text{as above.}$$

(2) Following the method outlined above, we have

$$2xx' - \frac{1}{y} x^2 = -y,$$

$$z' - \frac{1}{y} z = -y,$$

$$I = -\int \frac{dy}{y} = -\ell n\, y, \qquad e^I = \frac{1}{y},$$

$$\frac{z}{y} = -\int \frac{y\, dy}{y} = -y + C,$$

$$z = -y^2 + Cy \qquad \text{or} \qquad x^2 + y^2 - Cy = 0 \qquad \text{as before.}$$

Section 5

Note: In Section 6 you will find solutions of many more differential
equations like the ones here in Section 5 since the first step in
Section 6 problems is to solve the homogeneous equation.

1. We write the given differential equation as

$$(D^2 + D - 2)y = 0$$

and solve the auxiliary equation by factoring.

$$D^2 + D - 2 = 0, \qquad (D - 1)(D + 2) = 0,$$

$$D = 1, \ -2.$$

Thus, by text equation (5.11), the solution is

$$y = Ae^x + Be^{-2x}.$$

5. We factor the auxiliary equation to get

$$D^2 - 2D + 1 = (D - 1)^2 = 0.$$

Since the roots of the auxiliary equation are equal, the solution
is [see text equation (5.15) and Example 2, text page 355]

$$y = (Ax + B)e^x.$$

9. $(D^2 - 4D + 13)y = 0.$

Here the auxiliary equation does not factor using real numbers,
so we solve for D using the quadratic formula.

$$D = \frac{4 \pm \sqrt{16 - 52}}{2} = 2 \pm 3i.$$

Then by text equations (5.16), (5.17), and (5.18), the solution
may be written in any one of the three forms:

9.   (continued)

$$y = e^{2x}(Ae^{3ix} + Be^{-3ix}) \qquad \text{or}$$

$$y = e^{2x}(c_1 \sin 3x + c_2 \cos 3x) \qquad \text{or}$$

$$y = ce^{2x}\sin(3x + \gamma).$$

18.   We write and evaluate the Wronskian of the given functions (see text, page 131).

$$W = \begin{vmatrix} e^{ax} & xe^{ax} & x^2 e^{ax} \\ ae^{ax} & xae^{ax} + e^{ax} & x^2 ae^{ax} + 2xe^{ax} \\ a^2 e^{ax} & xa^2 e^{ax} + 2ae^{ax} & x^2 a^2 e^{ax} + 4xae^{ax} + 2e^{ax} \end{vmatrix}$$

Factor out e^{ax} from each row:

$$W = (e^{ax})^3 \begin{vmatrix} 1 & x & x^2 \\ a & ax+1 & x(ax+2) \\ a^2 & a(ax+2) & a^2 x^2 + 4ax + 2 \end{vmatrix}$$

Now row reduce (see text, Chapter 3, Sections 2 and 3): R2 - a R1, and R3 - a^2R1.

$$W = e^{3ax} \begin{vmatrix} 1 & x & x^2 \\ 0 & 1 & 2x \\ 0 & 2a & 4ax+2 \end{vmatrix}$$

$$= e^{3ax} \begin{vmatrix} 1 & 2x \\ 2a & 4ax+2 \end{vmatrix} = 2e^{3ax} \neq 0.$$

Since $W \neq 0$, the functions are linearly independent.

20.   Hint: See Chapter 2, Problem 10.18.

21.   Hints: Show that the solution of, say,

$$(D - a)(D - b)(D - c) \cdots (D - n)y = 0$$

is the sum of the solutions of

$$(D - a)y = 0, \quad (D - b)y = 0, \quad \cdots \quad (D - n)y = 0$$

if $a, b, \cdots, n$ are all different (see text, page 353). For the cases of equal roots, continue the method on text pages 354-355 to show that the general solution of $(D - a)(D - b)^2 y = 0$ is $y = Ae^{ax} + (Bx + C)e^{bx}$ and the general solution of $(D - a)^3 y = 0$ is $y = (Ax^2 + Bx + C)e^{ax}$.

23.   We solve the auxiliary equation by factoring it.

$$(D^2 + 1)(D^2 - 1) = (D + i)(D - i)(D + 1)(D - 1) = 0,$$

$$D = 1, \ -1, \ i, \ -i.$$

Then by Problem 21 above, the solution is

$$y = Ae^{x} + Be^{-x} + Ce^{ix} + De^{-ix}.$$

The last two terms may also be written as either

$$c_1 \sin x + c_2 \cos x \qquad \text{or} \qquad E \sin(x + \gamma).$$

24.   We write the differential equation as

$$(D^3 + 1)y = 0$$

and factor the auxiliary equation to get

$$D^3 + 1 = (D + 1)(D^2 - D + 1) = 0.$$

Then the values of D are -1 and (by quadratic formula)

$$D = \frac{1 \pm \sqrt{1 - 4}}{2} = \frac{1}{2}(1 \pm i\sqrt{3}).$$

Thus the solution of the differential equation is

$$y = Ae^{-x} + Be^{(1+i\sqrt{3})x/2} + Ce^{(1-i\sqrt{3})x/2}$$

$$= Ae^{-x} + Ee^{x/2}\sin\left(\frac{1}{2}x\sqrt{3} + \gamma\right).$$

29.   We factor the auxiliary equation to get

$$(D+1)^2(D-2)(D+2)(D-2i)(D+2i) = 0,$$

$$D = -1, -1, 2, -2, 2i, -2i.$$

Then by Problem 21, the solution is

$$y = (A+Bx)e^{-x} + Ce^{2x} + Ee^{-2x} + F\sin(2x+\gamma).$$

34.   Hint: See Problem 7.13.

35.   Hint: You should find a period of about 84 min, that is, 42 min through the earth and 42 min back.

39.   We first find the constants in text equations (5.30), (5.31), and (5.32) to make $y=1$ and $dy/dt=0$ at $t=0$.

$$(5.30) \quad y = Ae^{-\lambda t} + Be^{-\mu t}, \quad \begin{cases} \lambda = b + \sqrt{b^2 - \omega^2}, \\ \\ \mu = b - \sqrt{b^2 - \omega^2}. \end{cases}$$

$$1 = A + B,$$

$$0 = -\lambda A - \mu B,$$

$$A = -\mu/(\lambda - \mu), \qquad B = \lambda/(\lambda - \mu),$$

$$y = \frac{1}{\lambda - \mu}(-\mu e^{-\lambda t} + \lambda e^{-\mu t}).$$

$$(5.31) \quad y = (A+Bt)e^{-bt} = (A+Bt)e^{-\omega t} \qquad \text{since } b = \omega \text{ in this case.}$$

$$1 = A, \qquad 0 = A(-\omega) + B, \qquad \text{so} \qquad B = \omega.$$

$$y = (1 + \omega t)e^{-\omega t}.$$

$$(5.32) \quad y = ce^{-bt}\sin(\beta t + \gamma), \qquad \beta^2 = \omega^2 - b^2,$$

$$1 = c\sin\gamma,$$

$$0 = c[-b\sin\gamma + \beta\cos\gamma], \qquad \text{so} \qquad \tan\gamma = \frac{\beta}{b}.$$

39.   (continued)

Using the equations $\beta^2 = \omega^2 - b^2$ and $\tan \gamma = \frac{\beta}{b}$, we sketch a

right triangle showing the

relations among γ, β, ω,

and b.   Then

$$c = \frac{1}{\sin \gamma} = \frac{\omega}{\beta},$$

$$y = \frac{\omega}{\beta} e^{-bt} \sin(\beta t + \gamma), \text{ where } \tan \gamma = \frac{\beta}{b}.$$

We want to compare these three solutions as t increases.   Note

that ω is fixed; the kind of damping is determined by b (see

text, page 357).   In all these cases, $y \to 0$ as $t \to \infty$ because of

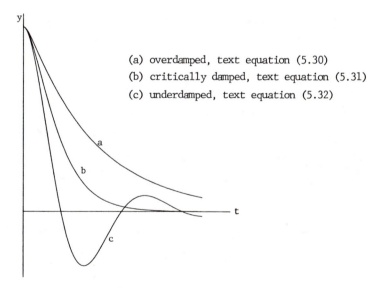

(a) overdamped, text equation (5.30)

(b) critically damped, text equation (5.31)

(c) underdamped, text equation (5.32)

39.   (continued)

the decreasing exponentials. It is clear from the figure,
however, that y in the critically damped case tends to zero
more rapidly than the other two. We can also see this by
comparing the exponents in the three cases; an exponential
e^{-at} tends to zero more rapidly for large $\underline{a}$ than for small
$\underline{a}$ $(a > 0)$. In (5.30), we show that $\mu < \omega$ as follows:

$$\mu - \omega = b - \omega - \sqrt{b^2 - \omega^2} = \sqrt{b - \omega}(\sqrt{b - \omega} - \sqrt{b + \omega}) < 0, \qquad \text{so} \qquad \mu < \omega.$$

In (5.32), we have $b < \omega$. Thus in both (5.30) and (5.32) we
have an exponential which decreases less rapidly than $e^{-\omega t}$
in (5.31). [The term $e^{-\lambda t}$ in (5.30) does decrease rapidly,
but we are concerned with the sum, which is kept from decreasing rapidly by the $e^{-\mu t}$ term.]

Section 6

In each of the following problems, we first solve the corresponding
homogeneous equation as in Section 5 in order to find the complementary
function y_c .   Then we choose an appropriate method to find a particular solution y_p .   The general solution is $y = y_c + y_p$.

3. $y'' + y' - 2y = e^{2x}$.

First solve the homogeneous equation as in Section 5.

$$D^2 + D - 2 = (D + 2)(D - 1) = 0, \qquad D = -2, \ 1.$$

$$y_c = Ae^{-2x} + Be^x.$$

To find y_p ,   we observe that the right-hand side, namely e^{2x},
is different from both the terms in y_c .   This is the case

3.   (continued)

c $\neq$ a, c $\neq$ b in text equation (6.18), so we assume a solution

y = Ce^{2x}.   Substituting this into the differential equation gives

$$y'' + y' - 2y = 4Ce^{2x} + 2Ce^{2x} - 2Ce^{2x} = 4Ce^{2x} \equiv e^{2x}.$$

This must be an identity (see text, page 339) so we have $4C = 1$,

$C = \frac{1}{4}$.   Then

$$y_p = \frac{1}{4} e^{2x}$$

and the general solution is

$$y = y_c + y_p = Ae^{-2x} + Be^x + \frac{1}{4} e^{2x}.$$

9. $(D^2 + 2D + 1)y = 2e^{-x}.$

First find y_c:

$$(D^2 + 2D + 1) = (D + 1)^2 = 0, \qquad D = -1, \; -1.$$

$$y_c = (Ax + B)e^{-x}. \qquad \text{[See text equation (5.15).]}$$

We use text equation (6.18) to find y_p .   In this problem,

c = a = b = -1, so we assume the solution $y = Cx^2 e^{-x}$ and substitute

this into the given differential equation.

$$y' = -Cx^2 e^{-x} + 2Cxe^{-x},$$
$$y'' = Cx^2 e^{-x} - 4Cxe^{-x} + 2Ce^{-x},$$
$$y'' + 2y' + y = 2Ce^{-x} \equiv 2e^{-x}, \qquad \text{so} \qquad C = 1.$$
$$y_p = x^2 e^{-x}.$$

Then the general solution is

$$y = y_c + y_p = (Ax + B)e^{-x} + x^2 e^{-x}.$$

13. (A) $(D^2 - 2D + 1)y = 2 \cos x.$

First find y_c.   We have $(D - 1)^2 = 0$ so by text equation (5.15)

$$y_c = (Ax + B)e^x.$$

To find a particular solution, we use the method of complex exponentials, text page 365.   First solve

(B) $Y'' - 2Y' + Y = 2e^{ix}.$

We assume the solution [text equation (6.18) with $c = i \neq a = 1$]

$$Y = Ce^{ix}.$$

Then

$$Y'' - 2Y' + Y = -Ce^{ix} - 2iCe^{ix} + Ce^{ix} \equiv 2e^{ix},$$

$$C = i,$$

(C) $Y = ie^{ix} = i(\cos x + i \sin x) = i \cos x - \sin x.$

Now the real part of equation (B) is equation (A), so the real part of equation (C) is a particular solution of equation (A). Thus

$$y_p = - \sin x$$

and the general solution of equation (A) is

$$y = y_c + y_p = (Ax + B)e^x - \sin x.$$

17. $(D^2 + 16)y = 16 \cos 4x.$

We first find $D = \pm 4i$, so

$$y_p = Ae^{4ix} + Be^{-4ix}$$

(or either of the other forms on text page 355).   Next we find a particular solution of

$$Y'' + 16Y = 16e^{4ix}.$$

17.   (continued)

Using text equation (6.18) with $c = a = 4i \neq b$, we assume

$$Y = Cxe^{4ix}.$$

Then

$$Y'' + 16Y = -16Cxe^{4ix} + 8iCe^{4ix} + 16Cxe^{4ix}$$

$$\equiv 16\, e^{4ix},$$

$$C = -2i,$$

$$Y = -2ixe^{4ix} = -2ix(\cos 4x + i \sin 4x)$$

$$= -2ix \cos 4x + 2x \sin 4x.$$

Now the real part of the Y equation is the y equation so $y_p = \operatorname{Re} Y$.

$$y_p = 2x \sin 4x.$$

The general solution is

$$y = Ae^{4ix} + Be^{-4ix} + 2x \sin 4x \qquad \text{or}$$

$$y = c_1 \cos 4x + c_2 \sin 4x + 2x \sin 4x.$$

18. $(D^2 + 2D + 17)y = 60e^{-4x} \sin 5x.$

We solve $D^2 + 2D + 17 = 0$ by the quadratic formula to get

$$D = \frac{-2 \pm \sqrt{4 - 4 \cdot 17}}{2} = -1 \pm 4i.$$

Thus

$$y_c = Ae^{(-1+4i)x} + Be^{(-1-4i)x}.$$

Next we solve

$$Y'' + 2Y' + 17Y = 60e^{(-4+5i)x}.$$

(The imaginary part of Y will be our y_p.)   By text equation

(6.18), with $c = -4 + 5i \neq a$ or b, we assume

$$Y = Ce^{(-4+5i)x}.$$

Substitute this into the Y equation to find C.

18.   (continued)

$$Y'' + 2Y' + 17Y = [(-4+5i)^2 + 2(-4+5i) + 17]Ce^{(-4+5i)x}$$
$$= (16 - 40i - 25 - 8 + 10i + 17)Ce^{(-4+5i)x}$$
$$= -30iCe^{(-4+5i)x} \equiv 60e^{(-4+5i)x},$$

$C = 2i,$

$$Y = 2ie^{(-4+5i)x} = 2ie^{-4x}(\cos 5x + i \sin 5x)$$
$$= 2e^{-4x}(i \cos 5x - \sin 5x),$$

$$y_p = \text{Im } Y = 2e^{-4x} \cos 5x.$$

Thus the general solution of the y equation is

$$y = Ae^{(-1+4i)x} + Be^{(-1-4i)x} + 2e^{-4x} \cos 5x \qquad \text{or}$$
$$y = e^{-x}(c_1 \cos 4x + c_2 \sin 4x) + 2e^{-4x} \cos 5x.$$

22. $2y'' + y' = 2x.$

We have $2D^2 + D = 0, \qquad D = 0, \ -\frac{1}{2}, \qquad$ so

$$y_c = c_1 + c_2 e^{-x/2}.$$

Next we use text equation (6.24). We see that $c = 0$ (the right-hand side is just a polynomial); $c = a \neq b$ since we found $D = 0, \ -\frac{1}{2}$; and $n = 1$. Thus we assume

$$y_p = x(A + Bx) = Ax + Bx^2.$$

Then

$$2y_p'' + y_p' = 4B + A + 2Bx \equiv 2x,$$
$$4B + A = 0 \qquad \text{and} \qquad 2B = 2 \qquad \text{so}$$
$$B = 1, \qquad A = -4,$$
$$y_p = x^2 - 4x.$$

The general solution is

$$y = c_1 + c_2 e^{-x/2} + x^2 - 4x.$$

22.   (continued)

It is interesting to note that we could also solve this equation by integrating once to get $2y' + y = x^2 + C$, and then solve the first order linear equation (text, Section 3). You might like to try it, but you may decide that the necessary integration of $x^2 e^{x/2}$ is no easier than the method above!

25. $(D - 3)(D + 1)y = 16x^2 e^{-x}$.

We see that $D = 3$, -1, so $y_c = c_1 e^{3x} + c_2 e^{-x}$. In text equation (6.24), we have $c = -1 = a \neq b$ and $n = 2$. Thus we assume
$$y_p = xe^{-x}(A + Bx + Cx^2) = e^{-x}(Ax + Bx^2 + Cx^3).$$
We substitute this into the differential equation and simplify the algebra to get
$$y'' - 2y' - 3y = e^{-x}[(2B - 4A) + (6C - 8B)x - 12Cx^2] \equiv 16x^2 e^{-x},$$
$$C = -\frac{4}{3}, \qquad B = -1, \qquad A = -\frac{1}{2}.$$

Then
$$y = c_1 e^{3x} + c_2 e^{-x} - \left(\frac{1}{2}x + x^2 + \frac{4}{3}x^3\right)e^{-x}.$$
Comment: See Problem 31 for a method of simplifying the algebra in this problem.

30.   (c) Let us find y_p for Problem 9 using an inverse operator. We have $(D + 1)^2 y = 2e^{-x}$ or
$$y_p = \frac{1}{(D + 1)^2} 2e^{-x}.$$
Since -1 is a double root of the auxiliary equation, we use part 5 of (b) to get

---

30.   (continued)
$$y_p = 2 \cdot \frac{1}{2} x^2 e^{-x} = x^2 e^{-x}$$

as we found before in the solution of Problem 9.

   In Problem 17, we use parts 1 and 4 of (b) to find

$$Y_p = 16 \frac{1}{D - 4i} \frac{1}{D + 4i} e^{4ix} = 16 \cdot \frac{1}{D - 4i} \frac{e^{4ix}}{8i}$$

$$= \frac{16}{8i} x e^{4ix} = -2ixe^{4ix}$$

as we had before.

31.   (d) Let us solve Problem 25 using the inverse exponential shift.
We multiply both sides of the equation by e^x and use (c) to get

$$e^x (D - 3) (D + 1) y = (D - 1 - 3) (D - 1 + 1) e^x y$$
$$= (D - 4) (D) e^x y = 16x^2 .$$

Call $e^x y = u$; then

$$u'' - 4u' = 16x^2 .$$

By text equation (6.24) with $c = a = 0 \neq b$, and $n = 2$, we assume

$$u = x (A + Bx + Cx^2) = Ax + Bx^2 + Cx^3 .$$

Then

$$u'' - 4u' = 2B + 6Cx - 4 (A + 2Bx + 3Cx^2) \equiv 16x^2 .$$

We find the same values of A, B, C that we found in the solution
of Problem 25 but with less algebra.   Having found u, then
$y = e^{-x} u$ and we have the same solution as before.

34. $y'' - 5y' + 6y = 2e^x + 6x - 5$.

   $(D - 3) (D - 2) = 0; \qquad y_c = c_1 e^{3x} + c_2 e^{2x} .$

To find y_p we solve the two differential equations corresponding
to the exponentials e^x and e^0 on the right-hand side, namely

34.   (continued)

(a) $y'' - 5y' + 6y = 2e^x$.

(b) $y'' - 5y' + 6y = 6x - 5$.

(a) Assume $y_p = Ce^x$ by text equation (6.18). Then

$$y'' - 5y' + 6y = Ce^x - 5Ce^x + 6Ce^x = 2Ce^x \equiv 2e^x,$$

$$C = 1, \quad y_p = e^x.$$

(b) Assume $y_p = Ax + B$ using text equation (6.24) with $c = 0$.

$$y'' - 5y' + 6y = -5A + 6Ax + 6B \equiv 6x - 5,$$

$$-5A + 6B = -5, \quad 6A = 6, \quad \text{so} \quad A = 1, \quad B = 0,$$

$$y_p = x.$$

Then the general solution is

$$y = c_1 e^{3x} + c_2 e^{2x} + e^x + x.$$

35. $(D^2 - 1)y = \sinh x = \frac{1}{2}(e^x - e^{-x})$.   (See text page 69 for discussion of $\sinh x$.) $(D - 1)(D + 1) = 0$; $y_c = c_1 e^x + c_2 e^{-x}$.

To find y_p we solve the two differential equations

(a) $y'' - y = \frac{1}{2} e^x$

(b) $y'' - y = -\frac{1}{2} e^{-x}$

(a)   Using text equation (6.18) with $c = a \neq b$, we assume $y = Cxe^x$. Then

$$y'' - y = 2Ce^x \equiv \frac{1}{2} e^x, \qquad C = \frac{1}{4},$$

$$y_p = \frac{1}{4} xe^x.$$

35.   (continued)

(b)   Similarly assume $y = Cxe^{-x}$ and find

$$y'' - y = -2Ce^{-x} \equiv -\frac{1}{2}e^{-x}, \qquad C = \frac{1}{4},$$

$$y_p = \frac{1}{4}xe^{-x}.$$

Then the general solution is

$$y = c_1 e^x + c_2 e^{-x} + \frac{1}{4}x(e^x + e^{-x}).$$

Now

$$\frac{1}{4}(e^x + e^{-x}) = \frac{1}{2}\frac{e^x + e^{-x}}{2} = \frac{1}{2}\cosh x.$$

Also

$$e^x = \sinh x + \cosh x, \qquad \text{and} \qquad e^{-x} = \sinh x - \cosh x$$

so a linear combination of e^x and e^{-x} is equivalent to a linear combination of $\sinh x$ and $\cosh x$. Thus the solution may also be written in the form

$$y = A\cosh x + B\sinh x + \frac{1}{2}x\cosh x.$$

36.   Hint: Although this problem looks as if it should be done by superposition (and can be done that way) it is actually simpler to write

$$2\sin x + 4x\cos x = \text{Re}(-2i + 4x)e^{ix}$$

and solve the equation (see text, page 365)

$$Y'' + Y = (-2i + 4x)e^{ix}.$$

Then $y_p = \text{Re } Y$.

41.   See text, Chapter 7, Problem 12.7.

Section 7

1. Since the independent variable x is missing, we use text
    equation (7.3) to get

    $$p \frac{dp}{dy} + yp = 0, \quad \text{or} \quad p\left(\frac{dp}{dy} + y\right) = 0.$$

    Thus either (a) $p = 0$, or (b) $\frac{dp}{dy} + y = 0$.

    (a) From $p = 0 = \frac{dy}{dx}$, we get the solution $y = C$.

    (b) From $\frac{dp}{dy} + y = 0$, we get by integration

    $$p + \frac{y^2}{2} = \text{const.}, \quad \text{or} \quad \frac{dy}{dx} = \text{const.} - \frac{y^2}{2}.$$

    There are three cases: (i) const. $= 0$, (ii) const. > 0,
    and (iii) const. < 0.

    (i)   const. $= 0$: $\frac{dy}{dx} = -\frac{y^2}{2}$ or $y^{-2} dy = -\frac{1}{2} dx$.

           Integrate to get

           $$-y^{-1} = -\frac{1}{2}(x + a),$$

           $$y(x + a) = 2.$$

    (ii)   const. $= \frac{K^2}{2}$: $\frac{dy}{dx} = \frac{K^2 - y^2}{2}$ or $\frac{2\,dy}{K^2 - y^2} = dx$.

           Integrate to get:

           $$x + a = 2 \int \frac{dy}{K^2 - y^2} = \frac{2}{K} \tanh^{-1} \frac{y}{K},$$

           $$y = K \tanh \frac{K}{2}(x + a), \quad \text{or with } K = 2A \text{ and } Ka/2 = B,$$

           $$y = 2A \tanh(Ax + B).$$

    (iii) const. $= \frac{-K^2}{2}$: $\frac{dy}{dx} = -\frac{K^2 + y^2}{2}$ or $\frac{2\,dy}{K^2 + y^2} = -dx$.

           Integrate to get:

1.   (continued)

$$-x + a = 2\int \frac{dy}{K^2 + y^2} = \frac{2}{K} \tan^{-1} \frac{y}{K},$$

$$y = K \tan \frac{K}{2}(-x + a) \quad \text{or with } K = -2A \text{ and } Ka/2 = B,$$

$$y = -2A \tan (Ax + B).$$

Thus we find four solutions of the given (nonlinear) equation:

$$y = C, \quad y(x + a) = 2, \quad y = 2A \tanh(Ax + B), \quad y = -2A \tan(Ax + B)$$

2.   Here the dependent variable y is missing so we use text
equation (7.2) to get

$$p' + 2xp = 0.$$

This equation is separable:

$$\frac{dp}{p} = -2x \, dx,$$

$$\ell n \, p = -x^2 + C,$$

$$p = e^{-x^2 + C} = A e^{-x^2}.$$

We now substitute back $p = \frac{dy}{dx}$ and integrate again.

$$\frac{dy}{dx} = p = A e^{-x^2},$$

(1) $$y = A \int e^{-x^2} dx + \text{const.}$$

This indefinite integral cannot be evaluated in terms of elementary functions. It is however tabulated as a function of the
upper limit and called the _error_ _function_. By definition (see
text, page 467)

$$\text{erf}(x) = \frac{2}{\sqrt{\pi}} \int_0^x e^{-t^2} dt.$$

In terms of the error function, we can write the solution (1) as

$$y = c_1 \text{erf}(x) + c_2.$$

8. Following method (c) on text page 377, multiply both sides of
the given differential equation by $\frac{dr}{dt}$:

(1) $m \frac{dr}{dt} \frac{d^2r}{dt^2} = \frac{-mgR^2}{r^2} \frac{dr}{dt}$.

We can integrate this with respect to t to get

(2) $\frac{1}{2} m \left(\frac{dr}{dt}\right)^2 = \frac{mgR^2}{r} + \text{const.}$

However, you may prefer to substitute $v = \frac{dr}{dt}$ and $\frac{dv}{dt} = \frac{d^2r}{dt^2}$ on the
left-hand side of equation (1) to get

(3) $mv \frac{dv}{dt} = \frac{-mgR^2}{r^2} \frac{dr}{dt}$ or $v \, dv = -\frac{gR^2}{r^2} dr$

Integrating (3) gives

(4) $\frac{1}{2} v^2 = \frac{gR^2}{r} + \text{const.}$

as in (2). Now we are given $v = v_0$ when $r = R$, so the constant
in (4) is

$$\frac{1}{2} v_0^2 - \frac{gR^2}{R} = \frac{1}{2} v_0^2 - gR$$

and we have

$$v^2 = \frac{2gR^2}{r} + v_0^2 - 2gR.$$

The mass m reaches its highest point $r = r_{max}$ when $v = 0$.

$$0 = \frac{2gR^2}{r_{max}} + v_0^2 - 2gR \quad \text{or} \quad r_{max} = \frac{2gR^2}{2gR - v_0^2} .$$

If the denominator of r_{max} is zero, then the mass m never returns
to earth. The corresponding value of v_0 is called the escape
velocity. We have

escape velocity = $\sqrt{2gR}$.

13.   Hints: See text, page 466. Dots over letters mean time
       derivatives (text, page 397).

16.   This is an Euler or Cauchy equation. Following text method (d),
       we let $x = e^z$ or $z = \ln x$. Then by text equation (7.19), the
       given differential equation

       (1) $x^2 \dfrac{d^2y}{dx^2} + x \dfrac{dy}{dx} + y = x$

       becomes

       (2) $\dfrac{d^2y}{dz^2} - \dfrac{dy}{dz} + \dfrac{dy}{dz} + y = e^z$ or $D^2 y + y = e^z$,

       where D now means a derivative with respect to z. This is a
       linear equation with constant coefficients which we solve as
       in text Sections 5 and 6.

       $$D^2 + 1 = 0, \qquad D = \pm i,$$

       $$y_c = A \cos z + B \sin z.$$

By text equation (6.18), a particular solution is $y_p = Ce^z$; we
substitute this into the differential equation (2) to find
$2Ce^z \equiv e^z$, so $C = 1/2$. Thus

       $$y = y_c + y_p = A \cos z + B \sin z + \tfrac{1}{2} e^z.$$

Putting back $x = e^z$ or $z = \ln x$ gives the solution of equation (1):

       $$y = A \cos(\ln x) + B \sin(\ln x) + \tfrac{1}{2} x.$$

21.   We want to solve the differential equations

(1) $r \dfrac{d}{dr}\left(r \dfrac{dR}{dr}\right) = n^2 R.$

(2) $\dfrac{d}{dr}\left(r^2 \dfrac{dR}{dr}\right) = \ell(\ell+1)R$ or $r^2 \dfrac{d^2R}{dr^2} + 2r \dfrac{dR}{dr} = \ell(\ell+1)R.$

(1)   With $r = e^z$, (1) becomes, by text (7.19):

$\dfrac{d}{dz}\left(\dfrac{dR}{dz}\right) = n^2 R$ or $(D^2 - n^2)R = 0$

where D means d/dz. Then $D = \pm n$, so

$R = Ae^{nz} + Be^{-nz} = Ar^n + Br^{-n}.$

For $n = 0$, $D = \pm 0$ gives only one solution [see text equations (5.12) to (5.15)]. Thus, when $n = 0$, we want the solution of $D^2 R = 0$, which is

$R = Az + B = A \, \ell n \, r + B.$

(2)   For equation (2) with $r = e^z$ and $D = \dfrac{d}{dz}$,   we find by text equation (7.19):

$D^2 R - DR + 2DR = \ell(\ell+1)R$ or $[D^2 + D - \ell(\ell+1)]R = 0;$

$(D - \ell)(D + \ell + 1) = 0;$ $D = \ell, \; -(\ell+1);$

$R = Ae^{\ell z} + Be^{-(\ell+1)z} = Ar^\ell + Br^{-\ell-1}.$

Here we do not have to consider $\ell = 0$ separately since we have two independent solutions (const. and r^{-1}) when $\ell = 0$.

Section 8

3. We write the differential equation as

$$(D^3 + 2D^2 + 2D)y = 0$$

and factor the auxiliary equation to get

$$D(D^2 + 2D + 2) = 0.$$

Then $D = 0$, and (by quadratic formula)

$$D = \frac{-2 \pm \sqrt{4 - 8}}{2} = -1 \pm i.$$

Thus (see text, page 355 and Problem 5.21)

$$y = Ae^0 + Be^{(-1+i)x} + Ce^{(-1-i)x} \qquad \text{or}$$

$$y = A + Ee^{-x}\sin(x + \gamma).$$

7. Here are two methods of solving the equation

$$3x^3 y^2 y' - x^2 y^3 = 1.$$

(A) Divide by $3x^3 y^2$:

$$y' - \frac{1}{3x} y = \frac{1}{3x^3} y^{-2}.$$

This is a Bernoulli equation. By text, page 350, we let $z = y^3$.
Then $z' = 3y^2 y'$ and the equation becomes

$$z' - \frac{1}{x} z = \frac{1}{x^3}.$$

This is a linear first order equation which we solve by text
equation (3.9):

$$I = -\int \frac{dx}{x} = -\ell n\, x, \qquad e^I = e^{-\ell n\, x} = \frac{1}{x},$$

$$\frac{z}{x} = \int \frac{1}{x^3} \frac{1}{x} dx = \frac{x^{-3}}{-3} + C,$$

$$z = -\frac{1}{3} x^{-2} + Cx = y^3 \qquad \text{or, with } 3C = A,$$

$$3x^2 y^3 + 1 = Ax^3.$$

7.   (continued)

(B) Divide by x^4 and multiply by dx:

$$\frac{3y^2}{x}\,dy - \left(\frac{y^3}{x^2} + \frac{1}{x^4}\right)dx = 0.$$

This equation is exact (that is, $\frac{1}{x^4}$ is an integrating factor

for the given equation) since

$$\frac{\partial}{\partial x}\left(\frac{3y^2}{x}\right) = -\frac{3y^2}{x^2} = \frac{\partial}{\partial y}\left(-\frac{y^3}{x^2} + \frac{1}{x^4}\right).$$

Integration gives

$$\frac{y^3}{x} + \frac{1}{3x^3} = \text{const.} \qquad \text{or} \qquad 3x^2y^3 + 1 = Ax^3$$

as above.

12.   We show several ways of solving $xy'' + y' = 4x$.

(A) This is a Cauchy equation if we multiply by x. By text,

page 378, we have

$$(D^2 - D)y + Dy = 4e^{2z} \qquad \text{or} \qquad D^2y = 4e^{2z},$$

where D means d/dz. By simply integrating twice, we get

$$y = e^{2z} + Cz + C'.$$

Put back $x = e^z$, $z = \ln x$ to get

$$y = x^2 + C\,\ln x + C'.$$

(B) We can also treat the equation as a "y-missing" problem.

By text equation (7.2), we let $y' = p$, $y'' = p'$.

$$xp' + p = 4x \qquad \text{or} \qquad xdp + (p - 4x)\,dx = 0.$$

This equation is exact and it is also a first order linear

equation: $p' + \frac{1}{x}p = 4$.

Integrating it by either method, we get

12.   (continued)

$$xp - 2x^2 = C.$$

Put back $p = y'$ and integrate again:

$$y' = 2x + \frac{C}{x},$$

$$y = x^2 + C \ln x + C',$$

as above.

18.   Hint: Write the equation in the form $x' + P(y)x = Q(y)$ and solve
       for $x(y)$ using text equation (3.9) with x and y interchanged.

19.   Hint: Let $u = x + y$, that is, $y = u - x$. The resulting equation
       is both separable and exact. Try solving it both ways.

20. $(D^2 - 2D + 5)y = 5x + 4e^x(1 + \sin 2x).$

       We first solve the homogeneous equation.

$$D = \frac{2 \pm \sqrt{4 - 20}}{2} = 1 \pm 2i,$$

$$y_c = Ae^x \sin(2x + \gamma).$$

We use the principle of superposition (text, page 369) to find
a particular solution. We solve the three equations

(1) $(D^2 - 2D + 5)y_{p_1} = 5x.$

(2) $(D^2 - 2D + 5)y_{p_2} = 4e^x.$

(3) $(D^2 - 2D + 5)y_{p_3} = 4e^x \sin 2x.$

For the first two of these equations, we have $c \neq a$ or b, so we
assume [see text equation (6.24)]

$$y_{p_1} = Ax + B,$$

$$y_{p_2} = Ce^x.$$

20.   (continued)

Substitute these into equations (1) and (2) to find

$$-2A + 5(Ax + B) \equiv 5x, \qquad A = 1, \quad B = 2/5$$

$$C(1 - 2 + 5) = 4, \qquad C = 1.$$

Thus

$$y_{p_1} = x + \frac{2}{5} ,$$

$$y_{p_2} = e^x.$$

Now to solve equation (3), we first solve (see text, page 365)

(4) $(D^2 - 2D + 5)Y = 4e^{(1+2i)x}$

and then take the imaginary part of Y:

(5) $y_{p_3} = \text{Im } Y .$

Since $1 + 2i$ is a root of the auxiliary equation, we assume [see text, equation (6.24), with $c = a \neq b$]

(6) $Y = Kxe^{(1+2i)x}.$

We want to substitute (6) into (4). We can simplify our work by remembering that $f = e^{(1+2i)x}$ is a solution of the homogeneous equation. Thus we write $Y = Kxf$ and find

$$(D^2 - 2D + 5)Y = K[x(D^2 - 2D + 5)f + 2f' - 2f] = 2K(f' - f)$$

$$= 2K(1 + 2i - 1)e^{(1+2i)x}.$$

Substituting this into equation (4) gives

$$4Ki = 4 \qquad \text{or} \qquad K = -i.$$

Then (6) becomes

$$Y = -ixe^{(1+2i)x} = -ixe^x(\cos 2x + i \sin 2x)$$

$$= xe^x(-i \cos 2x + \sin 2x)$$

20.   (continued)

and so (5) gives

$$y_{p_3} = -xe^x\cos 2x.$$

Now the solution of the original equation is obtained by adding the three particular solutions and the complementary function

$$y = Ae^x\sin(2x+\gamma) + x + \frac{2}{5} + e^x - xe^x\cos 2x.$$

23.   We write the given equation as

$$r' - \frac{\cos\theta}{\sin\theta}r = \frac{\sin\theta}{\cos\theta}$$

to recognize it as a linear first order equation. By text (3.9) we find

$$I = -\int\frac{\cos\theta}{\sin\theta}\,d\theta = -\ell n\sin\theta, \qquad e^I = \frac{1}{\sin\theta},$$

$$\frac{r}{\sin\theta} = \int\frac{\sin\theta}{\cos\theta}\cdot\frac{1}{\sin\theta}\,d\theta = \int\sec\theta\,d\theta = \ell n(\sec\theta + \tan\theta) + C,$$

$$r = \sin\theta\,\ell n(\sec\theta + \tan\theta) + C\sin\theta.$$

28. $yy'' + y'^2 + 4 = 0,\qquad y = 3,\ y' = 0$ when $x = 1$.

By text equation (7.3), we have

$$yp\frac{dp}{dy} + p^2 + 4 = 0.$$

This equation is separable.

$$\frac{p\,dp}{p^2+4} + \frac{dy}{y} = 0,$$

$$\frac{1}{2}\ell n(p^2+4) + \ell n\,y = \ell n\,C,$$

$$(p^2+4) = \frac{C^2}{y^2}.$$

Since $y = 3$ and $y' = p = 0$ when $x = 1$, we have $C = 6$. Put back $p = y'$

28.   (continued)

and integrate again.

$$y'^2 + 4 = \frac{36}{y^2} ,$$

$$y' = \frac{\sqrt{36 - 4y^2}}{y^2} = \frac{2}{y}\sqrt{9 - y^2} ,$$

$$\int \frac{y \, dy}{\sqrt{9 - y^2}} = 2\int dx ,$$

$$-\sqrt{9 - y^2} = 2x + K .$$

Since $y = 3$ when $x = 1$, we find $K = -2$. Thus

$$4(x - 1)^2 + y^2 = 9 .$$

32.   Let $N =$ number of pages covered,

$T =$ temperature,

$t =$ time in hours.

Then

(1) $\dfrac{dN}{dt} = \dfrac{k}{T - 75} .$

Note that it is implicit in the problem that $T > 75°$ during all
of your study period since dN/dt is defined only for temperatures
above $75°$. Also, if $T = 75$ when $t = t_0$,   then

(2) $T = 75 + C(t - t_0) .$

Again note that $C > 0$ (temperature is rising) and that $t - t_0 > 0$
for $t \geq 2$ since $T > 75$. We combine equations (1) and (2) to get

$$\frac{dN}{dt} = \frac{K}{t - t_0} .$$

Then

32.   (continued)
$$N = K \int \frac{dt}{t - t_0} = K \ln(t - t_0) + K'.$$

We use the given data to write three equations which determine K, K' and t_0 .

		multiply by
At $t = 2$, $N = 0$,	$0 = K \ln(2 - t_0) + K'$.	1
At $t = 3$, $N = 20$,	$20 = K \ln(3 - t_0) + K'$.	-3
At $t = 4$, $N = 30$,	$30 = K \ln(4 - t_0) + K'$.	2

Combine these equations using the multipliers indicated in the last column to get

$$K[\ln(2 - t_0) - 3 \ln(3 - t_0) + 2 \ln(4 - t_0)] = 0.$$

Then

$$(2 - t_0)(4 - t_0)^2 = (3 - t_0)^3.$$

This equation is simpler if we let $u = 2 - t_0$.

$$u(u + 2)^2 = (u + 1)^3,$$
$$u(u^2 + 4u + 4) = u^3 + 3u^2 + 3u + 1,$$
$$u^2 + u - 1 = 0, \qquad u = \frac{-1 + \sqrt{5}}{2} = 0.618 \text{ hr} = 37 \text{ min}.$$

(We use only the positive solution because $u = 2 - t_0 > 0$ as we saw above.)

$$t_0 = 2 \text{ hr} - 37 \text{ min} = 1:23 \text{ pm}.$$

35.   At time t, let

y = distance raindrop has fallen,

v = speed of raindrop,

r = radius of raindrop,

m = mass of raindrop $= \frac{4}{3} \pi r^3 \rho$.

35.   (continued)

We are to assume that

$$dm = k\pi r^2 dy \qquad \text{or} \qquad 4\pi r^2 \rho \, dr = k\pi r^2 dy,$$

that is, dr and dy are proportional, or $r = Cy$ if $r = 0$ when $y = 0$. Now by Newton's second law

$$\frac{d}{dt}(mv) = mg \qquad \text{or} \qquad \frac{d}{dt}(r^3 v) = r^3 g.$$

Since $r = Cy$, we have

(1) $\frac{d}{dt}(y^3 v) = y^3 g.$

This is an "independent variable missing" equation (no t). Following the method of text equation (7.3) we can write for any function f

$$\frac{df}{dt} = \frac{df}{dy}\frac{dy}{dt} = \frac{df}{dy}v.$$

With $f = y^3 v$, equation (1) becomes

(2) $\frac{d}{dt}(y^3 v) = v\frac{d}{dy}(y^3 v) = y^3 g.$

Multiply both sides of (2) by y^3 to get

(3) $(y^3 v)\frac{d}{dy}(y^3 v) = y^6 g.$

The left side of (3) is now the derivative of $(y^3 v)^2/2$, so we can integrate (3) to get

(4) $(y^3 v)^2/2 = y^7 g/7.$

The integration constant is zero since $y = v = 0$ at $t = 0$.

Solve (4) for $v = dy/dt$ and integrate again.

$$v = dy/dt = (2g/7)^{1/2}y^{1/2} \qquad \text{or} \qquad y^{-1/2}dy = (2g/7)^{1/2}dt,$$

$$2y^{1/2} = (2g/7)^{1/2}t.$$

Again the integration constant is zero. We find

$$y = \frac{g}{14} t^2.$$

Then

35.   (continued)

$$v = \frac{dy}{dt} = \frac{g}{7} t,$$

$$a = \frac{dv}{dt} = \frac{g}{7} ,$$

as claimed.

Chapter 9

Section 1

1. The distance from A to P is

$$AP = \sqrt{(x - x_1)^2 + y_1^2}$$

and the distance from P to B is

$$PB = \sqrt{(x_2 - x)^2 + y_2^2} \; .$$

A=(x_1, y_1)

B=(x_2, y_2)

P=$(x, 0)$

Then

$$D = APB = AP + PB,$$

$$t = (n/c)D = \frac{n}{c}\left(\sqrt{(x - x_1)^2 + y_1^2} + \sqrt{(x_2 - x)^2 + y_2^2} \right),$$

$$\frac{dt}{dx} = \frac{n}{c}\left(\frac{1}{2}\frac{2(x - x_1)}{\sqrt{(x - x_1)^2 + y_1^2}} + \frac{1}{2}\frac{2(x_2 - x)(-1)}{\sqrt{(x_2 - x)^2 + y_2^2}} \right)$$

$$= \frac{n}{c}\left(\frac{x - x_1}{AP} - \frac{x_2 - x}{PB} \right) \; .$$

We want $dt/dx = 0$ to make t stationary; thus

$$\frac{x - x_1}{AP} = \frac{x_2 - x}{PB} \; .$$

From the text diagram we see that

$$\sin\theta = \frac{x - x_1}{AP} \; , \qquad \sin\phi = \frac{x_2 - x}{PB} \; .$$

Then $\sin\theta = \sin\phi$, so $\theta = \phi$ (since, from the diagram, both are first quadrant angles).

3.   (a) By definition of an ellipse, the sum of the distances from
      two fixed points (the foci) to a point of the curve is a con-
      stant. Thus if C_1D_1 is an arc of an ellipse with foci A and B
      (see figure), then APB = AP'B. A reflecting surface in the shape
      of an ellipse has the property that every ray of light from one
      focus is reflected to the other focus.

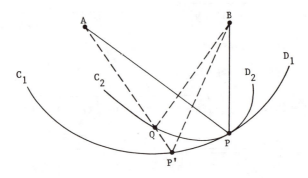

      (c) See figure. Let C_2D_2 be a reflecting surface tangent to the
      ellipse at P. Then APB is the actual path of the light ray and
      AQB is an example of a varied path. From the geometry we see
      that AQB is shorter than AP'B so [using APB = AP'B from part (a)
      above] we have AQB < APB. Thus the actual path is longer than the
      varied paths so the time is a maximum.

Section 2

2.   With $ds = \sqrt{1 + y'^2}\, dx$ [text page 209, equation (3.1)], we want
to make stationary

$$\int_{x_1}^{x_2} \frac{1}{x}\sqrt{1 + y'^2}\ dx.$$

Then

$$F = \frac{1}{x}\sqrt{1 + y'^2}\,,$$

$$\frac{\partial F}{\partial y'} = \frac{1}{x}\ \frac{y'}{\sqrt{1 + y'^2}}\,,$$

$$\frac{\partial F}{\partial y} = 0\,,$$

and the Euler equation, text equation (2.16), is

$$\frac{d}{dx}\left(\frac{1}{x}\ \frac{y'}{\sqrt{1 + y'^2}}\right) = 0 \qquad \text{or} \qquad \frac{1}{x}\ \frac{y'}{\sqrt{1 + y'^2}} = \text{const.} = C.$$

We square both sides of this equation and solve for y':

$$\frac{y'^2}{x^2(1 + y'^2)} = C^2\,, \qquad y'^2(1 - C^2 x^2) = C^2 x^2\,,$$

$$y' = \frac{Cx}{\sqrt{1 - C^2 x^2}}\ .$$

Then integrate both sides of this equation to get:

$$y + b = \int \frac{Cx\ dx}{\sqrt{1 - C^2 x^2}} = -\frac{1}{C}\sqrt{1 - C^2 x^2}\ .$$

Square both sides and simplify:

$$(y + b)^2 = \frac{1}{C^2}(1 - C^2 x^2) \qquad \text{or} \qquad x^2 + (y + b)^2 = 1/C^2.$$

This is a circle.

5.   Here $F = y'^2 + y^2$, so

$$\frac{\partial F}{\partial y'} = 2y' \qquad \text{and} \qquad \frac{\partial F}{\partial y} = 2y.$$

The Euler equation, text equation (2.16), is

$$\frac{d}{dx}(2y') - 2y = 0 \qquad \text{or} \qquad y'' = y.$$

The general solution of this differential equation is (see text, Chapter 8, Section 5, or page 339, Example 2)

$$y = ae^x + be^{-x}$$

[or other forms such as $y = A \cosh x + B \sinh x$, $y = C \cosh(x + b)$, etc.]

Section 3

3.   With y as the integration variable, the Euler equation is text equation (3.3). Since the integrand contains x, the second term in text equation (3.3) is not zero. We change to x as the integration variable by making the substitutions [compare text equations (3.4)]:

$$x' = \frac{1}{y'}, \qquad dy = y'\, dx,$$

in the given integral to get:

$$\int_{y_1}^{y_2} \frac{x'^2\, dy}{\sqrt{x'^2 + x^2}} = \int_{x_1}^{x_2} \frac{(1/y'^2)y'\, dx}{\sqrt{(1/y'^2) + x^2}} = \int_{x_1}^{x_2} \frac{dx}{\sqrt{1 + x^2 y'^2}}.$$

The Euler equation is now text equation (2.16). We have

$$F = \frac{1}{\sqrt{1 + x^2 y'^2}}, \qquad \frac{\partial F}{\partial y'} = -\frac{1}{2}(1 + x^2 y'^2)^{-3/2}(2x^2 y'), \qquad \frac{\partial F}{\partial y} = 0,$$

so text equation (2.16) becomes

$$\frac{d}{dx}\left[x^2 y'(1 + x^2 y'^2)^{-3/2}\right] = 0 \qquad \text{or} \qquad x^2 y'(1 + x^2 y'^2)^{-3/2} = C,$$

$$x^4 y'^2 = C^2 (1 + x^2 y'^2)^3.$$

7. This is the same as Problem 2.5 above but here we solve it by
    a different method. Since the integrand contains y, we change
    to y as the integration variable [text equations (3.4)] and use
    text equation (3.3).

$$\int (y'^2 + y^2)\,dx = \int \left(\frac{1}{x'^2} + y^2\right) x'\,dy = \int \left(\frac{1}{x'} + y^2 x'\right) dy.$$

$$F = \frac{1}{x'} + y^2 x', \qquad \frac{\partial F}{\partial x'} = -x'^{-2} + y^2, \qquad \frac{\partial F}{\partial x} = 0,$$

$$\frac{d}{dy}(-x'^{-2} + y^2) = 0, \qquad x'^{-2} - y^2 = y'^2 - y^2 = \text{const.}$$

The constant might be positive, negative, or zero and y' might
be either positive or negative, so we write

$$y' = \frac{dy}{dx} = \pm \sqrt{y^2 \pm K^2},$$

$$\pm \int dx = \int \frac{dy}{\sqrt{y^2 \pm K^2}}.$$

See Appendix A to these Solutions for the integrals needed.
The various possible forms are:

$$\pm x + C = \begin{cases} \int \frac{dy}{y} = \ell n\, y & \text{if} \quad K = 0, \\[2mm] \int \frac{dy}{\sqrt{y^2 - K^2}} = \cosh^{-1} \frac{y}{K}, \\[2mm] \int \frac{dy}{\sqrt{y^2 + K^2}} = \sinh^{-1} \frac{y}{K}, \\[2mm] \int \frac{dy}{\sqrt{y^2 \pm K^2}} = \ell n\left(y + \sqrt{y^2 \pm K^2}\right). \end{cases}$$

From any of these we find as in Problem 2.5 that y is a solution
of y'' = y, that is,

7.   (continued)

$$y = ae^x \quad \text{or} \quad be^{-x} \quad \text{or}$$

$$y = K \cosh(\pm x + C) \quad \text{or}$$

$$y = K \sinh(\pm x + C) \quad \text{or}$$

$$y = ae^x + be^{-x}.$$

Note that, in this problem, solving the second order equation as in Problem 2.5 was actually simpler than obtaining and solving the first order equation. However this is not usually true.

10.   Since the integrand contains s but not t, we change to s as the integration variable.

$$dt = t' \, ds, \qquad t' = \frac{dt}{ds}, \qquad s' = \frac{1}{t'},$$

$$\int s^{-1} \sqrt{s^2 + s'^2} \, dt = \int s^{-1} \sqrt{s^2 + (t')^{-2}} \, t' \, ds = \int s^{-1} \sqrt{s^2 t'^2 + 1} \, ds.$$

The Euler equation is

$$\frac{d}{ds} \frac{\partial F}{\partial t'} - \frac{\partial F}{\partial t} = 0 \quad \text{where } F = s^{-1} \sqrt{s^2 t'^2 + 1}.$$

Since $\frac{\partial F}{\partial t} = 0$, we have

(1) $$\frac{d}{ds}\left[s^{-1} \frac{s^2 t'}{\sqrt{s^2 t'^2 + 1}} \right] = 0 \quad \text{or} \quad \frac{st'}{\sqrt{s^2 t'^2 + 1}} = \text{const.}$$

Note that this is a function of (st'); thus

$$st' = \text{const.} = s\frac{dt}{ds} = \frac{1}{a},$$

(2) $$\frac{ds}{s} = a \, dt, \qquad \ln s = at + \ln b,$$

$$s = b \, e^{at}.$$

Comments: Observe the choice of constants to write this answer in a simple form. Suppose that we had solved (1) as follows:

10.   (continued)
$$s^2 t'^2 = K^2 (s^2 t'^2 + 1), \qquad s^2 t'^2 (1 - K^2) = K^2,$$

(3) $s \dfrac{dt}{ds} = \dfrac{K}{\sqrt{1 - K^2}}$.

Now the expression $\dfrac{K}{\sqrt{1 - K^2}}$ is a constant so we may as well

call it C.   Then (3) becomes

$$s \frac{dt}{ds} = C \qquad \text{or} \qquad \frac{ds}{s} = \frac{1}{C} dt,$$

$$\ell n\, s = \frac{1}{C} t + C', \qquad s = e^{t/C} e^{C'},$$

which is the same as (2) with $C = \dfrac{1}{a}$ and $e^{C'} = b$. Always simplify

constants as much as possible. However do <u>not</u> introduce extra

constants. For example, the expression $(K - 2)x + 3K$ contains

only one arbitrary constant K and must not be written as $ax + b$

which contains two arbitrary constants a and b.

11.   Hint: The path is a catenary. See text page 391.

13.   See Fermat's principle, text Section 1. We want to make

stationary

$$\int n\, ds \qquad \text{or} \qquad \int \sqrt{y}\, ds = \int \sqrt{y} \sqrt{1 + x'^2}\, dy.$$

[Note that we write ds so that y is the integration variable

because the integrand contains y but not x; the Euler equation

is then text equation (3.3) instead of (2.16).] We have:

$$F = \sqrt{y}\sqrt{1 + x'^2}, \qquad \frac{\partial F}{\partial x'} = \frac{x'\sqrt{y}}{\sqrt{1 - x'^2}}, \qquad \frac{\partial F}{\partial x} = 0;$$

$$\frac{d}{dy}\left(\frac{\partial F}{\partial x'}\right) = \frac{d}{dy}\left(\frac{x'\sqrt{y}}{\sqrt{1 + x'^2}}\right) = 0, \qquad \frac{x'\sqrt{y}}{\sqrt{1 + x'^2}} = K.$$

Solve for x' and integrate again:

13.   (continued)

$$x'^2 y = K^2(1 + x'^2), \qquad x'^2(y - K^2) = K^2,$$

$$\frac{dx}{dy} = x' = \frac{K}{\sqrt{y - K^2}},$$

$$x = K \int \frac{dy}{\sqrt{y - K^2}} = 2K\sqrt{y - K^2} + a, \qquad \text{or}$$

$$(x - a)^2 = 4K^2(y - K^2).$$

This is a parabola.

14.   Hints: Write ds in polar coordinates (text page 218).   Also
      see Problem 10 above.

18.   In spherical coordinates [see text, page 219, equation (4.5)],
      the arc length on a sphere of radius $\underline{a}$ is given by

$$ds^2 = a^2 d\theta^2 + a^2 \sin^2\theta \, d\phi^2.$$

Thus we want to minimize

$$\int ds = a \int \sqrt{1 + \sin^2\theta \, \phi'^2} \; d\theta.$$

The Euler equation is

$$\frac{d}{d\theta}\left(\frac{\partial F}{\partial \phi'}\right) - \frac{\partial F}{\partial \phi} = 0 \qquad \text{where} \qquad F = \sqrt{1 + \phi'^2 \sin^2\theta}.$$

$$\frac{\partial F}{\partial \phi'} = \frac{\phi' \sin^2\theta}{\sqrt{1 + \phi'^2 \sin^2\theta}}, \qquad \frac{\partial F}{\partial \phi} = 0,$$

$$\frac{d}{d\theta}\left(\frac{\phi' \sin^2\theta}{\sqrt{1 + \phi'^2 \sin^2\theta}}\right) = 0 \qquad \text{or} \qquad \frac{\phi' \sin^2\theta}{\sqrt{1 + \phi'^2 \sin^2\theta}} = \text{const.} = K.$$

We solve for ϕ' and integrate again:

$$\phi'^2 \sin^4\theta = K^2(1 + \phi'^2 \sin^2\theta), \qquad \phi'^2(\sin^4\theta - K^2 \sin^2\theta) = K^2,$$

$$\phi' = \frac{d\phi}{d\theta} = \frac{K}{\sqrt{\sin^4\theta - K^2 \sin^2\theta}} \quad .$$

18.   (continued)

$$\phi = \int \frac{K\,d\theta}{\sqrt{\sin^4\theta - K^2\sin^2\theta}} = \int \frac{K\csc^2\theta\,d\theta}{\sqrt{1 - K^2\csc^2\theta}} = \int \frac{\csc^2\theta\,d\theta}{\sqrt{K^{-2} + (1 + \cot^2\theta)}}.$$

To evaluate this integral, let $w = \cot\theta$ and $K^{-2} + 1 = A^2$; then $dw = -\csc^2\theta\,d\theta$, and we have

$$\phi = \int \frac{-dw}{\sqrt{A^2 - w^2}} = \arccos\frac{w}{A} + \alpha,$$

(1) $\cot\theta = w = A\cos(\phi - \alpha)$,

where A and α are constants. We want to show that, on the surface of the sphere, this is a great circle. Multiply (1) by $\sin\theta$ and also expand the cosine to get:

$\cos\theta = A\sin\theta(\cos\alpha\cos\phi + \sin\alpha\sin\phi)$.

Multiply by r and change to rectangular coordinates [text, page 219, equations (4.5)]:

$r\cos\theta = (A\cos\alpha)(r\sin\theta\cos\phi) + (A\sin\alpha)(r\sin\theta\sin\phi)$,

$z = (A\cos\alpha)x + (A\sin\alpha)y$.

This is the equation of a plane through the origin; its intersection with the sphere is a great circle. Thus the geodesics (1) on the sphere are great circles.

Section 4

3.   Hints: The initial energy is now $\frac{1}{2}mv_0^2$ instead of zero. Write the energy equation and show that y in the text solution is now replaced by $y + \left(\frac{1}{2}v_0^2/g\right)$.

4. Hints: Find v from the energy equation (compare text, top of
    page 394). The energy at r is equal to the energy at r = R, and
    v = 0 at r = R. Thus

    $$\frac{1}{2}mv^2 + \frac{mg}{2R}(r^2 - 3R^2) = \frac{mg}{2R}(-2R^2).$$

    From the first integral of the Euler equation you should find

    $$(r\theta')^2 = \frac{r_0^2(R^2 - r^2)}{R^2(r^2 - r_0^2)}.$$

    To evaluate the final integral for T, let $u = r^2$. You should
    find about 42 min for the transit time through the center of
    the earth as in Chapter 8, Problem 5.35.

7. We want to make stationary

    $$\int (2x + 5)^{-1/2} ds = \int (2x + 5)^{-1/2} \sqrt{1 + y'^2}\, dx.$$

    Then

    $$F = (2x + 5)^{-1/2}(1 + y'^2)^{1/2}, \qquad \frac{\partial F}{\partial y} = 0,$$

    and the Euler equation is

    $$\frac{d}{dx}\left(\frac{\partial F}{\partial y'}\right) = 0 \qquad \text{or} \qquad \frac{\partial F}{\partial y'} = \text{const.}$$

    $$\frac{\partial F}{\partial y'} = (2x + 5)^{-1/2}(1 + y'^2)^{-1/2}y' = \text{const.} = K, \qquad \text{or}$$

    $$(1 + y'^2)^{-1}y'^2 = K^2(2x + 5).$$

    To simplify this, let $u = K^2(2x + 5)$. Then solve for y' to get

    $$y' = \sqrt{\frac{u}{1 - u}} = \frac{dy}{dx} = 2K^2 \frac{dy}{du}.$$

    This can be integrated using tables:

    $$2K^2 y = \int \sqrt{\frac{u}{1 - u}}\, du = -\sqrt{u(1 - u)} - \frac{1}{2} \arccos(2u - 1) + C.$$

    As in text equations (4.2) and (4.4), we show that this is a
    cycloid by letting $\theta = \arccos(2u - 1)$. Then

7.   (continued)

$$u = \frac{1}{2}(1 - \cos \theta) = K^2(2x + 5),$$

$$x = \frac{1}{4K^2}(1 - \cos \theta) - \frac{5}{2},$$

$$y = \frac{1}{2K^2}\int\sqrt{\frac{(1/2)(1 - \cos \theta)}{(1/2)(1 + \cos \theta)}}\,\frac{1}{2}\sin \theta \; d\theta = \frac{1}{4K^2}\int\sqrt{\frac{(1 - \cos \theta)^2}{1 - \cos^2\theta}}\sin \theta \; d\theta$$

$$= \frac{1}{4K^2}\int(1 - \cos \theta)\,d\theta = \frac{1}{4K^2}(\theta - \sin \theta) + C.$$

Putting $\frac{1}{4K^2} = a$, we have the parametric equations of a cycloid:

$$x = a(1 - \cos \theta) - \frac{5}{2},$$

$$y = a(\theta - \sin \theta) + C.$$

We might have anticipated this result by comparing the integral to be made stationary, namely $\int(2x + 5)^{-1/2}\,ds$, with the integral $\int y^{-1/2}\,ds$ to be made stationary in the text discussion of the brachistochrone. If we replace x in the text by y, and replace y by $x + \frac{5}{2}$, we have the problem just solved. Note that our result is of the form of text equations (4.6) with these replacements and with the added constant of integration because here we do not have the condition (of the brachistochrone problem) that the curve is to pass through the origin.

Section 5

4.   See text, page 360, Problem 5.34; for the pendulum shown there,
      the velocity of the mass m is $v = \ell\,\dot{\theta}$, so the kinetic energy is
      $T = \frac{1}{2}\,m\,\ell^2\dot{\theta}^2$. The potential energy is mg times the height of m
      above its equilibrium position ($\theta = 0$); this height is $\ell - \ell\cos\theta$.
      Thus we have

$$V = mg\ell(1 - \cos\theta),$$

$$L = T - V = \frac{1}{2}\,m\,\ell^2\dot{\theta}^2 - mg\ell(1 - \cos\theta).$$

Then the Lagrange equation of motion is

$$\frac{d}{dt}\left(\frac{\partial L}{\partial \dot{\theta}}\right) - \frac{\partial L}{\partial \theta} = 0 \qquad \text{or}$$

$$m\ell^2\ddot{\theta} - \left[-mg\,\ell\,\frac{d}{d\theta}(-\cos\theta)\right] = m\,\ell^2\ddot{\theta} + mg\,\ell\sin\theta = 0.$$

[Note the product of <u>four</u> minus signs in the second term. Always
be careful about the sign in this term -- it is a frequent source
of error. There is always a minus sign in L = T - V, another minus
sign in the Lagrange equation ($-\frac{\partial L}{\partial \theta}$ in this problem) and then
possibly, as here, other minus signs in V or its derivative.]
Thus we have the equation of motion of a simple pendulum:

$$\ell\ddot{\theta} + g\sin\theta = 0.$$

7.   Constraint forces are normal to the surface and do no work on the
      particle; thus its velocity v is constant. Since both the poten-
      tial energy V and the kinetic energy $T = \frac{1}{2}\,mv^2$ are constant, the
      Lagrangian L = T - V is a constant. Hamilton's principle says that
      the particle follows a path for which $\int L\,dt$ is stationary; if L is
      constant then $\int dt$ is to be stationary. Now $v = \frac{ds}{dt}$ and v is con-
      stant, so $\int ds = \int v\,dt = v\int dt$. The requirement for a geodesic is

7.   (continued)

$\int ds$ stationary; Hamilton's principle (in this case) requires $\int dt$ stationary. We see that these are equivalent problems (for a particle subject <u>only</u> to constraint forces -- <u>not</u> in general).

8.   The kinetic energy of the particle on the table is, in polar coordinates,

$$T_1 = \frac{1}{2} m(\dot{r}^2 + r^2 \dot{\theta}^2)$$

and its potential energy is constant, say $V = 0$. The kinetic energy of the lower particle is $T_2 = \frac{1}{2} m \dot{z}^2$, and its potential energy relative to the table is $V = mgz$; note that z is negative. Then for the system of two particles, we have

$$L = T - V = \frac{1}{2} m(\dot{r}^2 + r^2 \dot{\theta}^2 + \dot{z}^2) - mgz.$$

Now r and z are not independent; they are related by $r + |z| = \text{length of string} = \ell$. We must use this constraint equation to eliminate one variable, say z. We have

$$z = -|z| = -(\ell - r), \qquad \dot{z}^2 = \dot{r}^2,$$

$$L = \frac{1}{2} m(\dot{r}^2 + r^2 \dot{\theta}^2 + \dot{r}^2) + mg(\ell - r) = \frac{1}{2} m(2\dot{r}^2 + r^2 \dot{\theta}^2) + mg(\ell - r),$$

$$\begin{cases} \dfrac{d}{dt} \dfrac{\partial L}{\partial \dot{r}} - \dfrac{\partial L}{\partial r} = 2m\ddot{r} - (mr\dot{\theta}^2 - mg) = 0 \\[2mm] \dfrac{d}{dt} \dfrac{\partial L}{\partial \dot{\theta}} - \dfrac{\partial L}{\partial \theta} = \dfrac{d}{dt}(mr^2 \dot{\theta}) = 0, \end{cases} \quad \text{or} \quad \begin{cases} 2\ddot{r} - r\dot{\theta}^2 + g = 0, \\[2mm] r^2 \dot{\theta} = \text{const.} \end{cases}$$

<u>Warning</u>: In the last equation, note carefully that $r^2 \ddot{\theta} \neq 0$. Since r is a variable, we have

$$\frac{d}{dt}(r^2 \dot{\theta}) = r^2 \ddot{\theta} + 2r\dot{r}\dot{\theta} = 0.$$

However, it is usually unnecessary to do this differentiation; instead, we write the first integral of the equation, namely $r^2 \dot{\theta} = \text{const.}$

Section 6

3. Let the area under the curve $y = y(x)$ between $x = 0$ and $x = 1$ be
    revolved about the x axis to form a volume of revolution. We
    are given that the volume is 10 cc. Thus (see text, page 212,
    Figure 3.4)

    $$V = \int_0^1 \pi y^2 \, dx = 10.$$

    The element of mass is $dM = \rho \pi y^2 \, dx$ where ρ is the density. The
    moment of inertia of this circular slab about the x axis is
    $\frac{1}{2} y^2 \, dM = \frac{1}{2} \rho \pi y^4 \, dx$ (see solution of Chapter 5, Problem 4.3a). We
    want to minimize

    $$I = \int_0^1 \frac{1}{2} \rho \pi y^4 \, dx$$

    subject to the condition that $V = 10$. Thus we write the Euler
    equation to make stationary the integral

    $$I + \lambda V = \int_0^1 \left(\frac{1}{2} \rho \pi y^4 + \lambda \pi y^2 \right) dx.$$

    We find (see text notation, Section 6)

    $$F + \lambda G = \frac{1}{2} \rho \pi y^4 + \lambda \pi y^2,$$

    $$\frac{\partial}{\partial y'}(F + \lambda G) = 0, \qquad \frac{\partial}{\partial y}(F + \lambda G) = 2\rho\pi y^3 + 2\lambda\pi y.$$

    Then the Euler equation is just

    $$\frac{\partial}{\partial y}(F + \lambda G) = 2\rho\pi y^3 + 2\lambda\pi y = 0.$$

    Since $y \neq 0$ (we are given $V = 10$), we have

    (A) $2\rho\pi y^2 + 2\lambda\pi = 0$ or $y = \text{const.}$

3.   (continued)

Revolving the area under y = const. about the x axis produces a circular cylinder. Thus, in order to minimize the moment of inertia, we should form the 10 cc of lead into a circular cylinder of radius y and height 1 so that $\pi y^2 \cdot 1 = 10$, or $y = \sqrt{10/\pi} = 1.8$ cm.   [We do not need the value of the Lagrange multiplier λ, but we can find it if we like. From (A), we find $\lambda = -\rho y^2 = -10\rho/\pi$.]

6.   Hint: As in Problem 6.5, assume that y = 0 satisfies the equation of the curve.

Section 8

1.   (a) We want to show that

$$S = F - y' \frac{\partial F}{\partial y'} = \text{const.}$$

if the Euler equation holds. We differentiate S with respect to x remembering that F is a function of y and y' where y and y' are functions of x.

$$\frac{dS}{dx} = \frac{dF}{dx} - \frac{d}{dx}\left(y' \frac{\partial F}{\partial y'}\right)$$

$$= \frac{\partial F}{\partial y} y' + \frac{\partial F}{\partial y'} y'' - \left[y'' \frac{\partial F}{\partial y'} + y' \frac{d}{dx}\left(\frac{\partial F}{\partial y'}\right)\right]$$

$$= - y'\left[\frac{d}{dx}\left(\frac{\partial F}{\partial y'}\right) - \frac{\partial F}{\partial y}\right] = 0$$

since the bracket is zero if the Euler equation holds. Thus S is a constant; the equation S = const. is a first integral of the Euler equation.

1. (continued)

    (b) Let us use (a) to solve Problem 3.13. First we must note
    that F is a function only of y and y'. Then we may use the
    formula in (a).

    $$F = \sqrt{y}\ \sqrt{1+y'^2},$$

    $$\frac{\partial F}{\partial y'} = \frac{y'\sqrt{y}}{\sqrt{1+y'^2}},$$

    $$F - y'\frac{\partial F}{\partial y'} = \sqrt{y}\ \sqrt{1+y'^2} - \frac{y'^2\sqrt{y}}{\sqrt{1+y'^2}}$$

    (1) $$= \frac{\sqrt{y}}{\sqrt{1+y'^2}} = \text{const.}$$

    In Problem 3.13 we had $\dfrac{x'\sqrt{y}}{\sqrt{1+x'^2}} = K$, where $x' = 1/y'$. If we
    multiply numerator and denominator by y' we get (1) above, so
    both methods lead to the same first integral of the Euler
    equation.

    The disadvantage of method (a) is that you must memorize (or
    look up) a formula, whereas the method of changing the variable
    of integration requires no formula. Note carefully that the
    formula in (a) is valid <u>only</u> if x is the integration variable and
    F is a function of y and y' only. You can, of course, write a
    similar formula if y is the integration variable and $F = F(x,x')$
    and so on for other variables.

1. (continued)

    (c) For $\int L \, dt = \int \left[\frac{1}{2} m\dot{x}^2 - V(x) \right] dt$, the formula corresponding to (a)
    is

    $$L - \dot{x} \frac{\partial L}{\partial \dot{x}} = \text{const.}$$

    We find

    $$\frac{1}{2} m\dot{x}^2 - V(x) - \dot{x}m\dot{x} = -\left[\frac{1}{2} m\dot{x}^2 + V(x) \right] = \text{const.}$$

    This says that the total energy is constant.

8. In cylindrical coordinates

    $$ds^2 = dr^2 + r^2 d\theta^2 + dz^2.$$

    We eliminate r and dr using the given equation of the cylinder.

    $$r = 1 + \cos \theta,$$

    $$dr = -\sin \theta \, d\theta,$$

    $$ds^2 = \sin^2 \theta \, d\theta^2 + (1 + \cos \theta)^2 d\theta^2 + dz^2$$

    $$= 2(1 + \cos \theta) d\theta^2 + dz^2.$$

    To find the geodesics we want to minimize $\int ds$.

    $$\int ds = \int \sqrt{2(1 + \cos \theta) d\theta^2 + dz^2} = \int_{\theta_1}^{\theta_2} \sqrt{2(1 + \cos \theta) + z'^2} \, d\theta$$

    where $z' = dz/d\theta$. The Euler equation is

    $$\frac{d}{d\theta} \left(\frac{\partial F}{\partial z'} \right) - \frac{\partial F}{\partial z} = 0,$$

    $$\frac{\partial F}{\partial z'} = \frac{z'}{\sqrt{2(1 + \cos \theta) + z'^2}} = K.$$

    We solve for z' and integrate again.

8.   (continued)

$$z'^2 = K^2(2 + 2\cos\theta + z'^2),$$

$$z'^2 = \frac{2K^2}{1 - K^2}(1 + \cos\theta) = A^2\left(\frac{1 + \cos\theta}{2}\right) \qquad \text{where } \frac{A^2}{2} = \frac{2K^2}{1 - K^2}$$

$$= A^2\cos^2(\theta/2),$$

$$z' = \frac{dz}{d\theta} = A\cos(\theta/2),$$

$$z = a + 2A\sin(\theta/2) = a + b\sin(\theta/2) \qquad \text{where } b = 2A.$$

The geodesics are the curves of intersection of the given
cylinder $r = 1 + \cos\theta$ with the surface $z = a + b\sin(\theta/2)$.

11.   We want to make stationary the integral

$$\int r^{-1/2}\, ds = \int r^{-1/2}\sqrt{dr^2 + r^2 d\theta^2} = \int_{r_1}^{r_2} r^{-1/2}\sqrt{1 + r^2\theta'^2}\, dr$$

where $\theta' = d\theta/dr$.   The Euler equation is

$$\frac{d}{dr}\left(\frac{\partial F}{\partial\theta'}\right) - \frac{\partial F}{\partial\theta} = 0,$$

$$\frac{\partial F}{\partial\theta'} = r^{-1/2}(1 + r^2\theta'^2)^{-1/2}r^2\theta' = K.$$

We solve for θ' and integrate again.

$$r^3\theta'^2 = K^2(1 + r^2\theta'^2), \qquad \theta'^2(r^3 - K^2 r^2) = K^2,$$

$$d\theta = \frac{K\,dr}{r\sqrt{r - K^2}},$$

(1) $$\theta + C = K\int \frac{dr}{r\sqrt{r - K^2}} = K \cdot \frac{2}{K}\tan^{-1}\sqrt{(r - K^2)/K^2}$$

from tables or the substitution $u = \sqrt{r - K^2}$.   (Another substitution which works well here is $r = K^2\sec^2\theta$; then we get

11. (continued)

$\theta + c = 2 \sec^{-1} \sqrt{r/K^2}$ which yields the result below immediately.)
From (1) we find

$$\frac{r - K^2}{K^2} = \tan^2 \frac{\theta + c}{2} \, ,$$

$$\frac{r}{K^2} = 1 + \tan^2 \frac{\theta + c}{2} = \sec^2 \frac{\theta + c}{2} \, ,$$

$$r = K^2 \sec^2 \frac{\theta + c}{2} \, .$$

This is a parabola. Using the trigonometric identity

$$\cos \frac{\alpha}{2} = \sqrt{\frac{1 + \cos \alpha}{2}} \, ,$$

we can also write the equation in the form

$$r \cos^2 \frac{\theta + c}{2} = K^2 = \frac{r}{2}[1 + \cos(\theta + c)] \qquad \text{or}$$

$$r = \frac{2K^2}{1 + \cos(\theta + c)} \, .$$

12. In the last integration, let $u = e^y$.

18. We want to make stationary the integral

$$(1) \qquad \int \sqrt{x + y} \, ds = \int \sqrt{x + y} \sqrt{dx^2 + dy^2} \, .$$

We make the change of variables indicated in the text. We find

$$dX^2 + dY^2 = \frac{1}{2}(dx^2 + 2 \, dx \, dy + dy^2 + dx^2 - 2 \, dx \, dy + dy^2) = dx^2 + dy^2 .$$

Then the integral in (1) becomes

$$\int \sqrt{\sqrt{2} \, X} \sqrt{dX^2 + dY^2} = 2^{1/4} \int \sqrt{X} \sqrt{1 + Y'^2} \, dX .$$

The Euler equation is

18.   (continued)

$$\frac{d}{dx}\left(\frac{\sqrt{X}\,Y'}{\sqrt{1+Y'^2}}\right) = 0 \qquad \text{or} \qquad \frac{\sqrt{X}\,Y'}{\sqrt{1+Y'^2}} = K.$$

We solve for Y' and integrate again.

$$XY'^2 = K^2(1+Y'^2),$$

$$Y'^2(X - K^2) = K^2 \qquad \text{so} \qquad Y' = K/\sqrt{X - K^2},$$

$$Y = K\int \frac{dX}{\sqrt{X - K^2}} = 2K\sqrt{X - K^2} + C,$$

$$(Y - C)^2 = 4K^2(X - K^2).$$

In the original variables, that is, relative to the (x,y) axes, the equation is

$$\left(\frac{x-y}{\sqrt{2}} - C\right)^2 = 4K^2\left(\frac{x+y}{\sqrt{2}} - K^2\right).$$

If we multiply both sides by $(\sqrt{2})^2$ and let $a = C\sqrt{2}$, $b = K^2\sqrt{2}$, we have

$$(x - y - a)^2 = 4b(x + y - b).$$

This is a parabola; it is easiest to see this from the (X,Y) equation.

22.   See the solution of Chapter 10, Problem 9.4.

23.   Hint: Let the (x,y) plane be vertical with the x axis pointing down. Use polar coordinates and let the equation of the circle be r = a.

26.   Hint: The electrostatic potential is proportional to $\int \frac{ds}{r}$.

Chapter 10

Section 1

4. We use the notation of text equation (1.1) to write element ij of ABC:

$$(ABC)_{ij} = \sum_{k\ell} A_{ik} B_{k\ell} C_{\ell j} .$$

The trace of the matrix ABC is

$$\sum_{i} (ABC)_{ii} = \sum_{i} \sum_{k\ell} A_{ik} B_{k\ell} C_{\ell i} = \sum_{ik\ell} A_{ik} B_{k\ell} C_{\ell i} .$$

The product $A_{ik} B_{k\ell} C_{\ell i}$ is a product of matrix <u>elements</u> (numbers); we can rearrange the order of a product of numbers any way we like. Let us rearrange this product in two ways with the i's next to each other so that we recognize the sum as the trace of a matrix product, namely

$$\sum_{ik\ell} A_{ik} B_{k\ell} C_{\ell i} = \sum_{\ell} \sum_{ik} C_{\ell i} A_{ik} B_{k\ell} = \sum_{\ell} (CAB)_{\ell\ell} = Tr(CAB) \qquad \text{or}$$

$$\sum_{ik\ell} A_{ik} B_{k\ell} C_{\ell i} = \sum_{k} \sum_{\ell i} B_{k\ell} C_{\ell i} A_{ik} = \sum_{k} (BCA)_{kk} = Tr(BCA) .$$

Thus $Tr(ABC) = Tr(CAB) = Tr(BCA)$. Note carefully, however that $Tr(ABC)$ is <u>not</u> in general equal to $Tr(ACB)$, etc. Only cyclic permutations leave the trace unchanged.

5. Hint: Note that the trace of a matrix (sum of diagonal elements) is not changed by transposing the matrix. That is,

$$tr\, M = Tr(M^T) .$$

Section 3

2.

$$r = \begin{pmatrix} x \\ y \\ z \end{pmatrix}, \qquad\qquad r^T = (x \quad y \quad z),$$

$$r^T r = (x \quad y \quad z) \begin{pmatrix} x \\ y \\ z \end{pmatrix} = (x^2 + y^2 + z^2).$$

$$r' = \begin{pmatrix} x' \\ y' \\ z' \end{pmatrix}, \qquad\qquad r'^T = (x' \quad y' \quad z'),$$

$$r'^T r' = (x' \quad y' \quad z') \begin{pmatrix} x' \\ y' \\ z' \end{pmatrix} = x'^2 + y'^2 + z'^2.$$

If $r' = Mr$, then $r'^T r' = (Mr)^T (Mr) = r^T M^T Mr$ by text equation (1.2). If M is orthogonal (that is, $M^T = M^{-1}$) this becomes

$$r'^T r' = r^T M^{-1} Mr = r^T r$$

since $M^{-1} M = 1$ (where 1 means the unit matrix). Thus

$$x'^2 + y'^2 + z'^2 = x^2 + y^2 + z^2,$$

which says that an orthogonal transformation does not change the length of a vector.

9. As in text page 130, equation (8.2), write the matrix A whose rows are the $n+1$ vectors; this matrix then has $n+1$ rows and n columns. If we row reduce the n by n matrix consisting of the first n rows of A, either there will be at least one row of zeros, or the reduced echelon form (text pages 121-122) will be the unit n by n matrix. In the latter case, we can apply further row operations to make row $n+1$ a row of zeros. In both cases, we find only n linearly independent vectors.

Section 4

2. To show that $\vec{r}_1$ and $\vec{r}_2$ are perpendicular, we find

$$\vec{r}_1 \cdot \vec{r}_2 = \left(\frac{1}{\sqrt{5}}\right)\left(\frac{-2}{\sqrt{5}}\right) + \left(\frac{2}{\sqrt{5}}\right)\left(\frac{1}{\sqrt{5}}\right) = 0.$$

To verify that C is orthogonal, that is, that $C^T = C^{-1}$, we show that $CC^T = $ unit matrix:

$$CC^T = \frac{1}{\sqrt{5}}\begin{pmatrix} 1 & -2 \\ 2 & 1 \end{pmatrix}\frac{1}{\sqrt{5}}\begin{pmatrix} 1 & 2 \\ -2 & 1 \end{pmatrix} = \frac{1}{5}\begin{pmatrix} 5 & 0 \\ 0 & 5 \end{pmatrix} = \begin{pmatrix} 1 & 0 \\ 0 & 1 \end{pmatrix}.$$

7. Let C be an n by n matrix whose columns are the components of n mutually perpendicular unit vectors $\vec{r}_1$, $\vec{r}_2$, $\cdots$, $\vec{r}_n$. Then

$$C^T C = \begin{pmatrix} x_1 & y_1 & \cdots \\ x_2 & y_2 & \cdots \\ \cdot & \cdot & \\ \cdot & \cdot & \\ \cdot & \cdot & \\ x_n & y_n & \cdots \end{pmatrix}\begin{pmatrix} x_1 & x_2 & \cdots & x_n \\ y_1 & y_2 & \cdots & y_n \\ \cdot & \cdot & & \cdot \\ \cdot & \cdot & & \cdot \\ \cdot & \cdot & & \cdot \end{pmatrix}$$

$$= \begin{pmatrix} \vec{r}_1 \cdot \vec{r}_1 & \vec{r}_1 \cdot \vec{r}_2 & \cdots & \vec{r}_1 \cdot \vec{r}_n \\ \vec{r}_2 \cdot \vec{r}_1 & \vec{r}_2 \cdot \vec{r}_2 & \cdots & \vec{r}_2 \cdot \vec{r}_n \\ \cdot & \cdot & & \cdot \\ \cdot & \cdot & & \cdot \\ \vec{r}_n \cdot \vec{r}_1 & \vec{r}_n \cdot \vec{r}_2 & \cdots & \vec{r}_n \cdot \vec{r}_n \end{pmatrix} = \begin{pmatrix} 1 & 0 & \cdots & 0 \\ 0 & 1 & \cdots & 0 \\ \cdot & \cdot & & \cdot \\ \cdot & \cdot & & \cdot \\ 0 & 0 & \cdots & 1 \end{pmatrix}$$

so $C^T = C^{-1}$. Thus C is an orthogonal matrix.

10.   Given: A and B are orthogonal matrices. This means

$$A^{-1} = A^T, \qquad\qquad B^{-1} = B^T.$$

Then

$$(AB)^T = B^T A^T \qquad\qquad \text{by text equation (1.2)},$$
$$= B^{-1} A^{-1} \qquad\qquad \text{since A and B are orthogonal},$$
$$= (AB)^{-1} \qquad\qquad \text{by text equation (1.3)}.$$

Thus AB is an orthogonal matrix since its inverse equals its transpose.

13.   To find the eigenvalues of the given matrix, we subtract μ from the elements of the main diagonal and set the resulting determinant equal to zero [see text equation (4.3) and the shaded box below it].

$$\begin{vmatrix} 2-\mu & 2 \\ 2 & -1-\mu \end{vmatrix} = 0.$$

Then we evaluate this determinant and solve the resulting equation (called the characteristic equation of the matrix).

$$\mu^2 - \mu - 6 = 0,$$
$$(\mu - 3)(\mu + 2) = 0, \qquad \mu = 3, \qquad \mu = -2.$$

Now an eigenvector (corresponding to each value of μ) is $\hat{i}x + \hat{j}y$ where (x,y) is a solution of the matrix equation

$$\begin{pmatrix} 2-\mu & 2 \\ 2 & -1-\mu \end{pmatrix} \begin{pmatrix} x \\ y \end{pmatrix} = 0.$$

For $\mu = 3$, we find

$$\begin{pmatrix} -1 & 2 \\ 2 & -4 \end{pmatrix} \begin{pmatrix} x \\ y \end{pmatrix} = 0, \qquad \text{or} \qquad \begin{cases} -x + 2y = 0, \\ 2x - 4y = 0. \end{cases}$$

13.   (continued)

These two equations both represent the same straight line, say
x = 2y.   [Note that this kind of problem is self-checking; you
<u>must</u> get equations representing one straight line corresponding
to any (non-repeated) eigenvalue.   If you don't, then hunt for
your mistake.]   A solution of x = 2y is   x = 2, y = 1, so an eigenvector corresponding to the eigenvalue $\mu = 3$ is $2\vec{i} + \vec{j}$ or (2,1).
A corresponding unit eigenvector is $(2\vec{i} + \vec{j})/\sqrt{5}$.   For $\mu = -2$,
we find

$$\begin{pmatrix} 4 & 2 \\ 2 & 1 \end{pmatrix} \begin{pmatrix} x \\ y \end{pmatrix} = 0 \qquad \text{or} \qquad 2x + y = 0.$$

A solution of this equation is x = 1, y = -2, so an eigenvector
corresponding to $\mu = -2$ is (1,-2) and the corresponding unit
eigenvector is $(1,-2)/\sqrt{5}$.

17.   To find the eigenvalues of the given matrix, we write and solve
the characteristic equation:

$$\begin{vmatrix} 5 - \mu & 0 & 2 \\ 0 & 3 - \mu & 0 \\ 2 & 0 & 5 - \mu \end{vmatrix} = 0. \qquad \mu = 3, \quad (5 - \mu)^2 = 4, \qquad \text{so}$$

$$\mu = 7, \ 3, \ 3.$$

We note that $\mu = 3$ is an eigenvalue since, if $\mu = 3$, we have a
column (or row) of zeros in the determinant.   In that case, do
<u>not</u> multiply the determinant out to obtain a cubic equation for
μ, since you must then do unnecessary work to factor it!
Instead, evaluate the determinant by a Laplace development using
the row (or column) containing $(3 - \mu)$; this gives
$(3 - \mu)[(5 - \mu)^2 - 4] = 0$, or $\mu = 3$, $(5 - \mu)^2 = 4$, as above.

To find the eigenvectors, we find (for each value of μ) a
solution (x,y,z) of the matrix equation

17.   (continued)

$$\begin{pmatrix} 5-\mu & 0 & 2 \\ 0 & 3-\mu & 0 \\ 2 & 0 & 5-\mu \end{pmatrix} \begin{pmatrix} x \\ y \\ z \end{pmatrix} = 0.$$

For $\mu = 7$, we have

$$\begin{cases} -2x & +2z = 0, \\ -4y & = 0, \\ 2x & -2z = 0, \end{cases} \qquad \text{or} \qquad x = z, \quad y = 0.$$

Thus, an eigenvector corresponding to $\mu = 7$ is $(1,0,1)$ or $\vec{i} + \vec{k}$.

The corresponding unit eigenvector is $\dfrac{\vec{i} + \vec{k}}{\sqrt{2}}$.

For $\mu = 3$, we find

$$\begin{cases} 2x & + & 2z = 0, \\ & & 0 = 0, \\ 2x & + & 2z = 0, \end{cases} \qquad \text{or} \qquad z = -x.$$

This is the equation of a plane; because $\mu = 3$ was a double eigenvalue, we find a whole plane of eigenvectors. We choose any two
perpendicular eigenvectors in this plane, for example $(1,1,-1)$
and $(-1,2,1)$. (Note that both these eigenvectors have $z = -x$, and
that the dot product of the two vectors is zero.) A simpler set
of two perpendicular eigenvectors in the plane $z = -x$ is $(1,0,-1)$
and $(0,1,0)$; the corresponding unit eigenvectors are $\dfrac{\vec{i} - \vec{k}}{\sqrt{2}}$ and $\vec{j}$.
Note that all the $\mu = 3$ eigenvectors are perpendicular to the
eigenvector $\vec{i} + \vec{k}$ found above for $\mu = 7$.

22.   To find the eigenvalues of the given matrix, we subtract μ from
the elements of the main diagonal, and set the resulting determinant equal to zero.

22.   (continued)

$$\begin{vmatrix} -3 - \mu & 2 & 2 \\ 2 & 1 - \mu & 3 \\ 2 & 3 & 1 - \mu \end{vmatrix} = 0.$$

We evaluate the determinant by a Laplace development (see text, Chapter 3, Section 3).

$$(-3 - \mu) \begin{vmatrix} 1 - \mu & 3 \\ 3 & 1 - \mu \end{vmatrix} - 2 \begin{vmatrix} 2 & 2 \\ 3 & 1 - \mu \end{vmatrix} + 2 \begin{vmatrix} 2 & 2 \\ 1 - \mu & 3 \end{vmatrix} = 0.$$

After a little algebra, we get

(1) $\mu^3 + \mu^2 - 22\mu - 40 = 0.$

To solve for μ we think of factoring this polynomial:

$(\mu - a)(\mu - b)(\mu - c) = 0.$ Then we see that $abc = 40$, so if there are integral roots they must divide 40.   The possibilities are ± 1, ± 2, ± 4, ± 5, etc.   We see quickly, by inspection, that $\mu = 1$ and $\mu = -1$ do not satisfy the equation.   By synthetic division (see Appendix B to these Solutions) we try $\mu = 2$:

```
1   1   -22   -40    |2
        2     6    -32
    _____
    1   3   -16   -72
```

Try $\mu = -2$:

```
1    1   -22   -40     |-2
         -2    2    40
    _____
    1   -1   -20    0
```

Then one solution of (1) is $\mu = -2$, and the other solutions are the roots of

$$\mu^2 - \mu - 20 = 0 = (\mu - 5)(\mu + 4),$$

so $\mu = 5$, -4.

22.   (continued)

To find the eigenvectors, we write the matrix equation (equivalent to three equations):

$$(2) \quad \begin{pmatrix} -3-\mu & 2 & 2 \\ 2 & 1-\mu & 3 \\ 2 & 3 & 1-\mu \end{pmatrix} \begin{pmatrix} x \\ y \\ z \end{pmatrix} = 0.$$

If $\mu = -2$, equation (2) gives:

$$-x + 2y + 2z = 0,$$
$$2x + 3y + 3z = 0,$$
$$2x + 3y + 3z = 0.$$

Eliminate x from the first two equations:

$$7y + 7z = 0, \quad \text{or} \quad y = -z.$$

If $z = 1$ then $y = -1$, and from the first equation $x = 2y + 2z = 0$. An eigenvector in the form $\vec{i}x + \vec{j}y + \vec{k}z$ is $-\vec{j} + \vec{k}$, or in the form (x,y,z) is $(0,-1,1)$; the corresponding unit eigenvector is $\frac{1}{\sqrt{2}}(-\vec{j} + \vec{k})$.

If $\mu = 5$, equation (2) gives:

$$-8x + 2y + 2z = 0,$$
$$2x - 4y + 3z = 0,$$
$$2x + 3y - 4z = 0.$$

Subtract the second equation from the third: $7y - 7z = 0$, $y = z$. Then $2x = 4y - 3z = y$. If $x = 1$, then $y = z = 2$. (Verify that these values satisfy all three equations.) Then $(1,2,2)$ is an eigen-

22.   (continued)

vector and $(1,2,2)/3$ is a unit eigenvector corresponding to
$\mu = 5$.

If $\mu = -4$, equation (2) gives

$$x + 2y + 2z = 0,$$
$$2x + 5y + 3z = 0,$$
$$2x + 3y + 5z = 0.$$

From the last two equations, $2y - 2z = 0$; $y = z$. If $y = z = 1$,
then $x = -4$. Then $(-4,1,1)$ is an eigenvector and $(-4,1,1)/(3\sqrt{2})$
is a unit eigenvector corresponding to $\mu = -4$.

The given matrix is symmetric, so the eigenvectors are
orthogonal. Verify this by finding the dot product.

$$\text{If} \qquad D = \begin{pmatrix} -2 & 0 & 0 \\ 0 & 5 & 0 \\ 0 & 0 & -4 \end{pmatrix} \qquad \text{then} \qquad C = \begin{pmatrix} 0 & \dfrac{1}{3} & \dfrac{-4}{3\sqrt{2}} \\ -\dfrac{1}{\sqrt{2}} & \dfrac{2}{3} & \dfrac{1}{3\sqrt{2}} \\ \dfrac{1}{\sqrt{2}} & \dfrac{2}{3} & \dfrac{1}{3\sqrt{2}} \end{pmatrix}$$

C is an orthogonal matrix.

29. $\begin{vmatrix} 3 - \mu & 4 \\ 4 & 9 - \mu \end{vmatrix} = \mu^2 - 12\mu + 11 = 0 = (\mu - 1)(\mu - 11).$

For $\mu = 1$: $\begin{cases} 2x + 4y = 0, & x = 2, \ y = -1 \\ 4x + 8y = 0, & \text{Eigenvector: } 2\vec{i} - \vec{j}. \end{cases}$

For $\mu = 11$: $\begin{cases} -8x + 4y = 0, & x = 1, \ y = 2, \\ 4x - 2y = 0, & \text{Eigenvector: } \vec{i} + 2\vec{j}. \end{cases}$

Relative to new axes x' along $2\vec{i} - \vec{j}$, and y' along $\vec{i} + 2\vec{j}$, the deformation is given by

$$\begin{pmatrix} X' \\ Y' \end{pmatrix} = \begin{pmatrix} 1 & 0 \\ 0 & 11 \end{pmatrix} \begin{pmatrix} x' \\ y' \end{pmatrix},$$

that is, X' = x', Y' = 11y'. This means that the plane is stretched by a factor 11 away from the x' axis parallel to the ±y' directions, with no deformation in the x' direction.

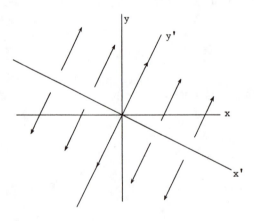

33. $\begin{vmatrix} A - \mu & H \\ H & B - \mu \end{vmatrix} = 0,$ $\qquad\qquad \mu^2 - (A+B)\mu + AB - H^2 = 0,$

$$\mu = \frac{A+B \pm \sqrt{(A+B)^2 - 4(AB - H^2)}}{2} = \frac{A+B \pm \sqrt{(A-B)^2 + 4H^2}}{2}.$$

We note that μ is real since $(A-B)^2 + 4H^2 \geqslant 0$. Let μ_1 and μ_2 be the two eigenvalues. The eigenvector equations are:

$$(A - \mu_1)x_1 + Hy_1 = 0, \qquad (A - \mu_2)x_2 + Hy_2 = 0,$$
$$Hx_1 + (B - \mu_1)y_1 = 0, \qquad Hx_2 + (B - \mu_2)y_2 = 0.$$

Let us choose the following as eigenvectors:

$$x_1 = -H, \qquad y_1 = A - \mu_1,$$
$$x_2 = B - \mu_2, \qquad y_2 = -H.$$

(You might verify that these values satisfy both equations.) Then the dot product of the eigenvectors is

$$x_1 x_2 + y_1 y_2 = -H(B - \mu_2 + A - \mu_1) = -H[A + B - (\mu_1 + \mu_2)] = 0$$

since $\mu_1 + \mu_2 = A + B$. Thus the eigenvectors are perpendicular. You might like to verify that other choices of eigenvectors (for example, $x_2 = -H$, $y_2 = A - \mu_2$) give the same result.

36. (b) By the definition of antisymmetric (see Problem 1.2), $M^T = -M$, and by text equation (3.4), $C^T = C^{-1}$. We want to show that $(C^{-1}MC)^T = -C^{-1}MC$. By text equation (1.2),

$$(C^{-1}MC)^T = C^T M^T (C^{-1})^T = C^{-1}(-M)C = -C^{-1}MC.$$

37. Hint: See text, page 117, equation (6.2c).

Section 5

2. We write the given equation in matrix form [see text equations
    (5.1) and (5.2)].

$$2x^2 + 4xy - y^2 = (x \quad y) \begin{pmatrix} 2 & 2 \\ 2 & -1 \end{pmatrix} \begin{pmatrix} x \\ y \end{pmatrix} = 24.$$

In Problem 4.13 above, we diagonalized this matrix and found
eigenvalues $\mu = 3$, -2. Thus relative to principal axes the
equation of this conic is (see text, page 421):

$$3x'^2 - 2y'^2 = 24.$$

We see that the curve is a hyperbola.

6. We write the given equation in matrix form:

$$x^2 + y^2 + z^2 + 4xy + 2xz - 2yz = 12 = (x \quad y \quad z) \begin{pmatrix} 1 & 2 & 1 \\ 2 & 1 & -1 \\ 1 & -1 & 1 \end{pmatrix} \begin{pmatrix} x \\ y \\ z \end{pmatrix} .$$

We want the eigenvalues of this matrix.

$$\begin{vmatrix} 1-\mu & 2 & 1 \\ 2 & 1-\mu & -1 \\ 1 & -1 & 1-\mu \end{vmatrix} = 0 = (1-\mu) \begin{vmatrix} 1-\mu & -1 \\ -1 & 1-\mu \end{vmatrix} - 2 \begin{vmatrix} 2 & 1 \\ -1 & 1-\mu \end{vmatrix} + \begin{vmatrix} 2 & 1 \\ 1-\mu & -1 \end{vmatrix} .$$

This simplifies to

$$\mu^3 - 3\mu^2 - 3\mu + 9 = 0.$$

Possible integral roots must divide 9 so we try ± 1, ± 3, ± 9. (See
solution of Problem 4.22.) It is clear by inspection that ± 1 do
not satisfy the equation. Try -3:

6.   (continued)

$$
\begin{array}{rrrr|r}
1 & -3 & -3 & 9 & -3 \\
  & -3 & 18 & -45 & \\
\hline
1 & -6 & 15 & -36 &
\end{array}
$$

Try +3:

$$
\begin{array}{rrrr|r}
1 & -3 & -3 & 9 & 3 \\
  & 3 & 0 & -9 & \\
\hline
1 & 0 & -3 & 0 &
\end{array}
\qquad \text{so } \mu = 3,
$$

$$\mu^2 - 3 = 0, \qquad \mu = \pm\sqrt{3}.$$

Then relative to principal axes, the equation is

$$3x'^2 + \sqrt{3}y'^2 - \sqrt{3}z'^2 = 12 \qquad \text{or}$$

$$\frac{x'^2}{2^2} + \frac{y'^2}{(2\sqrt[4]{3})^2} - \frac{z'^2}{(2\sqrt[4]{3})^2} = 1.$$

This is a hyperboloid of one sheet (note the one minus sign) with semi-axes 2, $2\sqrt[4]{3}$, $2\sqrt[4]{3}$. See the first figure.

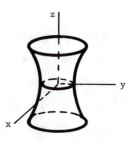

Comment: You may also see a hyperboloid of two sheets, for example $y^2 - x^2 - 4z^2 = 1$. Note the two minus signs and see the second figure.

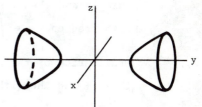

13. The Lagrangian for a single pendulum (Chapter 9, Problem 5.4) is
$L = \frac{1}{2} m\ell^2 \dot{\theta}^2 + mg\ell \cos \theta$.   Here we have a second pendulum and also a
spring with potential energy $\frac{1}{2} k(x - y)^2 = \frac{1}{2} k\ell^2 (\sin \theta - \sin \phi)^2$.
Thus the total Lagrangian is

$$L = \frac{1}{2} m\ell^2 \dot{\theta}^2 + \frac{1}{2} m\ell^2 \dot{\phi}^2 + mg\ell \cos \theta + mg\ell \cos \phi$$
$$- \frac{1}{2} k\ell^2 (\sin \theta - \sin \phi)^2 .$$

Using the suggested approximations and value of k, we get

$$L = \frac{1}{2} m\ell^2 \dot{\theta}^2 + \frac{1}{2} m\ell^2 \dot{\phi}^2 + mg\ell \left(1 - \frac{\theta^2}{2}\right) + mg\ell \left(1 - \frac{\phi^2}{2}\right)$$
$$- \frac{1}{2} \frac{mg\ell^2}{\ell} (\theta - \phi)^2 .$$

This simplifies to

$$L = \frac{1}{2} m\ell^2 (\dot{\theta}^2 + \dot{\phi}^2) - \frac{mg\ell}{2} (2\theta^2 + 2\phi^2 - 2\theta\phi) + 2mg\ell .$$

We diagonalize $2\theta^2 + 2\phi^2 - 2\theta\phi$ exactly as we diagonalized
$2x^2 + 2y^2 - 2xy$ (text page 424).   The eigenvalues and eigenvectors
are the same as in text Example 3.   Thus the two characteristic
modes of vibration are:

$\begin{cases} \text{Oscillation of two pendulums in the same direction like} \\ \text{this } \rightrightarrows \text{ and then like this } \leftleftarrows \text{ (spring not stretched)} \\ \text{with frequency } \omega = \sqrt{\dfrac{mg\ell}{m\ell^2}} = \sqrt{\dfrac{g}{\ell}} . \end{cases}$

$\begin{cases} \text{Oscillation of two pendulums in opposite directions like} \\ \text{this } \rightleftarrows \text{ and then like this } \leftrightarrows \text{ with frequency } \omega = \sqrt{\dfrac{3g}{\ell}} . \end{cases}$

Section 8

1. For spherical coordinates r, θ, φ,
    we have (see figure, or text,
    page 219)

    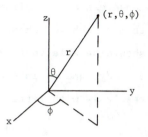

    $$x = r \sin \theta \cos \phi,$$
    $$y = r \sin \theta \sin \phi,$$
    $$z = r \cos \theta.$$

    Then

    $$dx = \sin \theta \cos \phi \, dr + r \cos \theta \cos \phi \, d\theta - r \sin \theta \sin \phi \, d\phi,$$
    $$dy = \sin \theta \sin \phi \, dr + r \cos \theta \sin \phi \, d\theta + r \sin \theta \cos \phi \, d\phi,$$
    $$dz = \qquad \cos \theta \, dr \qquad - r \sin \theta \, d\theta.$$

    To find $dx^2 + dy^2 + dz^2$, it is not necessary (or even desirable) to
    write out all the terms in dx^2, dy^2, and dz^2. Instead we con-
    sider the terms in groups and do the algebra mentally. We find
    the coefficient of dr^2, then of $dr \, d\theta$, then of $d\theta^2$, etc. The
    coefficient of dr^2 is $\sin^2\theta(\cos^2\phi + \sin^2\phi) + \cos^2\theta = 1$. (With a
    little practice, you should not need to write even this much.
    Note that the terms in dx, dy, and dz were carefully written with
    the dr terms in one column, the dθ terms in another, etc., to
    make the combinations easy to do.) Next, for the coefficient of
    $dr \, d\theta$ we find, from dz^2: $-2r \sin \theta \cos \theta$, and from $dx^2 + dy^2$:
    $+2r \sin \theta \cos \theta (\cos^2\phi + \sin^2\phi)$. Thus the dr dθ terms all cancel.
    Now I have written more than is necessary so that you can see
    what to do. Try the coefficient of dθ dφ -- you should see that
    there are only two terms, which are negatives of each other, so

1.   (continued)

these cancel. Similarly consider the $dr\,d\phi$ terms, and the $d\phi^2$ terms which you should easily find give $r^2\sin^2\theta\,d\phi^2$. Now you should be able to tackle the $d\theta^2$ terms: $r^2\cos^2\theta\,d\theta^2$ from $dx^2 + dy^2$ and $r^2\sin^2\theta\,d\theta^2$ from dz^2, giving just $r^2d\theta^2$. Thus our result -- with little or no writing -- is

$$ds^2 = dr^2 + r^2d\theta^2 + r^2\sin^2\theta\,d\phi^2.$$

The scale factors are 1, r, r sin θ, and the volume element (product of scale factors and differentials) is

$$dV = dr\cdot r\,d\theta\cdot r\sin\theta\,d\phi = r^2\sin\theta\,dr\,d\theta\,d\phi.$$

Then

$$\text{vector } \vec{ds} = \vec{e}_r\,dr + \vec{e}_\theta r\,d\theta + \vec{e}_\phi r\sin\theta\,d\phi$$

The $\vec{a}$ vectors are:

$$\vec{a}_r = \vec{i}\,\frac{\partial x}{\partial r} + \vec{j}\,\frac{\partial y}{\partial r} + \vec{k}\,\frac{\partial z}{\partial r} = \vec{i}\sin\theta\cos\phi + \vec{j}\sin\theta\sin\phi + \vec{k}\cos\theta = \vec{e}_r\,,$$

$$\vec{a}_\theta = \vec{i}\,\frac{\partial x}{\partial\theta} + \vec{j}\,\frac{\partial y}{\partial\theta} + \vec{k}\,\frac{\partial z}{\partial\theta} = \vec{i}r\cos\theta\cos\phi + \vec{j}r\cos\theta\sin\phi - \vec{k}r\sin\theta = r\vec{e}_\theta\,,$$

$$\vec{a}_\phi = \vec{i}\,\frac{\partial x}{\partial\phi} + \vec{j}\,\frac{\partial y}{\partial\phi} + \vec{k}\,\frac{\partial z}{\partial\phi} = -\vec{i}r\sin\theta\sin\phi + \vec{j}r\sin\theta\cos\phi = r\sin\theta\,\vec{e}_\phi\,.$$

Note that the $\vec{a}$ vectors are the scale factors times the $\vec{e}$ vectors. However, it is not necessary to know the scale factors in advance. For example, given $\vec{a}_\phi = -\vec{i}r\sin\theta\sin\phi + \vec{j}r\sin\theta\cos\phi$, we can just find its magnitude $= r\sin\theta$; this will be the scale factor, and we divide $\vec{a}_\phi$ by it giving $\vec{e}_\phi = -\vec{i}\sin\phi + \vec{j}\cos\phi$ which is a unit vector.

3.   In problem 1 we found

$$(1) \quad \begin{cases} \vec{e}_r = \vec{i} \sin\theta \cos\phi + \vec{j} \sin\theta \sin\phi + \vec{k} \cos\theta, \\ \vec{e}_\theta = \vec{i} \cos\theta \cos\phi + \vec{j} \cos\theta \sin\phi - \vec{k} \sin\theta, \\ \vec{e}_\phi = -\vec{i} \sin\phi + \vec{j} \cos\phi. \end{cases}$$

Then

$$(2) \quad \frac{d\vec{e}_r}{dt} = \vec{i}(\cos\theta \cos\phi \, \dot\theta - \sin\theta \sin\phi \, \dot\phi)$$
$$+ \vec{j}(\cos\theta \sin\phi \, \dot\theta + \sin\theta \cos\phi \, \dot\phi) + \vec{k}(-\sin\theta \, \dot\theta)$$
$$= \vec{e}_\theta \dot\theta + \vec{e}_\phi \sin\theta \, \dot\phi.$$

(We get the last step by comparing the terms we have in $d\vec{e}_r/dt$ with $\vec{e}_\theta$ and $\vec{e}_\phi$. Remember that dots over letters mean time derivatives: $\dot\theta = d\theta/dt$, $\dot{\vec{e}}_r = d\vec{e}_r/dt$.)

$$(3) \quad \frac{d\vec{e}_\theta}{dt} = i(-\sin\theta \cos\phi \, \dot\theta - \cos\theta \sin\phi \, \dot\phi)$$
$$+ \vec{j}(-\sin\theta \sin\phi \, \dot\theta + \cos\theta \cos\phi \, \dot\phi) - \vec{k} \cos\theta \, \dot\theta$$
$$= -\vec{e}_r \dot\theta + \vec{e}_\phi \cos\theta \, \dot\phi.$$

$$(4) \quad \frac{d\vec{e}_\phi}{dt} = -\vec{i} \cos\phi \, \dot\phi - \vec{j} \sin\phi \, \dot\phi = -(\vec{e}_r \sin\theta + \vec{e}_\theta \cos\theta)\dot\phi.$$

You may wonder how to get this last expression in (4). First observe that the derivative of a unit vector $d\vec{e}/dt$ cannot have an $\vec{e}$ component. If it did this would mean that part of the change in $\vec{e}$ is <u>along</u> $\vec{e}$, which would mean that $\vec{e}$ is changing in length; it can't change in length since it always has unit length. Thus $d\vec{e}/dt$ must be perpendicular to $\vec{e}$ [note that this was true in (2) and (3)]. Then $d\vec{e}_\phi/dt$ must be some combination of $\vec{e}_r$ and $\vec{e}_\theta$. Since $\vec{e}_r$ and $\vec{e}_\theta$ [equation (1)] contain $\vec{k}$ terms and $d\vec{e}_\phi/dt$ does not, we want the combination $\sin\theta \, \vec{e}_r + \cos\theta \, \vec{e}_\theta$ so that the $\vec{k}$ terms drop out. This gives

3.   (continued)

$$\vec{i} \cos\phi(\sin^2\theta + \cos^2\theta) + \vec{j}\,\sin\phi(\sin^2\theta + \cos^2\theta)$$

$$= \vec{i}\,\cos\phi + \vec{j}\,\sin\phi,$$

and we see from the middle expression in (4) that we need $-\dot\phi$

times this.

Using (2), (3) and (4) we can evaluate $\vec{v} = d\vec{s}/dt$, and
$\vec{a} = d^2\vec{s}/dt^2 = d\vec{v}/dt$. We differentiate $\vec{s} = r\vec{e}_r$ to get

(5) $\dfrac{d\vec{s}}{dt} = \dot{r}\vec{e}_r + r\dot{\vec{e}}_r = \dot{r}\vec{e}_r + r(\vec{e}_\theta\dot\theta + \vec{e}_\phi \sin\theta\,\dot\phi).$

We could have written (5) immediately by dividing vector $d\vec{s}$ in

Problem 1 by the scalar dt:

(6) $\dfrac{d\vec{s}}{dt} = \vec{e}_r \dfrac{dr}{dt} + \vec{e}_\theta r \dfrac{d\theta}{dt} + \vec{e}_\phi r \sin\theta \dfrac{d\phi}{dt}.$

(Note that we are not differentiating in this step, but merely

changing from differential to derivative notation. By definition,

$dr = \dfrac{dr}{dt}\,dt$; thus the formal division of dr by dt gives $\dfrac{dr}{dt}$.

Similarly $r \sin\theta\,d\phi$ means $r \sin\theta \dfrac{d\phi}{dt}\,dt$; dividing this by dt merely

cancels dt and has no effect on $r \sin\theta$.)

Next we differentiate (5) or (6) to get the acceleration:

(7) $\dfrac{d^2\vec{s}}{dt^2} = \vec{e}_r \ddot{r} + \dot{r}\dot{\vec{e}}_r + \vec{e}_\theta(r\ddot\theta + \dot{r}\dot\theta) + r\dot\theta\dot{\vec{e}}_\theta$

$\qquad\qquad + \vec{e}_\phi(r \sin\theta\,\ddot\phi + r \cos\theta\,\dot\theta\dot\phi + \sin\theta\,\dot{r}\dot\phi) + r \sin\theta\,\dot\phi\dot{\vec{e}}_\phi.$

Substituting from (2), (3), (4) for the derivatives of the unit

basis vectors, we find for those terms:

3.   (continued)

$$\dot{r}\dot{\vec{e}}_r + r\dot{\theta}\dot{\vec{e}}_\theta + r \sin\theta\ \dot{\phi}\dot{\vec{e}}_\phi =$$

$$\dot{r}(\vec{e}_\theta\dot{\theta} + \vec{e}_\phi \sin\theta\ \dot{\phi}) + r\dot{\theta}(-\vec{e}_r\dot{\theta} + \vec{e}_\phi \cos\theta\ \dot{\phi})$$

$$+ r \sin\theta\ \dot{\phi}(-\vec{e}_r \sin\theta\ \dot{\phi} - \vec{e}_\theta \cos\theta\ \dot{\phi})$$

$$= \vec{e}_r(-r\dot{\theta}^2 - r\sin^2\theta\ \dot{\phi}^2) + \vec{e}_\theta(r\dot{\theta} - r\sin\theta\cos\theta\ \dot{\phi}^2)$$

$$+ \vec{e}_\phi(\sin\theta\ r\dot{\phi} + r\cos\theta\ \dot{\theta}\dot{\phi}).$$

Now we combine these terms with the rest of the terms in (7)
to get

(8) $$\frac{d^2\vec{s}}{dt} = \vec{e}_r(\ddot{r} - r\dot{\theta}^2 - r\sin^2\theta\ \dot{\phi}^2) + \vec{e}_\theta(r\ddot{\theta} + 2\dot{r}\dot{\theta} - r\sin\theta\cos\theta\ \dot{\phi}^2)$$

$$+ \vec{e}_\phi(r\sin\theta\ \ddot{\phi} + 2r\cos\theta\ \dot{\theta}\dot{\phi} + 2\sin\theta\ \dot{r}\dot{\phi}).$$

9.   We find dx and dy from the given equations and simplify the
algebra using $\sin^2 v + \cos^2 v = 1$ and $\cosh^2 u - \sinh^2 u = 1$ (text,
page 70) to get

$$dx = \frac{a(1 + \cos v \cosh u)du + a \sinh u \sin v\, dv}{(\cosh u + \cos v)^2}$$

(1)

$$dy = \frac{-a \sinh u \sin v\, du + a(1 + \cosh u \cos v)dv}{(\cosh u + \cos v)^2}.$$

In finding $ds^2 = dx^2 + dy^2$, we first note that the coefficients of
du dv in dx^2 and dy^2 are of opposite sign but the same magnitude,
so there is no du dv term in ds^2 (orthogonal system).   Next we
note that the coefficient of du^2 and the coefficient of dv^2 are
the same, namely

9.   (continued)

(2) $\dfrac{a^2\sinh^2 u\,\sin^2 v + a^2(1+\cosh u\,\cos v)^2}{(\cosh u + \cos v)^4}$.

If we put $\sinh^2 u = \cosh^2 u - 1$ and $\sin^2 v = 1 - \cos^2 v$, this simplifies

to

$$\dfrac{a^2(\cosh u + \cos v)^2}{(\cosh u + \cos v)^4} = \dfrac{a^2}{(\cosh u + \cos v)^2}\ .$$

Remember that this is the coefficient of du^2 and of dv^2. Then

$$ds^2 = \dfrac{a^2}{(\cosh u + \cos v)^2}(du^2 + dv^2)\,.$$

Scale factors: $h_u = h_v = \dfrac{a}{\cosh u + \cos v}$.

Surface area element $dS = \dfrac{a^2}{(\cosh u + \cos v)^2}\,du\,dv.$

If also $z = z$ (that is, a 3-dimensional transformation), then the

volume element is

$$dV = \dfrac{a^2}{(\cosh u + \cos v)^2}\,du\,dv\,dz,$$

Vector $d\vec{s} = h_u\vec{e}_u du + h_v\vec{e}_v dv = \dfrac{a(\vec{e}_u du + \vec{e}_v dv)}{\cosh u + \cos v}$.

To find the $\vec{a}$ vectors, note that we can read $\dfrac{\partial x}{\partial u}$, etc. from (1),

since $dx = \dfrac{\partial x}{\partial u}du + \dfrac{\partial x}{\partial v}dv$; thus $\dfrac{\partial x}{\partial u}$ is the coefficient of du in the

equation for dx in (1), and similarly for the other partial

derivatives. The calculation of the lengths of $\vec{a}_u$ and $\vec{a}_v$ is the

same calculation we did in finding ds^2; thus we do not have to

find h_u and h_v again.

9.   (continued)

$$\vec{a}_u = \vec{i}\,\frac{\partial x}{\partial u} + \vec{j}\,\frac{\partial y}{\partial u} = \frac{a(1 + \cosh u \cos v)\vec{i} - a \sinh u \sin v\,\vec{j}}{(\cosh u + \cos v)^2} = h_u \vec{e}_u \;,$$

$$\vec{a}_v = \vec{i}\,\frac{\partial x}{\partial v} + \vec{j}\,\frac{\partial y}{\partial v} = \frac{a \sinh u \sin v\,\vec{i} + a(1 + \cosh u \cos v)\vec{j}}{(\cosh u + \cos v)^2} = h_v \vec{e}_v \;.$$

14.   We divide $d\vec{s}$ in Problem 9 above by dt to get

(1) $$\vec{v} = \frac{d\vec{s}}{dt} = \frac{a(\vec{e}_u \dot{u} + \vec{e}_v \dot{v})}{\cosh u + \cos v}\;.$$

[If you don't understand this, see the solution of Problem 3, just after equation (6).]   To find $d\vec{v}/dt$, we need the derivatives of the unit basis vectors $\vec{e}_u$ and $\vec{e}_v$.   From Problem 9, we have

(2)
$$\vec{e}_u = \frac{(1 + \cosh u \cos v)\vec{i} - \sinh u \sin v\,\vec{j}}{\cosh u + \cos v}\;,$$

$$\vec{e}_v = \frac{\sinh u \sin v\,\vec{i} + (1 + \cosh u \cos v)\vec{j}}{\cosh u + \cos v}\;.$$

Differentiating these functions involves rather complicated algebra; let us try to simplify our work.   We observe that $\vec{e}_u$ and $\vec{e}_v$ contain just three different functions of u and v; we define

(3)
$$\begin{cases} C = C(u,v) = 1 + \cosh u \cos v, \\ P = P(u,v) = \sinh u \sin v, \\ S = S(u,v) = \cosh u + \cos v. \end{cases}$$

Then

(4) $$\vec{e}_u = \frac{\vec{i}C - \vec{j}P}{S}\;, \qquad\qquad \vec{e}_v = \frac{\vec{i}P + \vec{j}C}{S}\;.$$

We want to differentiate these with respect to t.   Now for any function f(u,v), $\frac{df}{dt} = \frac{\partial f}{\partial u}\dot{u} + \frac{\partial f}{\partial v}\dot{v}$; we use this formula to find various partial derivatives we shall need.

14.   (continued)

From (3)

$$\frac{\partial}{\partial u}\left(\frac{C}{S}\right) = \frac{1}{S^2}\left(S\,\frac{\partial C}{\partial u} - C\,\frac{\partial S}{\partial u}\right)$$

$$= \frac{1}{S^2}[(\cosh u + \cos v)\sinh u \cos v - (1 + \cosh u \cos v)\sinh u]$$

$$= \frac{1}{S^2}\sinh u(\cos^2 v - 1) = -\frac{1}{S^2}\sinh u \sin^2 v = -\frac{P}{S^2}\sin v.$$

Similarly, by differentiating and simplifying, we find

$$\frac{\partial}{\partial v}\left(\frac{C}{S}\right) = -\frac{P}{S^2}\sinh u,$$

$$\frac{\partial}{\partial u}\left(\frac{P}{S}\right) = \frac{C}{S^2}\sin v,$$

$$\frac{\partial}{\partial v}\left(\frac{P}{S}\right) = \frac{C}{S^2}\sinh u.$$

Now we can find the time derivatives of the unit basis vectors in (4):

$$\dot{\vec{e}}_u = \vec{i}\left[\frac{\partial}{\partial u}\left(\frac{C}{S}\right)\dot{u} + \frac{\partial}{\partial v}\left(\frac{C}{S}\right)\dot{v}\right] - \vec{j}\left[\frac{\partial}{\partial u}\left(\frac{P}{S}\right)\dot{u} + \frac{\partial}{\partial v}\left(\frac{P}{S}\right)\dot{v}\right]$$

(5) $$= -\vec{i}\,\frac{P}{S^2}(\dot{u}\sin v + \dot{v}\sinh u) - \vec{j}\,\frac{C}{S^2}(\dot{u}\sin v + \dot{v}\sinh u)$$

$$= -(\dot{u}\sin v + \dot{v}\sinh u)\left(\frac{\vec{i}P + \vec{j}C}{S^2}\right) = -\frac{\dot{u}\sin v + \dot{v}\sinh u}{S}\vec{e}_v.$$

We could work out $\dot{\vec{e}}_v$ in the same way using the same partial derivatives we have calculated. However, if we observe that we can change $\vec{e}_u$ into $\vec{e}_v$ by changing $\vec{i}$ to $\vec{j}$, and $\vec{j}$ to $-\vec{i}$, we can make these same substitutions in $\dot{\vec{e}}_u$; then $\vec{e}_v$ in (5) becomes $(-\vec{e}_u)$, and we have

(6) $$\dot{\vec{e}}_v = \frac{\dot{u}\sin v + \dot{v}\sinh u}{\cosh u + \cos v}\vec{e}_u.$$

14.   (continued)

Finally, we find the acceleration by differentiating (1).

(7) $\frac{d\vec{v}}{dt} = \frac{a}{S}(\vec{e}_u\ddot{u} + \dot{u}\dot{\vec{e}}_u + \vec{e}_v\ddot{v} + \dot{v}\dot{\vec{e}}_v) - \frac{a}{S^2}(\vec{e}_u\dot{u} + \vec{e}_v\dot{v})(\dot{u}\sinh u - \dot{v}\sin v).$

Using (5) and (6) we find

$$\dot{u}\dot{\vec{e}}_u + \dot{v}\dot{\vec{e}}_v = \frac{1}{S}(\dot{u}\sin v + \dot{v}\sinh u)(-\dot{u}\vec{e}_v + \dot{v}\vec{e}_u).$$

Combining this with the other terms in (7) and simplifying, we find

(8) $\frac{d\vec{v}}{dt} = \vec{e}_u a\left(\frac{\ddot{u}}{S} + \frac{2\dot{u}\dot{v}\sin v + (\dot{v}^2 - \dot{u}^2)\sinh u}{S^2}\right)$

$\qquad\qquad + \vec{e}_v a\left(\frac{\ddot{v}}{S} - \frac{2\dot{u}\dot{v}\sinh u + (\dot{u}^2 - \dot{v}^2)\sin v}{S^2}\right)$,

where $S = \cosh u + \cos v$. Compare this problem with Problem 9.9 below where we find $\frac{d\vec{v}}{dt}$ with much less work. If the derivatives of the unit basis vectors are not needed for other purposes, finding them is an unnecessarily complicated way of finding the acceleration components.

Section 9

1. By text equation (9.2), $\nabla x_1 = \vec{e}_1/h_1$, and similarly for ∇x_2 and ∇x_3. If $\vec{e}_1 \times \vec{e}_2 = \vec{e}_3$, then

    (1) $(\nabla x_1) \times (\nabla x_2) = \dfrac{\vec{e}_3}{h_1 h_2}$.

    We now use identity (h) on text page 296, with $\vec{U} = \nabla x_1$ and $\vec{V} = \nabla x_2$.

    $$\nabla \cdot [(\nabla x_1) \times (\nabla x_2)] = (\nabla x_2) \cdot (\nabla \times \nabla x_1) - (\nabla x_1) \cdot (\nabla \times \nabla x_2) = 0$$

    since by identity (b) on text page 296,

    $$\nabla \times \nabla x_1 = 0, \qquad\qquad \nabla \times \nabla x_2 = 0.$$

    Thus we have

    $$\nabla \cdot \left(\frac{\vec{e}_3}{h_1 h_2} \right) = 0.$$

    Prove the other parts in a similar way.

4. In spherical coordinates,

    $$ds^2 = dr^2 + r^2 d\theta^2 + r^2 \sin^2\theta \, d\phi^2,$$

    $$\left(\frac{ds}{dt} \right)^2 = \dot{r}^2 + r^2 \dot{\theta}^2 + r^2 \sin^2\theta \, \dot{\phi}^2.$$

    If V is a function of r, θ, ϕ, then the Lagrangian is

    $$L = \frac{1}{2} m (\dot{r}^2 + r^2 \dot{\theta}^2 + r^2 \sin^2\theta \, \dot{\phi}^2) - V(r, \theta, \phi).$$

    Lagrange's equations are:

    $$\frac{d}{dt}(m\dot{r}) - mr\dot{\theta}^2 - mr \sin^2\theta \, \dot{\phi}^2 + \frac{\partial V}{\partial r} = 0,$$

    $$\frac{d}{dt}(mr^2\dot{\theta}) - mr^2 \sin\theta \cos\theta \, \dot{\phi}^2 + \frac{\partial V}{\partial \theta} = 0,$$

    $$\frac{d}{dt}(mr^2 \sin^2\theta \, \dot{\phi}) + \frac{\partial V}{\partial \phi} = 0.$$

4.   (continued)

We find the indicated time derivatives, and write each equation in the form mass · acceleration = force. Remember that $\vec{F} = -\nabla V$, and the components of the gradient in spherical coordinates are $\frac{\partial V}{\partial r}$, $\frac{1}{r}\frac{\partial V}{\partial \theta}$, $\frac{1}{r \sin \theta}\frac{\partial V}{\partial \phi}$. Thus we divide the θ Lagrange equation by r and the ϕ Lagrange equation by r sin θ.

r: $m(\ddot{r} - r\dot{\theta}^2 - r\sin^2\theta\,\dot{\phi}^2) = -\frac{\partial V}{\partial r} = F_r$.

θ: $\begin{cases} m(r^2\ddot{\theta} + 2r\dot{r}\dot{\theta} - r^2\sin\theta\cos\theta\,\dot{\phi}^2) = -\frac{\partial V}{\partial \theta} \qquad \text{or} \\[2mm] m(r\ddot{\theta} + 2\dot{r}\dot{\theta} - r\sin\theta\cos\theta\,\dot{\phi}^2) = -\frac{1}{r}\frac{\partial V}{\partial \theta} = F_\theta. \end{cases}$

ϕ: $\begin{cases} m(r^2\sin^2\theta\,\ddot{\phi} + 2r^2\sin\theta\cos\theta\,\dot{\theta}\dot{\phi} + 2r\sin^2\theta\,\dot{r}\dot{\phi}) = -\frac{\partial V}{\partial \phi} \qquad \text{or} \\[2mm] m(r\sin\theta\,\ddot{\phi} + 2r\cos\theta\,\dot{\theta}\dot{\phi} + 2\sin\theta\,\dot{r}\dot{\phi}) = -\frac{1}{r\sin\theta}\frac{\partial V}{\partial \phi} = F_\phi. \end{cases}$

Thus the acceleration components are:

$$\ddot{r} - r\dot{\theta}^2 - r\sin^2\theta\,\dot{\phi}^2,$$

$$r\ddot{\theta} + 2\dot{r}\dot{\theta} - r\sin\theta\cos\theta\,\dot{\phi}^2,$$

$$r\sin\theta\,\ddot{\phi} + 2r\cos\theta\,\dot{\theta}\dot{\phi} + 2\sin\theta\,\dot{r}\dot{\phi},$$

as in Problem 8.3, but with less calculation.

9.   In bipolar coordinates u,v we have (see solution of Problem 8.9)

$$\left(\frac{ds}{dt}\right)^2 = \frac{a^2}{(\cosh u + \cos v)^2}(\dot{u}^2 + \dot{v}^2).$$

Then

$$L = \frac{1}{2}ma^2\,\frac{\dot{u}^2 + \dot{v}^2}{(\cosh u + \cos v)^2} - V(u,v)$$

and the Lagrange equations are:

9.   (continued)

$$\frac{d}{dt}\left(\frac{ma^2\dot{u}}{(\cosh u + \cos v)^2}\right) + \frac{ma^2(\dot{u}^2 + \dot{v}^2)\sinh u}{(\cosh u + \cos v)^3} + \frac{\partial V}{\partial u} = 0,$$

$$\frac{d}{dt}\left(\frac{ma^2\dot{v}}{(\cosh u + \cos v)^2}\right) + \frac{ma^2(\dot{u}^2 + \dot{v}^2)(-\sin v)}{(\cosh u + \cos v)^3} + \frac{\partial V}{\partial v} = 0.$$

Differentiation with respect to t gives

$$\frac{ma^2\ddot{u}}{(\cosh u + \cos v)^2}$$

$$+ \frac{-2ma^2\dot{u}(\sinh u\,\dot{u} - \sin v\,\dot{v}) + ma^2\sinh u(\dot{u}^2 + \dot{v}^2)}{(\cosh u + \cos v)^3} = -\frac{\partial V}{\partial u},$$

$$\frac{ma^2\ddot{v}}{(\cosh u + \cos v)^2}$$

$$+ \frac{-2ma^2\dot{v}(\sinh u\,\dot{u} - \sin v\,\dot{v}) - ma^2\sin v(\dot{u}^2 + \dot{v}^2)}{(\cosh u + \cos v)^3} = -\frac{\partial V}{\partial v}.$$

We want to divide both equations by the scale factor

$$h_u = h_v = \frac{a}{\cosh u + \cos v}$$

so that the right-hand sides will be components of $\vec{F} = -\nabla V$. We
get

$$m\left[\frac{a\ddot{u}}{\cosh u + \cos v} + \frac{a(\dot{v}^2 - \dot{u}^2)\sinh u + 2a\dot{u}\dot{v}\sin v}{(\cosh u + \cos v)^2}\right] = -\frac{1}{h_u}\frac{\partial V}{\partial u} = F_u,$$

$$m\left[\frac{a\ddot{v}}{\cosh u + \cos v} + \frac{a(\dot{v}^2 - \dot{u}^2)\sin v - 2a\dot{u}\dot{v}\sinh u}{(\cosh u + \cos v)^2}\right] = -\frac{1}{h_v}\frac{\partial V}{\partial v} = F_v.$$

The expressions in brackets are the acceleration components.
Compare the (much longer) solution of Problem 8.14.

13.   From the solution of Problem 8.9 we have

$$h_u = h_v = \frac{a}{\cosh u + \cos v}, \qquad h_z = 1.$$

[The given transformation was in the (x,y) plane. Thus z was not changed and $ds^2 = dx^2 + dy^2 + dz^2 = h_u^2\,du^2 + h_v^2\,dv^2 + dz^2$. Then $h_z = 1$ and $\vec{e}_z = \vec{k}$. We need these values in order to use the text equations which are given in 3 dimensions. If U and $\vec{V}$ below are functions only of u and v, then the z terms simply drop out.]

By text equation (9.2):

$$\nabla U = \frac{\vec{e}_u}{h_u}\,\frac{\partial U}{\partial u} + \frac{\vec{e}_v}{h_v}\,\frac{\partial U}{\partial v} + \vec{e}_z\,\frac{\partial U}{\partial z}.$$

By text equation (9.8):

$$\nabla \cdot \vec{V} = \frac{1}{h_u h_v}\left[\frac{\partial}{\partial u}(h_v V_u) + \frac{\partial}{\partial v}(h_u V_v)\right] + \frac{\partial V_z}{\partial z}.$$

By text equation (9.10):

$$\nabla^2 U = \frac{1}{h_u h_v}\left[\frac{\partial}{\partial u}\!\left(\frac{h_v}{h_u}\,\frac{\partial U}{\partial u}\right) + \frac{\partial}{\partial v}\!\left(\frac{h_u}{h_v}\,\frac{\partial U}{\partial v}\right)\right] + \frac{\partial^2 U}{\partial z^2},$$

or, since $h_u = h_v$ in this example,

$$\nabla^2 U = \frac{1}{h_u h_v}\left[\frac{\partial^2 U}{\partial u^2} + \frac{\partial^2 U}{\partial v^2}\right] + \frac{\partial^2 U}{\partial z^2}.$$

Similarly, substituting in text equation (9.11), we can find $\nabla \times \vec{V}$.

Comment about text equations (9.2), (9.8), and (9.10):

These equations should be remembered, not as formulas, but using words, like this:

(9.2) Each component of the gradient is a partial derivative times the reciprocal of the corresponding scale factor.

(Informally: one over scale factor times partial derivative.)

Comment, continued:

(9.8) The divergence $\nabla \cdot \vec{V}$ starts with the reciprocal of the product of the scale factors times a bracket. Inside the bracket are 3 terms, each of which is a partial derivative of the corresponding component of $\vec{V}$ multiplied by the other two scale factors. (Informally: one over product of scale factors times sum of partials, each of its $\vec{V}$ component times other two scale factors.)

(9.10) Combine (9.2) and (9.8). (Informally: same as divergence, with component of $\vec{V}$ replaced by component of gradient.)

16. To find $\nabla \cdot \vec{e}_r$ in cylindrical coordinates, we use text equation (9.9) with $\vec{V} = V_r \vec{e}_r + V_\theta \vec{e}_\theta + V_z \vec{e}_z = \vec{e}_r$, that is, $V_r = 1$, $V_\theta = V_z = 0$:

$$\nabla \cdot \vec{e}_r = \frac{1}{r}\left[\frac{\partial}{\partial r}(r \cdot 1) + 0\right] = \frac{1}{r} .$$

Similarly, evaluate the other expressions using the appropriate text equations in cylindrical coordinates. Alternatively, use the general equations, text (9.2), (9.8), (9.10) and (9.11) with:
$x_1, x_2, x_3 = r, \theta, z$; $h_1, h_2, h_3 = 1, r, 1$; $V_1, V_2, V_3 = V_r, V_\theta, V_z$;
$\vec{e}_1, \vec{e}_2, \vec{e}_3 = \vec{e}_r, \vec{e}_\theta, \vec{e}_z$.

19. In spherical coordinates, we have:

$$x_1, x_2, x_3 = r, \theta, \phi \qquad \vec{e}_1, \vec{e}_2, \vec{e}_3 = \vec{e}_r, \vec{e}_\theta, \vec{e}_\phi$$

$$h_1, h_2, h_3 = 1, r, r \sin \theta \qquad V_1, V_2, V_3 = V_r, V_\theta, V_\phi$$

Then to find $\nabla \times (r\vec{e}_\theta)$ we use text equation (9.11) with $V_1 = V_3 = 0$, $V_2 = r$; we find

19.   (continued)

$$\frac{\partial}{\partial x_3}(h_2 V_2) = \frac{\partial}{\partial \phi}(r^2) = 0,$$

$$\frac{\partial}{\partial x_1}(h_2 V_2) = \frac{\partial}{\partial r}(r^2) = 2r,$$

and all other terms are zero since $V_1 = V_3 = 0$. Thus

$$\nabla \times (r\vec{e}_\theta) = \frac{\vec{e}_\phi}{r} 2r = 2\vec{e}_\phi .$$

Similarly evaluate the other expressions.

21.   Hint for $\nabla^2 e^{ikr \cos \theta}$: Since $r \cos \theta = z$, it is much easier to use rectangular coordinates.

Section 11

5.   A third order (Cartesian) tensor (in 3 dimensions) consists of $3^3 = 27$ components in every rectangular coordinate system with the components in two systems related by

(1) $$T'_{\ell mn} = \sum_i \sum_j \sum_k a_{\ell i} a_{mj} a_{nk} T_{ijk}$$

where A = matrix of the a_{ij}'s is the rotation matrix such that $r' = Ar$.

If also S_{ijk} is a third order (Cartesian) tensor, then

(2) $$S'_{\ell mn} = \sum_i \sum_j \sum_k a_{\ell i} a_{mj} a_{nk} S_{ijk} .$$

If we add the T equation and the S equation, we get

(3) $$T'_{\ell mn} + S'_{\ell mn} = \sum_i \sum_j \sum_k a_{\ell i} a_{mj} a_{nk} (T_{ijk} + S_{ijk}).$$

5.   (continued)

The expression on the right is correct since the coefficients of T_{ijk} and S_{ijk} are the same for each set of indices. Equation (3) says that $T + S$ is a third order tensor. Thus the sum of two third order tensors is a third order tensor.

The proof is almost identical for general tensors in Section 13. We just need to replace each a_{ij} by the appropriate partial derivative $\dfrac{\partial x_i'}{\partial x_j}$ or $\dfrac{\partial x_j}{\partial x_i'}$.

6.   (d) In 3 dimensions, $a_{ij}b_{jk} = a_{ij}c_{jk}$ means (according to the summation convention)

$$a_{i1}b_{1k} + a_{i2}b_{2k} + a_{i3}b_{3k} = a_{i1}c_{1k} + a_{i2}c_{2k} + a_{i3}c_{3k}$$

where i and k have some fixed values. Suppose $i = 3$ and $k = 1$; then the equation is:

$$a_{31}b_{11} + a_{32}b_{21} + a_{33}b_{31} = a_{31}c_{11} + a_{32}c_{11} + a_{33}c_{31}.$$

This is like saying

$$2 \cdot 5 + 1 \cdot 3 + 7 \cdot 1 = 2 \cdot 1 + 1 \cdot 4 + 7 \cdot 2 = 20.$$

Observe that no factor can be cancelled from every term in the equation. Thus, in $a_{ij}b_{jk} = a_{ij}c_{jk}$, we cannot cancel a_{ij} .

Section 12

1.

$$UV^T = \begin{pmatrix} U_1 \\ U_2 \\ U_3 \end{pmatrix} (V_1 \quad V_2 \quad V_3) = \begin{pmatrix} U_1V_1 & U_1V_2 & U_1V_3 \\ U_2V_1 & U_2V_2 & U_2V_3 \\ U_3V_1 & U_3V_2 & U_3V_3 \end{pmatrix}.$$

If $U' = AU$ and $V' = AV$, then $V'^T = V^T A^{-1}$, so

$$U'V'^T = AUV^T A^{-1} = A(UV^T)A^{-1}$$

as in (12.5).

5.   In matrix form, $U = TV$, or

$$\begin{pmatrix} U_1 \\ U_2 \\ U_3 \end{pmatrix} = \begin{pmatrix} T_{11} & T_{12} & T_{13} \\ T_{21} & T_{22} & T_{23} \\ T_{31} & T_{32} & T_{33} \end{pmatrix} \begin{pmatrix} V_1 \\ V_2 \\ V_3 \end{pmatrix}$$

In dyadic form

$$\vec{i}U_1 + \vec{j}U_2 + \vec{k}U_3 = (\vec{i}\vec{i}T_{11} + \vec{i}\vec{j}T_{12} + \cdots) \cdot (\vec{i}V_1 + \vec{j}V_2 + \vec{k}V_3).$$

In summation form

$$U_k = \sum_i T_{ki} V_i .$$

You should satisfy yourself that all these forms give the same result for U_1, U_2, and U_3, in terms of the nine components of T and the three components of V.

8.   In Problem 4.18, the eigenvalues and eigenvectors are:

$$\mu_1 = 4 \qquad \vec{a}_1 = 2\vec{i} + \vec{j} + 3\vec{k},$$
$$\mu_2 = 2, \qquad \vec{a}_2 = \qquad -3\vec{j} + \vec{k},$$
$$\mu_3 = -3, \qquad \vec{a}_3 = 5\vec{i} - \vec{j} - 3\vec{k}.$$

Then, if we interpret the matrix elements as components of a stress tensor, the stress consists of

tension of magnitude 4 in the direction $\vec{a}_1$,
tension of magnitude 2 in the direction $\vec{a}_2$,
compression of magnitude 3 in the direction $\vec{a}_3$.

11.   (a) Expand the triple vector product by text page 240, equation (3.9).

$$\vec{L} = m\vec{r} \times (\vec{\omega} \times \vec{r}) = m(\vec{r}\cdot\vec{r}\vec{\omega} - \vec{r}\cdot\vec{\omega}\vec{r}).$$

Now we can write

$$\vec{L} = m(r^2 \mathbf{1} - \vec{r}\vec{r})\cdot\vec{\omega}$$

where $\mathbf{1}$ is the unit dyadic (see Problem 3, text page 443). Then the inertial tensor is (for $m = 1$)

$$\mathbf{I} = r^2\mathbf{1} - \vec{r}\vec{r} \qquad \text{in dyadic notation,}$$
$$I = r^2 U - rr^T \qquad \text{in matrix notation}$$
$$\text{(where } U = \text{unit matrix),}$$
$$I_{ij} = r^2 \delta_{ij} - x_i x_j \qquad \text{in component notation.}$$

(See text, page 139, for δ_{ij}.) Write these three different forms out in detail to see that they are equivalent. Compare $L = \mathbf{I}\cdot\vec{\omega}$ with Problem 5, assuming $U = L$, $T = I$, $V = \omega$.

11.   (continued)

(b) Using the component form of I from part (a), we write

$$I'_{k\ell} = r'^2 \delta'_{k\ell} - x'_k x'_\ell .$$

Satisfy yourself that

$$\delta'_{k\ell} = \sum_{i=1}^{3} \sum_{j=1}^{3} a_{ki} a_{\ell j} \delta_{ij} .$$

(Hint: If $i = j$, then we have the dot product of rows k and ℓ of matrix A; recall that A is orthogonal.) As in text equation (11.11)

$$x'_k = \sum_{i=1}^{3} a_{ki} x_i \qquad \text{and} \qquad x'_\ell = \sum_{j=1}^{3} a_{\ell j} x_j .$$

Also for an orthogonal transformation, $r'^2 = r^2$. Thus

$$I'_{k\ell} = r^2 \sum_{i=1}^{3} \sum_{j=1}^{3} a_{ki} a_{\ell j} \delta_{ij} - \sum_{i=1}^{3} \sum_{j=1}^{3} a_{ki} a_{\ell j} x_i x_j$$

$$= \sum_{i=1}^{3} \sum_{j=1}^{3} a_{ki} a_{\ell j} \left(r^2 \delta_{ij} - x_i x_j \right) = \sum_{i=1}^{3} \sum_{j=1}^{3} a_{ki} a_{\ell j} I_{ij}$$

so by text equation (11.13), I is a second order tensor.
Suggestion: Try the proof in matrix notation.

(c) Hint: Let the x' axis be along $\vec{n}$. Write the moment of inertia about the x' axis and compare with $I_{x'x'} = \vec{n} \cdot \mathbf{I} \cdot \vec{n}$.

Section 13

4. By text (13.4)

$$dr' = J\, dr.$$

Then

$$(dr')^T = (dr)^T J^T.$$

Now

$$ds^2 = dx_1'^2 + dx_2'^2 + dx_3'^2 = (dx_1' \quad dx_2' \quad dx_3') \begin{pmatrix} dx_1' \\ dx_2' \\ dx_3' \end{pmatrix}$$

$$= (dr')^T (dr') = (dr)^T J^T J\, dr.$$

We are given

$$ds^2 + h_1^2 dx_1^2 + h_2^2 dx_2^2 + h_3^2 dx_3^2 \ .$$

Thus

$$J^T J = \begin{pmatrix} h_1^2 & 0 & 0 \\ 0 & h_2^2 & 0 \\ 0 & 0 & h_3^2 \end{pmatrix}.$$

Then

$$\det(J^T J) = (h_1 h_2 h_3)^2$$

and

$$\det(J^T J) = (\det J^T)(\det J) = (\det J)^2,$$

so

$$\det J = h_1 h_2 h_3 \ .$$

7.   By the chain rule for differentiation [text, page 162, equation (7.2)]

$$\frac{\partial u}{\partial x_i'} = \frac{\partial u}{\partial x_1}\frac{\partial x_1}{\partial x_i'} + \frac{\partial u}{\partial x_2}\frac{\partial x_2}{\partial x_i'} + \frac{\partial u}{\partial x_3}\frac{\partial x_3}{\partial x_i'} = \sum_{j=1}^{3}\frac{\partial u}{\partial x_j}\frac{\partial x_j}{\partial x_i'} \ .$$

We compare this with text equation (13.6) to see that ∇u with components $\frac{\partial u}{\partial x_j}$ is a covariant vector.

11. $$ds^2 = \sum_{i,j}g_{ij}\,dx_i\,dx_j = \sum_{k,\ell}g_{k\ell}'\,dx_k'\,dx_\ell' \ .$$

We know that dx is a contravariant vector, that is

$$dx_i = \sum_k \frac{\partial x_i}{\partial x_k'}\,dx_k'$$

so

$$\sum_{k,\ell}g_{k\ell}'\,dx_k'\,dx_\ell' \equiv \sum_{i,j}g_{ij}\sum_k\frac{\partial x_i}{\partial x_k'}\,dx_k'\sum_\ell\frac{\partial x_j}{\partial x_\ell'}\,dx_\ell'$$

$$\equiv \sum_{k,\ell}\left(\sum_{i,j}g_{ij}\frac{\partial x_i}{\partial x_k'}\frac{\partial x_j}{\partial x_\ell'}\right)dx_k'\,dx_\ell' \ .$$

Because this is an identity, the coefficients of $dx_k'\,dx_\ell'$ on the two sides of the equation must be equal.

$$g_{k\ell}' = \sum_{i,j}g_{ij}\frac{\partial x_i}{\partial x_k'}\frac{\partial x_j}{\partial x_\ell'} \ .$$

This is the transformation equation for a second order covariant tensor.

14. We are given

(1) $U_i = \sum_j T_{ij} V_j$

where U and V are vectors, that is, by text equation (11.12)

(2) $U_k' = \sum_i a_{ki} U_i$ and $V_m' = \sum_j a_{mj} V_j$.

In the primed coordinate system, let the relation between U and V be

(3) $U_k' = \sum_m T_{km}' V_m'$.

We want to find the relation between T' and T. Substitute (2) and (1) into (3) to get

$$\sum_i a_{ki} U_i = \sum_i a_{ki} \sum_j T_{ij} V_j = \sum_m T_{km}' \sum_j a_{mj} V_j \qquad \text{or}$$

(4) $\sum_j \left(\sum_i a_{ki} T_{ij} - \sum_m a_{mj} T_{km}' \right) V_j = 0 .$

Since V_j is an arbitrary vector, we can choose it to be, say, (1,0,0). Then equation (4) says that the parenthesis is zero when $j = 1$. Similarly we can choose V_j to have all zero components except for $j = 2$, and so on. Thus the parenthesis in (4) is zero for all j. Multiply the parenthesis by $a_{\ell j}$ and sum on j to get:

(5) $\sum_j \sum_i a_{ki} a_{\ell j} T_{ij} - \sum_j \sum_m a_{mj} a_{\ell j} T_{km}' = 0 .$

Now

(6) $\sum_j a_{mj} a_{\ell j} = \delta_{m\ell}$

14.   (continued)

because this is a dot product of rows m and ℓ of the A matrix
and A is orthogonal.   Then the second term in (5) is

$$\sum_m \delta_{m\ell} T'_{km} = T'_{k\ell}$$

so (5) gives

$$T'_{k\ell} = \sum_j \sum_i a_{ki} a_{\ell j} T_{ij} \; .$$

This is text equation (11.13); thus T_{ij} is a Cartesian tensor.

To give the proof in a general coordinate system, replace the
a_{ij} by appropriate partial derivatives according to text equations (13.6) and (13.7).

Section 15

1.   We find the eigenvalues and eigenvectors as in Section 4.

$$\begin{vmatrix} 1 - \mu & 0 \\ 3 & -2 - \mu \end{vmatrix} = \mu^2 + \mu - 2 = (\mu - 1)(\mu + 2) = 0$$

so the eigenvalues are $\mu = 1, -2$.

For $\mu = 1$: $\begin{pmatrix} 0 & 0 \\ 3 & -3 \end{pmatrix} \begin{pmatrix} x \\ y \end{pmatrix} = 0,$ $x - y = 0,$ $x = y,$

so an eigenvector is $(1,1)$ or $\vec{i} + \vec{j}$.

For $\mu = -2$: $\begin{pmatrix} 3 & 0 \\ 3 & 0 \end{pmatrix} \begin{pmatrix} x \\ y \end{pmatrix} = 0,$ $x = 0,$ $y = $ any number,

so an eigenvector is $(0,1)$ or $\vec{j}$.

8. From Problem 1, the unit eigenvectors are $(1,1)/\sqrt{2}$, and $(0,1)$.
    Recall from text, Section 4, that the columns of the C matrix
    are the unit eigenvectors. Thus

$$C = \begin{pmatrix} \dfrac{1}{\sqrt{2}} & 0 \\ \dfrac{1}{\sqrt{2}} & 1 \end{pmatrix} = \frac{1}{\sqrt{2}}\begin{pmatrix} 1 & 0 \\ 1 & \sqrt{2} \end{pmatrix}.$$

By text, page 123, equation (6.24), we find C^{-1}:

$$C^{-1} = \sqrt{2}\begin{pmatrix} 1 & 0 \\ -\dfrac{1}{\sqrt{2}} & \dfrac{1}{\sqrt{2}} \end{pmatrix} = \begin{pmatrix} \sqrt{2} & 0 \\ -1 & 1 \end{pmatrix}.$$

Note that C is <u>not</u> orthogonal $(C^{-1} \neq C^{T})$; we expected this since
M is not symmetric. However, we can verify by multiplying the
matrices that $C^{-1}MC = D$:

$$\begin{pmatrix} \sqrt{2} & 0 \\ -1 & 1 \end{pmatrix}\begin{pmatrix} 1 & 0 \\ 3 & -2 \end{pmatrix}\frac{1}{\sqrt{2}}\begin{pmatrix} 1 & 0 \\ 1 & \sqrt{2} \end{pmatrix} = \begin{pmatrix} 1 & 0 \\ 0 & -2 \end{pmatrix}.$$

10. As in Section 5, we write the equation in matrix form:

$$x^2 + y^2 - 5z^2 + 4xy = \begin{pmatrix} x & y & z \end{pmatrix}\begin{pmatrix} 1 & 2 & 0 \\ 2 & 1 & 0 \\ 0 & 0 & -5 \end{pmatrix}\begin{pmatrix} x \\ y \\ z \end{pmatrix} = 15.$$

We diagonalize the square matrix:

$$\mu = -5, \quad \text{and}$$

$$\begin{vmatrix} 1 - \mu & 2 \\ 2 & 1 - \mu \end{vmatrix} = 0, \quad \mu^2 - 2\mu - 3 = 0 = (\mu - 3)(\mu + 1), \quad \mu = 3,\ -1.$$

Relative to principal axes, the equation is

10.   (continued)

$$3x'^2 - y'^2 - 5z'^2 = 15 \quad \text{or} \quad \frac{x'^2}{(\sqrt{5})^2} - \frac{y'^2}{(\sqrt{15})^2} - \frac{z'^2}{(\sqrt{3})^2} = 1.$$

This is a hyperboloid of two sheets (see solution of Problem 5.6) and the shortest distance from the origin to the surface is $d = \sqrt{5}$ along the x' axis.

14.   From text Figure 5.1 with spring constants k, 3k, k, we have

$$V = \frac{1}{2} kx^2 + \frac{1}{2} \cdot 3k(x+y)^2 + \frac{1}{2} ky^2$$

$$= \frac{1}{2} k(4x^2 + 6xy + 4y^2).$$

We diagonalize the V matrix:

$$\begin{vmatrix} 4 - \mu & 3 \\ 3 & 4 - \mu \end{vmatrix} = \mu^2 - 8\mu + 7 = 0, \qquad \mu = 1, \ 7.$$

Then in terms of the primed variables (see text, page 424)

$$V = \frac{1}{2} k(x'^2 + 7y'^2),$$

$$T = \frac{1}{2} m(\dot{x}'^2 + \dot{y}'^2),$$

$$L = T - V = \frac{1}{2} m(\dot{x}'^2 + \dot{y}'^2) - \frac{1}{2} k(x'^2 + 7y'^2).$$

Lagrange's equations are

$$m\ddot{x}' = -kx',$$

$$m\ddot{y}' = -7ky',$$

so the characteristic frequencies are $\omega = \sqrt{k/m}, \ \sqrt{7k/m}.$

19.   In terms of the scale factors

$$ds^2 = h_u^2 \, du^2 + h_v^2 \, dv^2 .$$

Divide this by dt^2 to get

$$v^2 = \left(\frac{ds}{dt}\right)^2 = h_u^2 \, \dot{u}^2 + h_v^2 \, \dot{v}^2 .$$

Then the Lagrangian and Lagrange's equations are:

$$L = \frac{1}{2} m \left(h_u^2 \, \dot{u}^2 + h_v^2 \, \dot{v}^2 \right) - V(u,v)$$

$$m \frac{d}{dt} \left(h_u^2 \, \dot{u} \right) - m \left(\dot{u}^2 h_u \frac{\partial h_u}{\partial u} + \dot{v}^2 h_v \frac{\partial h_v}{\partial u} \right) + \frac{\partial V}{\partial u} = 0 ,$$

$$m \frac{d}{dt} \left(h_v^2 \, \dot{v} \right) - m \left(\dot{u}^2 h_u \frac{\partial h_u}{\partial v} + \dot{v}^2 h_v \frac{\partial h_v}{\partial v} \right) + \frac{\partial V}{\partial v} = 0 .$$

Evaluate the time derivatives, simplify, and divide each equation by the corresponding scale factor to write the equations in the form $m\vec{a} = -\nabla V = \vec{F}$.

(1) $$m \left(h_u \, \ddot{u} + \dot{u}^2 \frac{\partial h_u}{\partial u} + 2 \dot{u}\dot{v} \frac{\partial h_u}{\partial v} - \dot{v}^2 \frac{h_v}{h_u} \frac{\partial h_v}{\partial u} \right) = - \frac{1}{h_u} \frac{\partial V}{\partial u} = F_u ,$$

(2) $$m \left(h_v \, \ddot{v} + \dot{v}^2 \frac{\partial h_v}{\partial v} + 2 \dot{u}\dot{v} \frac{\partial h_v}{\partial u} - \dot{u}^2 \frac{h_u}{h_v} \frac{\partial h_u}{\partial v} \right) = - \frac{1}{h_v} \frac{\partial V}{\partial v} = F_v .$$

The expressions in parentheses are then the acceleration components.

The scale factors (see answer to Problem 15.17, text, page 765) are given by

$$h_u^2 = \frac{u^2 + v^2 - 1}{u^2 - 1} = 1 + \frac{v^2}{u^2 - 1} ,$$

$$h_v^2 = \frac{u^2 + v^2 - 1}{1 - v^2} = \frac{u^2}{1 - v^2} - 1 .$$

Then

19.   (continued)

$$2h_u \frac{\partial h_u}{\partial u} = - \frac{2v^2 u}{(u^2 - 1)^2} , \qquad\qquad 2h_v \frac{\partial h_v}{\partial u} = \frac{2u}{1 - v^2} ,$$

$$2h_u \frac{\partial h_u}{\partial v} = \frac{2v}{u^2 - 1} , \qquad\qquad 2h_v \frac{\partial h_v}{\partial v} = \frac{2u^2 v}{(1 - v^2)^2} .$$

Substitute these into equations (1) and (2) to get the acceleration components:

$$h_u \ddot{u} - \frac{1}{h_u} \left[\frac{v^2 u \dot{u}^2}{(u^2 - 1)^2} - \frac{2v \dot{u} \dot{v}}{u^2 - 1} + \frac{u \dot{v}^2}{1 - v^2} \right] = \frac{F_u}{m} ,$$

$$h_v \ddot{v} + \frac{1}{h_v} \left[\frac{u^2 v \dot{v}^2}{(1 - v^2)^2} + \frac{2u \dot{u} \dot{v}}{1 - v^2} - \frac{v \dot{u}^2}{u^2 - 1} \right] = \frac{F_v}{m} .$$

23.   In dyadic form (see text, page 442):

$$\mathbf{T} \cdot \vec{v} = (2\vec{\imath}\vec{\imath} - 3\vec{\imath}\vec{\jmath} + 5\vec{\jmath}\vec{k}) \cdot (\vec{\imath} - \vec{k}) = 2\vec{\imath} - 5\vec{\jmath},$$

$$\vec{v} \cdot \mathbf{T} = (\vec{\imath} - \vec{k}) \cdot (2\vec{\imath}\vec{\imath} - 3\vec{\imath}\vec{\jmath} + 5\vec{\jmath}\vec{k}) = 2\vec{\imath} - 3\vec{\jmath}.$$

In matrix form

$$\mathbf{T} \cdot \vec{v} = \begin{pmatrix} 2 & -3 & 0 \\ 0 & 0 & 5 \\ 0 & 0 & 0 \end{pmatrix} \begin{pmatrix} 1 \\ 0 \\ -1 \end{pmatrix} = \begin{pmatrix} 2 \\ -5 \\ 0 \end{pmatrix} = 2\vec{\imath} - 5\vec{\jmath},$$

$$\vec{v} \cdot \mathbf{T} = (1 \quad 0 \quad -1) \begin{pmatrix} 2 & -3 & 0 \\ 0 & 0 & 5 \\ 0 & 0 & 0 \end{pmatrix} = (2 \quad -3 \quad 0) = 2\vec{\imath} - 3\vec{\jmath}.$$

26.   The characteristic equation of $M = \begin{pmatrix} a & b \\ c & d \end{pmatrix}$ is

$$\begin{vmatrix} a - \mu & b \\ c & d - \mu \end{vmatrix} = 0 = \mu^2 - (a + d)\mu + (ad - bc) = 0.$$

Now $(a + d)$ is the trace of M (see text, page 409, Problem 4). Also $(ad - bc) = \det M$. Thus we can write the characteristic equation of M as

(1) $\mu^2 - (\text{Tr } M)\mu + \det M = 0.$

Let $M_1 = AB$ and $M_2 = BA$. By text, page 117, equation (6.2c),

$$\det M_1 = \det(AB) = (\det A)(\det B) = \det(BA) = \det M_2$$

and by text, page 409, Problem 4,

$$\text{Tr } M_1 = \text{Tr}(AB) = \text{Tr}(BA) = \text{Tr } M_2 \,.$$

Thus (1) above is the same equation for $M = M_1$ and $M = M_2$, so the eigenvalues μ are the same for $M_1 = AB$ and $M_2 = BA$.

33.   Verify that the given matrix C times its transpose is the unit matrix. Thus C is orthogonal [text equation (3.4)]. To find the rotation angle, we find tr C = sum of elements on the main diagonal, and use the given formula for $\cos \theta$.

$$\text{Tr } C = 2 \cos \theta + 1 = 0 - \frac{1}{3}\sqrt{\frac{1}{2}} + \frac{2}{3} \,,$$

$$\cos \theta = -0.2845, \qquad \theta = 106.5°.$$

Next we find the eigenvector corresponding to $\mu = 1$:

33.   (continued)

$$
\begin{vmatrix}
0-1 & -\sqrt{\tfrac{1}{2}} & \sqrt{\tfrac{1}{2}} \\
\tfrac{4}{3}\sqrt{\tfrac{1}{2}} & -\tfrac{1}{3}\sqrt{\tfrac{1}{2}}-1 & -\tfrac{1}{3}\sqrt{\tfrac{1}{2}} \\
\tfrac{1}{3} & \tfrac{2}{3} & \tfrac{2}{3}-1
\end{vmatrix}
\begin{pmatrix} x \\ y \\ z \end{pmatrix} = 0,
$$

$$
\begin{cases}
-x & - & \sqrt{\tfrac{1}{2}}\,y & + & \sqrt{\tfrac{1}{2}}\,z & = 0, \\
\tfrac{4}{3}\sqrt{\tfrac{1}{2}}\,x & - & \left(\tfrac{1}{3}\sqrt{2}+1\right)y & - & \tfrac{1}{3}\sqrt{\tfrac{1}{2}}\,z & = 0, \\
\tfrac{1}{3}x & + & \tfrac{2}{3}y & - & \tfrac{1}{3}z & = 0,
\end{cases}
\quad \text{or} \quad
\begin{cases}
x\sqrt{2} + y - z = 0, \\
4x - (1+3\sqrt{2})y - z = 0, \\
x + 2y - z = 0.
\end{cases}
$$

From the first and third equations, we find

$$y = x(\sqrt{2} - 1),$$
$$z = x(2\sqrt{2} - 1).$$

If $x = 1$, then $y = \sqrt{2} - 1$, $z = 2\sqrt{2} - 1$. As a check, verify that these values satisfy the second equation. Then the rotation axis is along the eigenvector

$$(1, \quad \sqrt{2} - 1, \quad 2\sqrt{2} - 1).$$

36.   See text, page 140, for definitions. Hermitian ($A^\dagger = A$) is the complex analogue of symmetric ($A^T = A$) and unitary ($A^\dagger = A^{-1}$) is the complex analogue of orthogonal ($A^T = A^{-1}$). If we take the transpose conjugate of the given matrix, we have the matrix we started with, so it is Hermitian. We find the eigenvalues and eigenvectors just as we did in Section 4 except that now x and y are allowed to be complex numbers.

36.   (continued)

$$\begin{vmatrix} 3 - \mu & 1 - i \\ 1 + i & 2 - \mu \end{vmatrix} = 0; \qquad \mu^2 - 5\mu + 4 = 0, \qquad \mu = 1, \ 4.$$

For $\mu = 1$: $\begin{pmatrix} 2 & 1 - i \\ 1 + i & 1 \end{pmatrix}\begin{pmatrix} x \\ y \end{pmatrix} = 0$. If $x = 1$, $y = -(1 + i) = -\sqrt{2}\, e^{i\pi/4}$.

Unit eigenvector: $(1, -1 - i)/\sqrt{3}$.

Note carefully that the length or norm of a complex eigenvector is defined as $(|x|^2 + |y|^2)^{1/2}$. Note also that we can multiply the unit eigenvector by any $e^{i\alpha}$ and still have a unit eigenvector. For example, multiply by $e^{3\pi i/4}$ to get the unit eigenvector

$$(e^{3\pi i/4}, -\sqrt{2}e^{i\pi})/\sqrt{3} = (\sqrt{2}e^{3\pi i/4}, 2)/\sqrt{6} = (-1 + i, 2)/\sqrt{6}$$

which we might have chosen originally.

For $\mu = 4$: $\begin{pmatrix} -1 & 1 - i \\ 1 + i & -2 \end{pmatrix}\begin{pmatrix} x \\ y \end{pmatrix} = 0$. If $y = 1$, $x = 1 - i$.

Unit eigenvector: $(1 - i, 1)/\sqrt{3}$.

Then one choice for C is

$$C = \frac{1}{\sqrt{3}}\begin{pmatrix} 1 & 1 - i \\ -1 - i & 1 \end{pmatrix}.$$

(All other choices for C are obtained by multiplying the first eigenvector by any $e^{i\alpha}$ and the second eigenvector by any $e^{i\beta}$.)
By text page 123, equation (6.24), we find

$$C^{-1} = \frac{1}{\sqrt{3}}\begin{pmatrix} 1 & -1 + i \\ 1 + i & 1 \end{pmatrix}.$$

We observe that $C^{-1} = C^\dagger$ so C is unitary.

Chapter 11

Section 1

For handbooks and other references on the functions in this chapter, see text, page 741 ff.

Section 2

See Solutions, Chapter 4.

Section 3

4. By repeated use of the recursion relation [text equation (3.4)], we find

    $$\Gamma(5.7) = 4.7\Gamma(4.7) = (4.7)(3.7)\Gamma(3.7)$$
    $$= (4.7)(3.7)(2.7)(1.7)\Gamma(1.7).$$

    From the table on the next page, $\Gamma(1.7) = 0.90864$. Then by calculator,

    $$\Gamma(5.7) = 72.528.$$

7. We use text equation (3.4) and solve for $\Gamma(p)$; this is also text equation (4.1). Then

    $$\Gamma(0.3) = \frac{1}{0.3}\Gamma(1.3) = \frac{1}{0.3}(0.89747) = 2.9916.$$

10. Compare the given integral with text equation (3.1). We see that $p = 1/2$, so the integral is $\Gamma(1/2)$. Either from the table on the next page [we find $\Gamma(0.5) = \frac{1}{0.5}\Gamma(1.5) = 2\Gamma(1.5)$], or by looking ahead to text equation (5.3), we find

    $$\Gamma(1/2) = \sqrt{\pi} = 1.7725.$$

15. Let $y = 8x$, $dy = 8\ dx$, $x^{-1/3} = (y/8)^{-1/3} = 2y^{-1/3}$. Then

$$\int_0^\infty x^{-1/3}\, e^{-8x}\, dx = \int_0^\infty 2y^{-1/3}\, e^{-y}\, \frac{1}{8}\, dy = \frac{1}{4}\int_0^\infty y^{-1/3}\, e^{-y}\, dy.$$

Compare this integral with text equation (3.1) to find $p = 2/3$.
Now $\Gamma(2/3) = \frac{1}{(2/3)}\, \Gamma\left(\frac{2}{3}+1\right) = \frac{3}{2}\,\Gamma\left(\frac{5}{3}\right)$. Thus the desired integral is

$$\frac{1}{4}\,\Gamma(2/3) = \frac{3}{8}\,\Gamma(5/3) \cong \frac{3}{8}\,\Gamma(1.67) \cong 0.3387.$$

A more accurate calculation (use a larger table, or interpolate,
or both) gives 0.3385.

Gamma function table for $1 \leqslant p \leqslant 2$.

p	$\Gamma(p)$	p	$\Gamma(p)$	p	$\Gamma(p)$	p	$\Gamma(p)$
1.00	1.00000	1.25	.90640	1.50	.88623	1.75	.91906
1.01	.99433	1.26	.90440	1.51	.88659	1.76	.92137
1.02	.98884	1.27	.90250	1.52	.88704	1.77	.92376
1.03	.98355	1.28	.90072	1.53	.88757	1.78	.92623
1.04	.97844	1.29	.89904	1.54	.88818	1.79	.92877
1.05	.97350	1.30	.89747	1.55	.88887	1.80	.93138
1.06	.96874	1.31	.89600	1.56	.88964	1.81	.93408
1.07	.96415	1.32	.89464	1.57	.89049	1.82	.93685
1.08	.95973	1.33	.89338	1.58	.89142	1.83	.93969
1.09	.95546	1.34	.89222	1.59	.89243	1.84	.94261
1.10	.95135	1.35	.89115	1.60	.89352	1.85	.94561
1.11	.94740	1.36	.89018	1.61	.89468	1.86	.94869
1.12	.94359	1.37	.88931	1.62	.89592	1.87	.95184
1.13	.93993	1.38	.88854	1.63	.89724	1.88	.95507
1.14	.93642	1.39	.88785	1.64	.89864	1.89	.95838
1.15	.93304	1.40	.88726	1.65	.90012	1.90	.96177
1.16	.92980	1.41	.88676	1.66	.90167	1.91	.96523
1.17	.92670	1.42	.88636	1.67	.90330	1.92	.96877
1.18	.92373	1.43	.88604	1.68	.90500	1.93	.97240
1.19	.92089	1.44	.88581	1.69	.90678	1.94	.97610
1.20	.91817	1.45	.88566	1.70	.90864	1.95	.97988
1.21	.91558	1.46	.88560	1.71	.91057	1.96	.98374
1.22	.91311	1.47	.88563	1.72	.91258	1.97	.98768
1.23	.91075	1.48	.88575	1.73	.91467	1.98	.99171
1.24	.90852	1.49	.88595	1.74	.91683	1.99	.99581
1.25	.90640	1.50	.88623	1.75	.91906	2.00	1.00000

Section 4

5.   By text equation (4.1) with p = -2.3, we find

$$\Gamma(-2.3) = \frac{1}{-2.3} \Gamma(-1.3) = \frac{1}{-2.3} \frac{1}{-1.3} \Gamma(-0.3)$$

$$= \frac{1}{(-2.3)(-1.3)(-0.3)(0.7)} \Gamma(1.7) = -1.4471$$

by table above and calculator.

7.   Graph of the gamma function, $\Gamma(p)$.

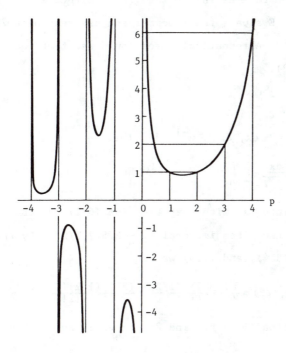

Section 5

1. Hint: Use text equations (3.4) and (5.3).

2. (b) Hint: $z! = \Gamma(z+1) = z\Gamma(z)$.

Section 6

3. Hint: Integrate text equation (6.1) by parts repeatedly (each time differentiating the factor with the smaller exponent) until only one factor remains in the integrand [either x^{n+m-2} or $(1-x)^{n+m-2}$]. Evaluate the integral and use text page 701, equation (4.5). Alternatively, you could use text equations (7.1) and (3.2).

Section 7

1. Let $u = x^2$, $du = 2x\,dx$, $dx = \frac{1}{2}u^{-1/2}\,du$; then

$$\int_0^1 \frac{x^4\,dx}{\sqrt{1-x^2}} = \frac{1}{2}\int_0^1 u^{3/2}(1-u)^{-1/2}\,du.$$

Compare this integral with text equation (6.1) to find $p = 5/2$, $q = 1/2$. Thus the given integral is $\frac{1}{2}B(5/2,1/2)$. By text equations (7.1), (3.4), and (5.3) we find

$$\frac{1}{2}B(5/2,1/2) = \frac{1}{2}\frac{\Gamma(5/2)\Gamma(1/2)}{\Gamma(3)} = \frac{(3/2)(1/2)[\Gamma(1/2)]^2}{2(2!)} = \frac{3\pi}{16}.$$

7. Use text equation (6.4) to find $2p - 1 = -1/2$, $2q - 1 = 0$ so $p = 1/4$, $q = 1/2$. Then

$$\int_0^{\pi/2} \frac{d\theta}{\sqrt{\sin\theta}} = \frac{1}{2}B(1/4,1/2) = \frac{\Gamma(1/4)\Gamma(1/2)}{2\Gamma(3/4)}$$

$$= \frac{4\Gamma(5/4)\sqrt{\pi}}{2(4/3)\Gamma(7/4)} = 2.6221.$$

12.   By text page 212 (with x and y interchanged), the volume is

$$V = \int \pi x^2 \, dy = \pi \int_0^2 (8 - y^3)^{2/3} \, dy.$$

Let $u = y^3$, $du = 3y^2 \, dy$, $dy = \frac{1}{3} u^{-2/3} \, du$; then

$$V = \frac{\pi}{3} \int_0^8 (8 - u)^{2/3} u^{-2/3} \, du.$$

We use text equation (6.3) to find $p = 1/3$, $q = 5/3$, $a^{p+q-1} = 8$, so

$$V = \frac{8\pi}{3} B(1/3, 5/3) = \frac{8\pi}{3} \frac{\Gamma(1/3)\Gamma(5/3)}{\Gamma(2)} = \frac{8\pi}{3} \Gamma(1/3) \frac{2}{3} \Gamma(2/3).$$

By text equation (5.4),

$$\Gamma(1/3)\Gamma(2/3) = \frac{\pi}{\sin(\pi/3)} = 2\pi/\sqrt{3}.$$

Thus

$$V = \frac{32\pi^2 \sqrt{3}}{27}.$$

Section 8

3.   Hints: $t = \int dt = \int \frac{ds}{v}$.   (See text, pages 384 and 394.) Here the
    energy equation is $\frac{1}{2} mv^2 + mgy = mgy_1$. Write $ds = \sqrt{1 + (dx/dy)^2} \, dy$,
    and show that $ds = \sqrt{2a/y} \, dy$.

Section 9

2. To prove (9.2b), let $t = u\sqrt{2}$. Then $t^2/2 = u^2$, $dt = \sqrt{2}\,du$, and
    $u = x/\sqrt{2}$ when $t = x$. Thus

$$P(0,x) = \frac{1}{\sqrt{2\pi}} \int_0^x e^{-t^2/2}\, dt = \frac{1}{\sqrt{2\pi}} \int_0^{x/\sqrt{2}} e^{-u^2} \sqrt{2}\, du$$

$$= \frac{1}{\sqrt{\pi}} \int_0^{x/\sqrt{2}} e^{-u^2}\, du = \tfrac{1}{2} \operatorname{erf}(x/\sqrt{2})$$

by text equation (9.1). Similarly prove (9.2a) and (9.3b),
noting that $\int_{-\infty}^0 e^{-t^2}\, dt = \int_0^\infty e^{-t^2}\, dt = \tfrac{1}{2}\sqrt{\pi}$ as in text equation (9.5).
To prove (9.4), replace x in (9.2b) by $x\sqrt{2}$.

Section 10

Comment: To evaluate the integrals in Problems 3 to 12, use power
series [text equation (9.6)] for $|x| \ll 1$, say $|x| < 10^{-2}$; use tables
of the error function or of the normal distribution [see text equations (9.2), (9.3), and (9.4)] for x in the middle range (say from
$x = 10^{-2}$ to $x = 2$ or 3); and use asymptotic series [text equation (10.4)]
for $|x| \gg 1$, say $|x| > 3$.

3. See text equation (9.1) to get

$$\int_0^2 e^{-x^2}\, dx = \tfrac{1}{2}\sqrt{\pi}\, \operatorname{erf}(2) = \tfrac{1}{2}\sqrt{\pi}\,(0.9953) = 0.8821,$$

using a table of the error function. If you are using a table of
the normal distribution, then by text equation (9.4), find
$x\sqrt{2} \cong 2.83$. From tables,

$$P(0, 2.83) = 0.49766, \qquad P(-\infty, 2.83) = 0.99766.$$

Thus by text equation (9.4), $\operatorname{erf}(2) = 2(0.49766) = 0.9953$, so we
obtain the same answer as above.

7.   Since 5 and 10 are both $\gg 1$, we use the asymptotic series. By
text equations (9.1) and (10.4)

$$\frac{2}{\sqrt{\pi}}\int_5^{10} e^{-x^2}\,dx = \mathrm{erf}(10) - \mathrm{erf}(5)$$

$$= \mathrm{erfc}(5) - \mathrm{erfc}(10)$$

$$= \frac{e^{-5^2}}{5\sqrt{\pi}}\left(1 - \frac{1}{50} + \frac{3}{50^2} - \frac{15}{50^3}\cdots\right) - \frac{e^{-10^2}}{10\sqrt{\pi}}(1 - \cdots)$$

$$= 1.537 \times 10^{-12} - 2 \times 10^{-45} = 1.537 \times 10^{-12}.$$

8.   Hint: To evaluate e^{-10^4} (which your calculator will probably
refuse to do), find the logarithm to the base 10 so that you
can write

$$e^{-10^4} = 10^{-4342.9448} = (10^{-4342})(10^{-0.9448}).$$

Continue the calculation using the second factor and finally
combine the powers of 10.

12.
$$\int_{-\infty}^{0.003} e^{-x^2}\,dx = \int_{-\infty}^{0} e^{-x^2}\,dx + \int_0^{0.003} e^{-x^2}\,dx = \frac{\sqrt{\pi}}{2}[1 + \mathrm{erf}(0.003)].$$

Since $0.003 \ll 1$, we use text equation (9.6).

$$\mathrm{erf}(0.003) = \frac{2}{\sqrt{\pi}}\left[(0.003) - \frac{(0.003)^3}{3} + \cdots\right] = \frac{2}{\sqrt{\pi}}(0.003).$$

Then

$$\int_{-\infty}^{0.003} e^{-x^2}\,dx = \frac{\sqrt{\pi}}{2} + 0.003 = 0.88923.$$

15.   (a) Hint: Integrate by parts repeatedly, integrating e^{-t} and
      differentiating the inverse powers of t.

      (c) Hint: Let $t = e^{-u}$.   What is u when $t = 0$?   When $t = x$?

Section 11

5.   By text equation (11.1),

$$\lim_{n \to \infty} \frac{\Gamma\left(n + \frac{3}{2}\right)}{\sqrt{n}\,\Gamma(n+1)} = \lim_{n \to \infty} \frac{\left(n + \frac{1}{2}\right)^{n+\frac{1}{2}} e^{-(n+\frac{1}{2})} \sqrt{2\pi\left(n + \frac{1}{2}\right)}}{\sqrt{n}\,n^n e^{-n}\,\sqrt{2\pi n}}$$

$$= e^{-1/2} \lim_{n \to \infty} \frac{\left(n + \frac{1}{2}\right)^{n+1}}{n^{n+1}} = e^{-1/2} \lim_{n \to \infty} \left(1 + \frac{1}{2n}\right)^{n+1}.$$

We can evaluate this limit as follows:

$$\ell n\left(1 + \frac{1}{2n}\right)^{n+1} = (n+1)\ell n\left(1 + \frac{1}{2n}\right) = (n+1)\left(\frac{1}{2n} - \frac{1}{8n^2} + \cdots\right)$$

for large n, by text, page 24, equation (13.4).   Then

$$\lim_{n \to \infty} \ell n\left(1 + \frac{1}{2n}\right)^{n+1} = \lim_{n \to \infty} \left(\frac{n+1}{2n} - \frac{n+1}{8n^2} + \cdots\right) = \frac{1}{2}.$$

Thus

$$e^{-1/2} \lim_{n \to \infty} \left(1 + \frac{1}{2n}\right)^{n+1} = e^{-1/2}\, e^{1/2} = 1.$$

8.

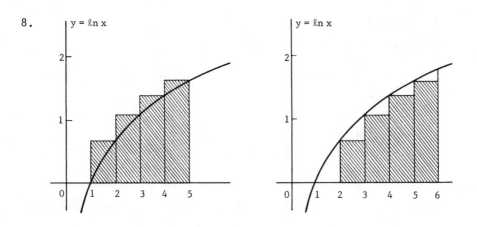

From the first graph:

$$\text{Area of shaded rectangles} > \int_1^5 \ell n\, x\, dx.$$

From the second graph:

$$\text{Area of shaded rectangles} < \int_2^6 \ell n\, x\, dx.$$

Each rectangle has width 1 and height $\ell n\, n$ for $n = 2, 3, \cdots$, so the total area of the shaded rectangles in either figure is

$$\ell n\, 2 + \ell n\, 3 + \ell n\, 4 + \ell n\, 5 = \ell n\, (5!).$$

Thus we have

$$\int_1^5 \ell n\, x\, dx < \ell n\, (5!) < \int_2^6 \ell n\, x\, dx.$$

Similarly, by extending the figures, we can show that

$$\int_1^n \ell n\, x\, dx < \ell n\, (n!) < \int_2^{n+1} \ell n\, x\, dx.$$

8.   (continued)

Evaluate the integrals to get

(1) $n \ln n - n + 1 < \ln(n!) < (n+1)\ln(n+1) - (n+1) + 2(1 - \ln 2)$.

Consider $n = 10^6$. By calculator,

$$n \ln n - n \cong 12{,}815{,}510$$
$$(n+1)\ln(n+1) - (n+1) \cong 12{,}815{,}524$$

so $\ln(10^6!) = 1.28155 \times 10^7$

with 5 decimal place accuracy. Note that, for large n, the terms
1 and $2(1 - \ln 2)$ in equation (1) above are negligible and also
$n+1$ and n are nearly equal. For $n \cong 10^{26}$ as in statistical
mechanics, the formula

$$\ln(n!) = n \ln n - n$$

gives about 24 decimal place accuracy!

9.   Hint: Note that $np + u = npx$ and $nq - u = nqy$.

Section 12

Comments about tables of elliptic integrals: Compare the notation of
the table you are using with text equations (12.1) and (12.4). The
CRC tables use the same notation as in the text. In the NBS tables,
you will find the following correspondence:

text:	ϕ	θ	k^2	$k = \sin\theta$
NBS:	ϕ	α	m	$\sqrt{m} = \sin\alpha$

The NBS tables give the complete elliptic integrals as functions of m
as well as of α. For other references, see text, page 741 ff.

5.   As discussed in the text (pages 475-6), we write

$$\int_0^{7\pi/3} = \int_0^{2\pi} + \int_{2\pi}^{7\pi/3} = 4\int_0^{\pi/2} + \int_0^{\pi/3}.$$

Thus

(1) $E(0.13, 7\pi/3) = 4E(0.13) + E(0.13, \pi/3).$

We find [see text equation (12.1)]

$$\theta = \sin^{-1}k = \sin^{-1}0.13 = 7.5°, \quad \text{and} \quad \phi = \pi/3 = 60°.$$

From a table of complete elliptic integrals, we find (using
$\theta = 7.5°$)

(2) $E(0.13) \cong 1.564.$

(You can interpolate more carefully for higher accuracy. Or for
this small k, you can use the power series for E in Problem 1.)
From a table of elliptic integrals of the second kind, we find
(using $\theta = 7.5°$ and $\phi = 60°$):

(3) $E(0.13, \pi/3) \cong 1.04.$

Note from the tables that the function does not change very much
between $\theta = 5°$ and $\theta = 10°$ so you can approximate it at 7.5°
although it is not tabulated there. Again you can interpolate
for more accuracy or use series. Putting (2) and (3) into (1)
gives

$$E(0.13, 7\pi/3) \cong 7.30.$$

11.   Following the text discussion of Figure 12.1, we write:

$$\int_{-7\pi/8}^{11\pi/4} = \int_{0}^{7\pi/8} + \int_{0}^{11\pi/4} = \int_{0}^{\pi} - \int_{0}^{\pi/8} + \int_{0}^{3\pi} - \int_{0}^{\pi/4}$$

$$= 8\int_{0}^{\pi/2} - \int_{0}^{\pi/8} - \int_{0}^{\pi/4}.$$

Compare text equations (12.1) to see that these integrals of $\sqrt{1 - 0.64 \sin^2\phi}$ are of the form $E(k,\phi)$ with $k = 0.8$. Thus our result is

(1) $8E(0.8) - E(0.8, 22.5°) - E(0.8, 45°)$.

From tables, we find [with $\theta = \sin^{-1} 0.8 = 53.1°$; or more accurately for $E(0.8)$, from the NBS tables with $m = k^2 = 0.64$]:

(2) $E(0.8) \cong 1.276$, $E(0.8, 22.5°) \cong 0.38$, $E(0.8, 45°) \cong 0.74$.

Substitute (2) into (1) to get the result 9.09.

13.   By text equations (12.1) and (12.4), we have

$$\int_{-1/2}^{1/2} \frac{dx}{\sqrt{(1 - x^2)(4 - 3x^2)}} = 2\int_{0}^{1/2} \frac{dx}{2\sqrt{(1 - x^2)(1 - \frac{3}{4}x^2)}} = F(k,\phi)$$

where

$$k^2 = 3/4, \qquad k = \sqrt{3}/2 = \sin\theta, \qquad \theta = 60°,$$

$$x = 1/2 = \sin\phi, \qquad \phi = 30°.$$

Then from tables

$$F(\sqrt{3}/2, 30°) = 0.542.$$

16.   Hint: Write $\int ds$ in terms of y.

17. Following the text hints, we write

$$\cos \theta = 1 - 2 \sin^2(\theta/2), \qquad \cos \alpha = 1 - 2 \sin^2(\alpha/2),$$

(1) $\cos \theta - \cos \alpha = 2[\sin^2(\alpha/2) - \sin^2(\theta/2)] = 2[\sin^2(\alpha/2)][1 - x^2]$

where

$$x = \frac{\sin(\theta/2)}{\sin(\alpha/2)} .$$

Then

$$dx = \frac{\cos(\theta/2)\,d\theta/2}{\sin(\alpha/2)} \qquad \text{or}$$

(2) $d\theta = \dfrac{2\sin(\alpha/2)\,dx}{\cos(\theta/2)} = \dfrac{2\sin(\alpha/2)\,dx}{\sqrt{1 - x^2 \sin^2(\alpha/2)}}$

We substitute (1) and (2) into the integral of text equation
(12.7) to get

$$\int_0^\alpha \frac{d\theta}{\sqrt{\cos\theta - \cos\alpha}} = \sqrt{2} \int_0^1 \frac{dx}{\sqrt{(1 - x^2)[1 - x^2\sin^2(\alpha/2)]}}$$

$$= \sqrt{2}\, K[\sin(\alpha/2)] .$$

19. Hint: For the first part, use $\tan\dfrac{x}{2} = \dfrac{1 - \cos x}{\sin x}$ with $x = \dfrac{\pi}{2} + \phi$.
For the later parts, find e^u and e^{-u}.

23. Hints: Recall that the buoyant force is equal to the weight of
displaced water. Use cylindrical coordinates to find the volume
of the sphere $r^2 + z^2 = a^2$ between $z = 0$ and $z = h$ (see text, Chapter
5, Section 4). Thus find the differential equation for the displacement z from equilibrium:

$$\ddot{z} = -\frac{3g}{2}\left[\frac{z}{a} - \frac{1}{3}\left(\frac{z}{a}\right)^3\right] .$$

23.   (continued)

Integrate once and evaluate the integration constant using $\dot{z} = 0$
when $z = a$. Then integrate again to find the period as an
elliptic integral. For small vibrations, neglect the z^3 term.

Section 13

3.   Hints: Use text equation (7.1) first; you now have $\Gamma(x)$ as a
factor, so your problem is to show that the rest of the expression tends to 1. For this use Stirling's formula. Also
see the solution of Problem 11.5.

10.   See the discussion of Figure 12.1, text pages 475-6, to see that

$$\int_{-\pi/4}^{3\pi/4} = \int_0^{\pi/4} + \int_0^{3\pi/4} = \int_0^{\pi/4} + \int_0^{\pi} - \int_0^{\pi/4} = \int_0^{\pi} = 2\int_0^{\pi/2} .$$

Also

$$1 + \cos^2\phi = 2 - \sin^2\phi = 2\left(1 - \tfrac{1}{2}\sin^2\phi\right).$$

Thus, using these results, text equation (12.2), and tables,
we have

$$\int_{-\pi/4}^{3\pi/4} \frac{d\phi}{\sqrt{1 + \cos^2\phi}} = \frac{2}{\sqrt{2}} \int_0^{\pi/2} \frac{d\phi}{\sqrt{1 - \tfrac{1}{2}\sin^2\phi}}$$

$$= \sqrt{2}\,K(1/\sqrt{2}) = \sqrt{2}(1.854) = 2.622.$$

13.   Hint: See text, page 479.

21.   Comments: You will need to evaluate $\Gamma(13/3)\Gamma(2/3)$. While this
      can be done numerically from tables, it is interesting to see
      that you can write an exact formula for it. We use text equa-
      tion (5.4) to find

$$\Gamma\left(\tfrac{1}{3}\right)\Gamma\left(\tfrac{2}{3}\right) = \frac{\pi}{\sin(\pi/3)} = \frac{2\pi}{\sqrt{3}}\,.$$

Then

$$\Gamma\left(\tfrac{13}{3}\right)\Gamma\left(\tfrac{2}{3}\right) = \frac{10}{3}\,\frac{7}{3}\,\frac{4}{3}\,\frac{1}{3}\,\Gamma\left(\tfrac{1}{3}\right)\Gamma\left(\tfrac{2}{3}\right) = \frac{280}{3^4}\,\frac{2\pi}{\sqrt{3}}\,.$$

Use this result to finish the problem.

24.   By text equation (5.4) with $p = -K$,

$$\Gamma(-K)\Gamma(1+K) = -\frac{\pi}{\sin K\pi} \qquad \text{or}$$

$$\Gamma(-K) = -\frac{\pi}{\Gamma(1+K)\sin K\pi}\,.$$

If K is a large positive number, we can find $\Gamma(1+K)$ by
Stirling's formula [text equation (11.1)] and so find $\Gamma(-K)$ for
large K more easily than by using the recursion relation [text
equation (4.1)]. Thus we can find $\Gamma(-54.5)$ as follows. Let
$K = 54.5$. Then

$$\Gamma(1+K) = K^K e^{-K}\sqrt{2\pi K} = (54.5)^{54.5}e^{-54.5}\sqrt{2\pi(54.5)}$$

$$= 1.71 \times 10^{72}$$

$$\sin K\pi = \sin 54.5\pi = \sin\tfrac{\pi}{2} = 1$$

$$\Gamma(-54.5) = -\frac{\pi}{(1.71 \times 10^{72})(1)} = -1.84 \times 10^{-72}.$$

If we correct this by using 2 terms of the asymptotic expansion,
text equation (11.5), we find

24.   (continued)

$$\Gamma(1+K) = 1.708 \times 10^{72}$$

$$\Gamma(-54.5) = -1.839 \times 10^{-72}.$$

Chapter 12

Section 1

3. $xy' = xy + y$ is a separable equation (text, page 341) so we can solve it as follows.

$$\frac{dy}{y} = \frac{x+1}{x}\,dx,$$

$$\ln y = x + \ln x + C,$$

$$y = Kxe^{x}.$$

Although it is completely unnecessary to use power series to solve this simple equation, we do it to demonstrate the method and to verify that the power series solution gives the same result. Substitute

$$y = a_0 + a_1 x + a_2 x^2 + a_3 x^3 + \cdots = \sum_{n=0}^{\infty} a_n x^n,$$

$$y' = a_1 + 2a_2 x + 3a_3 x^2 + \cdots = \sum_{n=0}^{\infty} n a_n x^{n-1}$$

into the differential equation and collect coefficients of powers of x:

	const.	x	x^2	$\cdots$	x^n	$\cdots$
xy'		a_1	$2a_2$		na_n	
xy		a_0	a_1		a_{n-1}	
y	a_0	a_1	a_2		a_n	

For each power of x, the coefficients must be the same on the two sides of the given differential equation. Thus we find

$$a_0 = 0, \qquad a_1 = a_0 + a_1, \qquad 2a_2 = a_1 + a_2, \cdots, \qquad na_n = a_{n-1} + a_n$$

361

3.   (continued)

or

$$a_0 = 0, \quad a_1 \text{ is arbitrary,} \quad a_2 = a_1, \cdots, \quad a_n = a_{n-1}/(n-1), \text{ so}$$

$$a_3 = a_1/2, \quad a_4 = a_1/3!, \quad a_5 = a_1/4!, \cdots, \quad a_n = a_1/(n-1)! \ .$$

Then

$$y = a_1\left(x + x^2 + \frac{x^3}{2} + \frac{x^4}{3!} + \cdots + \frac{x^n}{(n-1)!} + \cdots\right)$$

$$= a_1 x \left(1 + x + \frac{x^2}{2} + \frac{x^3}{3!} + \cdots + \frac{x^{n-1}}{(n-1)!} + \cdots\right) = a_1 x e^x,$$

as we found above.

9.   By elementary methods (text, Chapter 8, Sections 5 and 6), we
first solve the homogeneous equation to find y_c .

$$(D^2 + 1)y = 0, \qquad\qquad D = \pm i.$$

$$y_c = C_1 e^{ix} + C_2 e^{-ix} \qquad \text{or} \qquad a \sin x + b \cos x.$$

Next solve $(D^2 + 1)Y = 4xe^{ix}$; $y_p = \text{Im } Y$ (see text, page 365).
Since i is a simple root of the auxiliary equation, we assume
[see text, page 366, equation (6.24)]

$$Y = x(Ax + B)e^{ix} = (Ax^2 + Bx)e^{ix}.$$

Then

$$Y'' + Y = e^{ix}[2A + 2i(2Ax + B) - (Ax^2 + Bx) + Ax^2 + Bx] \equiv 4xe^{ix}.$$

$$2A + 2iB = 0 \qquad \text{and} \qquad 4iA = 4,$$

$$A = -i, \qquad B = 1,$$

$$Y = (-ix^2 + x)e^{ix},$$

$$y_p = \text{Im } Y = -x^2 \cos x + x \sin x,$$

$$y = y_c + y_p = a \sin x + b \cos x - x^2 \cos x + x \sin x.$$

9.   (continued)

Next we solve the equation by power series.

$$y = a_0 + a_1 x + a_2 x^2 + a_3 x^3 + a_4 x^4 + a_5 x^5 + \cdots,$$

$$y'' = 2a_2 + 3 \cdot 2a_3 x + 4 \cdot 3a_4 x^2 + 5 \cdot 4a_5 x^3 + \cdots,$$

$$4x \sin x = 4x \left[x - \frac{x^3}{3!} + \frac{x^5}{5!} - \frac{x^7}{7!} + \cdots \right].$$

Equate coefficients of powers of x in $y'' + y = 4x \sin x$:

const. $a_0 + 2a_2 = 0$ $\qquad a_2 = -\frac{1}{2} a_0$

$x \qquad a_1 + 3 \cdot 2a_3 = 0 \qquad a_3 = -\frac{1}{3!} a_1$

$x^2 \qquad a_2 + 4 \cdot 3a_4 = 4 \qquad a_4 = \frac{1}{3} - \frac{1}{12} a_2 = \frac{1}{3} + \frac{1}{4!} a_0$

$x^3 \qquad a_3 + 5 \cdot 4a_5 = 0 \qquad a_5 = -\frac{1}{20} a_3 = \frac{1}{5!} a_1$

$x^4 \qquad a_4 + 6 \cdot 5a_6 = -\frac{4}{3!} \qquad a_6 = -\frac{4}{6 \cdot 5 \cdot 3!} - \frac{1}{6 \cdot 5} a_4 = -\frac{1}{30} - \frac{1}{6!} a_0$,

and so on. Then

$$y = a_0 \left(1 - \frac{1}{2} x^2 + \frac{1}{4!} x^4 - \frac{x^6}{6!} \cdots \right)$$

$$+ a_1 \left(x - \frac{x^3}{6} + \frac{x^5}{5!} \cdots \right) + \frac{1}{3} x^4 - \frac{1}{30} x^6 \cdots .$$

We easily recognize the a_0 and a_1 series as $\sin x$ and $\cos x$. The rest of the solution should be y_p from the elementary solution:

$$y_p = x \sin x - x^2 \cos x = x \left(x - \frac{x^3}{3!} + \frac{x^5}{5!} \cdots \right) - x^2 \left(1 - \frac{x^2}{2} + \frac{x^4}{4!} \cdots \right)$$

$$= x^4 \left(-\frac{1}{3!} + \frac{1}{2!} \right) + x^6 \left(\frac{1}{5!} - \frac{1}{4!} \right) + \cdots = \frac{1}{3} x^4 - \frac{1}{30} x^6 \cdots$$

which checks the first few terms. It is possible, although somewhat messy, to write general formulas (for the coefficient of x^n) and compare the two solutions, but we shall not go this far.

11. To find a power series solution of the differential equation
     $y'' - x^2 y' - xy = 0$, we substitute text equations (2.2) into the
     equation and collect coefficients of powers of x.

	const	x	x^2	x^3	x^4	$\cdots$	x^n	$\cdots$
y''	$2a_2$	$3 \cdot 2a_3$	$4 \cdot 3a_4$	$5 \cdot 4a_5$	$6 \cdot 5a_6$		$(n+2)(n+1)a_{n+2}$	
$-x^2 y'$			$-a_1$	$-2a_2$	$-3a_3$		$-(n-1)a_{n-1}$	
$-xy$		$-a_0$	$-a_1$	$-a_2$	$-a_3$		$-a_{n-1}$	

Each column must add to zero, so we find

$$a_2 = 0, \qquad a_3 = \frac{1}{3!} a_0, \qquad a_4 = \frac{1}{6} a_1, \qquad a_5 = \frac{3a_2}{20} = 0, \cdots,$$

(A) $a_{n+2} = \dfrac{n}{(n+2)(n+1)} a_{n-1}$ or $a_{n+3} = \dfrac{n+1}{(n+3)(n+2)} a_n$.

We can write this in the useful form

(B) $a_{n+3} = \dfrac{(n+1)^2}{(n+3)(n+2)(n+1)} a_n$.

We can use (A) or (B) to find all the coefficients in terms of
a_0, a_1, or a_2. Since $a_2 = 0$, we also have

$$a_5 = a_8 = a_{11} = \cdots = 0.$$

With $n = 0$, 3, 6, 9, $\cdots$, we find from (B)

$$a_3 = \frac{1}{3!} a_0,$$

$$a_6 = \frac{4^2}{6 \cdot 5 \cdot 4} a_3 = \frac{4^2}{6!} a_0,$$

$$a_9 = \frac{7^2}{9 \cdot 8 \cdot 7} a_6 = \frac{7^2 \cdot 4^2}{9!} a_0, \cdots.$$

11. (continued)

Similarly we find a_4, a_7, $\cdots$ in terms of a_1.

$$a_4 = \frac{2^2}{4!} a_1 \, ,$$

$$a_7 = \frac{5^2}{7 \cdot 6 \cdot 5} a_4 = \frac{5^2 \cdot 2^2}{7!} a_1 \, ,$$

$$a_{10} = \frac{8^2}{10 \cdot 9 \cdot 8} a_7 = \frac{(8 \cdot 5 \cdot 2)^2}{10!} a_1 \, , \cdots \, .$$

Thus the solution of the given differential equation is

$$y = a_0 \left[1 + \frac{1}{3!} x^3 + \frac{4^2}{6!} x^6 + \frac{(7 \cdot 4)^2}{9!} x^9 + \cdots \right]$$

$$+ a_1 \left[x + \frac{2^2}{4!} x^4 + \frac{(5 \cdot 2)^2}{7!} x^7 + \frac{(8 \cdot 5 \cdot 2)^2}{10!} x^{10} + \cdots \right] \, .$$

16. $(x^2 + 1)y'' - 2xy' + 2y = 0.$

We substitute text equations (2.2) into the differential equation and collect coefficients of powers of x:

	const	x	x^2	x^3	$\cdots$	x^n
$x^2 y''$			$2a_2$	$6a_3$		$n(n-1)a_n$
y''	$2a_2$	$6a_3$	$12a_4$	$20a_5$		$(n+2)(n+1)a_{n+2}$
$-2xy'$		$-2a_1$	$-4a_2$	$-6a_3$		$-2na_n$
$2y$	$2a_0$	$2a_1$	$2a_2$	$2a_3$		$2a_n$

16.   (continued)

Each column must add to zero; thus we find

$$2(a_2 + a_0) = 0, \quad 6a_3 = 0, \quad 12a_4 = 0, \quad 20a_5 + 2a_3 = 0, \quad \cdots$$

$$(n+2)(n+1)a_{n+2} + (n-1)(n-2)a_n = 0.$$

Since $a_3 = a_4 = 0$, then $a_5 = 0$, and all $a_n = 0$ for $n > 4$ by the general formula relating a_n and a_{n+2}. The coefficients a_0 and a_1 are arbitrary and $a_2 = -a_0$. Thus

$$y = a_0(1 - x^2) + a_1 x.$$

Section 2

1.   Let us find P_2 using text equation (2.7). For $\ell = 2$, we see that the coefficient of x^4 is zero and, by text equation (2.6), all the following coefficients of even powers of x are zero also. Thus the a_0 series is just $1 - 3x^2$. Remember that a_0 and a_1 are arbitrary. We set $a_1 = 0$ here since the a_1 series is an infinite series and we want a polynomial. Then (with $\ell = 2$, $a_1 = 0$)

$$y = a_0(1 - 3x^2).$$

Legendre polynomials are required to be 1 when $x = 1$. We get

$$1 = a_0(1 - 3) \qquad \text{or} \qquad a_0 = -1/2, \qquad \text{so}$$

$$P_2(x) = -\frac{1}{2}(1 - 3x^2) = \frac{1}{2}(3x^2 - 1)$$

as in text equation (2.8). Similarly, to find $P_3(x)$ from text equation (2.7), let $\ell = 3$, $a_0 = 0$, and find a_1 to make $y = 1$ when $x = 1$. To find P_4, let $\ell = 4$, etc. To check your answers, see Problems 4.3 and 5.3 below.

3. Graphs of Legendre polynomials.

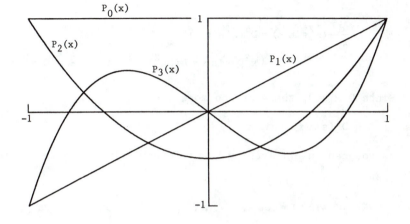

Note that the graphs agree with the following facts:

$$P_\ell(1) = 1 \qquad \text{for all } \ell.$$

$$P_\ell(0) = 0 \qquad \text{for odd } \ell \text{ (but not for even } \ell).$$

$$P_\ell(-1) = (-1)^\ell = \begin{cases} 1, & \ell \text{ even,} \\ -1, & \ell \text{ odd.} \end{cases}$$

Section 3

3.   By Leibniz' rule (using $D = \frac{d}{dx}$)

$$\frac{d^6}{dx^6}(x^2 \sin x) = D^6(x^2 \sin x)$$

$$= x^2 D^6 \sin x + 6(Dx^2)(D^5 \sin x) + \frac{6 \cdot 5}{2!}(D^2 x^2)(D^4 \sin x).$$

Since $D^3 x^2 = 0$, there are no more terms.

$$D^4(\sin x) = \sin x$$

as you can quickly verify by differentiating $\sin x$ four times.
Then

$$D^5(\sin x) = \cos x, \qquad \text{and} \qquad D^6(\sin x) = - \sin x.$$

Thus

$$\frac{d^6}{dx^6}(x^2 \sin x) = x^2(-\sin x) + 6(2x)\cos x + 15(2)\sin x$$

$$= (30 - x^2)\sin x + 12x \cos x.$$

6.   Proof of a theorem by mathematical induction means the following:
(a) Show that the theorem is true for n <u>if</u> it is true for $n-1$.
(b) Show that the theorem is true for $n = 1$.
Then by (a), the theorem is true for $n = 2$, thus for $n = 3$, etc.
Here we want to prove that

(1) $$D^n(uv) = \sum_{k=0}^{n} \binom{n}{k}(D^k u)(D^{n-k} v)$$

(see text, pp. 700-702 for binomial coefficients).   We assume
the theorem for $n-1$:

(2) $$D^{n-1}(uv) = \sum_{k=0}^{n-1} \binom{n-1}{k}(D^k u)(D^{n-1-k} v).$$

6.   (continued)

Differentiate (2) once:

(3) $D^n(uv) = \sum_{k=0}^{n-1} \binom{n-1}{k} [(D^{k+1}u)(D^{n-1-k}v) + (D^k u)(D^{n-k}v)]$

$= \sum_{k=1}^{n} \binom{n-1}{k-1} (D^k u)(D^{n-k}v) + \sum_{k=0}^{n-1} \binom{n-1}{k} (D^k u)(D^{n-k}v).$

(We have replaced k by $k-1$ in the first summation.) Now the
coefficient of $(D^k u)(D^{n-k}v)$ in (3) is (for $k = 1$ to $n-1$)

$\binom{n-1}{k-1} + \binom{n-1}{k} = \frac{(n-1)!}{(k-1)!(n-k)!} + \frac{(n-1)!}{k!(n-1-k)!}$

$= \frac{(n-1)!}{k!(n-k)!}(k+n-k) = \frac{n!}{k!(n-k)!} = \binom{n}{k}$

which is the desired coefficient in (1). The first and last
terms in (1) for $k = 0$ and $k = n$ are $u(D^n v)$ and $v(D^n u)$. These are
the same as the $k = 0$ term in the second series in (3) and the
$k = n$ term in the first series in (3). Thus we have shown that
(2) implies (1). Now for $n = 1$, the theorem says that

(4) $D(uv) = u(Dv) + v(Du)$

which is true by elementary calculus. With $n = 2$ in (1) and (2),
(4) and our proof above show that (1) is true when $n = 2$, and so
on by induction.

You may want to write out in detail several terms of each of
the series above.

<u>Section 4</u>

3.   To find P_3 from Rodrigues' formula, we set $\ell = 3$ in text
      equation (4.1).

$$P_3(x) = \frac{1}{2^3 3!} \frac{d^3}{dx^3} (x^2 - 1)^3 = \frac{1}{48} \frac{d^3}{dx^3} (x^6 - 3x^4 + 3x^2 - 1)$$

$$= \frac{1}{48}(6 \cdot 5 \cdot 4x^3 - 3 \cdot 4 \cdot 3 \cdot 2x) = \frac{1}{2}(5x^3 - 3x).$$

Similarly find the other Legendre polynomials and check your
results with the text answers to Problem 5.3, page 766.

<u>Section 5</u>

3.   The easiest way to find P_4 (if we know P_2 and P_3) is to use
      text equation (5.8a) with $\ell = 4$.

$$4P_4 = 7xP_3 - 3P_2 = 7x \cdot \frac{1}{2}(5x^3 - 3x) - 3 \cdot \frac{1}{2}(3x^2 - 1)$$

$$= \frac{1}{2}(35x^4 - 21x^2 - 9x^2 + 3),$$

$$P_4 = \frac{1}{8}(35x^4 - 30x^2 + 3).$$

Similarly find P_5 and P_6 using P_3 and P_4; check your results
with the text answers on page 766.

5.   We differentiate text equation (5.8a) to get

$$\ell P'_\ell = (2\ell - 1)(xP'_{\ell-1} + P_{\ell-1}) - (\ell - 1)P'_{\ell-2}.$$

Text equation (5.8b) with ℓ replaced by $\ell - 1$ is

$$xP'_{\ell-1} - P'_{\ell-2} = (\ell - 1)P_{\ell-1}.$$

We want to prove text equation (5.8c); we observe that there is
no $P_{\ell-2}$ term in (5.8c). Thus we eliminate the $P_{\ell-2}$ term between

5.   (continued)

our two equations above.

$$\ell P_\ell' = (2\ell - 1)(xP_{\ell-1}' + P_{\ell-1}) - (\ell - 1)[xP_{\ell-1}' - (\ell - 1)P_{\ell-1}]$$

$$= (2\ell - \ell - \ell + 1)xP_{\ell-1}' + (2\ell - \ell + \ell^2 - 2\ell + 1)P_{\ell-1},$$

$$P_\ell' = xP_{\ell-1}' + \ell P_{\ell-1}$$

which is text equation (5.8c).

12.   To write $f(x) = 7x^4 - 3x + 1$ in terms of Legendre polynomials, we
start by writing x^4 in terms of P_4 and lower powers of x.

$$P_4 = (35x^4 - 30x^2 + 3)/8,$$

$$x^4 = (8P_4 + 30x^2 - 3)/35.$$

Substitute this into $f(x)$:

$$f(x) = \frac{7}{35}(8P_4 + 30x^2 - 3) - 3x + 1$$

$$= \frac{8}{5}P_4 + 6x^2 - \frac{3}{5} - 3x + 1$$

$$= \frac{8}{5}P_4 + 6x^2 - 3x + \frac{2}{5}.$$

Next we write the x^2 term using $P_2 = \frac{1}{2}(3x^2 - 1)$,

$$x^2 = \frac{2P_2 + 1}{3},$$

$$6x^2 - 3x + \frac{2}{5} = 4P_2 + 2 - 3x + \frac{2}{5} = 4P_2 - 3x + \frac{12}{5}.$$

Finally $x = P_1(x)$ and $1 = P_0(x)$, so

$$f(x) = \frac{8}{5}P_4 + 4P_2 - 3P_1 + \frac{12}{5}P_0.$$

14.   As in Problem 12, we can write any polynomial of degree n in
      terms of Legendre polynomials with $\ell \leqslant n$. We start by writing
      x^n in terms of $P_n(x)$ and a polynomial of degree $n - 1$ or less.
      We combine terms and then write the x^{n-1} term (or the highest
      power remaining) in terms of its P_ℓ and so on until we reach
      P_1 and P_0.

Section 6

1.   Hint: Take real and imaginary parts of both equations.

6.   We want to show that $\int_{-1}^{1} P_\ell(x) P_\ell'(x) \, dx = 0$. Here are three
      different methods:

      (a) $\int_{-1}^{1} P_\ell(x) P_\ell'(x) \, dx = \frac{1}{2}[P_\ell(x)]^2 \Big|_{-1}^{1} = \frac{1}{2}[P_\ell(1)]^2 - \frac{1}{2}[P_\ell(-1)]^2$

                              $= \frac{1}{2}(1)^2 - \frac{1}{2}(\pm 1)^2 = 0$

               since $P_\ell(1) = 1$ and $P_\ell(-1) = (-1)^\ell$; see Problems, Section 2.

      (b)   Note that each Legendre polynomial contains just even
               powers of x or else just odd powers of x depending on
               whether ℓ is even or odd. Derivatives of odd powers are
               even and derivatives of even powers are odd. Thus if ℓ
               is even, $P_\ell(x)$ is an even function and $P_\ell'(x)$ is an odd
               function, so $P_\ell(x)P_\ell'(x)$ is an odd function. (What is the
               corresponding statement if ℓ is odd?) Recall from text,
               page 322, that the integral of an odd function over a
               symmetric interval is zero. Thus $\int_{-1}^{1} P_\ell(x)P_\ell'(x) \, dx = 0$.

6.   (continued)

(c) $P_\ell'(x)$ is a polynomial of degree $\ell - 1$. Thus the terms in $P_\ell'(x)$ are constants times various powers of x, with the power always $< \ell$. By Problem 4.4, $\int_{-1}^{1} x^m P_\ell(x)\, dx = 0$ if $m < \ell$, so $\int_{-1}^{1} P_\ell'(x) P_\ell(x)\, dx = 0$.

9.   Hint for one method: Use Problem 4.4 with $m = 0$.

Section 7

1.   Write the differential equations whose solutions are y_n and y_k:

(1) $y_n'' = -n^2 y_n$,

(2) $y_k'' = -k^2 y_k$.

Multiply (1) by y_k and (2) by y_n and subtract. Then integrate from $-\pi$ to π.

$$\int_{-\pi}^{\pi} (y_k y_n'' - y_n y_k'')\, dx = (k^2 - n^2) \int_{-\pi}^{\pi} y_n y_k\, dx.$$

Now

$$\frac{d}{dx}(y_k y_n' - y_n y_k') = y_k y_n'' + \cancel{y_k' y_n'} - y_n y_k'' - \cancel{y_n' y_k'} \qquad \text{so}$$

$$\int_{-\pi}^{\pi} (y_k y_n'' - y_n y_k'')\, dx = (y_k y_n' - y_n y_k') \Big|_{-\pi}^{\pi} .$$

The functions y_k, y_n, y_k', and y_n' are sines and cosines (or complex exponentials). These functions all have period 2π (we are assuming that n and k are integers). Thus $y_k y_n' - y_n y_k'$ has the same value at π and $-\pi$, so the integrated term is zero. Then

$$(k^2 - n^2) \int_{-\pi}^{\pi} y_n y_k\, dx = 0, \qquad \text{so}$$

$$\int_{-\pi}^{\pi} y_n y_k\, dx = 0 \qquad \text{if} \qquad n \neq k.$$

3.   Hint: See the solution (c) of Problem 6.3.

Section 8

1.   Hint: See text, page 306.

5.   We want to evaluate

$$N^2 = \int_0^\infty \left(x\, e^{-x^2/2} \right)^2 dx = \int_0^\infty x^2 e^{-x^2}\, dx.$$

From a table of definite integrals we find $N^2 = \frac{1}{4}\pi^{1/2}$.   Then the
norm of the function $x\, e^{-x^2/2}$ on $(0, \infty)$ is $N = \frac{1}{2}\pi^{1/4}$, and the
normalized function is $2\pi^{-1/4} x\, e^{-x^2/2}$.   [Compare dividing a
vector by its length (norm) to make it a unit vector.]

Comment on evaluation of the integral above: Instead of using
tables, we can proceed as follows. First evaluate

(1) $\displaystyle\int_0^\infty e^{-ax^2}\, dx = \frac{1}{2}\sqrt{\frac{\pi}{a}}$

(see text page 461 and text page 196, Problem 16). Then (see
text page 194) differentiate both sides of (1) with respect to a,
and set $a = 1$:

$$\int_0^\infty -x^2 e^{-ax^2}\, dx = \frac{1}{2}\sqrt{\pi}\left(-\frac{1}{2} a^{-3/2} \right)$$

$$\int_0^\infty x^2 e^{-x^2}\, dx = \frac{1}{4}\sqrt{\pi}.$$

Section 9

1. Hint: Note that $f(x)$ is an odd function (see text, page 322). Also note that $P_\ell(x)$ is an even function when ℓ is even and $P_\ell(x)$ is an odd function when ℓ is odd. Thus, in this problem, $c_\ell = 0$ when ℓ is even; when ℓ is odd, we may (as for Fourier series) replace $\int_{-1}^{1}$ by $2\int_{0}^{1}$.

3. Hint: Many coefficients are zero here. Don't waste time computing a coefficient you know is zero. Further hints:
   (a) Consider text equation (7.6).   (b) Is $f(x)$ even or odd?

4. We want the coefficients c_ℓ so that

$$f(x) = \text{arc sin} \, x = \sum_{\ell=0}^{\infty} c_\ell P_\ell(x).$$

As in the text, page 502, we find

$$\int_{-1}^{1} (\text{arc sin} \, x) P_m(x) \, dx = c_m \int_{-1}^{1} \left[P_m(x)\right]^2 dx = \frac{2c_m}{2m+1}.$$

Since arc sin x is an odd function (see text, Chapter 7, Section 9), and $P_m(x)$ for even m is an even function, we see that all c_m with m even are zero. For odd m, we have

$$\int_{-1}^{1} (\text{arc sin} \, x) P_m(x) \, dx = 2 \int_{0}^{1} (\text{arc sin} \, x) P_m(x) \, dx = \frac{2c_m}{2m+1},$$

(1) $c_m = (2m+1) \int_{0}^{1} (\text{arc sin} \, x) P_m(x) \, dx.$

We shall consider the details of evaluating these integrals below. We find

4.   (continued)

$$c_1 = 3 \int_0^1 (\text{arc } \sin x) P_1(x) \, dx = 3\pi/8,$$

$$c_3 = 7 \int_0^1 (\text{arc } \sin x) P_3(x) \, dx = 7\pi/2^7,$$

$$c_5 = 11 \int_0^1 (\text{arc } \sin x) P_5(x) \, dx = 11\pi/2^9, \qquad \text{etc.}$$

Then

$$\text{arc } \sin x = \frac{\pi}{8}\left[3P_1(x) + \frac{7}{16}P_3(x) + \frac{11}{64}P_5(x) + \cdots\right].$$

Evaluation of the integrals above: We first find the integrals of $x^n \text{arc } \sin x$ for odd n. Either from tables or by integrating by parts (integrate x^n and differentiate $\text{arc } \sin x$) we get

(2) $$\int_0^1 x^n \text{arc } \sin x \, dx = \frac{x^{n+1} \text{arc } \sin x}{n+1}\bigg|_0^1 - \frac{1}{n+1}\int_0^1 \frac{x^{n+1} \, dx}{\sqrt{1-x^2}}$$

$$= \frac{1}{n+1}\left[\frac{\pi}{2} - \int_0^1 \frac{x^{n+1} \, dx}{\sqrt{1-x^2}}\right].$$

To evaluate this last integral, let $x = \sin \theta$; then

$$\frac{dx}{\sqrt{1-x^2}} = \frac{\cos \theta \, d\theta}{\cos \theta} = d\theta, \qquad \theta \text{ goes from 0 to } \pi/2.$$

(3) $$\int_0^1 \frac{x^{n+1} \, dx}{\sqrt{1-x^2}} = \int_0^{\pi/2} (\sin \theta)^{n+1} \, d\theta = \frac{1\cdot3\cdot5\cdots n}{2\cdot4\cdot6\cdots(n+1)}\frac{\pi}{2}$$

by tables (look at a table of definite integrals) or by text equations (6.4) and (7.1) on page 463 and (5.3) on page 461. From equations (2) and (3), we have

4.   (continued)

(4) $\int_0^1 x \text{ arc sin } x \, dx = \frac{1}{2}\left(\frac{\pi}{2} - \frac{1}{2}\frac{\pi}{2}\right) = \frac{\pi}{8}$,

$\int_0^1 x^3 \text{arc sin } x \, dx = \frac{1}{4}\left(\frac{\pi}{2} - \frac{1 \cdot 3}{2 \cdot 4}\frac{\pi}{2}\right) = \frac{5\pi}{4}$,

$\int_0^1 x^5 \text{arc sin } x \, dx = \frac{1}{6}\left(\frac{\pi}{2} - \frac{1 \cdot 3 \cdot 5}{2 \cdot 4 \cdot 6}\frac{\pi}{2}\right) = \frac{11\pi}{192}$,   etc.

Now we use the Legendre polynomials (text, page 766, Problem 5.3) with equations (1) and (4) above to find:

$c_1 = 3\int_0^1 x \text{ arc sin } x \, dx = 3\pi/8$,

$c_3 = 7\int_0^1 \frac{1}{2}(5x^3 - 3x)\text{arc sin } x \, dx = \frac{7}{2}\left(5 \cdot \frac{5\pi}{64} - 3 \cdot \frac{\pi}{8}\right) = 7\pi/2^7$,

$c_5 = 11\int_0^1 \frac{1}{8}(63x^5 - 70x^3 + 15x)\text{arc sin } x \, dx = 11\pi/2^9$.

(You might like to work out the details of the c_5 computation.)

6.   Hint: To evaluate the integrals, let $x = e^{-u}$ and use the definite integral $\int_0^\infty u^n e^{-au} \, du = n!/a^{n+1}$.

11. We are going to do Problem 5.12 over again, this time using the more general method of getting Legendre series.

$$f(x) = 7x^4 - 3x + 1 = \sum c_\ell P_\ell(x).$$

Multiply both sides of the equation by $P_m(x)$ and integrate from -1 to 1:

(1) $$\int_{-1}^{1} f(x) P_m(x)\,dx = \sum c_\ell \int_{-1}^{1} P_\ell(x) P_m(x)\,dx = c_m \int_{-1}^{1} [P_m(x)]^2 dx = \frac{2c_m}{2m+1}$$

because $\int_{-1}^{1} P_\ell P_m\,dx = 0$ unless $\ell = m$ (orthogonality), and by text equation (8.1), $\int_{-1}^{1} [P_m(x)]^2 dx = 2/(2m+1)$ (normalization). We use (1) repeatedly to find the coefficients c_m. For $m = 0$, (1) gives (remember $P_0 = 1$):

$$\int_{-1}^{1} (7x^4 - 3x + 1)\cdot 1\, dx = 2c_0 , \qquad c_0 = \frac{1}{2}\left[\frac{7}{5}\cdot 2 - \frac{3}{2}(1-1) + 2\right] = \frac{12}{5} .$$

For $m = 1$, $P_1 = x$, and (1) gives:

$$\int_{-1}^{1} (7x^4 - 3x + 1) x\, dx = \frac{2c_1}{3} , \qquad c_1 = \frac{3}{2}\left[0 - \frac{3\cdot 2}{3} + 0\right] = -3.$$

(Remember from text, page 322, that the integral of an odd function over a symmetric interval is zero.)

For $m = 2$, $P_2 = \frac{1}{2}(3x^2 - 1)$, and (1) gives:

$$\int_{-1}^{1} (7x^4 - 3x + 1)\frac{1}{2}(3x^2 - 1)\, dx = \frac{2c_2}{5} ,$$

$$c_2 = \frac{5}{2}\int_{-1}^{1} \frac{1}{2}(21x^6 - 7x^4 - 9x^3 + 3x^2 + 3x - 1)\, dx$$

$$= \frac{5}{4}\left[21\cdot \frac{2}{7} - 7\cdot \frac{2}{5} - 0 + 3\cdot \frac{2}{3} + 0 - 2\right] = \frac{5}{4}\cdot \frac{16}{5} = 4.$$

11.   (continued)

For $m = 3$, we can conclude that $c_3 = 0$ without computing it:

$$\int_{-1}^{1} (7x^4 - 3x + 1)P_3\, dx = 7\int_{-1}^{1} x^4 P_3\, dx + \int_{-1}^{1} (-3P_1 + P_0)P_3\, dx.$$

The first integral is zero because $x^4 P_3$ is an odd function. The second integral is zero because P_3 is orthogonal to P_1 and P_0. Thus $c_3 = 0$.

For $m = 4$, $P_4(x) = \frac{1}{8}(35x^4 - 30x^2 + 3)$. Since P_4 is orthogonal to P_0 and P_1, we can omit all but $7x^4$ in $f(x)$.

$$\int_{-1}^{1} 7x^4 \cdot \frac{1}{8}(35x^4 - 30x^2 + 3)\, dx = \frac{2c_4}{9},$$

$$c_4 = \frac{9}{2} \cdot \frac{7}{8}\left(\frac{35 \cdot 2}{9} - \frac{30 \cdot 2}{7} + \frac{3 \cdot 2}{5}\right) = 1.6 \qquad \text{(by calculator)}$$

$$= \frac{8}{5} \qquad \text{as in Problem 5.12.}$$

There is an easier way to find c_4. Solve the P_4 equation to get $7x^4 = \frac{8}{5}P_4 +$ quadratic terms; multiply by P_4 and integrate. Since P_4 is orthogonal to the quadratic terms, we have

$$\int_{-1}^{1} 7x^4 P_4\, dx = \frac{8}{5}\int_{-1}^{1} (P_4)^2 dx = c_4 \int_{-1}^{1} (P_4)^2 dx,$$

so $c_4 = \frac{8}{5}$ as before.

Note, however, that the solution of Problem 9.11 was longer than the solution of Problem 5.12. The method of Problem 5.12 is usually easier than the general method for <u>polynomial</u> $f(x)$. However, what we have done here, using equation (1) repeatedly for $m = 0, 1, 2, \cdots$, is the general method to use for any $f(x)$ which is not a polynomial.

13. Note that x^4 is an even function so for a second-degree polynomial approximation, we need only c_0 and c_2. We write

$$x^4 = c_0 P_0(x) + c_2 P_2(x) \cdots .$$

Multiply this equation by $P_\ell(x)$ and integrate from -1 to 1.
For even $P_\ell(x)$, we get

$$\int_{-1}^{1} x^4 P_\ell(x) \, dx = 2\int_{0}^{1} x^4 P_\ell(x) \, dx = c_\ell \int_{-1}^{1} [P_\ell(x)]^2 \, dx = c_\ell \cdot \frac{2}{2\ell+1}$$

as in text equation (9.3). Then

$$c_\ell = (2\ell + 1)\int_{0}^{1} x^4 P_\ell(x) \, dx,$$

$$c_0 = \int_{0}^{1} x^4 \, dx = 1/5,$$

$$c_2 = 5\int_{0}^{1} x^4 \cdot \frac{1}{2}(3x^2 - 1) \, dx = \frac{5}{2}\left(\frac{3}{7} - \frac{1}{5}\right) = \frac{4}{7}.$$

Thus we have the approximation

$$x^4 \cong \frac{1}{5}P_0(x) + \frac{4}{7}P_2(x) = \frac{1}{5} + \frac{4}{7} \cdot \frac{1}{2}(3x^2 - 1) = \frac{6}{7}x^2 - \frac{3}{35}.$$

Note, in this last step, that we no longer have a Legendre
series and in general you should <u>not</u> do this. Here we simply
used Legendre series (as proved in Problem 16) to find a second-
degree polynomial approximation to x^4 and our answer is the
polynomial, not the Legendre series.

Section 10

2.   If $x = \cos \theta$, then by the chain rule (text, Chapter 4, Section 5)

(1) $\dfrac{df}{d\theta} = \dfrac{df}{dx}\dfrac{dx}{d\theta} = -\sin \theta \dfrac{df}{dx}$ or $\dfrac{1}{\sin \theta}\dfrac{df}{d\theta} = -\dfrac{df}{dx}$.

Write the first term of the given differential equation as

$$\dfrac{1}{\sin \theta}\dfrac{d}{d\theta}\left[(\sin^2\theta)\dfrac{1}{\sin \theta}\dfrac{dy}{d\theta}\right].$$

Then

$$\dfrac{1}{\sin \theta}\dfrac{dy}{d\theta} = -\dfrac{dy}{dx} \qquad \text{and} \qquad \dfrac{1}{\sin \theta}\dfrac{d}{d\theta}[\quad] = -\dfrac{d}{dx}[\quad].$$

Also replace $\sin^2\theta$ by $1 - \cos^2\theta = 1 - x^2$. Then

$$\dfrac{1}{\sin \theta}\dfrac{d}{d\theta}\left[(\sin^2\theta)\dfrac{1}{\sin \theta}\dfrac{dy}{d\theta}\right] = -\dfrac{d}{dx}\left[(1 - x^2)\left(-\dfrac{dy}{dx}\right)\right]$$

$$= \dfrac{d}{dx}\left[(1 - x^2)\dfrac{dy}{dx}\right]$$

$$= (1 - x^2)y'' - 2xy'.$$

These are the first two terms of text equation (10.1). The rest of the terms are the same as text equation (10.1) if we replace $\sin^2\theta$ by $(1 - x^2)$ as above.

5.   By text equation (10.6) with $m = 1$, $\ell = 4$:

$$P_4^1(x) = (1 - x^2)^{1/2}\dfrac{d}{dx}P_4(x) = (1 - x^2)^{1/2}\dfrac{d}{dx}(35x^4 - 30x^2 + 3)/8$$

$$= (1 - x^2)^{1/2}(35 \cdot 4x^3 - 30 \cdot 2x)/8$$

$$= (1 - x^2)^{1/2}(35x^3 - 15x)/2.$$

Now we put $x = \cos \theta$, $1 - x^2 = 1 - \cos^2\theta = \sin^2\theta$ to get

$$P_4^1(\cos \theta) = \tfrac{1}{2}(\sin \theta)(35 \cos^3\theta - 15 \cos \theta).$$

Section 11

4. Substitute text equations (11.3) into $x^2 y'' - 6y = 0$ and tabulate powers of x as follows:

	x^s	x^{s+1}	$\cdots$	x^{n+s}
$x^2 y''$	$s(s-1)a_0$	$(s+1)sa_1$		$(n+s)(n+s-1)a_n$
$-6y$	$-6a_0$	$-6a_1$		$-6a_n$

Each column (coefficients of a power of x) must add to zero. Thus the indicial equation (first column) gives

$$s^2 - s - 6 = 0, \qquad (s-3)(s+2) = 0, \qquad s = 3 \text{ or } s = -2.$$

For $s = 3$, we find from the second column, $a_1 = 0$. Similarly, from the x^{n+s} column, we find

$$[(n+3)(n+2) - 6]a_n = (n^2 + 5n)a_n = 0$$

so $a_n = 0$ for all $n \neq 0$. The solution corresponding to $s = 3$ is just $a_0 x^s = a_0 x^3$, or

$$y = Ax^3.$$

For $s = -2$, we also find $a_n = 0$, $n \neq 0$ (verify this), so the corresponding solution is $y = Bx^{-2}$, and the general solution of the given differential equation is

$$y = Ax^3 + Bx^{-2}.$$

(Note that the given differential equation is an Euler or Cauchy equation and so can also be solved as on page 378 of the text.)

8. Comment: Compare Problem 21.10.

11.   Use text equations (11.3) and tabulate powers of x.

	x^s	x^{s+1}	x^{s+2}	$\cdots$	x^{n+s}
$36x^2y''$	$36s(s-1)a_0$	$36(s+1)sa_1$	$36(s+2)(s+1)a_2$		$36(n+s)(n+s-1)a_n$
$5y$	$5a_0$	$5a_1$	$5a_2$		$5a_n$
$-9x^2y$			$-9a_0$		$-9a_{n-2}$

We find the values of s from the indicial equation.

$$36s^2 - 36s + 5 = 0 = (6s - 1)(6s - 5), \qquad s = 1/6 , 5/6 .$$

For either value of s, the second column gives $a_1 = 0$. From the x^{n+s} column we get

(1) $$a_n = \frac{9}{36(n+s)(n+s-1)+5} a_{n-2} .$$

Thus for either value of s, we have (since $a_1 = 0$),

$$a_3 = a_5 = a_7 = \cdots = 0.$$

For s = 1/6, use (1) to find

$$a_2 = \frac{9}{36\left(2+\frac{1}{6}\right)\left(1+\frac{1}{6}\right)+5} a_0 = \frac{9}{96} a_0 = \frac{3}{2^5} a_0 ,$$

$$a_4 = \frac{9}{36\left(4+\frac{1}{6}\right)\left(3+\frac{1}{6}\right)+5} a_2 = \frac{3}{5\cdot2^5} a_2 = \frac{3^2}{5\cdot2^{10}} a_0 ,$$

and so on.   For s = 5/6, we find from (1):

$$a_2' = \frac{9}{36\left(2+\frac{5}{6}\right)\left(1+\frac{5}{6}\right)+5} a_0' = \frac{9}{192} a_0' = \frac{3}{2^6} a_0' ,$$

$$a_4' = \frac{9}{672} a_2' = \frac{3}{7\cdot2^5} a_2' = \frac{3^2}{7\cdot2^{11}} a_0'$$

11.   (continued)

and so on. The primes emphasize the fact that these coefficients are not related to the first set found above. Then with $a_0 = A$ and $a_0' = B$, the general solution of the given differential equation is

$$y = Ax^{1/6}\left(1 + \frac{3}{2^5}x^2 + \frac{3^2}{5 \cdot 2^{10}}x^4 + \cdots\right)$$

$$+ Bx^{5/6}\left(1 + \frac{3}{2^6}x^2 + \frac{3^2}{7 \cdot 2^{11}}x^4 + \cdots\right).$$

13.   Hint: What is a_0 ?

14.   Hint: Look at the second column of your tabulation of coefficients. What is b_1 ? Compare the corresponding equation for a_1.

Section 12

1.   Using text page 12, equations (6.2) and (6.3), and text page 512, equation (12.9), we find

$$\rho_n = \left| \frac{(-1)^{n+1}}{\Gamma(n+2)\Gamma(n+p+2)} \div \frac{(-1)^n}{\Gamma(n+1)\Gamma(n+p+1)} \right| = \frac{1}{(n+1)(n+p+1)}$$

since $\Gamma(n+2) = (n+1)\Gamma(n+1)$ and $\Gamma(n+p+1) = (n+p+1)\Gamma(n+p+1)$ [see text page 459, equation (3.4)]. Then

$$\rho = \lim_{n \to \infty} \rho_n = 0$$

so the series for $J_p(x)$ converges for all x.

5. We write text equation (12.9) with $p = 1$, multiply by x, and differentiate to get

$$\frac{d}{dx}[x\,J_1(x)] = \frac{d}{dx}\sum_{n=0}^{\infty}\frac{(-1)^n}{\Gamma(n+1)\Gamma(n+2)}\left(\frac{1}{2}\right)^{2n+1}x^{2n+2}$$

(A)
$$= \sum_{n=0}^{\infty}\frac{(-1)^n(2n+2)}{\Gamma(n+1)\Gamma(n+2)}\left(\frac{1}{2}\right)^{2n+1}x^{2n+1}.$$

Now $\Gamma(n+2) = (n+1)\Gamma(n+1)$ [text, page 459, equation (3.4)]. Then

$$\frac{2n+2}{\Gamma(n+2)}\left(\frac{1}{2}\right)^{2n+1} = \frac{2(n+1)}{(n+1)\Gamma(n+1)}\left(\frac{1}{2}\right)^{2n+1} = \frac{1}{\Gamma(n+1)}\left(\frac{1}{2}\right)^{2n}.$$

Using this result in (A), we have

$$\frac{d}{dx}[x\,J_1(x)] = \sum_{n=0}^{\infty}\frac{(-1)^n}{\Gamma(n+1)\Gamma(n+1)}\left(\frac{1}{2}\right)^{2n}x^{2n+1}$$

$$= x\sum_{n=0}^{\infty}\frac{(-1)^n}{\Gamma(n+1)\Gamma(n+1)}\left(\frac{x}{2}\right)^{2n}.$$

From text equation (12.9) with $p = 0$, we see that this is $x\,J_0(x)$.

9. Hints: Remember that $\Gamma(p+1) = p\Gamma(p)$, and $\Gamma(1/2) = \sqrt{\pi}$. See text page 24 for the $\sin x$ series.

Section 13

2.   Let us write a few terms of $J_3(x)$ and of $J_{-3}(x)$ to see that
     $J_{-3}(x) = -J_3(x)$.   First we use text equation (12.9) with $p = 3$.
     Remember that the gamma function $\Gamma(n)$ is just $(n-1)!$ for inte-
     gral n, and that $\Gamma(1) = 0! = 1$ (text, page 459).   Then

$$J_3(x) = \frac{1}{\Gamma(1)\Gamma(4)}\left(\frac{x}{2}\right)^3 - \frac{1}{\Gamma(2)\Gamma(5)}\left(\frac{x}{2}\right)^5$$

$$+ \frac{1}{\Gamma(3)\Gamma(6)}\left(\frac{x}{2}\right)^7 - \frac{1}{\Gamma(4)\Gamma(7)}\left(\frac{x}{2}\right)^9 + \cdots$$

$$= \frac{1}{3!}\left(\frac{x}{2}\right)^3 - \frac{1}{4!}\left(\frac{x}{2}\right)^5 + \frac{1}{2!5!}\left(\frac{x}{2}\right)^7 - \frac{1}{3!7!}\left(\frac{x}{2}\right)^9 + \cdots .$$

To find J_{-3} we will need Γ of negative numbers.   See the text,
page 460, and look at the graph of the Γ function in the solution
of Chapter 11 Problem 4.7, to see that $\Gamma(n)$ is infinite for all
negative integers and for $n = 0$.   Using text equation (12.9) with
$p = -3$, we find that the first three terms have $\Gamma(-2)$, $\Gamma(-1)$,
$\Gamma(0)$, in their denominators, and so are zero.   Then starting with
the 4th term, we find

$$J_{-3}(x) = \frac{-1}{\Gamma(4)\Gamma(1)}\left(\frac{x}{2}\right)^{6-3} + \frac{1}{\Gamma(5)\Gamma(2)}\left(\frac{x}{2}\right)^{8-3}$$

$$- \frac{1}{\Gamma(6)\Gamma(3)}\left(\frac{x}{2}\right)^{10-3} + \frac{1}{\Gamma(7)\Gamma(4)}\left(\frac{x}{2}\right)^{12-3} \cdots$$

$$= -\frac{1}{3!}\left(\frac{x}{2}\right)^3 + \frac{1}{4!}\left(\frac{x}{2}\right)^5 - \frac{1}{5!2!}\left(\frac{x}{2}\right)^7 + \frac{1}{6!3!}\left(\frac{x}{2}\right)^9 \cdots$$

which is just the negative of J_3 as claimed.

2.   (continued)

Now to show in general that, for integral p,

$J_{-p}(x) = (-1)^P J_p(x)$, we use text equation (13.1). Since

$n - p + 1 \leqslant 0$ for all $n \leqslant p - 1$, then $\dfrac{1}{\Gamma(n - p + 1)} = 0$ for $n \leqslant p - 1$,

so the series for $J_{-p}(x)$ starts with the term $n = p$. We have

$$J_{-p}(x) = \sum_{n=p}^{\infty} \frac{(-1)^n}{\Gamma(n+1)\Gamma(n-p+1)}\left(\frac{x}{2}\right)^{2n-p}.$$

Now change the summation index to $m = n - p$; then the sum over m

is from 0 to ∞.   Replace n by $m + p$ to get

$$J_{-p}(x) = \sum_{m=0}^{\infty} \frac{(-1)^{m+p}}{\Gamma(m+p+1)\Gamma(m+1)}\left(\frac{x}{2}\right)^{2(m+p)-p}$$

$$= (-1)^P \sum_{m=0}^{\infty} \frac{(-1)^m}{\Gamma(m+1)\Gamma(m+p+1)}\left(\frac{x}{2}\right)^{2m+p} = (-1)^P J_p(x)$$

by text equation (12.9). Note carefully that p is an integer

in this problem; the result is not true for non-integral p.

3.   Let $p = 1/2$ in text equation (13.1) or $p = -1/2$ in text equation
      (12.9).   Then

$$J_{-1/2}(x) = \sum_{n=0}^{\infty} \frac{(-1)^n}{\Gamma(n+1)\Gamma\left(n+\frac{1}{2}\right)} \left(\frac{x}{2}\right)^{2n} \left(\frac{x}{2}\right)^{-1/2} .$$

Factor out $\left(\frac{x}{2}\right)^{-1/2} = \sqrt{\frac{2}{x}}$ and put $\Gamma(n+1) = n!$ to get

$$J_{-1/2}(x) = \sqrt{\frac{2}{x}} \sum_{n=0}^{\infty} \frac{(-1)^n}{n!\,\Gamma\left(n+\frac{1}{2}\right)} \left(\frac{x}{2}\right)^{2n}$$

$$= \sqrt{\frac{2}{x}} \left(\frac{1}{\Gamma(1/2)} - \frac{1}{\Gamma(3/2)}\left(\frac{x}{2}\right)^2 + \frac{1}{2\Gamma(5/2)}\left(\frac{x}{2}\right)^4 - \frac{1}{3!\,\Gamma(7/2)}\left(\frac{x}{2}\right)^6 + \cdots \right) .$$

Use repeatedly $\Gamma(p+1) = p\Gamma(p)$ to write the Γ functions as
multiples of $\Gamma(1/2)$ and then use $\Gamma(1/2) = \sqrt{\pi}$ (see text, pages
459 and 461).

$$J_{-1/2}(x) = \sqrt{\frac{2}{\pi x}} \left(1 - \frac{x^2}{2} + \frac{x^4}{4!} - \frac{x^6}{6!} \cdots \right) = \sqrt{\frac{2}{\pi x}} \cos x .$$

Section 14

See the graphs in Problem 15.5 below. Note from text equation (12.9) that $J_p(x)$ for small x is (a constant times) x^p. Thus J_1 starts off like x, J_2 like x^2, J_3 like x^3, $J_{1/2}$ like $\sqrt{x}$, and so on (times a numerical factor in each case). $J_0(0) = 1$, but for all other p, $J_p(0) = 0$. All the functions $N_p(x)$ go to $-\infty$ at the origin (for $x \geqslant 0$ and $p \geqslant 0$). Away from the origin, note the oscillation of both J's and N's. Note that the zeros (values of x where a graph crosses the x axis) are not evenly spaced. Find and learn to use tables of values of Bessel functions and their zeros. (The CRC tables give J_0 and J_1; the NBS tables are very extensive. See References, text, page 742.) You should find the following approximate values:

$$J_0(x) = 0 \quad \text{for} \quad x = 2.4, \ 5.5,$$
$$J_1(x) = 0 \quad \text{for} \quad x = 3.8, \ 7.0,$$
$$J_2(x) = 0 \quad \text{for} \quad x = 5.1, \ 8.4,$$
$$J_3(x) = 0 \quad \text{for} \quad x = 6.4, \ 9.8.$$

Section 15

3. From text equation (15.1), we find

$$x^p J_p' + px^{p-1} J_p = x^p J_{p-1} .$$

Solving for J_p', we get

$$J_p' = J_{p-1} - \frac{p}{x} J_p$$

which is part of (15.5). Similarly, use equation (15.2) to prove the rest of (15.5).

4.   To do Problem 12.4, we use text equation (15.2) with $p = 0$.

$$\frac{d}{dx} J_0(x) = -x^0 J_1(x) = -J_1(x).$$

5.   From text equation (15.4) with $p = 1$, we find $J_0 - J_2 = 2J_1'$, so $J_0 - J_2 = 0$ if $J_1' = 0$, that is,

$$J_0 = J_2 \quad \text{at every maximum or minimum of } J_1.$$

From text equation (15.5) with $p = 1$, we find

$$J_1' = -\frac{1}{x} J_1 + J_0 = \frac{1}{x} J_1 - J_2, \qquad \text{so}$$

$$J_1' = J_0 = -J_2 \quad \text{when} \quad J_1 = 0.$$

Since text equations (15.1) to (15.5) hold for the N's as well as the J's, the results we have just proved for J_0, J_1, and J_2 also hold for N_0, N_1, and N_2. Study the following graphs to see that these theorems are true.

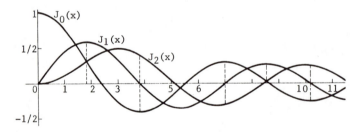

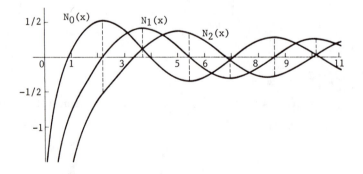

Section 16

5. We compare the given equation with text equation (16.1) to get

$$-1 = 1 - 2a, \qquad \left(bcx^{c-1}\right)^2 = 4, \qquad a^2 - p^2c^2 = 1, \qquad \text{so}$$

$$a = 1, \qquad c = 1, \qquad b = 2, \qquad p = 0.$$

By text equation (16.2), the solution is

$$y = x\, J_0(2x) \qquad \text{or} \qquad x\, N_0(2x)$$

or any linear combination of them. (This is like saying that the solution of $y'' = -y$ is $\sin x$ or $\cos x$ or e^{ix} or e^{-ix} or $\sinh ix$, etc.) For Bessel functions, it is rather common to use the letter Z to mean any of the various Bessel functions J, N, $H^{(1)}$ or $H^{(2)}$ [see text equation (17.1)]. We write the solution as

$$y = x\, Z_0(2x).$$

This means that the general solution is

$$y = Ax\, J_0(2x) + Bx\, N_0(2x)$$

or any other linear combination of the various Bessel functions.

7. Divide the equation by x to get:

$$y'' + \frac{3}{x}\, y' + x^2 y = 0.$$

We now try to fit this to text equation (16.1). We must have

$$1 - 2a = 3, \qquad a^2 - p^2c^2 = 0, \qquad \left(bcx^{c-1}\right)^2 \equiv x^2,$$

$$a = -1, \qquad pc = 1, \qquad bc = 1, \qquad 2(c - 1) = 2.$$

Thus $a = -1$, $c = 2$, $p = 1/2$, $b = 1/2$, and from text equation (16.2) the solution of the differential equation is

7.   (continued)

$$y = x^{-1} Z_{1/2}\left(\tfrac{1}{2} x^2\right).$$

You may have wondered why we chose $p = 1/2$ rather than $p = -1/2$ which also satisfies the equation $1 - 4p^2 = 0$. Recall that if $J_{1/2}$ is one solution of a Bessel equation, $J_{-1/2}$ is another solution. However, we usually use $N_{1/2}$ rather than $J_{-1/2}$ as the second solution. Thus we ordinarily use a positive value for p. A negative value of b just gives a constant factor times the solution using positive b, so again, we use the positive value.

13.   Let's prove that text equation (16.2) gives a solution of text equation (16.1). We make the change of variables

(1) $y = x^a u,$ $u = Z_p(z),$ $z = bx^c.$

By the rules of partial differentiation (text, Chapter 4),

(2)
$$\begin{cases} \dfrac{dy}{dx} = ax^{a-1} u + x^a \dfrac{du}{dx}, \\[2mm] \dfrac{d^2 y}{dx^2} = a(a-1)x^{a-2} u + 2ax^{a-1}\dfrac{du}{dx} + x^a \dfrac{d^2 u}{dx^2}. \end{cases}$$

(3)
$$\begin{cases} \dfrac{du}{dx} = \dfrac{du}{dz}\dfrac{dz}{dx} = \dfrac{du}{dz} bcx^{c-1}, \\[2mm] \dfrac{d^2 u}{dx^2} = \dfrac{d^2 u}{dz^2}\left(bcx^{c-1}\right)^2 + \dfrac{du}{dz} bc(c-1)x^{c-2}. \end{cases}$$

Substitute equations (3) into equations (2) and then substitute equations (2) into text equation (16.1). We tabulate the coefficients of $\dfrac{d^2 u}{dz^2}$, $\dfrac{du}{dz}$, and u.

13.   (continued)

$$\frac{d^2u}{dz^2} : \quad x^a\left(bcx^{c-1}\right)^2$$

$$\frac{du}{dz} : \quad 2ax^{a-1}bcx^{c-1} + x^abc(c-1)x^{c-2} + \frac{1-2a}{x}x^abcx^{c-1}$$

$$u: \quad a(a-1)x^{a-2} + \frac{1-2a}{x}ax^{a-1} + \left[\left(bcx^{c-1}\right)^2 + \frac{a^2-p^2c^2}{x^2}\right]x^a$$

Simplify the algebra, divide each term by c^2x^{a-2}, and put $z = bx^c$
to get the u(z) equation

(4) $$z^2\frac{d^2u}{dz^2} + z\frac{du}{dz} + (z^2 - p^2)u = 0.$$

This is Bessel's equation [see text equation (12.1)] so we have
verified that text equation (16.2) gives a solution of text equation (16.1) if $u = Z_p$ is a solution of Bessel's equation.

Section 17

Table comparing Bessel functions with trigonometric and
hyperbolic functions.

$\sin x,\ \cos x$	$e^{\pm ix} = \cos x \pm i\sin x$	$\sinh x = i^{-1}\sin ix$
$J_p(x),\ N_p(x)$ Bessel functions of the first and second kind	$H_p^{(1)(2)}(x) = J_p(x) \pm iN_p(x)$ Bessel functions of the third kind or Hankel functions	$I_p(x) = i^{-p}J_p(ix)$ $K_p(x)$ Hyperbolic Bessel functions

3. Note that text equations (15.1) to (15.5) hold for $Y_p(x)$ as well
    as for $J_p(x)$. Then by (15.2) we have

    $$Y_{p+1}(x) = -x^p \frac{d}{dx}\left[x^{-p}Y_p(x)\right].$$

    We let $p = 1/2$, and then $p = 3/2$, to get

    (1) $Y_{3/2}(x) = -x^{1/2}\frac{d}{dx}\left[x^{-1/2}Y_{1/2}(x)\right]$,

    (2) $Y_{5/2}(x) = -x^{3/2}\frac{d}{dx}\left[x^{-3/2}Y_{3/2}(x)\right].$

    From Problems 13.3 and 13.5 we have

    (3) $Y_{1/2}(x) = -J_{-1/2}(x) = -\sqrt{\dfrac{2}{\pi x}}\cos x.$

    (Remember that Y and N are the same.)   Substitute (3) into (1)
    and the result into (2) to get

    $$Y_{3/2}(x) = -x^{1/2}\frac{d}{dx}\left[x^{-1/2}\left(-\sqrt{\frac{2}{\pi x}}\cos x\right)\right] = \sqrt{\frac{2}{\pi}}\,x^{1/2}\frac{d}{dx}\left(\frac{\cos x}{x}\right),$$

    $$Y_{5/2}(x) = -x^{3/2}\frac{d}{dx}\left[x^{-3/2}\sqrt{\frac{2}{\pi}}\,x^{1/2}\frac{d}{dx}\left(\frac{\cos x}{x}\right)\right]$$

    $$= -\sqrt{\frac{2}{\pi}}\,x^{3/2}\frac{d}{dx}\left[\frac{1}{x}\frac{d}{dx}\left(\frac{\cos x}{x}\right)\right].$$

    Then by the definition of $y_n(x)$ in text equation (17.4),

    $$y_0(x) = \sqrt{\frac{\pi}{2x}}\,Y_{1/2}(x) = -\sqrt{\frac{\pi}{2x}}\sqrt{\frac{2}{\pi x}}\cos x = -\frac{\cos x}{x},$$

    $$y_1(x) = \sqrt{\frac{\pi}{2x}}\,Y_{3/2}(x) = \frac{d}{dx}\left(\frac{\cos x}{x}\right) = -x\left(-\frac{1}{x}\frac{d}{dx}\right)\left(\frac{\cos x}{x}\right),$$

    $$y_2(x) = \sqrt{\frac{\pi}{2x}}\,Y_{5/2}(x) = -x\frac{d}{dx}\left(\frac{1}{x}\frac{d}{dx}\right)\left(\frac{\cos x}{x}\right)$$

    $$= -x^2\left(-\frac{1}{x}\frac{d}{dx}\right)\left(-\frac{1}{x}\frac{d}{dx}\right)\left(\frac{\cos x}{x}\right) = -x^2\left(-\frac{1}{x}\frac{d}{dx}\right)^2\left(\frac{\cos x}{x}\right),$$

    as in (17.4).

5.   We leave the problem as stated in the text for you to do and here
     do the very similar problem to find $h_n^{(2)}(x)$. Using text equation
     (17.4), we have

$$h_n^{(2)}(x) = j_n(x) - i\, y_n(x)$$

$$= x^n \left(-\frac{1}{x}\frac{d}{dx}\right)^n \frac{\sin x}{x} + i\, x^n \left(-\frac{1}{x}\frac{d}{dx}\right)^n \frac{\cos x}{x}$$

$$= x^n \left(-\frac{1}{x}\frac{d}{dx}\right)^n \left(\frac{\sin x + i \cos x}{x}\right)$$

$$= i\, x^n \left(-\frac{1}{x}\frac{d}{dx}\right)^n \left(\frac{\cos x - i \sin x}{x}\right)$$

$$= i\, x^n \left(-\frac{1}{x}\frac{d}{dx}\right)^n \left(\frac{1}{x} e^{-ix}\right).$$

11. Again we leave the text problem to you and do the similar problem
     to find $h_0^{(2)}(ix)$. From Problem 5 above, we have

$$h_0^{(2)}(x) = i\, e^{-ix}/x.$$

Replacing x by ix gives the desired result:

$$h_0^{(2)}(ix) = i\, e^{x}/(ix) = e^{x}/x.$$

14. Using the definition of $j_n(x)$ in text equation (17.4), we see
     that we want to prove

$$\frac{d}{dx}\left(\sqrt{\frac{2}{\pi x}}\, J_{n+\frac{1}{2}}\right) = \sqrt{\frac{2}{\pi x}}\, J_{n-\frac{1}{2}} - (n+1)\frac{1}{x}\sqrt{\frac{2}{\pi x}}\, J_{n+\frac{1}{2}}.$$

We cancel the factor $\sqrt{2/\pi}$ and do the differentiation to get

$$x^{-1/2} J'_{n+\frac{1}{2}} - \frac{1}{2}x^{-3/2} J_{n+\frac{1}{2}} = x^{-1/2} J_{n-\frac{1}{2}} - (n+1)x^{-3/2} J_{n+\frac{1}{2}},$$

$$J'_{n+\frac{1}{2}} = J_{n-\frac{1}{2}} - \left(n+\frac{1}{2}\right)x^{-1} J_{n+\frac{1}{2}}.$$

Compare this with the first part of text equation (15.5); we see
that they are the same if $p = n + \frac{1}{2}$. Thus the original equation
is correct.

Section 18

3.   Following the text outline we find:

$$\left[x^2 J''_{-p} + x J'_{-p} + (x^2 - p^2) J_{-p}\right] J_p - \left[x^2 J''_p + x J'_p + (x^2 - p^2) J_p\right] J_{-p} = 0,$$

(A) $x^2 (J_p J''_{-p} - J_{-p} J''_p) + x(J_p J'_{-p} - J_{-p} J'_p) = 0.$

Cancel x and show that (A) is then a derivative:

$$\frac{d}{dx}\left[x(J_p J'_{-p} - J_{-p} J'_p)\right] = x \frac{d}{dx}(J_p J'_{-p} - J_{-p} J'_p) + (J_p J'_{-p} - J_{-p} J'_p)\frac{dx}{dx}$$

$$= x(J_p J''_{-p} - J_{-p} J''_p) + (J_p J'_{-p} - J_{-p} J'_p) = 0$$

from (A). Thus

$$x(J_p J'_{-p} - J_{-p} J'_p) = \text{const.} \quad \text{or} \quad J_p J'_{-p} - J_{-p} J'_p = \frac{c}{x}.$$

To find c we look at text equation (12.9). The lowest powers of
x are the first terms of each of the series. J_p starts with x^p
and J_{-p} starts with x^{-p} (for p ≠ integer). The product of either
of these times the derivative of the other gives a multiple of
1/x. Any higher terms in either or both series would give positive powers of x. Using the first terms in text equation (12.9),
we get

$$J_p J'_{-p} - J_{-p} J'_p = \frac{1}{\Gamma(1+p)\Gamma(1-p)}\left(\frac{1}{2}\right)^p\left(\frac{1}{2}\right)^{-p}\left[x^p \frac{d}{dx}(x^{-p}) - x^{-p} \frac{d}{dx}(x^p)\right]$$

$$= \frac{1}{\Gamma(1+p)\Gamma(1-p)}\left[x^p(-px^{-p-1}) - x^{-p}(px^{p-1})\right]$$

$$= \frac{-2px^{-1}}{\Gamma(1+p)\Gamma(1-p)}.$$

Using equation (3.4) on text page 459 and equation (5.4) on text
page 462, we get

$$J_p J'_{-p} - J_{-p} J'_p = \frac{-2px^{-1}}{p\Gamma(p)\Gamma(1-p)} = \frac{-2\sin \pi p}{\pi x}.$$

5.   Although Problems 3 and 4 assumed $p \neq$ integer, the result in
     Problem 4 is valid for $p =$ integer since in text equation (13.3)
     the functions N_p for integral p are defined as limits when
     $p \to$ integer.   Thus

(A) $J_n N_n' - J_n' N_n = \frac{2}{\pi x}$.

From text equation (15.2) with $p = 0$, we have $J_0' = -J_1$ and
similarly $N_0' = -N_1$.   Thus (A) with $n = 0$ gives

(B) $-J_0 N_1 + J_1 N_0 = \frac{2}{\pi x}$ or $J_0 N_1 - J_1 N_0 = -\frac{2}{\pi x}$.

For $p = 1$, text equation (15.4) gives

(C) $J_1' = \frac{1}{2}(J_0 - J_2)$ and $N_1' = \frac{1}{2}(N_0 - N_2)$.

Substitute (C) into (A) with $n = 1$ to get

$$\frac{1}{2} J_1 (N_0 - N_2) - \frac{1}{2}(J_0 - J_2) N_1 = \frac{2}{\pi x} \qquad \text{or}$$

$$J_1 N_2 - J_2 N_1 = J_1 N_0 - J_0 N_1 - \frac{4}{\pi x} = -\frac{2}{\pi x}$$

using (B).   In general, text equation (15.4) gives

(D) $J_n' = \frac{1}{2}(J_{n-1} - J_{n+1})$ and $N_n' = \frac{1}{2}(N_{n-1} - N_{n+1})$.

Substitute (D) into (A) to get

$$\frac{1}{2} J_n (N_{n-1} - N_{n+1}) - \frac{1}{2}(J_{n-1} - J_{n+1}) N_n = \frac{2}{\pi x} \qquad \text{or}$$

$$J_n N_{n+1} - J_{n+1} N_n = J_n N_{n-1} - J_{n-1} N_n - \frac{4}{\pi x} = -\frac{2}{\pi x}$$

assuming the result proved for the $n - 1$ case (mathematical
induction).

8. From text equations (18.10) and (18.7) with $B = 0$

$\theta = 0$ when $J_1(u) = 0,$ $u = (\text{zero of } J_1) = r_1$,

$\dot{\theta} = 0$ when $J_2(u) = 0,$ $u = (\text{zero of } J_2) = r_2$.

From text equation (18.5)

$$u^2 = \frac{4g}{v^2}(\ell_0 + vt) \qquad \text{or} \qquad t = \frac{v}{4g}u^2 - \frac{\ell_0}{v} .$$

Then the time for an outward swing from $\theta = 0$ to $\dot{\theta} = 0$ is

$$t_2 - t_1 = \frac{v}{4g}\left(r_2^2 - r_1^2\right)$$

and the time for an inward swing from $\dot{\theta} = 0$ to $\theta = 0$ is

$$t_1 - t_2 = \frac{v}{4g}\left(r_1^2 - r_2^2\right).$$

From tables (NBS, Jahnke-Emde, etc.; see references, text page 741), we have

$r_1 = 3.83, \ 7.02, \ 10.17,$

$r_2 = 5.14, \ 8.42, \ 11.62.$

Thus we find for successive quarter periods [as multiples of $v/(4g)$]:

out	in	out	in	out
11.7	23	21.6	32.6	31.5

Note that each inward swing takes longer than the outward swings before and after it.

9. Hint: As t increases, how is $u = b\sqrt{\ell}$ changing?

Section 19

1.   On text page 523, we have

$$u(x) = J_p(ax), \qquad\qquad v(x) = J_p(bx),$$

(1) $u'(x) = \dfrac{d}{dx} J_p(ax) = aJ_p'(ax), \qquad v(1) = J_p(b),$

$$u'(1) = aJ_p'(a).$$

If a is a zero of $J_p(x)$, then $J_p(a) = 0$, so at $x = 1$:

(2) $u(1) = J_p(a) = 0.$

Use (1) and (2) in text equation (19.7):

$$vxu' - uxv' \Big|_0^1 = J_p(b) \cdot 1 \cdot aJ_p'(a) - 0 = (b^2 - a^2) \int_0^1 xuv\, dx$$

or

(3) $\displaystyle \int_0^1 xuv\, dx = \dfrac{aJ_p(b)J_p'(a)}{b^2 - a^2}.$

Remember that a is a zero of J_p but b is any number. As $b \to a$,

the fraction becomes indeterminate. We evaluate it by L'Hôpital's

rule (text, page 35): Differentiate numerator and denominator

with respect to b and let $b \to a$.

$$\lim_{b \to a} \frac{a\, J_p(b)J_p'(a)}{b^2 - a^2} = \lim_{b \to a} \frac{a\, J_p'(b)J_p'(a)}{2b} = \frac{1}{2}[J_p'(a)]^2.$$

Thus (3) with $u = v = J_p(ax)$ becomes

(4) $\displaystyle \int_0^1 x[J_p(ax)]^2 dx = \frac{1}{2}[J_p'(a)]^2,$

as claimed in text equation (19.10). The other two forms of the

integral given in text equation (19.10) can be found from text

equation (15.5) with $x = a$. Remember that $J_p(a) = 0$ to get

$$J_p'(a) = J_{p-1}(a) = -J_{p+1}(a)$$

so

$$[J_p'(a)]^2 = [J_{p-1}(a)]^2 = [J_{p+1}(a)]^2.$$

3.   Hint: Let $p = (2n+1)/2$ in text equation (19.10) and then rewrite each J_p in terms of j_n using text equation (17.4).

6.   Hint: In Problem 6.8 you proved orthogonality for the cosine functions of order $n+\frac{1}{2}$ by direct integration of the product of two of them. Here prove it by a different method, that is, by using the orthogonality of the $N_{1/2}$ Bessel functions. For still another method, see the solution of Chapter 13 Problem 9.6.

Section 20

4.   From text page 525 we find the approximations for small x

$$J_p(x) \sim \frac{1}{\Gamma(p+1)}\left(\frac{x}{2}\right)^p,$$

$$N_p(x) \sim -\frac{\Gamma(p)}{\pi}\left(\frac{2}{x}\right)^p, \qquad p > 0.$$

Then for $p > 0$,

$$\lim_{x \to 0} J_p(x)N_p(x) = \lim_{x \to 0}\left[\frac{-\Gamma(p)}{\pi\Gamma(p+1)}\left(\frac{x}{2}\right)^p\left(\frac{2}{x}\right)^p\right]$$

$$= \frac{-\Gamma(p)}{\pi\Gamma(p+1)} = -\frac{1}{\pi p}$$

since $\Gamma(p+1) = p\Gamma(p)$. (See text, page 459.) For $p = 0$, we see from text page 525 that $J_0(x)N_0(x) \to -\infty$ as $x \to 0$ since $J_0 \to 1$ and $N_0 \sim \frac{2}{\pi}\ln x \to -\infty$.

7.   By text equation (17.4),

$$h_n^{(1)}(x) = j_n(x) + i\, y_n(x).$$

By text page 525, the asymptotic approximations for j_n and y_n are

$$j_n(x) \sim \frac{1}{x} \sin\left(x - \frac{n\pi}{2}\right), \qquad y_n(x) \sim -\frac{1}{x} \cos\left(x - \frac{n\pi}{2}\right).$$

Thus the asymptotic approximation for $h_n^{(1)}(x)$ is

$$h_n^{(1)}(x) \sim \frac{1}{x}\left[\sin\left(x - \frac{n\pi}{2}\right) - i \cos\left(x - \frac{n\pi}{2}\right)\right].$$

Now

$$\sin\theta - i\cos\theta = -i(\cos\theta + i\sin\theta) = -ie^{i\theta} = e^{-i\pi/2}e^{i\theta}.$$

Thus

$$h_n^{(1)}(x) \sim \frac{1}{x}e^{-i\pi/2}e^{i(x-n\pi/2)} = \frac{1}{x}e^{i[x-(n+1)\pi/2]}.$$

9 and 10.   Hints: Do Problems 7 and 8 and then replace x by ix. Remember that $e^{i\pi/2} = i$, so $e^{in\pi/2} = i^n$.

<u>Section 21</u>

6.   By Kamke, equation 2.225, we find

$$y = C_1 x + C_2 x \int x^{-2}\sqrt{x^2 + 1}\, dx.$$

However, the integral can be evaluated by tables:

$$\int x^{-2}\sqrt{x^2+1}\, dx = \ln\left(x + \sqrt{x^2+1}\right) - x^{-1}\sqrt{x^2+1}.$$

Thus the solution is

$$y = C_1 x + C_2\left[x \ln\left(x + \sqrt{x^2+1}\right) - \sqrt{x^2+1}\right].$$

Also see Problem 20 below.

7.   By Kamke, equation 2.328, we find

$$y = C_1 \frac{x}{x-1} + C_2 \left(x + 1 - \frac{x}{x-1} \ell n\, x^2 \right).$$

Also see Problem 14 below.

10.   By Kamke equation 2.204 with $a = 1$, $b = 0$, we find

$$y = C_1 \left(\frac{1}{x} - \frac{1}{2} \right) + C_2 \left(\frac{1}{x} + \frac{1}{2} \right) e^{-x}$$

$$= A(1 - 2x^{-1}) + B(1 + 2x^{-1}) e^{-x},$$

with $C_1 = -2A$, $C_2 = 2B$.

Comment: We can write the solution of Problem 11.8 in closed form by using this same Kamke equation with $a = 2$, $b = 0$.

14.   This is the same differential equation as in Problem 7 above. Here we solve it by using Frobenius series as outlined in the text directions. Substitute text equations (11.3) into the given differential equation

$$x(x-1)^2 y'' - 2y = 0 = (x^3 - 2x^2 + x) y'' - 2y = 0,$$

and collect coefficients of powers of x:

14.   (continued)

	x^{s-1}	x^s	x^{s+1}	...	x^{s+n} ...
$x^3 y''$			$s(s-1)a_0$		$(n+s-1)(n+s-2)a_{n-1}$
$-2x^2 y''$		$-2s(s-1)a_0$	$-2(s+1)sa_1$		$-2(n+s)(n+s-1)a_n$
xy''	$s(s-1)a_0$	$(s+1)sa_1$	$(s+2)(s+1)a_2$		$(n+s+1)(n+s)a_{n+1}$
$-2y$		$-2a_0$	$-2a_1$		$-2a_n$

The indicial equation (first column) gives $s = 0$ or $s = 1$. We see that $s = 0$ is not allowed by the second column since $a_0 \neq 0$ (remember that a_0 is, by definition, the coefficient of the first nonzero term in the solution). Thus, as the directions claim, we must use the larger value of s. For $s = 1$, we have

$$x^s: \quad 2a_1 - 2a_0 = 0, \qquad\qquad a_1 = a_0$$

$$x^{s+1}: \quad -4a_1 + 6a_2 - 2a_1 = 0, \qquad a_2 = a_1 = a_0$$

...

$$x^{s+n}: \quad n(n-1)a_{n-1} - 2(n+1)na_n + (n+2)(n+1)a_{n+1} - 2a_n = 0.$$

If all the a's up to a_n are equal to a_0, then the x^{s+n} column gives

$$a_{n+1} = \frac{a_0}{(n+2)(n+1)}[-n(n-1) + 2(n+1)n + 2] = a_0.$$

Thus, by mathematical induction, all the a's are equal to a_0 and we have a solution

$$(1) \qquad y = a_0(x + x^2 + x^3 + \cdots + x^n + \cdots) = a_0 \frac{x}{1-x}.$$

14.   (continued)

According to the text directions, a second solution is

$$y = \frac{x \, \ell n \, x}{1 - x} + \sum b_n x^{n+s} .$$

Substitute this into the differential equation to get

$$x(x - 1)^2 \left[\frac{1 - x^2 + 2x \, \ell n \, x}{x(1 - x)^3} + \sum b_{n+2} (n + s + 2) (n + s + 1) x^{n+s} \right]$$

$$- \frac{2x \, \ell n \, x}{1 - x} - 2 \sum b_n x^{n+s} = 0, \qquad \text{or}$$

$$1 + x + x(1 - x)^2 \sum b_{n+2} (n + s + 2) (n + s + 1) x^{n+s} - 2 \sum b_n x^{n+s} = 0.$$

The lowest order term is a constant so we must have $s = 0$ (as we expected -- see text directions) and

$$1 - 2b_0 = 0, \qquad b_0 = \frac{1}{2} .$$

We tabulate the coefficients of the powers of x with $s = 0$:

1	x	x^2	x^3	$\cdots$	x^n
1	1				
	$2b_2$	$6b_3$	$12b_4$		$(n + 1)nb_{n+1}$
		$-4b_2$	$-12b_3$		$-2n(n - 1)b_n$
			$2b_2$		$(n - 1)(n - 2)b_{n-1}$
$-2b_0$	$-2b_1$	$-2b_2$	$-2b_3$		$-2b_n$

From the x column we find a relation between b_2 and b_1. We can choose either b_1 or b_2 arbitrarily and solve for the other. Let us choose $b_2 = 0$ and find

$$1 - 2b_1 = 0, \qquad b_1 = \frac{1}{2} .$$

14.   (continued)

Then from the x^2 column we find $b_3 = 0$, and from the later columns all the rest of the b's are zero. Thus we have a solution

(2) $y = \dfrac{x \, \ell n \, x}{1 - x} + \dfrac{1}{2} + \dfrac{1}{2} x.$

Any arbitrary constant (say 2B) times the solution (2) is a solution of the differential equation, so from (1) and (2) the general solution is

$$y = A \frac{x}{1 - x} + B \left(\frac{2x \, \ell n \, x}{1 - x} + 1 + x \right).$$

17.   Given the solution $u = e^x$, we let $y = e^x v(x)$ and find the differential equation which $v(x)$ must satisfy.

$$y = e^x v,$$
$$y' = e^x (v' + v),$$
$$y'' = e^x (v'' + 2v' + v),$$
$$xy'' - 2(x + 1) y' + (x + 2) y = 0$$
$$= e^x [x(v'' + 2v' + v) - 2(x + 1)(v' + v) + (x + 2)v]$$
$$= e^x (xv'' - 2v') = 0.$$

We solve the differential equation for v:

$$xv'' - 2v' = 0 \qquad \text{or}$$
$$xp' - 2p = 0 \qquad \text{where } p = v'.$$
$$\frac{dp}{p} = \frac{2dx}{x}, \qquad \ell n \, p = 2 \, \ell n \, x + \ell n \, C,$$
$$p = \frac{dv}{dx} = Cx^2, \qquad v = Cx^3 / 3.$$

Thus a second solution of the differential equation is

$$y = \frac{1}{3} Cx^3 e^x = Ax^3 e^x.$$

20.   Let $y = xv$ in the given differential equation, and solve the v
       equation.

$$(x^2 + 1)y'' - xy' + y = 0$$
$$= (x^2 + 1)(xv'' + 2v') - x(xv' + v) + xv = 0,$$

$$(x^3 + x)v'' + (x^2 + 2)v' = 0,$$

$$(x^3 + x)\frac{dp}{dx} + (x^2 + 2)p = 0, \qquad \text{where } p = v'.$$

$$\int\frac{dp}{p} = -\int\frac{x^2 + 2}{x^3 + x}\,dx = -\int\left(\frac{2}{x} - \frac{x}{x^2 + 1}\right)dx,$$

$$\ln p = -2\ln x + \frac{1}{2}\ln(x^2 + 1)$$
$$= \ln\left[x^{-2}(x^2 + 1)^{1/2}\right],$$

$$p = \frac{dv}{dx} = x^{-2}(x^2 + 1)^{1/2},$$

$$v = \int\frac{1}{x^2}\sqrt{x^2 + 1}\,dx = \ln\left(x + \sqrt{x^2 + 1}\right) - x^{-1}\sqrt{x^2 + 1}.$$

Then a second solution of the y equation is

$$y = x\ln\left(x + \sqrt{x^2 + 1}\right) - \sqrt{x^2 + 1}.$$

Section 22

3. $$D\left[e^{-x^2/2} f(x)\right] = \frac{d}{dx}\left[e^{-x^2/2} f(x)\right] = -xe^{-x^2/2} f(x) + e^{-x^2/2}\frac{df}{dx}$$

$$= e^{-x^2/2}(D - x)f(x).$$

Multiply by $e^{x^2/2}$:

(1) $$e^{x^2/2} D\left[e^{-x^2/2} f(x)\right] = (D - x)f(x).$$

3.   (continued)

Write $f(x) = (D - x)g(x) = e^{x^2/2} D\left[e^{-x^2/2} f(x)\right]$; then

(2) $(D - x)f(x) = (D - x)(D - x)g(x) = (D - x)^2 g(x)$,   and

(3) $D\left\{e^{-x^2/2} f(x)\right\} = D\left\{e^{-x^2/2} e^{x^2/2} D\left[e^{-x^2/2} g(x)\right]\right\}$

$$= D^2\left[e^{-x^2/2} g(x)\right].$$

Substitute (2) and (3) into (1):

$$e^{x^2/2} D^2\left[e^{-x^2/2} g(x)\right] = (D - x)^2 g(x).$$

Repeat this process with

$$g(x) = (D - x)h(x) = e^{x^2/2} D\left[e^{-x^2/2} h(x)\right]$$

to get

$$e^{x^2/2} D^2\left\{e^{-x^2/2} e^{x^2/2} D\left[e^{-x^2/2} h(x)\right]\right\} = (D - x)^2 (D - x)h(x),   \text{or}$$

$$e^{x^2/2} D^3\left[e^{-x^2/2} h(x)\right] = (D - x)^3 h(x).$$

We can continue this process as many steps as we like to get

$$e^{x^2/2} D^n\left[e^{-x^2/2} F(x)\right] = (D - x)^n F(x)$$

for any $F(x)$ and any integer n.   If $F(x) = e^{-x^2/2}$, we have

$$e^{x^2/2} D^n\left(e^{-x^2}\right) = (D - x)^n e^{-x^2/2}.$$

Thus we can write the Hermite functions as

$$y_n = (D - x)^n e^{-x^2/2} = e^{x^2/2} D^n\left(e^{-x^2}\right).$$

7.   We find

$$e^{x^2} \frac{d}{dx}\left(e^{-x^2}y'\right) = e^{x^2}\left(e^{-x^2}y'' - 2xe^{-x^2}y'\right) = y'' - 2xy'.$$

Thus text equation (22.14) can be written as

(1) $e^{x^2} \frac{d}{dx}\left(e^{-x^2}y_n'\right) + 2ny_n = 0,$ or $\frac{d}{dx}\left(e^{-x^2}y_n'\right) + 2ne^{-x^2}y_n = 0,$

where $y_n = H_n(x)$. We also write (1) for $y_k = H_k(x)$:

(2) $\frac{d}{dx}\left(e^{-x^2}y_k'\right) + 2ke^{-x^2}y_k = 0.$

Multiply (1) by y_k and (2) by y_n; subtract and integrate from $-\infty$ to ∞.

(3) $\int_{-\infty}^{\infty}\left[y_k\left(e^{-x^2}y_n'\right)' - y_n\left(e^{-x^2}y_k'\right)'\right]dx + 2(n-k)\int_{-\infty}^{\infty}e^{-x^2}y_ny_k\,dx = 0.$

Now the bracket in (3) is the derivative of

(4) $y_k\left(e^{-x^2}y_n'\right) - y_n\left(e^{-x^2}y_k'\right).$

To check this, write $\frac{d}{dx}y_k\left(e^{-x^2}y_n'\right) = y_k'\left(e^{-x^2}y_n'\right) + y_k\left(e^{-x^2}y_n'\right)'$, and differentiate the other term similarly. Then the first integral in (3) is given by (4) evaluated at $\pm\infty$. This is zero because of the e^{-x^2} factors. Thus (3) gives

$$0 + 2(n-k)\int_{-\infty}^{\infty}e^{-x^2}y_ny_k\,dx = 0,$$

so for $n \neq k$,

$$\int_{-\infty}^{\infty}e^{-x^2}y_ny_k\,dx = 0.$$

We say that the functions $y_n(x)$ are orthogonal on $(-\infty,\infty)$ with respect to the weight function e^{-x^2} (see text, page 524).

11.   The differential equation $y'' + (\epsilon - x^2)y = 0$ is text equation (22.1)
       if $\epsilon = 2n + 1$. For each integer $n \geqslant 0$ we found a solution y_n [text
       equation (22.11)] which tends to zero as $x \to \pm\infty$. Thus the eigen-
       values are $\epsilon = 2n + 1$ and the eigenfunctions are the Hermite func-
       tions y_n in text equation (22.11).

15.   Substitute text equations (2.2) into the differential equation
       and collect powers of x.

	const	x	x^2	$\cdots$	x^n
xy''		$2a_2$	$6a_3$		$(n+1)na_{n+1}$
y'	a_1	$2a_2$	$3a_3$		$(n+1)a_{n+1}$
$-xy'$		$-a_1$	$-2a_2$		$-na_n$
py	pa_0	pa_1	pa_2		pa_n

The coefficients are:

$$a_1 = -pa_0 \,,$$

$$a_2 = -\tfrac{1}{4} a_1 (p - 1) = \tfrac{1}{4} p(p - 1)a_0 \,,$$

$$a_3 = -\tfrac{1}{9} a_2 (p - 2) = -\tfrac{1}{9} \cdot \tfrac{1}{4} p(p - 1)(p - 2)a_0 \,,$$

$$\cdots$$

$$a_{n+1} = \frac{-(p - n)}{(n + 1)^2} a_n \,.$$

If p is any integer n, then all coefficients from a_{n+1} on contain
the factor p - n and so are zero. Thus we have a polynomial solution if p = integer. Using the formulas above for the coefficients and setting $a_0 = 1$, we find the following solutions:

15.   (continued)

$$p = 0: \quad a_1 = 0, \quad L_0(x) = 1.$$

$$p = 1: \quad a_1 = -1, \quad L_1(x) = 1 - x.$$

$$p = 2: \quad a_1 = -2, \quad a_2 = \frac{1}{4} \cdot 2 \cdot 1 = \frac{1}{2}, \quad L_2(x) = 1 - 2x + \frac{1}{2}x^2.$$

$$p = 3: \quad a_1 = -3, \quad a_2 = \frac{1}{4} \cdot 3 \cdot 2 = \frac{3}{2}, \quad a_3 = -\frac{1}{9} \cdot \frac{1}{4} \cdot 3! = -\frac{1}{6},$$

$$L_3(x) = 1 - 3x + \frac{3}{2}x^2 - \frac{1}{6}x^3.$$

We have solved an eigenvalue problem, namely: Find the values of
p (eigenvalues) for which the given differential equation has
polynomial solutions, and find the corresponding solutions
(eigenfunctions).

18.   (c) From text equation (22.23) we have

$$\Phi = \frac{e^{-xh/(1-h)}}{1-h} = \sum L_n(x)h^n,$$

$$\frac{\partial \Phi}{\partial x} = \frac{-he^{-xh/(1-h)}}{(1-h)^2} = \sum L_n'(x)h^n,$$

$$\frac{\partial \Phi}{\partial h} = \frac{(1-h-x)e^{-xh/(1-h)}}{(1-h)^3} = \sum nL_n(x)h^{n-1},$$

$$x\frac{\partial \Phi}{\partial x} + h\Phi - h(1-h)\frac{\partial \Phi}{\partial h} = -\frac{xh\Phi}{(1-h)} + h\Phi - \frac{h(1-h-x)\Phi}{1-h} = 0$$

$$= \sum xL_n'(x)h^n + \sum L_n(x)h^{n+1} - \sum h(1-h)nL_n(x)h^{n-1}.$$

Now select the coefficient of h^n in each sum, to get

$$xL_n'(x) + L_{n-1}(x) - nL_n(x) + (n-1)L_{n-1}(x) = 0 \qquad \text{or}$$

$$xL_n'(x) - nL_n(x) + nL_{n-1}(x) = 0,$$

which is part (c) of text equation (22.24).

19.   We want to evaluate $I = \int_0^\infty e^{-x}\left[L_n(x)\right]^2 dx$. Use text equation
(22.18) to get

$$I = \int_0^\infty e^{-x} L_n(x)\frac{1}{n!} e^x D^n(x^n e^{-x}) dx$$

$$= \frac{1}{n!}\int_0^\infty L_n(x)D^n(x^n e^{-x}) dx.$$

Integrate once by parts:

$$(n!)I = L_n(x)D^{n-1}(x^n e^{-x})\Big|_0^\infty - \int_0^\infty D^{n-1}(x^n e^{-x})\frac{d}{dx}L_n(x) dx.$$

The integrated term is zero (satisfy yourself that after differentiating $n-1$ times you still have the factor xe^{-x} which
is zero at $x = 0$ and $x = \infty$). Repeat the integration by parts
a total of n times to get

$$(n!)I = (-1)^n \int_0^\infty D^{n-n}(x^n e^{-x})D^n L_n(x) dx$$

$$= (-1)^n \int_0^\infty x^n e^{-x}(-1)^n dx$$

since by text equation (22.19), $D^n L_n(x) = (-1)^n$. By text page
458, $\int_0^\infty x^n e^{-x} dx = n!$, so we find

$$(n!)I = n! , \qquad I = 1.$$

23.   (b) We write the result in Problem 18(c) above with n replaced
by $n+k$:

$$xL'_{n+k} - (n+k)L_{n+k} + (n+k)L_{n+k-1} = 0.$$

Now differentiate k times. Use Leibniz' rule (text page 488) to
differentiate the first term.

$$x\frac{d^k}{dx^k}L'_{n+k} + k\frac{dx}{dx}\frac{d^{k-1}}{dx^{k-1}}L'_{n+k} - (n+k)\frac{d^k}{dx^k}(L_{n+k} - L_{n+k-1}) = 0,$$

or

$$x\frac{d}{dx}\frac{d^k}{dx^k}L_{n+k} - n\frac{d^k}{dx^k}L_{n+k} + (n+k)\frac{d^k}{dx^k}L_{n+k-1} = 0.$$

Multiply by $(-1)^k$ and use the notation of text equation (22.25):

$$x\frac{d}{dx}L_n^k - nL_n^k + (n+k)L_{n-1}^k = 0,$$

which is part (b) of text equation (22.28).

27.   Hints: Be very careful about the notation. In this problem we
are using the notation of quantum mechanics and n does not have
the same meaning as in the text equations. For example, for
$f_3(x)$ with $\ell = 1$, $n = 3$, we have $2\ell + 1 = 3$, $n - \ell - 1 = 3 - 1 - 1 = 1$, so
we want the function L_1^3. First find $L_1^3(x)$. By text equation
(22.25), and text page 767, Problem 22.13, this is

$$-\frac{d^3}{dx^3}L_4(x) = \frac{1}{24}(16 \cdot 6 - 4!x) = 4 - x.$$

Now replace x by $\frac{x}{n} = \frac{x}{3}$ and write $f_n(x)$ for $n = 3$, $\ell = 1$:

$$f_3(x) = x^2 e^{-x/6}\left(4 - \frac{x}{3}\right).$$

For the orthogonality proof, evaluate the integrals to show that
$\int_0^\infty f_2 f_3 \, dx = 0$ and similarly for $f_2 f_4$ and $f_3 f_4$.

Section 23

1. From text equations (5.1) and (5.2), we have

$$\Phi = (1 - 2xh + h^2)^{-1/2} = \sum h^\ell P_\ell(x).$$

Then

(1)
$$\int_{-1}^{1} \Phi^2 \, dx = \int_{-1}^{1} \frac{dx}{1 - 2xh + h^2}$$

$$= \sum_{\ell,n} h^\ell h^n \int_{-1}^{1} P_\ell(x) P_n(x) \, dx = \sum_\ell h^{2\ell} \int_{-1}^{1} \left[P_\ell(x) \right]^2 dx$$

since (text page 499) the Legendre polynomials are orthogonal.
Evaluate the integral of Φ^2 and expand the logarithms using text
page 24, equation (13.4).

$$\frac{1}{-2h} \ell n (1 - 2xh + h^2) \bigg|_{-1}^{1} = \frac{1}{-2h} \ell n \, \frac{(1-h)^2}{(1+h)^2} = \frac{1}{h} \left[\ell n (1+h) - \ell n (1-h) \right]$$

$$= \frac{1}{h} \left[h - \frac{h^2}{2} + \frac{h^3}{3} \cdots - \left(-h - \frac{h^2}{2} - \frac{h^3}{3} \cdots \right) \right]$$

(2)
$$= 2 + \frac{2h^2}{3} + \frac{2h^4}{5} \cdots = \sum h^{2\ell} \frac{2}{2\ell+1} .$$

Equate coefficients of $h^{2\ell}$ in (1) and (2):

$$\int_{-1}^{1} [P_\ell(x)]^2 dx = \frac{2}{2\ell+1} .$$

8.   From text equations (5.8e) and (7.2) we have

    (1) $(2\ell + 1)P_\ell(x) = P'_{\ell+1}(x) - P'_{\ell-1}(x)$,

    (2) $\left[(1 - x^2)P'_\ell(x)\right]' + \ell(\ell + 1)P_\ell = 0$.

Integrate (1) from $x = -1$ to $x = b$:

    (3) $(2\ell + 1)\displaystyle\int_{-1}^{b} P_\ell(x)\,dx = P_{\ell+1}(x) - P_{\ell-1}(x)\Big|_{-1}^{b} = P_{\ell+1}(b) - P_{\ell-1}(b)$

since $P_{\ell+1}(-1) = P_{\ell-1}(-1)$ (see text Problem 2.2). Also
integrate (2) from -1 to b.

    (4) $(1 - x^2)P'_\ell(x)\Big|_{-1}^{b} + \ell(\ell + 1)\displaystyle\int_{-1}^{b} P_\ell(x)\,dx = 0$.

Combine (3) and (4) to get two expressions both equal to
$\displaystyle\int_{-1}^{b} P_\ell(x)\,dx$.

$$\frac{1}{2\ell + 1}\left[P_{\ell+1}(b) - P_{\ell-1}(b)\right] = -\frac{1}{\ell(\ell+1)}(1 - b^2)P'_\ell(b).$$

Thus $P_{\ell+1}(b) = P_{\ell-1}(b)$ when $b = \pm 1$ and when $P'_\ell(b) = 0$.

11.   We use text equations (17.4) and (15.2) to find

$$\frac{d}{dx}\left(x^{-n}j_n\right) = \frac{d}{dx}\left(x^{-n}\sqrt{\frac{\pi}{2x}}\,J_{(2n+1)/2}\right)$$

$$= \sqrt{\frac{\pi}{2}}\frac{d}{dx}\left(x^{-n-\frac{1}{2}}J_{n+\frac{1}{2}}\right) = -\sqrt{\frac{\pi}{2}}\,x^{-n-\frac{1}{2}}J_{n+\frac{3}{2}}$$

$$= -x^{-n}j_{n+1}\,.$$

Then

$$\int_{0}^{\infty} x^{-n}j_{n+1}(x)\,dx = -x^{-n}j_n(x)\Big|_{0}^{\infty} = 0 + \lim_{x\to 0}\frac{j_n(x)}{x^n}$$

$$= \frac{1}{(2n+1)!!}$$

by text page 525.

12, 13, 14. Warning hint: Text equations (15.1) to (15.5) hold for J_p
and N_p and consequently for $H_p^{(1)}$ and $H_p^{(2)}$. They do <u>not</u> in general hold for the other functions discussed in Section 17 of the
text. [For example, compare Problems 12 and 13 with text equation (15.4).] To prove formulas involving the hyperbolic Bessel
functions I_p and K_p or the spherical Bessel functions j_n, y_n,
$h_n^{(1)}$, $h_n^{(2)}$, write these functions in terms of the J's and N's
using the formulas in text Section 17 as we did in Problem 11
above.

16.   By text equation (15.2) solved for J_{p+1}

$$J_{p+1} = -x^p \frac{d}{dx}\left(x^{-p} J_p\right).$$

We write this repeatedly with $p = 0$, $p = 1$, $p = 2, \cdots$, to get

$$J_1 = -\frac{d}{dx}(J_1) = x\left(-\frac{1}{x}\frac{d}{dx}\right)J_0 \, ,$$

$$J_2 = -x\frac{d}{dx}\left(x^{-1} J_1\right) = x^2\left(-\frac{1}{x}\frac{d}{dx}\right)\left(x^{-1} J_1\right) = x^2\left(-\frac{1}{x}\frac{d}{dx}\right)^2 J_0 \, ,$$

$$J_3 = -x^2\frac{d}{dx}\left(x^{-2} J_2\right) = x^3\left(-\frac{1}{x}\frac{d}{dx}\right)\left(x^{-2} J_2\right) = x^3\left(-\frac{1}{x}\frac{d}{dx}\right)^3 J_0$$

and so on. By mathematical induction,

$$J_n(x) = -x^{n-1}\frac{d}{dx}\left(x^{-(n-1)} J_{n-1}\right) = x^n\left(-\frac{1}{x}\frac{d}{dx}\right)\left(x^{-(n-1)} J_{n-1}\right)$$

$$= x^n\left(-\frac{1}{x}\frac{d}{dx}\right)^n J_0 \, .$$

20.   Problem 19 with $h = e^{i\theta}$ gives

$$\frac{x}{2}(h - h^{-1}) = \frac{x}{2}(e^{i\theta} - e^{-i\theta}) = ix \sin \theta \qquad \text{(text page 67)},$$

$$\Phi(x,h) = e^{i\, x \sin\, \theta} = \cos(x \sin \theta) + i \sin(x \sin \theta)$$

$$= \sum_{-\infty}^{\infty} h^n J_n(x) = \sum_{-\infty}^{\infty} e^{in\theta} J_n(x)$$

$$= J_0(x) + \sum_{1}^{\infty} (\cos n\theta)(J_n + J_{-n}) + i \sum_{1}^{\infty} (\sin n\theta)(J_n - J_{-n})$$

$$= J_0(x) + 2 \sum_{\text{even } n} J_n(x) \cos n\theta + 2i \sum_{\text{odd } n} J_n(x) \sin n\theta$$

(see Problem 13.2). Take real and imaginary parts:

$$\cos(x \sin \theta) = J_0(x) + 2 \sum_{\text{even } n} J_n(x) \cos n\theta,$$

$$\sin(x \sin \theta) = 2 \sum_{\text{odd } n} J_n(x) \sin n\theta.$$

These are Fourier series; we write the formulas for the coefficients (text, page 323, $\ell = \pi$):

$$a_n = 2J_n(x) = \frac{2}{\pi} \int_0^\pi \cos(x \sin \theta) \cos n\theta \, d\theta, \qquad \text{even } n,$$

$$a_0/2 = J_0(x) = \frac{1}{\pi} \int_0^\pi \cos(x \sin \theta) d\theta,$$

$$b_0 = 2J_n(x) = \frac{2}{\pi} \int_0^\pi \sin(x \sin \theta) \sin n\theta \, d\theta, \qquad \text{odd } n.$$

To obtain an integral representing J_n for all n, let's use the
complex exponential Fourier series (see above)

$$e^{ix \sin \theta} = \sum_{-\infty}^{\infty} e^{in\theta} J_n(x).$$

Then the Fourier coefficients are (text page 316)

20. (continued)

$$c_n = J_n(x) = \frac{1}{2\pi} \int_{-\pi}^{\pi} e^{ix \sin \theta} e^{-in\theta} d\theta$$

$$= \frac{1}{2\pi} \int_{-\pi}^{\pi} e^{-i(n\theta - x \sin \theta)} d\theta$$

$$= \frac{1}{\pi} \int_{0}^{\pi} \cos(n\theta - x \sin \theta) d\theta.$$

(Write the integrand as the sum of its even and odd parts -- see
text, page 322.)

25. The solutions of the differential equation (Problem 22.26)

(1) $$y'' + \left(\frac{\lambda}{x} - \frac{1}{4} - \frac{\ell(\ell+1)}{x^2} \right) y = 0$$

are

(2) $$y = x^{\ell+1} e^{-x/2} L_{\lambda-\ell-1}^{2\ell+1}(x).$$

If we replace x by x/n, the differential equation becomes

$$n^2 y'' + \left(\frac{\lambda n}{x} - \frac{1}{4} - \frac{n^2 \ell(\ell+1)}{x^2} \right) y = 0$$

or, with $\lambda = n$,

(3) $$y'' + \left(\frac{1}{x} - \frac{1}{4n^2} - \frac{\ell(\ell+1)}{x^2} \right) y = 0.$$

The functions in Problem 22.27, namely

(4) $$f_n(x) = x^{\ell+1} e^{-x/(2n)} L_{n-\ell-1}^{2\ell+1}(x/n),$$

are then solutions of (3). [Replace x in (2) by x/n to get (4).
Since any constant times a solution of (3) is a solution of (3),
we do not need to include the factor $(1/n)^{\ell+1}$.]

25.   (continued)

It is easy to prove that the solutions of (3) (for fixed ℓ and different n) are orthogonal. As in the text, pages 499 and 523, and in the Solutions of Problems 7.1 and 22.7, we write (3) for y_n and for y_k, multiply the y_n equation by y_k and vice versa, and subtract the two equations. We get

$$(5) \qquad y_n'' y_k - y_k'' y_n = \left(\frac{1}{4n^2} - \frac{1}{4k^2}\right) y_n y_k .$$

Now

$$y_n'' y_k - y_k'' y_n = \frac{d}{dx}(y_n' y_k - y_k' y_n) .$$

Integrate (5) from 0 to ∞.

$$\int_0^\infty (y_n'' y_k - y_k'' y_n)\,dx = (y_n' y_k - y_k' y_n)\Big|_0^\infty = \left(\frac{1}{4n^2} - \frac{1}{4k^2}\right)\int_0^\infty y_n y_k\,dx .$$

The solutions y_n or y_k are the functions in (4) and we see that they are zero at $x = 0$ and at $x = \infty$. Thus we have

$$\int_0^\infty y_n y_k\,dx = 0, \qquad n \neq k.$$

Comment: Both equations (1) and (3) are special cases of a Sturm-Liouville equation discussed in Problem 24 so we could have just appealed to the general proof there to say that their solutions are a set of orthogonal functions.

Chapter 13

<u>Section 1</u>

2.   If $u = \sin(x - vt)$, then (see Chapter 4, Section 5)

$$\frac{\partial u}{\partial x} = \cos(x - vt), \qquad\qquad \frac{\partial u}{\partial t} = -v\cos(x - vt),$$

$$\frac{\partial^2 u}{\partial x^2} = -\sin(x - vt), \qquad\qquad \frac{\partial^2 u}{\partial t^2} = -v^2\sin(x - vt).$$

Thus $\dfrac{\partial^2 u}{\partial x^2} = \dfrac{1}{v^2}\dfrac{\partial^2 u}{\partial t^2}$ which is text equation (1.4) when $u = u(x,t)$.

Similarly if $u = f(x + vt)$, then

$$\frac{\partial^2 u}{\partial x^2} = f''(x + vt) \quad \text{and} \quad \frac{\partial^2 u}{\partial t^2} = v^2 f''(x + vt), \quad \text{so} \quad \frac{\partial^2 u}{\partial x^2} = \frac{1}{v^2}\frac{\partial^2 u}{\partial t^2},$$

and a similar calculation for $f(x - vt)$.

<u>Section 2</u>

1.   The basic solutions of Laplace's equation are given by text equation (2.7). We want $T \to 0$ as $y \to \infty$ and $T = 0$ when $x = 0$, so we use
the solution $e^{-ky}\sin kx$. We also want $T = 0$ when $x = 10$ so we set
$\sin 10k = 0$ or $10k = n\pi$. Thus our basic solutions are
$e^{-n\pi y/10}\sin(n\pi x/10)$, $n = 1,2,3\cdots$. Any linear combination of
these solutions is a solution of Laplace's equation satisfying
the given boundary conditions on three sides of the semi-infinite
strip. We write

(1) $T = \displaystyle\sum_n b_n e^{-n\pi y/10}\sin(n\pi x/10)$.

Now we want $T = x$ when $y = 0$:

(2) $T = x = \displaystyle\sum_n b_n \sin(n\pi x/10)$.

1.   (continued)

This means that we want to expand x in a Fourier sine series.
By text Chapter 7, Section 9, we find

$$b_n = \frac{2}{10}\int_0^{10} x \sin\frac{n\pi x}{10}\,dx = \frac{2}{10}\left(\frac{10}{n\pi}\right)^2\left(\sin\frac{n\pi x}{10} - \frac{n\pi x}{10}\cos\frac{n\pi x}{10}\right)\Bigg|_0^{10}$$

$$= \frac{20}{n^2\pi^2}(-n\pi\cos n\pi) = -\frac{20}{n\pi}(-1)^n .$$

We substitute the values of b_n into (1) [caution: not into (2),
which is just a step in our work and not the final answer] to
obtain T(x,y) satisfying Laplace's equation and all the boundary
conditions:

$$T(x,y) = \frac{20}{\pi}\sum\frac{(-1)^{n+1}}{n}e^{-n\pi y/10}\sin\frac{n\pi x}{10} .$$

Further comment: Note that we can now easily find the temperature distribution in a finite plate. Suppose that we cut the
semi-infinite plate off at height 15 cm and keep the top edge
at 0°. Then we replace (1) above by

(3) $T = \sum B_n \sinh\frac{n\pi}{10}(15 - y) \sin\frac{n\pi x}{10}$

so that $T = 0$ at $y = 15$ as well as at $x = 0$ and $x = 10$ (see text,
page 546). Then at $y \doteq 0$, we want

(4) $T = x = \sum B_n \sinh\frac{15n\pi}{10} \sin\frac{n\pi x}{10} = \sum b_n \sin\frac{n\pi x}{10}$

where $B_n \sinh\frac{3n\pi}{2} = b_n$. The Fourier coefficients b_n are the same
as above. We solve for B_n and substitute into (3) to find the
temperature distribution in the finite plate:

$$T = \frac{20}{\pi}\sum\frac{(-1)^{n+1}}{n\sinh(3n\pi/2)}\sinh\frac{n\pi}{10}(15 - y)\sin\frac{n\pi x}{10} .$$

2. Caution: Here the width is 20, so you want $\sin 20k = 0$, $k = n\pi/20$.

5. Comments: Here are two ways to verify that you have a solution
    of a differential equation:

    (a) Substitute the supposed solution into the differential
        equation.

    (b) Any linear combination of solutions of Laplace's equation is
        a solution of Laplace's equation. The same statement is true
        for <u>any</u> (homogeneous) linear differential equation, ordinary
        or partial. [A linear differential equation is one of the
        form $L(u) = f$ where $L(u)$ is a linear combination of (ordinary
        or partial) derivatives of u (such as $\frac{\partial^2 u}{\partial x^2} - \frac{1}{\alpha^2}\frac{\partial u}{\partial t}$) and f is a
        given function. If $f = 0$, the differential equation is called
        homogeneous. See text, Chapter 8, Sections 1 and 5.]

6. Using $\sin(n\pi x/10) = \operatorname{Im} e^{in\pi x/10}$, we write (2.12) as

$$T = \frac{400}{\pi} \sum_{\text{odd } n} \frac{1}{n} e^{-n\pi y/10} \sin\frac{n\pi x}{10} = \frac{400}{\pi} \operatorname{Im} \sum_{\text{odd } n} \frac{1}{n} e^{-n\pi y/10} e^{in\pi x/10}$$

$$= \frac{400}{\pi} \operatorname{Im} \sum_{\text{odd } n} \frac{1}{n}(e^{-\pi y/10} e^{i\pi x/10})^n = \frac{400}{\pi} \operatorname{Im} \sum_{\text{odd } n} \frac{1}{n} z^n$$

where $z = e^{(\pi/10)(-y+ix)}$. Now from text Chapter 1, equation (13.4)

$$\ell n \frac{1+z}{1-z} = \ell n(1+z) - \ell n(1-z) = 2\left(z + \frac{z^3}{3} + \frac{z^5}{5} \cdots\right) = 2\sum_{\text{odd } n} \frac{1}{n} z^n .$$

Then

$$T = \frac{400}{\pi} \operatorname{Im}\left(\frac{1}{2}\ell n \frac{1+z}{1-z}\right) = \frac{200}{\pi}\left(\text{angle of } \frac{1+z}{1-z}\right)$$

(see text, page 71). Now the angle of a complex number w is
$\arctan\frac{\operatorname{Im} w}{\operatorname{Re} w}$. Let $w = \frac{1+z}{1-z}$ and make the denominator of w real
(text, Chapter 2, Section 5). We find

6.   (continued)

$$w = \frac{1+z}{1-z}\frac{1-\bar{z}}{1-\bar{z}} = \frac{1+2i\ \text{Im}\ z - |z|^2}{1 - 2\ \text{Re}\ z + |z|^2},$$

(angle of w) $= \text{arc tan}\ \dfrac{2\ \text{Im}\ z}{1 - |z|^2} = \text{arc tan}\ \dfrac{2\ e^{-\pi y/10}\ \sin(\pi x/10)}{1 - e^{-2\pi y/10}}$

$$= \text{arc tan}\ \frac{2\ \sin(\pi x/10)}{e^{\pi y/10} - e^{-\pi y/10}} = \text{arc tan}\ \frac{\sin(\pi x/10)}{\sinh(\pi y/10)},$$

$$T = \frac{200}{\pi}\ \text{arc tan}\ \frac{\sin(\pi x/10)}{\sinh(\pi y/10)},$$

as claimed.

12.   We first want the temperature distribution when the boundary
temperatures are as shown in
Figure 1.   This is just the
problem solved in the text
with solution (2.17):

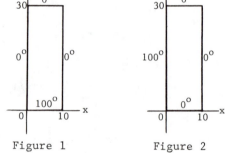

Figure 1   Figure 2

(1) $T_1 = \displaystyle\sum_{\text{odd }n} \frac{400}{n\pi\ \sinh 3n\pi}\ \sinh\frac{n\pi}{10}(30 - y)\ \sin\frac{n\pi x}{10}.$

Next we find the temperature distribution when the boundary temperatures are as in Figure 2.   Here we use the solutions of
Laplace's equation given in text equation (2.18).   Since we want
$T = 0$ when $y = 0$, we use the $\sin ky$ solution.   We also want $T = 0$
when $y = 30$, so $\sin 30k = 0$, $k = n\pi/30$.   For the x part of the solution, we need a linear combination of e^{kx} and e^{-kx} which is 0
when $x = 10$.   We use

12.   (continued)

$$\sinh k(10 - x) = \tfrac{1}{2}(e^{k(10-x)} - e^{-k(10-x)}).$$

(Verify that this is 0 when x = 10.)   Then the series

(2) $T_2 = \sum_n B_n \sinh \tfrac{n\pi}{30}(10 - x) \sin \tfrac{n\pi y}{30}$

is a solution of Laplace's equation which gives the correct
boundary temperatures of Figure 2 on the three T = 0 sides of the
rectangle.   Now we want T = 100 when x = 0, that is

(3) $T = 100 = \sum_n B_n \sinh \tfrac{n\pi}{3} \sin \tfrac{n\pi y}{30} = \sum_n b_n \sin \tfrac{n\pi y}{30}$

where $b_n = B_n \sinh \tfrac{n\pi}{3}$ or $B_n = b_n / \sinh \tfrac{n\pi}{3}$.

Now (3) says to expand 100 in a Fourier sine series on (0,30).
Thus

$$b_n = \frac{2}{30} \int_0^{30} 100 \sin \tfrac{n\pi y}{30} \, dy = - \frac{200}{30} \frac{30}{n\pi} \cos \tfrac{n\pi y}{30} \Big|_0^{30}$$

$$= \frac{200}{n\pi}(1 - \cos n\pi) = \begin{cases} 0 & , \quad \text{even } n, \\ \dfrac{400}{n\pi} & , \quad \text{odd } n. \end{cases}$$

Then for odd n,

$$B_n = 400 / \left(n\pi \sinh \tfrac{n\pi}{3} \right).$$

Substituting this into (2) gives the temperature distribution
for Figure 2.

(4) $T_2 = \displaystyle\sum_{\text{odd } n} \frac{400}{n\pi \sinh(n\pi/3)} \sinh \tfrac{n\pi}{30}(10 - x) \sin \tfrac{n\pi y}{30}$

Now the final solution when both sides of the rectangle along the
axes are held at 100° is the sum of the solutions for Figures 1
and 2, that is, the sum of equations (1) and (4).   This is true

12.   (continued)

because a sum of solutions of Laplace's equation $\nabla^2 T = 0$ is a solution, and the sum of the two temperatures on any one side from Figures 1 and 2 gives the result we want. Thus the final solution is:

$$T(x,y) = T_1 + T_2$$

$$= \frac{400}{\pi} \sum_{\text{odd } n} \frac{1}{n}\left(\frac{1}{\sinh 3n\pi} \sinh \frac{n\pi}{10}(30 - y) \sin \frac{n\pi x}{10}\right.$$

$$\left. + \frac{1}{\sinh \frac{n\pi}{3}} \sinh \frac{n\pi}{30}(10 - y) \sin \frac{n\pi y}{30}\right).$$

15.   The basic solutions of Laplace's equation are given by text equation (2.7). We want $\frac{\partial T}{\partial x} = 0$ at $x = 0$ because the sides of the rectangle are insulated (see text Problem 2.14). Thus we use solutions containing $\cos kx$. Also $\frac{\partial T}{\partial x} = 0$ at $x = 10$ if $\frac{\partial}{\partial x} \cos kx = -k \sin kx = 0$ at $x = 10$, that is, if $\sin 10k = 0$, or $k = n\pi/10$. The y part of the solution is $\sinh \frac{n\pi}{10}(30 - y)$ as in text (2.15) or Problem 12 above. Then we might expect to find a solution of the form:

(1) $T = \sum A_n \sinh \frac{n\pi}{10}(30 - y) \cos \frac{n\pi x}{10}$.

Given $T = f(x)$ when $y = 0$, we want to expand $f(x)$ in a Fourier cosine series on $(0,10)$. But note that there is no constant term in (1) since, for $n = 0$, we have $\sinh 0 = 0$. Now putting

15.   (continued)

$k = n\pi/10 = 0$ into $e^{\pm ky} \cos kx$ from text equation (2.7) gives a constant as a basic solution of Laplace's equation. However, we
cannot simply add a nonzero constant to (1) because then $T \neq 0$
when $y = 30$. We must go back to text equations (2.5) and solve
them when $k = 0$. We find

$$X'' = 0 \qquad\qquad X' = \text{const.} \qquad\qquad X = \begin{cases} x \\ \text{const.} \end{cases}$$

$$Y'' = 0 \qquad\qquad Y' = \text{const.} \qquad\qquad Y = \begin{cases} y \\ \text{const.} \end{cases}$$

Thus text equation (2.7) did not include all the basic solutions;
for $k = 0$ we must include x, y, and xy as well as a constant term.
(In previous problems we have not needed these because we were
using sine series which do not include an $n = 0$ term.) In this
problem we want $\frac{\partial T}{\partial x} = 0$ when $x = 10$ so we can use only the const.
solution for X. For the Y solution we take a linear combination
of y and const. which is zero for $y = 30$. Thus we add to (1) the
solution $a(30 - y)$ which [like each of the terms in (1)] satisfies
the boundary conditions on three sides of the rectangle.

(2) $T = a(30 - y) + \sum_{1}^{\infty} A_n \sinh \frac{n\pi}{10}(30 - y) \cos \frac{n\pi x}{10}$.

When $y = 0$, we are given $T = f(x)$; thus we want

(3) $f(x) = 30a + \sum_{1}^{\infty} A_n \sinh 3n\pi \cos \frac{n\pi x}{10} = a_0/2 + \sum_{1}^{\infty} a_n \cos \frac{n\pi x}{10}$

where

$$a_0/2 = 30a, \qquad\qquad a_n = A_n \sinh 3n\pi, \qquad \text{so}$$

$$a = a_0/60, \qquad\qquad A_n = a_n/\sinh 3n\pi.$$

15.   (continued)

For $f(x) = 100$, you should find

$$T = \frac{10}{3}(30 - y)$$

which is just the result you may know from elementary physics.

For $f(x) = x$, we expand x in a cosine series on $(0,10)$.

$$a_0 = \frac{2}{10}\int_0^{10} x\,dx = 10, \qquad a = a_0/60 = 1/6.$$

$$a_n = \frac{2}{10}\int_0^{10} x \cos\frac{n\pi x}{10}\,dx = \frac{2}{10}\left(\frac{10}{n\pi}\right)^2 \left(\cos\frac{n\pi x}{10} + \frac{n\pi x}{10}\sin\frac{n\pi x}{10}\right)\Bigg|_0^{10}$$

$$= \frac{20}{n^2\pi^2}(\cos n\pi - 1) = \begin{cases} 0 \ , & \text{even } n, \\ -\dfrac{40}{n^2\pi^2}, & \text{odd } n. \end{cases}$$

Then the final solution when $f(x) = x$ is

$$T = \frac{1}{6}(30 - y) - \frac{40}{\pi^2}\sum_{\text{odd }n}\frac{1}{n^2\sinh 3n\pi}\sinh\frac{n\pi}{10}(30 - y)\cos\frac{n\pi x}{10}\ .$$

Section 3

5.   Initially the temperatures are as shown in the diagram. For each
slab, the temperature is linear, $u_0 = ax + b$,
where a and b must be found for each slab.

   For the first slab:

   $u_0 = 0$ when $x = 0$ and $u_0 = 100$ when $x = 1$.

   For the second slab:

   $u_0 = 100$ when $x = 1$ and $u_0 = 0$ when $x = 2$.

5.   (continued)

Thus the initial temperature distribution is

$$u_0 = \begin{cases} 100x, & 0 < x < 1, \\ 100(2 - x), & 1 < x < 2. \end{cases}$$

The final temperature is $u_f = 100$. The temperature distribution as a function of x and t must be some linear combination of the basic solutions of the heat flow equation [text equation (3.1)]. These solutions are given by text equation (3.10) when $k > 0$. When $k = 0$, text equations (3.5), (3.6), (3.8) and (3.9) become

$$\nabla^2 F = 0 \quad \text{or} \quad \frac{d^2 F}{dx^2} = 0, \quad F = ax + b,$$

$$\frac{dT}{dt} = 0, \quad T = \text{const}.$$

Thus the basic solutions of the heat flow equation are

$$u = \begin{cases} e^{-k^2 \alpha^2 t} \sin kx, & k > 0, \\ e^{-k^2 \alpha^2 t} \cos kx, & k > 0, \\ ax + b, & k = 0, \end{cases}$$

and we can write the solution of our problem in the form

$$(1) \qquad u = \sum_k e^{-k^2 \alpha^2 t} (b_k \sin kx + a_k \cos kx) + ax + b.$$

As $t \to \infty$, we see from (1) that $u \to ax + b$; this must be the final steady state u_f. In our problem $u_f = 100$ so we can write (1) as

$$(2) \qquad u = \sum_k e^{-k^2 \alpha^2 t} (b_k \sin kx + a_k \cos kx) + 100.$$

Now we must satisfy the conditions $u = 100$ at $x = 0$ and at $x = 2$ for all t. From (2) we see that $u = 100$ if the terms in the series

5.   (continued)

are zero. This will be true for all t if we keep only the sine

terms and take $k = n\pi/2$. Thus we write (2) as

(3) $u = 100 + \sum\limits_{n=1}^{\infty} b_n \, e^{-(n\pi\alpha/2)^2 t} \, \sin\frac{n\pi x}{2}$.

[This is text equation (3.16) with $\ell = 2$ and $u_f = 100$.]

When $t = 0$, (3) becomes

$$u_0 = 100 + \sum\limits_{n=1}^{\infty} b_n \sin\frac{n\pi x}{2} \qquad \text{or}$$

(4) $u_0 - 100 = \sum\limits_{n=1}^{\infty} b_n \sin\frac{n\pi x}{2}$.

Equation (4) says to expand in a Fourier sine series the function

$$u_0 - u_f = \begin{cases} 100x - 100 = 100(x-1), & 0 < x < 1, \\ 100(2-x) - 100 = -100(x-1), & 1 < x < 2. \end{cases}$$

We find

$$\frac{b_n}{100} = \frac{2}{2}\left[\int_0^1 (x-1)\sin\frac{n\pi x}{2}\,dx - \int_1^2 (x-1)\sin\frac{n\pi x}{2}\,dx\right]$$

$$= \left[\left(\frac{2}{n\pi}\right)^2 \sin\frac{n\pi x}{2} - \frac{2}{n\pi}(x-1)\cos\frac{n\pi x}{2}\right]_0^1 - \left[\begin{array}{c}\text{same}\\\text{integral}\end{array}\right]_1^2$$

$$= \left(\frac{2}{n\pi}\right)^2 \left(2\sin\frac{n\pi}{2}\right) + \frac{2}{n\pi}(-1 + \cos n\pi) .$$

$$b_n = 100\begin{cases} 0, & \text{even } n \\ \dfrac{8}{n^2\pi^2} - \dfrac{4}{n\pi}, & n = 1 + 4k \\ -\dfrac{8}{n^2\pi^2} - \dfrac{4}{n\pi}, & n = 3 + 4k \end{cases} = 400\begin{cases} 0, & \text{even } n, \\ \dfrac{2}{n^2\pi^2} - \dfrac{1}{n\pi}, & n = 1 + 4k, \\ -\dfrac{2}{n^2\pi^2} - \dfrac{1}{n\pi}, & n = 3 + 4k. \end{cases}$$

Then the temperature distribution is given by (3) above with

these values for the b_n.

6.   Hint: Find $u_0 - u_f$ and compare with u_0 in text (3.13).

7.   When the ends of the bar are insulated, we want $\frac{\partial u}{\partial x} = 0$ at the
      ends. We use from text (3.10) the basic solution containing
      $\cos kx$ since then $\frac{\partial}{\partial x}(\cos kx) = -k \sin kx$ is zero at $x = 0$ and, with
      $k = n\pi/\ell$, it is also zero at $x = \ell$. When $k = 0$, text (3.8) is
      $F'' = 0$, so the $k = 0$ solutions are x and const. Since $\partial u/\partial x = 0$
      at the ends, we use only the constant which is already contained
      in $\cos kx$ with $k = 0$. Thus the temperature distribution in the
      bar is given by

$$u(x,t) = \sum_{n=0}^{\infty} a_n e^{-(n\pi\alpha/\ell)^2 t} \cos \frac{n\pi x}{\ell} .$$

We are given $u(x,0) = x$ so the a_n's are determined by

$$u(x,0) = x = \sum_{n=0}^{\infty} a_n \cos \frac{n\pi x}{\ell} .$$

This says to expand x on $(0,\ell)$ in a Fourier cosine series. We
find

$$a_n = \frac{2}{\ell} \int_0^{\ell} x \cos \frac{n\pi x}{\ell} dx = \frac{2}{\ell}\left(\frac{\ell}{n\pi}\right)^2 \left(\cos \frac{n\pi x}{\ell} + \frac{n\pi x}{\ell} \sin \frac{n\pi x}{\ell}\right)\bigg|_0^{\ell}$$

$$= \frac{2\ell}{n^2\pi^2}(\cos n\pi - 1) = \begin{cases} 0 , & \text{even } n \neq 0, \\ -\frac{4\ell}{n^2\pi^2} , & \text{odd } n. \end{cases}$$

The constant term is called a_0 in the series above instead of
$a_0/2$ as in Chapter 7 of the text; then the integral for a_0 from
text Chapter 7 must here be divided by 2 (or else simply derive
it using orthogonality as discussed in text Chapter 12). We find

$$a_0 = \frac{1}{\ell} \int_0^{\ell} x \, dx = \frac{1}{\ell}\frac{\ell^2}{2} = \frac{\ell}{2} .$$

Thus

$$u(x,t) = \frac{\ell}{2} - \frac{4\ell}{\pi^2} \sum_{\text{odd } n} \frac{1}{n^2} e^{-(n\pi\alpha/\ell)^2 t} \cos \frac{n\pi x}{\ell} .$$

Section 4

3.   The basic solutions of the vibrating string problem are given by
     text equation (4.5) or, for a string fastened at $x = 0$ and $x = \ell$,
     by text equation (4.6). In this problem we are given that the
     initial displacement of the string is $y_0 = f(x)$ and the initial
     velocity of points of the string is $V = (\partial y/\partial t)_{t=0} = 0$. Then text
     equation (4.7) gives the solution where the coefficients b_n are
     determined by text equation (4.8). Thus we want to expand the
     given $y_0(x)$ in a Fourier sine series. From the text figure for
     this problem we have

$$
y_0 = \begin{cases} \dfrac{8h}{\ell}x, & 0 < x < \dfrac{\ell}{8}, \\[2mm] \dfrac{8h}{\ell}\left(\dfrac{\ell}{4} - x\right), & \dfrac{\ell}{8} < x < \dfrac{\ell}{4}, \\[2mm] 0, & \dfrac{\ell}{4} < x < \ell. \end{cases}
$$

Then

$$
b_n = \frac{2}{\ell}\left[\int_0^{\ell/8} \frac{8h}{\ell} x \sin\frac{n\pi x}{\ell}\,dx + \int_{\ell/8}^{\ell/4} \frac{8h}{\ell}\left(\frac{\ell}{4} - x\right)\sin\frac{n\pi x}{\ell}\,dx\right]
$$

$$
= \frac{16h}{\ell^2}\left[\left(\frac{\ell}{n\pi}\right)^2\left(\sin\frac{n\pi x}{\ell} - \frac{n\pi x}{\ell}\cos\frac{n\pi x}{\ell}\right)\Big|_0^{\ell/8}\right.
$$

$$
\left. - \left(\begin{array}{c}\text{same}\\\text{integral}\end{array}\right)\Big|_{\ell/8}^{\ell/4} + \frac{\ell}{4}\frac{\ell}{n\pi}\left(-\cos\frac{n\pi x}{\ell}\right)\Big|_{\ell/8}^{\ell/4}\right]
$$

$$
= \frac{16h}{n^2\pi^2}\left(2\sin\frac{n\pi}{8} - \frac{2n\pi}{8}\cos\frac{n\pi}{8} - \sin\frac{n\pi}{4}\right.
$$

$$
\left. + \frac{n\pi}{4}\cos\frac{n\pi}{4} - \frac{n\pi}{4}\cos\frac{n\pi}{4} + \frac{n\pi}{4}\cos\frac{n\pi}{8}\right)
$$

$$
= \frac{16h}{n^2\pi^2}\left(2\sin\frac{n\pi}{8} - \sin\frac{n\pi}{4}\right).
$$

The displacement y is [text equation (4.7)]

$$
y = \frac{16h}{\pi^2}\sum_1^\infty B_n \sin\frac{n\pi x}{\ell}\cos\frac{n\pi vt}{\ell}
$$

where $B_n = \left(2\sin\frac{n\pi}{8} - \sin\frac{n\pi}{4}\right)/n^2$.

4. Hint: You may find $b_n = \frac{16h}{n^2\pi^2}\left(\sin\frac{n\pi}{4} - \sin\frac{3n\pi}{4}\right)$. This is correct

but unnecessarily complicated. Here are some ways of getting
a simpler answer.

(a) Use the identity

$$\sin x - \sin y = 2 \sin\tfrac{1}{2}(x - y) \cos\tfrac{1}{2}(x + y).$$

   Then $b_n = 0$ for odd n so let n = 2m.

(b) Note that $b_n = 0$ for odd n and let n = 2m for even n. Then use

$$\sin\tfrac{m\pi}{2} - \sin\tfrac{3m\pi}{2} = 2 \sin\tfrac{m\pi}{2}.$$

(c) Note to start with that ℓ is a period of the given function
    and expand in terms of the functions $2n\pi x/\ell$.

By any of these methods you should find

$$y = \frac{8h}{\pi^2}\sum_1^\infty \frac{1}{n^2} \sin\tfrac{n\pi}{2} \sin\tfrac{2n\pi x}{\ell} \cos\tfrac{2n\pi vt}{\ell}.$$

This can also be written as a sum over odd n with $\sin\frac{n\pi}{2}$ replaced
by $(-1)^{(n-1)/2}$.

6. By text (4.10) and (4.11), the displacement of a string fastened at 0 and ℓ and started with initial conditions $y = 0$ and $(\partial y/\partial t)_{t=0} = V(x)$ is given by

(1) $y = \sum B_n \sin \frac{n\pi x}{\ell} \sin \frac{n\pi vt}{\ell}$

where the B_n's are determined by

$$\left(\frac{\partial y}{\partial t}\right)_{t=0} = V(x) = \sum B_n \frac{n\pi v}{\ell} \sin \frac{n\pi x}{\ell} = \sum b_n \sin \frac{n\pi x}{\ell} \, .$$

Then $\frac{n\pi v}{\ell} B_n = b_n$, so $B_n = \frac{\ell}{n\pi v} b_n$. Thus we want to expand the given $V(x)$ in a Fourier sine series to find b_n, solve for B_n and substitute into (1). Using the text figure for this problem, we find

$$V(x) = \begin{cases} h, & \frac{\ell}{2} - w < x < \frac{\ell}{2} + w, \\ 0, & \text{otherwise.} \end{cases}$$

$$b_n = \frac{2}{\ell} \int_{(\ell/2)-w}^{(\ell/2)+w} h \sin \frac{n\pi x}{\ell} \, dx = \frac{2h}{\ell} \left(-\frac{\ell}{n\pi} \cos \frac{n\pi x}{\ell} \right) \Big|_{(\ell/2)-w}^{(\ell/2)+w}$$

$$= -\frac{2h}{n\pi} \left[\cos\left(\frac{n\pi}{2} + \frac{n\pi w}{\ell}\right) - \cos\left(\frac{n\pi}{2} - \frac{n\pi w}{\ell}\right) \right]$$

$$= \frac{4h}{n\pi} \sin \frac{n\pi}{2} \sin \frac{n\pi w}{\ell} \, .$$

The last step follows from the identity

$$\cos x - \cos y = -2 \sin\tfrac{1}{2}(x+y) \sin\tfrac{1}{2}(x-y) \, .$$

Then

$$B_n = \frac{\ell}{n\pi v} b_n = \frac{4h\ell}{n^2 \pi^2 v} \sin \frac{n\pi}{2} \sin \frac{n\pi w}{\ell}$$

6.   (continued)

and (1) gives

$$y = \frac{4h\ell}{\pi^2 v} \sum_1^\infty \frac{1}{n^2} \sin\frac{n\pi}{2} \sin\frac{n\pi w}{\ell} \sin\frac{n\pi x}{\ell} \sin\frac{n\pi vt}{\ell} .$$

[Comment: Since $\sin\frac{n\pi}{2} = (1,0,-1,0,$ and repeat), we could indicate a sum over odd n only.   Also, for odd n, we could write $\sin\frac{n\pi}{2} = (-1)^{(n-1)/2}.$]

9.   Recall (text, page 329) that the intensities of the various harmonics are proportional to the squares of the Fourier coefficients.

For Problem 3, we compute $(B_n/B_1)^2$ for n = 1 to 9:

1, 3.2, 4.8, 4.6, 3.1, 1.3, 0.3, 0, 0.1, $\cdots$.

Thus n = 3 gives the most important harmonic; the corresponding frequency is $\omega = 3\pi v/\ell$ or $\nu = 3v/(2\ell)$.   The intensity of the n = 4 harmonic is almost as great and the intensities of the n = 2 and n = 5 harmonics are also fairly large.   The frequency in each case is $\omega = n\pi v/\ell$ or $\nu = nv/(2\ell)$.

For Problem 6, the even harmonics are not present because $\sin(n\pi/2) = 0$ for even n.   The intensities of the odd harmonics are proportional to $\frac{1}{n^4} \sin^2\frac{n\pi w}{\ell}$.   The n = 1 harmonic is the most important; its frequency is $\omega = \pi v/\ell$ or $\nu = v/(2\ell)$.

Section 5

1.   Hint: Note that $J_1 = -J_0'$ [by text page 514, equation (15.2) with
     $p = 0$]. You may find either J_1 or J_0' tabulated at the zeros of J_0.

2.   The basic solutions for this problem are given in text equation
     (5.10). Since we need a solution containing $\sin \theta$, we let $n = 1$
     and use the basic solution $J_1(kr) \sin \theta \, e^{-kz}$. The condition $u = 0$
     when $r = 1$ gives $J_1(k) = 0$. Thus the possible values of k are the
     zeros of J_1; call these k_m. Then we try to find the solution of
     our problem in the form

(1) $u(r,\theta,z) = \sum\limits_{m=1}^{\infty} c_m J_1(k_m r) \sin \theta \, e^{-k_m z}$.

We want $u = r \sin \theta$ when $z = 0$:

$$u(r,\theta,0) = r \sin \theta = \sum\limits_{m=1}^{\infty} c_m J_1(k_m r) \sin \theta$$

or, cancelling $\sin \theta$,

(2) $r = \sum\limits_{m=1}^{\infty} c_m J_1(k_m r)$.

To expand r in a Bessel series, we multiply both sides of equation (2) by $rJ_1(k_\ell r)$ [see text, page 561] and integrate from 0
to 1. By orthogonality, all terms drop out except the $m = \ell$ term;
we use text equation (19.10), page 523, to get:

(3) $\displaystyle\int_0^1 r^2 J_1(k_\ell r) \, dr = \sum\limits_{m=1}^{\infty} c_m \int_0^1 r J_1(k_\ell r) J_1(k_m r) \, dr$

$$= c_\ell \int_0^1 r [J_1(k_\ell r)]^2 \, dr = c_\ell \cdot \frac{1}{2} J_2^2(k_\ell).$$

2.   (continued)

To integrate $r^2 J_1$, we use text equation (15.1), page 514, with $p = 2$ and $x = k_\ell r$:

$$\frac{d}{d(k_\ell r)}\left[(k_\ell r)^2 J_2(k_\ell r)\right] = (k_\ell r)^2 J_1(k_\ell r).$$

Cancelling k_ℓ^2 and integrating gives:

(4) $\displaystyle\int_0^1 r^2 J_1(k_\ell r)\, dr = \frac{r^2}{k_\ell} J_2(k_\ell r)\Big|_0^1 = \frac{1}{k_\ell} J_2(k_\ell).$

Combine equations (3) and (4) to get

$$c_\ell = \frac{2}{J_2^2} \cdot \frac{1}{k_\ell} J_2 = \frac{2}{k_\ell J_2(k_\ell)}.$$

Substituting the value of the c's into (1) now gives the solution of our problem:

$$u(r,\theta,z) = \sum_{m=1}^\infty \frac{2}{k_m J_2(k_m r)} J_1(k_m r)\sin\theta\, e^{-k_m z}$$

where $k_m = $ zeros of J_1.

Further comment: In using this series for computation, it may be useful to recall (see text, Chapter 12, Problem 15.5) that at the zeros k_m of J_1, we have $J_2(k_m) = -J_0(k_m) = -J_1'(k_m)$. You may find either J_0 or J_1' tabulated at the zeros of J_1.

3.   Hint: Compare the finite plate problems in Section 2.

8. We need the basic solutions of the heat flow equation [text equa-
    tion (3.1)] in polar coordinates. By text equations (3.5) and
    (3.6), these are of the form $F(r,\theta)T(t)$ where

$$T(t) = e^{-(k\alpha)^2 t}$$

and F satisfies the Helmholtz equation $\nabla^2 F + k^2 F = 0$. This equation
in polar coordinates is text equation (5.5) with solutions

$$R(r)\Theta(\theta) = J_n(kr) \begin{Bmatrix} \sin n\theta \\ \cos n\theta \end{Bmatrix} .$$

We consider a circular cross section of the pipe. The temperature
is a function only of r (no θ dependence) so we use the $\cos n\theta$
solution with $n = 0$. Thus our basic solutions are

(1) $J_0(kr) e^{-(k\alpha)^2 t}$.

These solutions tend to zero as $t \to \infty$. As in text Section 3, we
must add to any solutions of the form (1), the final steady state
$u_f = 40$. Then we want solutions of the form (1) to be zero when
$r = 1$, for any $t > 0$. Thus we must have $J_0(k) = 0$ so $k = k_m$ are the
zeros of J_0. We write the solution of our problem in the form

(2) $u(r,t) = 40 + \sum_{m=1}^{\infty} c_m J_0(k_m r) e^{-(k_m \alpha)^2 t}$

and find the coefficients c_m so that $u = 100$ when $t = 0$.

$$u(r,0) = 100 = 40 + \sum_{m=1}^{\infty} c_m J_0(k_m r),$$

$$\sum_{m=1}^{\infty} c_m J_0(k_m r) = 60.$$

8.   (continued)

This is the same as text equation (5.12) if we change 100 to 60.
The coefficients c_m are found as on page 561 of the text, and so
by text equation (5.16) we have

$$c_m = \frac{2 \cdot 60}{k_m J_1(k_m)} = \frac{120}{k_m J_1(k_m)} \ .$$

We substitute these coefficients into (2) to obtain the solution.

$$u(r,t) = 40 + \sum_{m=1}^{\infty} \frac{120}{k_m J_1(k_m)} J_0(k_m r) \, e^{-(k_m \alpha)^2 t} \ .$$

11.   For one method of solving these equations, see page 271 of
these Solutions.

Another method is to assume a solution of the form $R = r^p$ and
find p to satisfy the equation. Substitute this into the first
equation:

$$\frac{dR}{dr} = p r^{p-1} \qquad \text{or} \qquad r\frac{dR}{dr} = p r^p,$$

$$\frac{d}{dr}\left(r\frac{dR}{dr}\right) = p^2 r^{p-1} \qquad \text{or} \qquad r\frac{d}{dr}\left(r\frac{dR}{dr}\right) = p^2 r^p = n^2 r^p.$$

Thus $p^2 = n^2$ so $p = \pm n$ and the solutions of the differential
equation are $R(r) = r^n$ or r^{-n}. When $n = 0$, both these solutions
give a constant as the solution and we must go back to the
differential equation to find the second solution. With $n = 0$,
the differential equation becomes:

$$\frac{d}{dr}\left(r\frac{dR}{dr}\right) = 0, \qquad \text{so} \qquad r\frac{dR}{dr} = \text{const.} = A.$$

Then $\frac{dR}{dr} = \frac{A}{r}$ $\qquad$ so $\qquad$ $R = A \ln r + B.$

Thus the solutions of the first differential equation are:

$$R(r) = \begin{cases} r^n, \ r^{-n}, & n > 0, \\ \ln r, \ \text{const.}, & n = 0. \end{cases}$$

You can solve the second equation in a similar way.

13. We need the solutions of Laplace's equation in polar coordinates.
Text equation (5.5) with $k = 0$ is Laplace's equation and we see
that the variables separate to give the θ solution as in text
equation (5.6) and the R equation

$$\frac{r}{R}\frac{d}{dr}\left(r\frac{dR}{dr}\right) = n^2 \qquad \text{or} \qquad r\frac{d}{dr}\left(r\frac{dR}{dr}\right) = n^2 R$$

which we just solved in Problem 11 above. Since $r = 0$ is a point
of the plate, we use only the r^n solutions, $n > 0$, because the
temperature is finite. Thus the basic solutions we need are:

$$\{r^n\} \begin{Bmatrix} \sin n\theta \\ \cos n\theta \end{Bmatrix}.$$

For our purposes, this notation is going to be confusing, so let
us write our solutions as

$$\{r^k\} \begin{Bmatrix} \sin k\theta \\ \cos k\theta \end{Bmatrix}.$$

From the text figure for this problem, we see that $T = 0$ when $\theta = 0$,
so we use the $\sin k\theta$ solution. Also $T = 0$ when $\theta = \pi/4$, so we want
$\sin \frac{k\pi}{4} = 0$. Then $k\pi/4$ must be a multiple of π, say $n\pi$; thus $k = 4n$
and our θ solutions are $\sin 4n\theta$. Then the temperature of the
plate is

$$u(r,\theta) = \sum_{n=1}^{\infty} B_n r^{4n} \sin 4n\theta.$$

13.   (continued)

The B_n's are determined by

$$u(10,\theta) = \sum_{n=1}^{\infty} B_n \, 10^{4n} \sin 4n\theta = \sum_{n=1}^{\infty} b_n \sin 4n\theta = 100$$

where $B_n = b_n/10^{4n}$. We expand 100 in a Fourier sine series
on $(0,\pi/4)$.

$$b_n = \frac{200}{\pi/4} \int_0^{\pi/4} \sin 4n\theta \, d\theta = -\frac{200}{\pi/4} \frac{1}{4n} \cos 4n\theta \Big|_0^{\pi/4} = -\frac{200}{n\pi}(\cos n\pi - 1)$$

$$= \begin{cases} 0, & \text{even } n, \\ \dfrac{400}{n\pi}, & \text{odd } n. \end{cases}$$

$$B_n = b_n/10^{4n}.$$

Then

$$u(r,\theta) = \frac{400}{\pi} \sum_{\text{odd } n} \frac{1}{n}\left(\frac{r}{10}\right)^{4n} \sin 4n\theta.$$

Further comment: We have also solved an electrostatics
problem: Consider a long cylinder of
radius 10 cut into eight pieces as
shown with alternate pieces held at
potentials +100 and -100. Then the
potential inside the cylinder is given
by our solution above. (Note that the
potential along the radial lines is
zero by symmetry.)

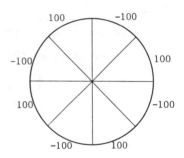

14. We again need the solutions of Laplace's equation in polar

coordinates (see Problems 12 and 13). Since $r = 0$ is not part

of the annulus, we do not discard the solutions for $n \leqslant 0$. The

basic solutions are then

$$\begin{Bmatrix} r^n \\ r^{-n} \end{Bmatrix} \begin{Bmatrix} \sin n\theta \\ \cos n\theta \end{Bmatrix} \quad \text{for } n > 0, \quad \text{and} \quad \begin{Bmatrix} \text{const.} \\ \ell n\, r \end{Bmatrix} \quad \text{for } n = 0.$$

For any θ, the temperature is zero when $r = 1$. Thus we use a

combination of r solutions which is zero when $r = 1$, namely,

$(r^n - r^{-n})$. The combination $(a + b\,\ell n\, r)$ also must be zero when

$r = 1$, so $a = 0$. Thus we write the solution of our problem in

the form

$$u(r,\theta) = b\,\ell n\, r + \sum_{n=1}^{\infty} (r^n - r^{-n})(a_n \cos n\theta + b_n \sin n\theta).$$

This satisfies $u = 0$ when $r = 1$. When $r = 2$, we have

$$u(2,\theta) = b\,\ell n\, 2 + \sum_{n=1}^{\infty} (2^n - 2^{-n})(a_n \cos n\theta + b_n \sin n\theta)$$

$$= \begin{cases} 100, & 0 < \theta < \pi, \\ 0, & \pi < \theta < 2\pi \quad (\text{or } -\pi < \theta < 0). \end{cases}$$

We can simplify the evaluation of the Fourier coefficients by

observing that the given function is $50 +$ an odd function, namely

$$50 + \begin{Bmatrix} 50, & 0 < \theta < \pi \\ -50, & -\pi < \theta < 0 \end{Bmatrix}.$$

14.   (continued)

Thus

$$b \, \ell n \, 2 = 50, \qquad a_n = 0,$$

$$(2^n - 2^{-n})b_n = 50 \cdot \frac{2}{\pi} \int_0^{\pi} \sin n\theta \, d\theta = -\left. \frac{100}{\pi} \frac{\cos n\theta}{n} \right|_0^{\pi}$$

$$= \begin{cases} \dfrac{200}{n\pi}, & n \text{ odd}, \\[2mm] 0, & n \text{ even}. \end{cases}$$

Then the temperature in the annular ring $1 < r < 2$ is

$$u(r, \theta) = \frac{50 \, \ell n \, r}{\ell n \, 2} + \frac{200}{\pi} \sum_{\text{odd } n} \frac{r^n - r^{-n}}{n(2^n - 2^{-n})} \sin n\theta.$$

Section 6

1.   Hint: Here is one more example; this is the k_{43} mode of vibration
      (corresponding to the fourth zero of J_3).
      The radii of the circles are $k_{13}/k_{43} \cong 0.4$,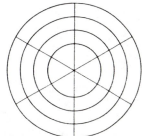
      $k_{23}/k_{43} \cong 0.6$, $k_{33}/k_{43} \cong 0.8$, and 1.   The
      formula for the displacement is

$$z = J_3(k_{43}r) \cos 3\theta \cos k_{43}vt.$$

2.   Hints: The lowest frequency is $\omega_0 = k_{10}v$. The others are
      $\omega_{mn} = k_{mn}v$. You want the ratio of the various frequencies to the
      fundamental (lowest) frequency, that is, you want

$$\omega_{mn}/\omega_0 = k_{mn}/k_{10}$$

      where $k_{10} = 2.4$ is the first zero of J_0 and k_{mn} is the mth zero of
      J_n (see Chapter 12, Section 14). Thus the first two frequencies
      are ω_0 and $1.59\omega_0$. Similarly find the next four frequencies as
      multiples of ω_0.

3.   The wave equation in two-dimensional rectangular coordinates is

(1) $\dfrac{\partial^2 z}{\partial x^2} + \dfrac{\partial^2 z}{\partial y^2} = \dfrac{1}{v^2}\dfrac{\partial^2 z}{\partial t^2}$

where z is the displacement of the membrane from the (x,y) plane.
We separate variables as follows (see text, pages 544, 554, 559,
and 564). Substitute $z = X(x)\,Y(y)\,T(t)$ into (1) and divide by XYT:

$$\frac{X''YT}{XYT} + \frac{XY''T}{XYT} = \frac{1}{v^2}\frac{XY\ddot{T}}{XYT} \qquad \text{or}$$

$$\frac{X''}{X} + \frac{Y''}{Y} = \frac{1}{v^2}\frac{\ddot{T}}{T}\,.$$

Each of the three terms is a function of only one variable;
therefore each of the terms must be constant. We put

$$\frac{X''}{X} = -k_x^2\,, \qquad\qquad \frac{Y''}{Y} = -k_y^2\,, \qquad\qquad \frac{\ddot{T}}{T} = -\left(k_x^2 + k_y^2\right)v^2\,.$$

The solutions of these equations are

$$X = \begin{cases} \sin k_x x \\ \cos k_x x \end{cases} \qquad Y = \begin{cases} \sin k_y y \\ \cos k_y y \end{cases} \qquad T = \begin{cases} \sin\left(vt\sqrt{k_x^2 + k_y^2}\right) \\ \cos\left(vt\sqrt{k_x^2 + k_y^2}\right) \end{cases}$$

Since the membrane is attached to supports along the sides, we
must have

$$X = 0 \text{ at } x = 0 \text{ and } x = a, \qquad Y = 0 \text{ at } y = 0 \text{ and } y = b.$$

We use the sine solution to make $X = 0$ when $x = 0$. At $x = a$ we want
$\sin k_x a = 0$ so $k_x = n\pi/a$, and $X = \sin\dfrac{n\pi x}{a}$. Similarly $Y = \sin\dfrac{m\pi y}{b}$,
and

$$T = \begin{cases} \sin\left(vt\pi\sqrt{(n/a)^2 + (m/b)^2}\right) \\ \cos\left(vt\pi\sqrt{(n/a)^2 + (m/b)^2}\right) \end{cases}\,.$$

3. (continued)

Thus the frequencies (text page 300) are given by

$$\nu_{nm} = \frac{v}{2}\sqrt{(n/a)^2 + (m/b)^2}.$$

The first few normal modes of vibration are:

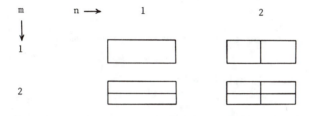

For a square membrane of side a, the frequencies are

$$\nu_{nm} = \frac{v}{2a}\sqrt{n^2 + m^2}.$$

The 5,5 mode and the 7,1 mode both
have the frequency

$$\nu_{55} = \nu_{71} = \frac{v}{2a}\sqrt{50}.$$

Can you find some more examples? Hint: There are two modes with
frequency $\frac{v}{2a}\sqrt{65}$ and two modes with frequency $\frac{v}{2a}\sqrt{85}$.

Section 7

1. The basic solutions for this problem are given by text equation
    (7.10). If the given surface temperature is independent of ϕ,
    then $m = 0$, and P_ℓ^m becomes just P_ℓ. Then we will try to find a
    solution of the form of text equation (7.11).

(1) $u(r, \theta) = \displaystyle\sum_{\ell=0}^{\infty} c_\ell r^\ell P_\ell(\cos \theta).$

When $r = 1$ (surface of the sphere), (1) becomes

(2) $u(1, \theta) = \displaystyle\sum_{\ell=0}^{\infty} c_\ell P_\ell(\cos \theta).$

That is, we want to expand the given surface temperature in a
Legendre series as in Chapter 12. We found in Chapter 12,
Problem 5.10, that

$$x^4 = \frac{8}{35} P_4 + \frac{4}{7} P_2 + \frac{1}{5} P_0 \qquad\qquad \text{or}$$

$$35x^4 = 8P_4(x) + 20P_2(x) + 7P_0(x) \qquad \text{or, if } x = \cos \theta,$$

$$35 \cos^4\theta = 8P_4(\cos \theta) + 20P_2(\cos \theta) + 7P_0(\cos \theta).$$

This is equation (2) above, so we find $c_0 = 7$, $c_2 = 20$, $c_4 = 8$, and
all other c's are zero. The temperature inside the sphere is
then given by (1) with these values of the c's:

$$u(r, \theta) = 8r^4 P_4(\cos \theta) + 20r^2 P_2(\cos \theta) + 7P_0(\cos \theta).$$

6.   This is much like Problem 1 above. The solution is given by
      equation (1) in Problem 1 and we use equation (2) to find the
      c_ℓ's.   Then we want to expand in a Legendre series the given
      surface temperature $\frac{\pi}{2} - \theta$. If $x = \cos \theta$, then

$$\frac{\pi}{2} - \theta = \frac{\pi}{2} - \text{arc cos } x = \text{arc sin } x.$$

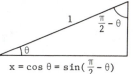

$$x = \cos \theta = \sin(\frac{\pi}{2} - \theta)$$

In Chapter 12, Problem 9.4, we found

$$\text{arc sin } x = \frac{\pi}{8}\left[3P_1(x) + \frac{7}{16}P_3(x) + \frac{11}{64}P_5(x) \cdots\right].$$

Then, with $x = \cos \theta$, we have

$$\frac{\pi}{2} - \theta = \frac{\pi}{8}\left[3P_1(\cos \theta) + \frac{7}{16}P_3(\cos \theta) + \frac{11}{64}P_5(\cos \theta) \cdots\right].$$

Thus, from equation (2) of Problem 1:

$$c_0 = 0, \quad c_1 = \frac{3\pi}{8}, \quad c_2 = 0, \quad c_3 = \frac{7}{16}\frac{\pi}{8}, \quad c_4 = 0, \quad c_5 = \frac{11}{64}\frac{\pi}{8}, \quad \cdots$$

and from equation (1) of Problem 1

$$u(r,\theta) = \frac{\pi}{8}\left[3r\,P_1(\cos \theta) + \frac{7}{16}r^3 P_3(\cos \theta) + \frac{11}{64}r^5 P_5(\cos \theta) \cdots\right].$$

10.   Here the surface temperature depends on ϕ as well as θ so the
        basic solutions are given by text equation (7.10). In general
        in a problem like this we write

(1) $u(r,\theta,\phi) = \sum_{\ell=0}^{\infty} \sum_{m=-\ell}^{\ell} r^{\ell} P_{\ell}^{m}(\cos\theta)(a_{m\ell}\cos m\phi + b_{m\ell}\sin m\phi).$

When $r = 1$ (surface of sphere)

(2) $u(1,\theta,\phi) = \sum_{\ell=0}^{\infty} \sum_{m=-\ell}^{\ell} P_{\ell}^{m}(\cos\theta)(a_{m\ell}\cos m\phi + b_{m\ell}\sin m\phi).$

In general, the coefficients are integrals. In this problem we
are given

(3) $u(1,\theta,\phi) = \sin^2\theta \cos\theta \cos 2\phi - \cos\theta.$

Since each of the terms in (3) is just one term of the series
in (2), we can easily determine the coefficients by inspection.
The last term in (3) is just $-P_1(\cos\theta)$, that is, $m = 0$, $\ell = 1$,
$a_{01} = -1$. In the first term of (3), the factor $\cos 2\phi$ tells us
that $m = 2$. The factor $\sin^2\theta \cos\theta$ with $x = \cos\theta$ becomes $(1 - x^2)x$.
Using text equation (10.6) on page 505 with $m = 2$, $\ell = 3$, and
$P_3(x)$ from text page 766, answer to Problem 5.3, we find

$$P_3^2(x) = (1 - x^2)^{2/2} \frac{d^2}{dx^2} P_3(x) = (1 - x^2)\frac{d^2}{dx^2}\frac{1}{2}(5x^3 - 3x)$$

$$= (1 - x^2)\cdot 15x.$$

10.   (continued)

Thus the first term in (3) is $\frac{1}{15} P_3^2(\cos \theta) \cos 2\phi$ so $a_{23} = 1/15$.
All the other $a_{m\ell}$ and all $b_{m\ell}$ are zero. Substituting these
values into (1) gives the desired steady state temperature:

(4) $u(r, \theta, \phi) = \frac{1}{15} r^3 P_3^2(\cos \theta) \cos 2\phi - r P_1(\cos \theta)$.

In a more complicated problem, we would need to find the
coefficients by integration. First let us substitute $\cos \theta = x$
in $u(1, \theta, \phi)$ to obtain a function of x and ϕ, say $f(x, \phi)$. Then
equation (2) above would be

(5) $f(x, \phi) = \sum\limits_{\ell=0}^{\infty} \sum\limits_{m=-\ell}^{\ell} P_\ell^m(x) (a_{m\ell} \cos m\phi + b_{m\ell} \sin m\phi)$.

Multiply this equation by $P_\ell^{m'}(x) \cos m'\phi$ and integrate as follows:

$$\int_{x=-1}^{1} \int_{\phi=0}^{2\pi} f(x, \phi) P_\ell^{m'}(x) \cos m'\phi \, dx \, d\phi$$

$$= \sum\limits_{\ell=0}^{\infty} \sum\limits_{m=-\ell}^{\ell} \int_{-1}^{1} P_\ell^{m'}(x) P_\ell^m(x) \, dx \int_{0}^{2\pi} (\cos m'\phi)(a_{m\ell} \cos m\phi + b_{m\ell} \sin m\phi) d\phi.$$

The ϕ integrals are all zero except for the $\cos m'\phi \cos m\phi$ integral
with $m = m'$ (text, page 308); this integral $= \pi$. Now with $m = m'$,
the functions $P_\ell^m(x)$ are orthogonal on $(-1,1)$ and the normalization
integral is given by text equation (10.8), page 505. Thus we
find the formula

$$a_{m\ell} = \frac{2\ell + 1}{2\pi} \frac{(\ell - m)!}{(\ell + m)!} \int_{x=-1}^{1} \int_{\phi=0}^{2\pi} f(x, \phi) P_\ell^m(x) \cos m\phi \, d\phi \, dx$$

and a similar formula for $b_{m\ell}$.

15.   The equation satisfied by the temperature inside the sphere is
       the heat flow equation $\nabla^2 u = \dfrac{1}{\alpha^2}\dfrac{\partial u}{\partial t}$;   we want ∇^2 in spherical
       coordinates. First we separate the space dependent and time
       dependent parts of the solution as in text equations (3.3)
       to (3.6), page 551. Thus we have

$$u = F(r,\theta,\phi)\,e^{-k^2\alpha^2 t} \qquad \text{and}$$

$$\nabla^2 F + k^2 F = 0 \qquad\qquad \text{where} \qquad\qquad F = R(r)\Theta(\theta)\Phi(\phi).$$

This is just text equation (7.3) plus the term $k^2 r^2$. The Φ
equation and its solutions are the same as text equation (7.4).
Then we have [compare text equation (7.5)]

$$\frac{1}{R}\frac{d}{dr}\left(r^2\frac{dR}{dr}\right) + \frac{1}{\Theta}\frac{1}{\sin\theta}\frac{d}{d\theta}\left(\sin\theta\,\frac{d\Theta}{d\theta}\right) - \frac{m^2}{\sin^2\theta} + k^2 r^2 = 0.$$

The first and last terms are functions of r only; the middle
terms are functions of θ only. Thus (see box, text page 559) we
can write the separated r and θ equations

$$\frac{1}{R}\frac{d}{dr}\left(r^2\frac{dR}{dr}\right) + k^2 r^2 = \ell(\ell+1),$$

$$\frac{1}{\Theta}\frac{1}{\sin\theta}\frac{d}{d\theta}\left(\sin\theta\,\frac{d\Theta}{d\theta}\right) - \frac{m^2}{\sin^2\theta} + \ell(\ell+1) = 0,$$

where we have called the separation constant $\ell(\ell+1)$. The
solutions of the θ equation are $P_\ell^m(\cos\theta)$ as in text equation
(7.8). We solve the r equation using text equations (16.1)

15.   (continued)

and (16.2) on page 516; the solutions are $R = r^{-1/2} J_p(kr)$

where $p = \ell + \frac{1}{2}$.   Then

$$u = r^{-1/2} J_p(kr)\, P_\ell^m(\cos\theta) \begin{Bmatrix} \sin m\phi \\ \cos m\phi \end{Bmatrix} e^{-k^2\alpha^2 t}.$$

Since there is no θ or ϕ dependence in this problem, we take

$m = 0$, $\ell = 0$, $p = 1/2$:

$$u = r^{-1/2} J_{1/2}(kr) e^{-k^2\alpha^2 t} \quad\text{or}\quad u = j_0(kr) e^{-k^2\alpha^2 t} \quad\text{or}$$

$$u = r^{-1} \sin kr\; e^{-k^2\alpha^2 t}.$$

(See the definition of spherical Bessel functions, text page 518.)

Following the hint, we want $u = 0$ for all $t > 0$ when $r = a$, and

$u = -100$ for $t = 0$.

$$j_0(ka) = \frac{1}{ka}\sin ka = 0, \qquad ka = n\pi, \qquad k = n\pi/a.$$

(1) $u = \sum_n c_n\, j_0\!\left(\frac{n\pi r}{a}\right) e^{-(n\pi\alpha/a)^2 t}$ or $u = \frac{1}{r}\sum_n b_n \sin\frac{n\pi r}{a} e^{-(n\pi\alpha/a)^2 t},$

$$-100 = \sum_n c_n\, j_0\!\left(\frac{n\pi r}{a}\right) \qquad\qquad\text{or}\qquad\qquad -100r = \sum_n b_n \sin\frac{n\pi r}{a}.$$

<u>Fourier series solution</u>: Expand $-100r$ in a Fourier sine series

on $(0,a)$. You should find $b_n = (-1)^n 200a/(n\pi)$. Substitute these

coefficients into (1) and add back the 100° we subtracted; this

is the answer as a Fourier series.

15.   (continued)

Bessel series solution:

$$-100 = \sum_n c_n j_0 \left(\frac{n\pi r}{a}\right) = \sum_n c_n \sqrt{\frac{\pi a}{2n\pi r}} \, J_{1/2}\left(\frac{n\pi r}{a}\right). \quad \text{Let } x = r/a.$$

$$-100\sqrt{x} = \sum_n \frac{c_n}{\sqrt{2n}} \, J_{1/2}(n\pi x).$$

Multiply by $x \, J_{1/2}(m\pi x)$ and integrate from 0 to 1.

$$(2) \qquad -100 \int_0^1 x^{3/2} \, J_{1/2}(m\pi x) \, dx = \frac{c_m}{\sqrt{2m}} \int_0^1 x \, J_{1/2}^2(m\pi x) \, dx.$$

We use formulas from text Chapter 12 to evaluate the integrals.
Right-hand side:

$$\int_0^1 x \, J_{1/2}^2(m\pi x) \, dx = \frac{1}{2} J_{3/2}^2(m\pi) \qquad \begin{array}{l} \text{by text equation (19.10),} \\ \text{page 523.} \end{array}$$

Left-hand side: Let $u = m\pi x$, so $x = \dfrac{u}{m\pi}$.

$$\int_0^1 x^{3/2} \, J_{1/2}(m\pi x) \, dx = \int_0^{m\pi} \left(\frac{u}{m\pi}\right)^{3/2} J_{1/2}(u) \frac{du}{m\pi}$$

$$= \frac{1}{(m\pi)^{5/2}} u^{3/2} J_{3/2}(u) \Big|_0^{m\pi} \qquad \begin{array}{l} \text{by text equation (15.1),} \\ \text{page 514, with } p = 3/2 \end{array}$$

$$= \frac{1}{m\pi} J_{3/2}(m\pi).$$

15.   (continued)

Substitute these results into (2):

$$\frac{-100}{m\pi} J_{3/2}(m\pi) = \frac{c_m}{\sqrt{2m}} \frac{1}{2} J_{3/2}^2(m\pi),$$

(3) $$c_m = \frac{-200\sqrt{2m}}{m\pi \, J_{3/2}(m\pi)} \, .$$

By text equation (17.4), page 518, we find

$$J_{3/2}(x) = \sqrt{\frac{2x}{\pi}} \, x \left(-\frac{1}{x} \frac{d}{dx} \right) \frac{\sin x}{x} = -\sqrt{\frac{2x}{\pi}} \frac{x \cos x - \sin x}{x^2} \, ,$$

$$J_{3/2}(m\pi) = -\sqrt{\frac{2m\pi}{\pi}} \frac{m\pi \cos m\pi - \sin m\pi}{(m\pi)^2} = -\sqrt{2m} \frac{(-1)^m}{m\pi} \, .$$

Thus from (3)

$$c_m = 200(-1)^m$$

and (1) plus 100° gives the result as a Bessel series.

$$u = 200 \sum_n (-1)^n \, j_0(n\pi r/a) \, e^{-(n\pi\alpha/a)^2 t} + 100.$$

If we use $j_0(x) = \frac{\sin x}{x}$ [text equation (17.4), page 518], that
is, $j_0(n\pi r/a) = \frac{a}{n\pi r} \sin(n\pi r/a)$, we find

$$u = \frac{200a}{\pi r} \sum_{n=1}^{\infty} \frac{(-1)^n}{n} \sin\left(\frac{n\pi r}{a}\right) e^{-(n\pi\alpha/a)^2 t} + 100$$

which is the Fourier series solution above.

Section 8

2.   Hint: Let $h = \dfrac{R^2}{ra}$.

4.   We first use Gauss's law to find the potential of a line charge.
      For simplicity in this problem we write Gauss's law in electro-
      static units [this is text equation (10.23) on page 278, but
      using esu instead of mks units].

$$\int_S \vec{E} \cdot \vec{n}\, d\sigma = 4\pi \cdot (\text{total charge inside } S)$$

where S is a closed surface. Take S to be a
cylinder of length 1 and radius r with axis
along the line charge as shown. Let K be the
line charge per unit length. Then K is the
charge inside S. By symmetry, $\vec{E}$ is radial,
so $\vec{E} \cdot \vec{n} = 0$ over the ends of the cylinder and
$\int \vec{E} \cdot \vec{n}\, d\sigma = 2\pi r \cdot 1 \cdot E$ over the curved surface
(remember that the length of the cylinder
is 1). Thus

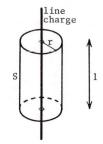

$$2\pi r E = 4\pi K, \qquad E = \frac{2K}{r} .$$

Then if V is the potential such that $\vec{E} = -\nabla V$, we have

(1) $V = -2K \ln r = -K \ln r^2$.

This is the potential of a uniform line charge at distance r
from the line.

4.   (continued)

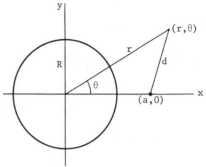

Now consider a line charge along
a line perpendicular to the (x,y)
plane passing through the point
(a,0), and a grounded cylinder of
radius R with axis along the z
axis. The diagram shows a cross
section of the line and cylinder.

The potential of the line charge at the point (r,θ) is, by (1),
$V_K = -K \ln d^2$, where d is the distance from (a,0) to (r,θ). By
the law of cosines, we have

(2) $V_K = -K \ln (r^2 - 2ra \cos \theta + a^2)$.

Now we want the potential after we introduce the grounded
cylinder. We must add to (2) solutions of Laplace's equation
so that the sum will be zero at $r = R$ (surface of the grounded
cylinder). The solutions of Laplace's equation in polar coordinates are (Problems 5.11 to 5.14)

$$\begin{Bmatrix} r^n \\ r^{-n} \end{Bmatrix} \begin{Bmatrix} \sin n\theta \\ \cos n\theta \end{Bmatrix}, \quad n \neq 0, \quad \text{and} \quad \begin{Bmatrix} \text{const.} \\ \ln r \end{Bmatrix}, \quad n = 0.$$

We use the cosine solutions because (see diagram) the problem
is symmetric with respect to the x axis. We use the r^{-n} solutions which $\to 0$ as $r \to \infty$. We also include the $n = 0$ solutions
(the constant is the $n = 0$ term in the cosine series). Then
we assume a potential of the form

4.   (continued)

(3) $V = V_K + A \ln r + \sum_{n=0}^{\infty} a_n r^{-n} \cos n\theta$

$= -K \ln (r^2 - 2ra \cos \theta + a^2) + A \ln r + \sum_{n=0}^{\infty} a_n r^{-n} \cos n\theta.$

If we agree to make the potential finite at ∞, then we must take $A = 2K$ so that the sum of the first two terms $\to 0$ as $r \to \infty$. At $r = R$ (surface of the grounded cylinder) we want $V = 0$. Thus (3) becomes

(4) $0 = -K \ln (R^2 - 2Ra \cos \theta + a^2) + 2K \ln R + \sum_{n=0}^{\infty} a_n R^{-n} \cos n\theta.$

For later convenience we write this as:

(5) $\sum_{n=0}^{\infty} a_n R^{-n} \cos n\theta = K \ln \left[1 - \frac{2R}{a} \cos \theta + \left(\frac{R}{a} \right)^2 \right] + 2K \ln \frac{a}{R}.$

To find the a_n's easily, we need a formula using complex numbers (text, Chapter 2, see pages 51 and 71). If $z = re^{i\theta}$ is any complex number, then $\ln z = \ln r + i\theta$ so $\ln|z| = \ln r = \text{Re} \ln z$. Now let $z = (1 - \rho e^{i\theta})$. Then

$\ln|1 - \rho e^{i\theta}| = \text{Re} \ln(1 - \rho e^{i\theta}),$

$2 \ln|1 - \rho e^{i\theta}| = \ln|1 - \rho e^{i\theta}|^2 = \ln[(1 - \rho e^{i\theta})(1 - \rho e^{-i\theta})]$

$= \ln(1 - 2\rho \cos \theta + \rho^2).$

4.   (continued)

For $\rho < 1$, we can expand the logarithm by equation (13.4), text page 24.

$$\ell n(1 - \rho e^{i\theta}) = -\sum_{n=1}^{\infty} \frac{1}{n}(\rho e^{i\theta})^n$$

$$= -\sum_{n=1}^{\infty} \frac{1}{n} \rho^n e^{in\theta},$$

$$\text{Re } \ell n(1 - \rho e^{i\theta}) = -\sum_{n=1}^{\infty} \frac{1}{n} \rho^n \cos n\theta.$$

Putting these equations together, we find

(6) $$\ell n(1 - 2\rho \cos \theta + \rho^2) = 2\ell n|1 - \rho e^{i\theta}| = 2 \text{ Re } \ell n(1 - \rho e^{i\theta})$$

$$= -2\sum_{n=1}^{\infty} \frac{1}{n} \rho^n \cos n\theta.$$

We use (6) with $\rho = R/a$ to write (5) as

$$\sum_{n=0}^{\infty} a_n R^{-n} \cos n\theta = K(-2)\sum_{n=1}^{\infty} \frac{1}{n}\left(\frac{R}{a}\right)^n \cos n\theta + 2K \ell n \frac{a}{R}.$$

Equate coefficients of $\cos n\theta$ to get

$$a_0 = 2K \ell n \frac{a}{R}, \quad a_n R^{-n} = -2K \cdot \frac{1}{n}\left(\frac{R}{a}\right)^n, \quad a_n = -2K \cdot \frac{1}{n}\left(\frac{R^2}{a}\right)^n.$$

Then the series in (3) becomes [use (6) again with $\rho = R^2/(ar)$]

$$\sum_{n=0}^{\infty} a_n r^{-n} \cos n\theta = 2K \ell n \frac{a}{R} - 2K\sum_{n=1}^{\infty} \frac{1}{n}\left(\frac{R^2}{ar}\right)^n \cos n\theta$$

$$= 2K \ell n \frac{a}{R} + K \ell n \left[1 - \frac{2R^2}{ar} \cos \theta + \left(\frac{R^2}{aR}\right)^2\right].$$

4.   (continued)

Combine this with the other terms in (3) to get

$$V = -K \, \ell n \, (r^2 - 2ra \cos \theta + a^2) + 2K \, \ell n \, r$$

$$+ 2K \, \ell n \frac{a}{R} + K \, \ell n \left[1 - \frac{2R^2}{ar} \cos \theta + \left(\frac{R^2}{ar} \right)^2 \right]$$

$$= -K \, \ell n \, (r^2 - 2ra \cos \theta + a^2)$$

$$+ K \, \ell n \left[r^2 - \frac{2R^2}{a} r \cos \theta + \left(\frac{R^2}{a} \right)^2 \right] + 2K \, \ell n \frac{a}{R} \, .$$

Section 9

6.   The basic solutions of the heat flow equation are given by text
equation (3.10). Here we use the cos kx solutions because we
want $\partial u / \partial x = 0$ at $x = 0$ (this is the mathematical requirement at
an insulated end -- see text, page 553). The final steady state
is $u_f = 20°$. Since the solutions in text equation (3.10) all
tend to zero as $t \to \infty$ (because of the factors $e^{-(k\alpha)^2 t}$) we must
add 20° to the series of basic solutions in order to get the
right final steady state. The series then is the solution of
the problem we would have if we subtract 20° from all temperatures. We first do this auxiliary problem and then add back
the 20°. After we subtract 20°, we have the $x = \ell$ end held at
0° for all $t > 0$. Then we want the cos kx factors to be zero at
$x = \ell$, that is

(1) $\cos k\ell = 0$,   so $k\ell = \frac{2n+1}{2} \pi$.

6.   (continued)

Thus we assume a solution of the form

(2) $u = 20 + \sum\limits_{n=0}^{\infty} a_n e^{-(k\alpha)^2 t} \cos\left(\frac{2n+1}{2\ell}\pi x\right)$

where k is given in (1). Initially (that is, at t = 0), u = 0, so from (2) with t = 0 and u = 0, we get

(3) $\sum\limits_{n=0}^{\infty} a_n \cos\left(\frac{2n+1}{2\ell}\pi x\right) = -20.$

The functions $\cos\left(\frac{2n+1}{2\ell}\pi x\right)$ are an orthogonal set on $(0,\ell)$ (see text, Chapter 12, Problems 6.8, 7.1 and 19.6, and the discussion below). To find the coefficients a_n we multiply (3) by $\cos\left(\frac{2m+1}{2\ell}\pi x\right)$ and integrate from 0 to ℓ. Since the cosine functions are orthogonal, all integrated terms are zero except the one with n = m, and we have

$$a_m \int_0^{\ell} \cos^2\left(\frac{2m+1}{2\ell}\pi x\right) dx = -20 \int_0^{\ell} \cos\left(\frac{2m+1}{2\ell}\pi x\right) dx.$$

The $\cos^2$ integral is $\ell/2$ (see text, page 306) so we have

$$a_m = -\frac{40}{\ell}\int_0^{\ell}\cos\left(\frac{2m+1}{2\ell}\pi x\right)dx = -\frac{40}{\ell}\frac{2\ell}{(2m+1)\pi}\sin\left(\frac{2m+1}{2\ell}\pi x\right)\Bigg|_0^{\ell}$$

$$= -\frac{80(-1)^m}{(2m+1)\pi}.$$

Substitute this result into (2) to find the temperature:

$$u = 20 - \frac{80}{\pi}\sum\limits_{n=0}^{\infty}\frac{(-1)^n}{2n+1}e^{-[(2n+1)\pi\alpha/(2\ell)]^2 t}\cos\left(\frac{2n+1}{2\ell}\pi x\right).$$

Comments on the orthogonality of the functions $\cos\left(\frac{2n+1}{2\ell}\pi x\right)$: It is straightforward to prove this by just integrating the pro-

6.   (continued)

duct of two functions. However, it is even easier to prove it using the differential equation satisfied by the functions (see text pages 499 and 523, Problem 24 on page 540, and the solution of Chapter 12, Problem 7.1). Let y_k be the solution of the differential equation $y_k'' = -k^2 y_k$ satisfying the conditions $y_k'(0) = 0$ and $y_k(\ell) = 0$. [Note that the functions y_k are the functions $\cos kx$ where k is given by (1) above.] We want to prove that $\int_0^{\ell} y_k y_j \, dx = 0$, $k \neq j$. Write the differential equations for y_k and y_j:

(A) $y_k'' = -k^2 y_k$,

(B) $y_j'' = -j^2 y_j$.

Multiply (A) by y_j and (B) by y_k, subtract, and integrate from 0 to ℓ.

$$\int_0^{\ell} (y_j y_k'' - y_k y_j'') \, dx = (j^2 - k^2) \int_0^{\ell} y_j y_k \, dx.$$

As in the solution of Chapter 12 Problem 7.1, the left-hand side is

$$\int_0^{\ell} (y_j y_k'' - y_k y_j'') \, dx = (y_j y_k' - y_k y_j') \Big|_0^{\ell} = 0$$

since the functions are zero at $x = \ell$ and their derivatives are zero at $x = 0$. Thus

$$(j^2 - k^2) \int_0^{\ell} y_j y_k \, dx = 0.$$

and so

$$\int_0^{\ell} y_j y_k \, dx = 0, \qquad j \neq k.$$

11.   Hint: Follow the solution of Problem 2.6 with $z = \frac{r}{a} e^{i\theta}$.

13.   Hint: See hint in Problem 11 above. What is z here?

18.   Consider a continuation of text Figure 6.1 and see Problem 6.1
      in these solutions. If $n = 3$, the radial lines where $\cos n\theta = 0$
      are 60° apart. Since there is no vibration of these lines, we
      could attach them to a rigid support (just as the circumference
      is attached) without affecting the motion. Then any one sector
      is a membrane of the shape desired. Similarly if $n = 6$, 9, 12, $\cdots$
      there are 60° sectors bounded by nodal lines. The value of n is
      the order of the Bessel function in the displacement [see text
      equation (6.6)] and the corresponding frequencies are given by
      $\omega_{mn} = k_{mn} v$ where k_{mn} is the mth zero of J_n. The frequencies of
      the 60° sector membrane as multiples of the lowest frequency of
      the corresponding circular membrane are (see solution of Problem
      6.2)

      $$\omega_{m,3n} = (k_{m,3n}/k_{10})\omega_0.$$

      The first few frequencies are (look up the zeros and verify
      these values):

      $$\omega_{13} = 2.65\omega_0, \qquad \omega_{23} = 4.06\omega_0, \qquad \omega_{16} = 4.13\omega_0, \quad \cdots .$$

22.   Hint: Subtract 1 from all temperatures and solve the problem;
      then add back the 1. See text, page 569.

26.   We separate variables in the equation

(1) $\frac{1}{r}\frac{\partial}{\partial r}\left(r\frac{\partial z}{\partial r}\right)+\frac{1}{r^2}\frac{\partial^2 z}{\partial\theta^2}=\frac{1}{v^2}\frac{\partial^2 z}{\partial t^2}+\lambda^2 z.$

Substitute $z=R(r)\Theta(\theta)T(t)$ into (1) and divide by $R\Theta T$ to get

$$\frac{1}{rR}\frac{d}{dr}\left(r\frac{dR}{dr}\right)+\frac{1}{r^2\Theta}\frac{d^2\Theta}{dt^2}=\frac{1}{v^2 t}\frac{d^2 T}{dt^2}+\lambda^2.$$

Separate variables (see text, page 559) and solve the separated
equations.

$$\frac{d^2 T}{dt^2}=-(k^2+\lambda^2)v^2 T,\qquad T=\begin{cases}\sin\left(\sqrt{k^2+\lambda^2}\,vt\right)\\[2mm]\cos\left(\sqrt{k^2+\lambda^2}\,vt\right)\end{cases}.$$

$$\frac{d^2\Theta}{d\theta^2}=-n^2\Theta,\qquad \Theta=\begin{cases}\sin n\theta\\[1mm]\cos n\theta\end{cases}.$$

$$r\frac{d}{dr}\left(r\frac{dR}{dr}\right)+(k^2 r^2-n^2)R=0,\qquad R=J_n(kr).$$

[See text, page 523, equation (19.2), or page 516, equation
(16.1).]   The membrane is rigidly attached to a support at $r=a$
so we must have $R=0$ at $r=a$.   Thus

$$J_n(ka)=0,\qquad ka=k_{mn}$$

where $k_{mn}=$ mth zero of J_n.   The frequencies [from the $T(t)$
solution] are

$$\omega=v\sqrt{k^2+\lambda^2}=v\sqrt{\lambda^2+(k_{mn}/a)^2}=2\pi\nu,$$

$$\nu_{mn}=\frac{v}{2\pi}\sqrt{\lambda^2+(k_{mn}/a)^2}.$$

Chapter 14

Section 1

3. $\bar{z} = x - iy = u(x,y) + iv(x,y)$,   so $u = x$ and $v = -y$.

4. $|z| = \sqrt{x^2 + y^2}$ so $u = \sqrt{x^2 + y^2}$ and $v = 0$.

6. $e^z = e^{x+iy} = e^x e^{iy} = e^x(\cos y + i \sin y)$,   so

   $u = e^x \cos y$ and $v = e^x \sin y$.

7. By Chapter 2, Problem 12.4,

$$\cosh z = \cosh(x + iy) = \cosh x \cos y + i \sinh x \sin y.$$

   Thus

$$u = \cosh x \cos y, \qquad v = \sinh x \sin y.$$

11. $\dfrac{2z - i}{iz + 2} = \dfrac{2(x + iy) - i}{i(x + iy) + 2} = \dfrac{2x + i(2y - 1)}{(2 - y) + ix} \cdot \dfrac{(2 - y) - ix}{(2 - y) - ix}$

$$= \frac{2x(2 - y) + x(2y - 1) + i[(2y - 1)(2 - y) - 2x^2]}{(2 - y)^2 + x^2}$$

$$= \frac{3x - i(2x^2 + 2y^2 - 5y + 2)}{x^2 + (y - 2)^2}.$$

   Thus

$$u = \frac{3x}{x^2 + (y - 2)^2}, \qquad\qquad v = -\frac{2x^2 + 2y^2 - 5y + 2}{x^2 + (y - 2)^2}.$$

18. $\sqrt{z} = \pm \sqrt{re^{i\theta}} = \pm\sqrt{r}\ e^{i\theta/2} = \pm\sqrt{r}\left(\cos\frac{\theta}{2} + i\ \sin\frac{\theta}{2}\right)$,

$\sqrt{r}\cos\frac{\theta}{2} = \pm\sqrt{r}\ \sqrt{\frac{1+\cos\theta}{2}} = \pm\ \sqrt{\frac{r+r\cos\theta}{2}} = \pm\sqrt{\frac{\sqrt{x^2+y^2}+x}{2}}$,

$\sqrt{r}\sin\frac{\theta}{2} = \pm\sqrt{r}\ \sqrt{\frac{1-\cos\theta}{2}} = \pm\ \sqrt{\frac{r-r\cos\theta}{2}} = \pm\sqrt{\frac{\sqrt{x^2+y^2}-x}{2}}$,

$u = \pm\sqrt{\frac{\sqrt{x^2+y^2}+x}{2}}$, $v = \pm\sqrt{\frac{\sqrt{x^2+y^2}-x}{2}}$.

The correct choice of signs is most easily seen by plotting z
and $\sqrt{z}$ in the complex plane. If z is in the first or second
quadrants ($\theta < \pi$), then the angle of $\sqrt{z}$ [which is $\theta/2$ or
$(\theta + 2\pi)/2$] is in the first or third quadrants, so u and v have
the same sign. If z is in the third or fourth quadrants, then
u and v have opposite signs. We can summarize this by saying
that uv has the sign of y.

19. $\ell n\ z = \ell n(re^{i\theta}) = \ell n\ r + i\theta = \ell n\sqrt{x^2+y^2} + i\ \tan^{-1}\frac{y}{x}$,   so

$u = \ell n\ r = \frac{1}{2}\ell n(x^2+y^2)$ and $v = \theta = \tan^{-1}\frac{y}{x}$.

21.   Hint: Set $z = x + iy$.

Section 2

3.   By Problem 1.3, $u = x$ and $v = -y$; then $\frac{\partial u}{\partial x} = 1 \neq \frac{\partial v}{\partial y} = -1$. Thus, by
the Cauchy-Riemann equations [text equation (2.2)], $f(z) = \bar{z}$ is
not an analytic function.

4. By Problem 1.4, $u = \sqrt{x^2 + y^2}$, $v = 0$; then $\frac{\partial u}{\partial x} \neq 0$, $\frac{\partial v}{\partial y} = 0$, so $|z|$
    is not analytic.

6. By Problem 1.6, $u = e^x \cos y$, $v = e^x \sin y$. Then the Cauchy-
    Riemann equations are

    $$\frac{\partial u}{\partial x} = e^x \cos y = \frac{\partial v}{\partial y}, \qquad \frac{\partial v}{\partial x} = e^x \sin y = -\frac{\partial u}{\partial y}.$$

    Since u, v, and their partial derivatives are continuous
    everywhere and satisfy the Cauchy-Riemann equations, e^z
    is analytic for all z.

7. By Problem 1.7, $u = \cosh x \cos y$, $\qquad v = \sinh x \sin y$,

    $$\frac{\partial u}{\partial x} = \sinh x \cos y = \frac{\partial v}{\partial y},$$

    $$\frac{\partial u}{\partial y} = -\cosh x \sin y = -\frac{\partial v}{\partial x}.$$

    Thus u and v satisfy the Cauchy-Riemann equations. Also u, v
    and their partial derivatives are continuous, so $\cosh z$ is
    analytic.

12. To do problems like this, we _can_ find u and v and then apply the
    Cauchy-Riemann conditions. However, a more efficient method is
    to use the results of Problems 25 to 28. We see that we can
    find derivatives of sums, products, and quotients (when the
    denominator $\neq 0$) of analytic functions. Then, if z and $z^2 + 1$
    are analytic, so is $z/(z^2 + 1)$ when $z^2 + 1 \neq 0$, that is, when $z \neq \pm i$.
    It is easy to verify that z and $z^2 + 1$ are both analytic.

18.   In Problem 48 below, we show that $\sqrt{z}$ satisfies the Cauchy-
      Riemann equations for $z \neq 0$. By properly choosing the branch of
      $\sqrt{z}$ (text pages 580, 607, 608), we can make u, v, and their
      derivatives continuous in any region not containing the origin
      and not crossing the branch cut. Then in this region, $\sqrt{z}$ is
      analytic. Since the branch cut can be moved, it is possible to
      define an analytic branch of $\sqrt{z}$ at any $z \neq 0$.

         Caution: If you do Problem 48 for $z^{3/2}$, you find that the
      Cauchy-Riemann equations are satisfied even at the origin.
      This does <u>not</u> mean that $z^{3/2}$ is analytic at the origin. An easy
      way to see this is to observe that the second and higher deriva-
      tives of $z^{3/2}$ do not exist for $z = 0$, whereas an analytic func-
      tion has derivatives of all orders (Theorem III, text page 584).
      The reason $z^{3/2}$ is not analytic at $z = 0$ is that u and v are not
      continuous across the branch cut so $z^{3/2}$ does not have a deriva-
      tive at all points of any small circle about $z = 0$ (see text
      definition of "analytic at a point", page 581).

24.   Hint: Show that $\dfrac{y - ix}{x^2 + y^2} = \dfrac{-i}{\bar{z}}$ and see Problem 12 and Problem 3 above.

25 to 31. Hint: See a calculus text for proofs of these formulas for
      a function of x. Since text equation (2.1) defining df/dz is of
      the same form as the definition of df/dx in calculus, the same
      methods apply to the problems here.

36.   See Problem 18; $\sqrt{1 + z^2}$ does not have a derivative if $1 + z^2 = 0$.

37. See text, pages 69, 70 (Problem 16), and 26 (Example 3). We
     could write

$$\tanh z = \frac{e^z - e^{-z}}{e^z + e^{-z}} = \frac{z + \frac{z^3}{3!} + \frac{z^5}{5!} \cdots}{1 + \frac{z^2}{2!} + \frac{z^4}{4!} \cdots}$$

and do the long division as for the $\tan x$ series in the text,
page 26. However, it is simpler to use the result for the $\tan x$
series with x replaced by iz, and Problem 16 on text page 70,
as follows:

$$\tanh z = -i \tan iz = -i\left[iz + \frac{1}{3}(iz)^3 + \frac{2}{15}(iz)^5 \cdots\right]$$

$$= z - \frac{1}{3}z^3 + \frac{2}{15}z^5 \cdots .$$

The denominator of $\tanh z$ is $\cosh z = \cos iz = 0$ for $iz = \pm\pi/2$, that
is, for $z = \pm i\pi/2$. The series converges inside the circle
$|z| = \pi/2$.

38. $\dfrac{1}{2i + z} = \dfrac{1}{2i\left(1 + \frac{z}{2i}\right)} = \dfrac{1}{2i}\left(1 + \frac{z}{2i}\right)^{-1}.$

We expand this in a binomial series [see text, page 24,
equation (13.5)].

$$\frac{1}{2i + z} = \frac{1}{2i}\left[1 - \frac{z}{2i} + \left(\frac{z}{2i}\right)^2 - \left(\frac{z}{2i}\right)^3 \cdots\right]$$

$$= -\frac{i}{2} + \frac{z}{4} + \frac{iz^2}{8} - \frac{z^3}{16} + \cdots .$$

The function $1/(z + 2i)$ is not analytic at
$z = -2i$, so the series converges inside the
circle with center at the origin and radius
2, that is, for $|z| < 2$.

44. In text equation (2.2), differentiate the first equation with respect to x and the second equation with respect to y, and add the equations to get

$$\frac{\partial^2 u}{\partial x^2} + \frac{\partial^2 v}{\partial y \partial x} = \frac{\partial^2 v}{\partial x \partial y} - \frac{\partial^2 u}{\partial y^2} \; .$$

Assuming that $\frac{\partial^2 v}{\partial y \partial x} + \frac{\partial^2 v}{\partial x \partial y}$ (see text, bottom of page 147), we have

$$\frac{\partial^2 u}{\partial x^2} + \frac{\partial^2 u}{\partial y^2} = 0$$

which is Laplace's equation. Similarly, show that $\nabla^2 v = 0$.

46. In polar coordinates, we write

$$f(z) = u(r,\theta) + iv(r,\theta) \quad \text{and} \quad z = re^{i\theta}.$$

Following the method of text equations (2.3) and (2.4), we find:

(1) $\dfrac{\partial f}{\partial r} = \dfrac{df}{dz}\dfrac{\partial z}{\partial r} = \dfrac{df}{dz}e^{i\theta} = \dfrac{\partial u}{\partial r} + i\dfrac{\partial v}{\partial r}$.

(2) $\dfrac{\partial f}{\partial \theta} = \dfrac{df}{dz}\dfrac{\partial z}{\partial \theta} = \dfrac{df}{dz}ire^{i\theta} = \dfrac{\partial u}{\partial \theta} + i\dfrac{\partial v}{\partial \theta}$.

(3) $e^{i\theta}\dfrac{df}{dz} = \dfrac{\partial u}{\partial r} + i\dfrac{\partial v}{\partial r} = \dfrac{1}{ir}\left(\dfrac{\partial u}{\partial \theta} + i\dfrac{\partial v}{\partial \theta}\right).$

                           from (1)   from (2)

Equate real parts and equate imaginary parts in (3) to get:

$$\frac{\partial u}{\partial r} = \frac{1}{r}\frac{\partial v}{\partial \theta}, \qquad\qquad \frac{\partial v}{\partial r} = -\frac{1}{r}\frac{\partial u}{\partial \theta} \; .$$

These are the Cauchy-Riemann equations in polar coordinates.

48.   In polar coordinates, we have

$$\sqrt{z} = r^{1/2} e^{i\theta/2} = r^{1/2}\left(\cos\frac{\theta}{2} + i\sin\frac{\theta}{2}\right),$$

$$u = r^{1/2}\cos\frac{\theta}{2}, \qquad\qquad v = r^{1/2}\sin\frac{\theta}{2},$$

$$\frac{\partial u}{\partial r} = \frac{1}{2}r^{-1/2}\cos\frac{\theta}{2}, \qquad\qquad \frac{\partial v}{\partial r} = \frac{1}{2}r^{-1/2}\sin\frac{\theta}{2},$$

$$\frac{\partial u}{\partial\theta} = -\frac{1}{2}r^{1/2}\sin\frac{\theta}{2}, \qquad\qquad \frac{\partial v}{\partial\theta} = \frac{1}{2}r^{1/2}\cos\frac{\theta}{2}.$$

We see that these partial derivatives satisfy the Cauchy-Riemann equations as given in Problem 46 above, for $r \neq 0$.

54.   If $u(x,y) = y$, then $\nabla^2 u = \dfrac{\partial^2 u}{\partial x^2} + \dfrac{\partial^2 u}{\partial y^2} = 0$, that is, u is harmonic. To find v so that $f(z) = u + iv$ is an analytic function, we follow the method of the Example on text page 585. We find

$$\frac{\partial v}{\partial y} = \frac{\partial u}{\partial x} = 0 \qquad\qquad \text{so } v = g(x);$$

$$\frac{\partial v}{\partial x} = -\frac{\partial u}{\partial y} = -1 \qquad\qquad \text{so } v = -x + \text{const.}$$

Then $f(z) = u + iv = y - ix + \text{const.} = -iz + \text{const.}$ We observe that $\nabla^2 v = -\nabla^2 x = 0$, so v is also harmonic.

60.   We first verify that $u = \ln(x^2 + y^2)$ is harmonic:

$$\frac{\partial u}{\partial x} = \frac{2x}{x^2 + y^2}, \qquad \frac{\partial^2 u}{\partial x^2} = \frac{2(x^2 + y^2) - 2x \cdot 2x}{(x^2 + y^2)^2} = \frac{2(y^2 - x^2)}{(x^2 + y^2)^2}.$$

Similarly

$$\frac{\partial^2 u}{\partial y^2} = \frac{2(x^2 - y^2)}{(x^2 + y^2)^2} \qquad \text{so} \qquad \nabla^2 u = 0.$$

60.   (continued)

To find $v(x,y)$, we use the method of the Example on text page 585.

$$\frac{\partial u}{\partial x} = \frac{2x}{x^2 + y^2} = \frac{\partial v}{\partial y} , \qquad v = 2 \tan^{-1} \frac{y}{x} + g(x),$$

$$\frac{\partial v}{\partial x} = \frac{-2y/x^2}{1 + y^2/x^2} + g'(x) = -\frac{\partial u}{\partial y} = \frac{-2y}{x^2 + y^2} \qquad \text{so} \qquad g' = 0. \quad \text{Then}$$

$$f(z) = u + iv = \ell n(x^2 + y^2) + 2i \tan^{-1} \frac{y}{x}$$

$$= \ell n \, r^2 + 2i\theta = \ell n(re^{i\theta})^2 = \ell n \, z^2.$$

This problem is easier in polar coordinates using Problem 46. Since $\ell n(x^2 + y^2) = \ell n \, r^2 = 2 \ell n \, r$, we find

$$\frac{\partial u}{\partial r} = \frac{2}{r} = \frac{1}{r} \frac{\partial v}{\partial \theta} , \qquad v = 2\theta,$$

$$u + iv = 2(\ell n \, r + i\theta) = 2 \ell n(re^{i\theta}) = 2 \ell n \, z = \ell n \, z^2$$

as we found above using rectangular coordinates.

It is also easier in polar coordinates to verify that $u = 2 \ell n \, r$ and $v = 2\theta$ are both harmonic. We have (see text page 433):

$$\nabla^2 = \frac{1}{r} \frac{\partial}{\partial r} \left(r \frac{\partial}{\partial r} \right) + \frac{1}{r^2} \frac{\partial^2}{\partial \theta^2} ,$$

$$\nabla^2 (2 \ell n \, r) = \frac{1}{r} \frac{\partial}{\partial r} \left(r \cdot \frac{2}{r} \right) + 0 = 0,$$

$$\nabla^2 (2\theta) = 0.$$

64.   We use the method outlined to do Problems 54 and 60 again.

Problem 54

Let $z_0 = 0$. Then the formula gives

$$f(z) = 2u\left(\frac{z}{2}, \frac{z}{2i}\right) + \text{const.}$$

This means replace x by $\frac{z}{2}$ and y by $\frac{z}{2i}$ in u(x,y). Since u(x,y) = y, this gives

$$f(z) = 2 \cdot \frac{z}{2i} = \frac{z}{i} = -iz + \text{const.}$$

Problem 60

Here we cannot use $z_0 = 0$ since $\ell n\, 0$ is not defined. Let $z_0 = 1$. Then we replace x by $\frac{z+1}{2}$ and y by $\frac{z-1}{2i}$ in u to get

$$f(z) = 2 \,\ell n\left[\left(\frac{z+1}{2}\right)^2 + \left(\frac{z-1}{2i}\right)^2\right] = 2\,\ell n\, z.$$

Section 3

1.   Integrating from z = i to z = 1 + i means integrating from (0,1) to (1,1). Thus y = 1, dy = 0, and x varies from 0 to 1. Then

$$\int_i^{1+i} z\, dz = \int_{(0,1)}^{(1,1)} (x + iy)(dx + i\, dy)$$

$$= \int_0^1 (x + i)\, dx = \frac{x^2}{2} + ix\Big|_0^1 = \frac{1}{2} + i.$$

3.   (b) Along the x axis, $z = x$, $dz = dx$. Along the semicircle,
$z = e^{i\theta}$, $dz = ie^{i\theta} d\theta$. Then

$$\oint_C z^2 \, dz = \int_{-1}^{1} x^2 \, dx + \int_0^{\pi} e^{2i\theta} \, ie^{i\theta} \, d\theta = \frac{x^3}{3}\Big|_{-1}^{1} + \frac{ie^{3i\theta}}{3i}\Big|_0^{\pi}$$

$$= \frac{2}{3} + \frac{1}{3}(e^{3i\pi} - 1) = \frac{2}{3} - \frac{2}{3} = 0.$$

We can check this by observing that z^2 is analytic on and inside
C and that C is "smooth" except for two corners; thus Cauchy's
theorem applies and the integral is zero.

5.   If $y = \pi$, then $dy = 0$, $z = x + i\pi$, $dz = dx$. Thus

$$\int_{i\pi}^{\infty + i\pi} e^{-z} \, dz = \int_0^{\infty} e^{-(x+i\pi)} \, dx = e^{-i\pi}(-e^{-x})\Big|_0^{\infty}$$

$$= e^{-i\pi}(0 + 1) = -1.$$

12.   (a) Along the given straight line, $y = 2x$, so

$$dy = 2dx, \qquad dz = dx + i \, dy = (1 + 2i) \, dx,$$

$$|z|^2 = x^2 + y^2 = x^2 + (2x)^2 = 5x^2,$$

and x varies from 0 to 1. Thus along the straight line path,

$$\int_0^{1+2i} |z|^2 dz = \int_0^1 5x^2 (1 + 2i) \, dx = (1 + 2i)\frac{5}{3} x^3 \Big|_0^1 = \frac{5}{3}(1 + 2i).$$

Recall from text Section 2 that $|z^2|$ does not have a derivative,
that is, $|z|^2 = x^2 + y^2$ is not analytic (or use the Cauchy-Riemann
equations). Thus we expect the integral to depend on the path
of integration. Evaluate the integral along path (b) and observe that the answer is different.

14. If $z = e^{i\theta}$, then $|z| = 1$; as θ varies from 0 to 2π, the point
     $z = e^{i\theta}$ traverses the unit circle $|z| = r = 1$. Along this circle C,
     we have

$$z = e^{i\theta}, \qquad\qquad dz = i e^{i\theta} \, d\theta,$$

$$\int_C z^{n-m-1} \, dz = \int_0^{2\pi} e^{(i\theta)(n-m-1)} i e^{i\theta} \, d\theta = i \int_0^{2\pi} e^{in\theta} e^{-im\theta} \, d\theta.$$

Now if $n > m$, then z^{n-m-1} is analytic on and inside C. Then,
by Cauchy's theorem [text equation (3.1)], the integral is zero.
Thus we have proved

(1) $\displaystyle\int_0^{2\pi} e^{in\theta} e^{-im\theta} \, d\theta = 0, \qquad n > m.$

If $n < m$, then the same proof starting with z^{m-n-1} gives

(2) $\displaystyle\int_0^{2\pi} e^{-in\theta} e^{im\theta} \, d\theta = 0, \qquad n < m.$

Remember that, if a complex expression is zero, its real and
imaginary parts are zero and so its complex conjugate is zero.
(In symbols, if $a + bi = 0$, then $a = 0$, $b = 0$, and so $a - bi = 0$.)
Taking the complex conjugate of (2) gives

(3) $\displaystyle\int_0^{2\pi} e^{in\theta} e^{-im\theta} \, d\theta = 0, \qquad n < m.$

[Alternatively, we could have taken the complex conjugate of (1)
and interchanged n and m.] Then (1) and (3) together give the
desired result for $n \neq m$.

17.   First we must write the denominator in the form $(z - a)$:

$$\oint_C \frac{\sin z \, dz}{2z - \pi} = \oint_C \frac{\sin z \, dz}{2\left(z - \frac{\pi}{2}\right)} = \frac{1}{2} \oint_C \frac{\sin z \, dz}{z - \frac{\pi}{2}} .$$

(a) Since $\pi/2 > 1$, $z = \pi/2$ lies outside the circle C of radius 1 and center at the origin. Thus $f(z) = \frac{\sin z}{z - (\pi/2)}$ is analytic on and inside C so by Cauchy's theorem [text equation (3.1)] the integral is zero.

(b) Since $\pi/2 < 2$, $z = \pi/2$ lies inside the circle C of radius 2 and center at the origin. Then by Cauchy's integral formula [text equation (3.9)] with $f(z) = \sin z$,

$$\frac{1}{2} \oint_C \frac{\sin z \, dz}{z - \frac{\pi}{2}} = \frac{1}{2} \cdot 2\pi i \sin \frac{\pi}{2} = \pi i.$$

22.   First we must write the denominator in the form $(z - a)^3$:

$$\oint_C \frac{\sin 2z \, dz}{(6z - \pi)^3} = \frac{1}{6^3} \oint_C \frac{\sin 2z \, dz}{\left(z - \frac{\pi}{6}\right)^3} .$$

Next we see that the point $z = a = \frac{\pi}{6} < 3$ is inside C (the circle $|z| = 3$). Then by Problem 21 with $f(z) = \sin 2z$ and $n = 2$

$$\frac{1}{6^3} \oint_C \frac{\sin 2z \, dz}{\left(z - \frac{\pi}{6}\right)^3} = \frac{1}{6^3} \left(\frac{2\pi i}{2!}\right) \frac{d^2}{dz^2}(\sin 2z) \Bigg|_{z = \pi/6}$$

$$= \frac{1}{6^3}(\pi i)\left(-4 \sin \frac{2\pi}{6}\right) = \frac{-4\pi i}{6^3} \frac{\sqrt{3}}{2} = \frac{-\pi i \sqrt{3}}{108} .$$

Section 4

4.   By the method of partial fractions (see Appendix C to these
Solutions), we write

$$\frac{1}{(z-1)(z-2)^2} = \frac{A}{z-1} + \frac{B}{z-2} + \frac{C}{(z-2)^2} \qquad \text{or}$$

$$1 = A(z-2)^2 + B(z-1)(z-2) + C(z-1).$$

For $z = 1$: $1 = A$.

For $z = 2$: $1 = C$.

For $z = 0$: $1 = 4A + 2B - C = 3 + 2B$, $B = -1$.   Then

(A) $$\frac{1}{z(z-1)(z-2)^2} = \frac{1}{z}\left[\frac{1}{z-1} - \frac{1}{z-2} + \frac{1}{(z-2)^2}\right].$$

We will want each of the fractions in the bracket in (A)
expanded in powers of z and in powers of $1/z$.   In powers of z
[see text, page 24, equation (13.5)]:

(B) $$\frac{1}{z-1} = -(1-z)^{-1} = -(1 + z + z^2 + z^3 + \cdots),$$

(C) $$\frac{1}{z-2} = -\frac{1}{2}\left(1 - \frac{z}{2}\right)^{-1} = -\frac{1}{2}\left(1 + \frac{z}{2} + \frac{z^2}{4} + \frac{z^3}{8} + \cdots\right),$$

(D) $$\frac{1}{(z-2)^2} = \frac{1}{4}\left(1 - \frac{z}{2}\right)^{-2} = \frac{1}{4}\left(1 + z + \frac{3}{4}z^2 + \frac{1}{2}z^3 + \cdots\right).$$

In powers of $\frac{1}{z}$:

(E) $$\frac{1}{z-1} = \frac{1}{z}\left(1 - \frac{1}{z}\right)^{-1} = \frac{1}{z}\left(1 + \frac{1}{z} + \frac{1}{z^2} + \frac{1}{z^3} + \frac{1}{z^4} + \cdots\right),$$

(F) $$\frac{1}{z-2} = \frac{1}{z}\left(1 - \frac{2}{z}\right)^{-1} = \frac{1}{z}\left(1 + \frac{2}{z} + \frac{4}{z^2} + \frac{8}{z^3} + \frac{16}{z^4} + \cdots\right),$$

(G) $$\frac{1}{(z-2)^2} = \frac{1}{z^2}\left(1 - \frac{2}{z}\right)^{-2} = \frac{1}{z^2}\left(1 + \frac{4}{z} + \frac{12}{z^2} + \frac{32}{z^3} + \cdots\right).$$

4.   (continued)

As in text Figure 4.1, there are three regions, each with a different Laurent series. Here the regions are $|z| < 1$, $1 < |z| < 2$, and $|z| > 2$. In the region $|z| < 1$, we use only powers of z. Substituting equations (B), (C), (D) into (A), we find

$$\frac{1}{z(z-1)(z-2)^2} = \frac{1}{z}\left[-(1 + z + z^2 + z^3 \cdots) + \frac{1}{2}\left(1 + \frac{z}{2} + \frac{z^2}{4} + \frac{z^3}{8} \cdots\right)\right.$$
$$\left. + \frac{1}{4}\left(1 + z + \frac{3}{4}z^2 + \frac{1}{2}z^3 \cdots\right)\right]$$

$$= \frac{1}{z}\left(-\frac{1}{4} - \frac{1}{2}z - \frac{11}{16}z^2 - \frac{13}{16}z^3 \cdots\right)$$

$$= -\frac{1}{4z} - \frac{1}{2} - \frac{11}{16}z - \frac{13}{16}z^2 \cdots .$$

This series converges near $z = 0$; the residue of the given function at the origin is the coefficient of $1/z$ in this series, that is, $R = -1/4$.

In the region $1 < |z| < 2$, we use the series in powers of z for $1/(z-2)$ and $1/(z-2)^2$ since these converge for $|z| < 2$, and the series in powers of $1/z$ for $1/(z-1)$ since this converges for $|z| > 1$. Then for $1 < |z| < 2$, we substitute (C), (D), and (E) into (A) to get:

$$\frac{1}{z(z-1)(z-2)^2} = \frac{1}{z}\left[\frac{1}{z}\left(1 + \frac{1}{z} + \frac{1}{z^2} + \cdots\right) + \frac{1}{2}\left(1 + \frac{z}{2} + \frac{z^2}{4} + \frac{z^3}{8} \cdots\right)\right.$$
$$\left. + \frac{1}{4}\left(1 + z + \frac{3}{4}z^2 + \frac{1}{2}z^3 \cdots\right)\right]$$

$$= \frac{1}{z}\left(\cdots \frac{1}{z^3} + \frac{1}{z^2} + \frac{1}{z} + \frac{3}{4} + \frac{1}{2}z + \frac{5}{16}z^2 + \frac{3}{16}z^3 \cdots\right)$$

$$= \left(\cdots + \frac{1}{z^4} + \frac{1}{z^3} + \frac{1}{z^2} + \frac{3}{4z} + \frac{1}{2} + \frac{5}{16}z + \frac{3}{16}z^2 \cdots\right)$$

4.   (continued)

This series is an expansion about $z = 0$ (that is, in powers of z).
Note carefully, however, that the point $z = 0$ is not in the region
of convergence; thus this series cannot be used to find the residue at the origin.   The coefficient of $1/z$ in this series is <u>not</u>
the residue at the origin.

In the region $|z| > 2$, we use only the series of powers of $1/z$
since all of these converge for $|z| > 2$ (at least).   We substitute
(E), (F), and (G) into (A) to get:

$$\frac{1}{z(z-1)(z-2)^2} = \frac{1}{z}\left[\frac{1}{z}\left(1 + \frac{1}{z} + \frac{1}{z^2} + \frac{1}{z^3} + \frac{1}{z^4} + \frac{1}{z^5}\cdots\right)\right.$$

$$-\frac{1}{z}\left(1 + \frac{2}{z} + \frac{4}{z^2} + \frac{8}{z^3} + \frac{16}{z^4} + \frac{32}{z^5}\cdots\right)$$

$$\left.+\frac{1}{z^2}\left(1 + \frac{4}{z} + \frac{12}{z^2} + \frac{32}{z^3} + \frac{80}{z^4}\cdots\right)\right]$$

$$= \frac{1}{z}\left(\frac{1}{z^3} + \frac{5}{z^4} + \frac{17}{z^5} + \frac{49}{z^6}\cdots\right) = \frac{1}{z^4} + \frac{5}{z^5} + \frac{17}{z^6} + \frac{49}{z^7}\cdots \ .$$

9.   (a) $\dfrac{\sin z}{z} = \dfrac{z - \frac{1}{3!}z^3 + \cdots}{z} = 1 - \frac{1}{3!}z^2 + \cdots \ .$

All the b's are zero, so $z = 0$ is a regular point of $\frac{\sin z}{z}$ if we
define $\frac{\sin z}{z} = 1$ at $z = 0$.   [Sometimes we say that $f(z)$ has a
"removable singularity" at z_0 when the numerator and the denominator have the same factor equal to zero at z_0.]

9.   (continued)

(b) $\dfrac{\cos z}{z^3} = \dfrac{1 - \dfrac{z^2}{2!} + \cdots}{z^3} = \dfrac{1}{z^3} - \dfrac{1}{2z} \cdots$.

Since $b_3 \neq 0$, but all b_n for $n > 3$ are zero, $\dfrac{\cos z}{z^3}$ has a pole of order 3 at $z = 0$. Another way to see this is to say that $\cos z$ is analytic and not zero at $z = 0$; then the z^3 in the denominator means that $\dfrac{\cos z}{z^3}$ has a pole of order 3 at $z = 0$.

10.   (b) Since $\cot\left(z - \dfrac{\pi}{2}\right) = -\tan z$, we can write

$$\tan^2 z = \cot^2\left(z - \dfrac{\pi}{2}\right) = \dfrac{\cos^2\left(z - \dfrac{\pi}{2}\right)}{\sin^2\left(z - \dfrac{\pi}{2}\right)} .$$

Now $\cos^2\left(z - \dfrac{\pi}{2}\right)$ is analytic and not zero at $z = \dfrac{\pi}{2}$, and

$$\sin^2\left(z - \dfrac{\pi}{2}\right) = \left[\left(z - \dfrac{\pi}{2}\right) \cdot \dfrac{1}{3!}\left(z - \dfrac{\pi}{2}\right)^3 + \cdots\right]^2$$

has a factor $\left(z - \dfrac{\pi}{2}\right)^2$ which is zero at $z = \pi/2$. Thus $\tan^2 z$ has a pole of order 2 at $z = \pi/2$.

12.   (b) $\dfrac{z^2 - 1}{(z^2 + 1)^2} = \dfrac{z^2 - 1}{(z + i)^2 (z - i)^2}$ has a pole of order 2 at $z = i$.

Section 5

1. If $z = z_0 + \rho e^{i\theta}$ on C, where $\rho = $ const., then, for $n \neq 1$:

$$\oint_C \frac{dz}{(z - z_0)^n} = \int_0^{2\pi} \frac{\rho i e^{i\theta} \, d\theta}{\rho^n e^{in\theta}} = \frac{i\rho}{\rho^n} \int_0^{2\pi} e^{i(1-n)\theta} \, d\theta$$

$$= \frac{i}{\rho^{n-1}} \frac{e^{i(1-n)\theta}}{i(1-n)} \Big|_0^{2\pi} = \frac{1}{\rho^{n-1}(1-n)}(e^{2\pi i(1-n)} - 1) = 0$$

since $2\pi(1 - n)$ is an integral multiple of 2π. If $n = 1$,

$$\oint_C \frac{dz}{z - z_0} = \int_0^{2\pi} \frac{\rho i e^{i\theta} \, d\theta}{\rho e^{i\theta}} = i \int_0^{2\pi} d\theta = 2\pi i.$$

Section 6

1. The Laurent series about $z = 0$ is

$$\frac{1}{z(z+1)} = \frac{1}{z}(1+z)^{-1} = \frac{1}{z}(1 - z + z^2 \cdots) = \frac{1}{z} - 1 + z \cdots .$$

(Residue at $z = 0$) = (coefficient of $\frac{1}{z}$) = 1.

If we want the residue at $z = -1$, we must expand in powers of $z + 1$:

$$\frac{1}{z(z+1)} = \frac{1}{[(z+1) - 1](z+1)} = \frac{-1}{z+1}[1 - (z+1)]^{-1}$$

$$= \frac{-1}{z+1}[1 + (z+1) + (z+1)^2 + \cdots]$$

$$= \frac{-1}{z+1} - 1 - (z+1) \cdots .$$

This series converges inside a circle of radius 1 with center at $z = -1$. Then the residue at $z = -1$ is the coefficient of $\frac{1}{z+1}$ in the series, that is, $R = -1$.

1. (continued)

Note that for a function with a simple pole, finding the Laurent series is not as easy as text Method B. Using Method B here:

$$\text{Residue of } \frac{1}{z(z+1)} \text{ at } z = 0 \text{ is } \left. \cancel{z}\frac{1}{\cancel{z}(z+1)} \right|_{z=0} = 1.$$

$$\text{Residue of } \frac{1}{z(z+1)} \text{ at } z = -1 \text{ is } \left. \cancel{(z+1)}\frac{1}{z\cancel{(z+1)}} \right|_{z=-1} = -1.$$

3. The Laurent series is

$$\frac{\sin z}{z^4} = \frac{z - \frac{z^3}{3!} + \frac{z^5}{5!} \cdots}{z^4} = \frac{1}{z^3} - \frac{1}{6}\frac{1}{z} + \frac{1}{5!} z + \cdots .$$

The residue at $z = 0$ is the coefficient of $1/z$ so $R = -1/6$. In this problem, the Laurent series method is about as easy as text Method C. Using Method C here with $m = 4$, $m - 1 = 3$, we find

$$\left. \frac{1}{3!}\left(\frac{d^3}{dz^3} \sin z\right) \right|_{z=0} = \frac{1}{3!}(- \cos 0) = -\frac{1}{6} .$$

10. Rule B applies to functions $f(z)$ which have a simple pole at $z = z_0$. Then the Laurent series for $f(z)$ [see text equation (4.1)] has a b_1 term but no other b terms. Multiply text equation (4.1) by $z - z_0$ and find the limit as $z \to z_0$.

$$\lim_{z \to z_0} (z - z_0)f(z)$$

$$= \lim_{z \to z_0} \left[a_0(z - z_0) + a_1(z - z_0)^2 + a_2(z - z_0)^3 + \cdots + b_1 \right] = b_1.$$

14. There are two simple poles; for each we use text Method B. To find the residue at $z = -2/3$, we multiply $f(z)$ by $\left(z + \frac{2}{3}\right)$ and evaluate at $z = -2/3$. [Note carefully: Factor out the 3 to get $3z + 2 = 3\left(z + \frac{2}{3}\right)$ and then cancel $\left(z + \frac{2}{3}\right)$.]

$$R\left(-\frac{2}{3}\right) = \left(z + \frac{2}{3}\right)\frac{1}{3\left(z + \frac{2}{3}\right)(2 - z)}\Bigg|_{z = -\frac{2}{3}} = \frac{1}{3 \cdot \frac{8}{3}} = \frac{1}{8}.$$

To find the residue at $z = 2$, we multiply by $(z - 2)$ and evaluate. [Again caution. Note the minus sign: $(2 - z) = -(z - 2)$.]

$$R(2) = (z - 2)\frac{1}{(3z + 2)(2 - z)}\Bigg|_{z = 2} = -\frac{1}{3z + 2}\Bigg|_{z = 2} = -\frac{1}{8}.$$

18. Simple pole; thus use text method B. We can use either text equation (6.1) or text equation (6.2).

Using (6.1):

$$(z - 3i)\frac{z + 2}{z^2 + 9} = (z - 3i)\frac{z + 2}{(z + 3i)(z - 3i)} = \frac{z + 2}{z + 3i},$$

$$R(3i) = \frac{z + 2}{z + 3i}\Bigg|_{z = 3i} = \frac{3i + 2}{6i} = \frac{1}{2} - \frac{1}{3}i.$$

Using (6.2)

$$R(3i) = \frac{z + 2}{\frac{d}{dz}(z^2 + 9)}\Bigg|_{z = 3i} = \frac{z + 2}{2z}\Bigg|_{z = 3i} = \frac{3i + 2}{6i}$$

$$= \frac{1}{2} - \frac{1}{3}i.$$

21.   Simple pole; use text method B. This problem can be done using
       text equation (6.1), but it is much easier to use text equation
       (6.2). We first verify that for

$$f(z) = \frac{z^2}{z^4 + 16} = \frac{g(z)}{h(z)} ,$$

we have at $z_0 = \sqrt{2}(1+i)$: $g(z_0)$ is finite and not equal to 0,
and $h(z_0) = 0$, $h'(z_0) \neq 0$. Then by text equation (6.2),

$$\frac{g(z)}{dh/dz} = \frac{z^2}{4z^3} = \frac{1}{4z} \qquad \text{so}$$

$$R[\sqrt{2}(1+i)] = \frac{1}{4\sqrt{2}(1+i)} = \frac{1-i}{8\sqrt{2}} \qquad \text{or} \qquad \frac{\sqrt{2}}{16}(1-i).$$

Note that in this problem it is much easier to differentiate the
denominator as we have done than to find the complex factors of
$z^4 + 16$.

31.   Multiple pole; use text method C with $m = 4$, $m - 1 = 3$.

$$R(0) = \frac{1}{3!} \frac{d^3}{dz^3}(e^{3z} - 3z - 1)\bigg|_{z=0} = \frac{1}{3!}(3^3 e^{3z})_{z=0} = \frac{9}{2}.$$

General comment: The problems above can be done by Laurent
series; however, the methods shown are usually easier. Problem
31 is not hard by Laurent series; let's try it that way:

$$\frac{e^{3z} - 3z - 1}{z^4} = \frac{1 + 3z + \frac{9z^2}{2!} + \frac{27z^3}{3!} + \cdots - 3z - 1}{z^4} ,$$

$$\frac{\frac{9}{2}z^2 + \frac{9}{2}z^3 + \cdots}{z^4} = \frac{9}{2z^2} + \frac{9}{2z} + \cdots .$$

31.   (continued)

The residue is the coefficient of $1/z$; $R = 9/2$. Note that the pole is of order 2 (because of the $1/z^2$ term). In our first solution, we multiplied by z^4 which is allowed since $m = 4 \geqslant 2$. Also note the residue of 0 in Problem 35 below although there is a double pole; this means that the Laurent series has a $1/z^2$ term but no $1/z$ term. Note that it is not possible to have a zero residue at a simple pole; it is not a simple pole if there is no $1/z$ term.

35.   Double pole; thus use text method C.

$$\frac{z}{(z^2+1)^2} = \frac{z}{(z+i)^2(z-i)^2}.$$

Multiply by $(z-i)^2$; since $m = 2$, we differentiate $\dfrac{z}{(z+i)^2}$ once and divide by $1!$.

$$\left(\frac{d}{dz}\frac{z}{(z+i)^2}\right)\bigg|_{z=i} = \frac{(z+i)^2 - z\cdot 2(z+i)}{(z+i)^4}\bigg|_{z=i} = \frac{z+i-2z}{(z+i)^3}\bigg|_{z=i} = 0.$$

Thus the residue is 0.

General comment about Problems 14' to 35':

By Cauchy's theorem or the residue theorem, the integral of
$f(z)$ around C, a circle of radius 3/2 with center at the origin,
is $2\pi i$ times the sum of the residues at the singular points
<u>inside</u> C. Thus we ignore any singular points which are farther
than 3/2 away from the origin.

14'. From Problem 14, we have the residues at the two poles. However,
note carefully that $z = 2$ is <u>outside</u> the circle $|z| = 3/2$ and so
does not contribute to the integral. The point $z = -2/3$ is inside
the circle; thus

$$\oint_C \frac{dz}{(3z+2)(2-z)} = 2\pi i R(-2/3) = 2\pi i \cdot \frac{1}{8} = \frac{\pi i}{4} .$$

18'. There are no singularities inside C (the points $\pm 3i$ are at the
distance $3 > 3/2$ from the origin). Thus

$$\oint_C \frac{z+2}{z^2+9} \, dz = 0 .$$

21'. At all of the singularities, we have $z^4 + 16 = 0$, and therefore
$|z| = 2 > 3/2$. Since all singularities are outside C, the integral
around C is zero.

31'. The only singularity is at the origin. From Problem 31 above,
$R(0) = 9/2$. Thus

$$\oint_C \frac{e^{3z} - 3z - 1}{z^4} \, dz = 2\pi i \cdot \frac{9}{2} = 9\pi i .$$

35'.   There are singularities at $z = i$ and $z = -i$, both at distance 1
       from the origin and so inside C.

$\qquad$ $R(i) = 0$ from Problem 35.

$\qquad$ $R(-i) = 0$ similarly, so $\displaystyle\oint_C \frac{z}{(z^2 + 1)^2}\, dz = 0$.

Section 7

1.   This is like text Example 1, Section 7. We let $z = e^{i\theta}$ and
     integrate around the unit circle, C. Then $dz = i\, e^{i\theta}\, d\theta$, or
     $d\theta = \dfrac{dz}{iz}$, and we find

$$\sin\theta = \frac{e^{i\theta} - e^{-i\theta}}{2i} = \frac{z - \dfrac{1}{z}}{2i} = \frac{z^2 - 1}{2iz},$$

$$13 + 5\sin\theta = 13 + \frac{5(z^2 - 1)}{2iz} = \frac{5z^2 + 26iz - 5}{2iz} = \frac{(5z + i)(z + 5i)}{2iz}$$

$$= \frac{5}{2iz}\left(z + \frac{i}{5}\right)(z + 5i),$$

$$\int_0^{2\pi} \frac{d\theta}{13 + 5\sin\theta} = \oint_C \frac{\dfrac{dz}{iz}}{\dfrac{5}{2iz}\left(z + \dfrac{i}{5}\right)(z + 5i)} = \frac{2}{5}\oint_C \frac{dz}{\left(z + \dfrac{i}{5}\right)(z + 5i)}.$$

The point $z = -i/5$ is inside the unit circle; $z = -5i$ is outside.
Thus we need the residue only at $z = -i/5$.

$$R\left(-\frac{i}{5}\right) = \frac{1}{z + 5i}\bigg|_{z = -i/5} = \frac{1}{-\dfrac{i}{5} + 5i} = \frac{5}{24i} = -\frac{5}{24}\,i.$$

Then the value of the integral is

$$2\pi i\left(\frac{2}{5}\right)\left(-\frac{5}{24}\,i\right) = \frac{\pi}{6}.$$

7.   Hint: If $z = e^{i\theta}$, then $\cos 2\theta = \frac{1}{2}(e^{2i\theta} + e^{-2i\theta}) = \frac{1}{2}(z^2 + z^{-2})$.

Also write $\cos \theta$ and $d\theta$ in terms of z and dz as in text Example 1.

13.
$$\int_0^\infty \frac{x^2 dx}{(x^2+4)(x^2+9)} = \frac{1}{2}\int_{-\infty}^\infty \frac{x^2 dx}{(x^2+4)(x^2+9)}$$

because the integrand is an even function. (See text page 322.)
The problem is now an example of $\int_{-\infty}^\infty \frac{P(x)}{Q(x)} dx$ discussed in text
Example 2. We integrate around the countour C of text Figure 7.2.
The degree of $Q(x)$ is 4 and the degree of $P(x)$ is 2; thus the
integral around the semicircle tends to zero as $\rho \to \infty$. $Q(x)$ has
no zeros on the x axis. Then the desired integral along the x
axis is equal to $2\pi i$ times the sum of the residues of
$\frac{z^2}{(z^2+4)(z^2+9)}$ in the upper half plane, that is, at $z = 2i$
and $z = 3i$.

$$R(2i) = \frac{z^2}{\frac{d}{dz}(z^4 + 13z^2 + 36)}\bigg|_{z=2i} = \frac{z^2}{4z^3 + 26z}\bigg|_{2i}$$

$$= \frac{z}{4z^2 + 26}\bigg|_{2i} = \frac{2i}{10} = \frac{i}{5} .$$

$$R(3i) = \frac{z}{4z^2 + 26}\bigg|_{z=3i} = \frac{3i}{-10} = -\frac{3}{10} i.$$

Then

$$\int_0^\infty \frac{x^2 dx}{(x^2+4)(x^2+9)} = \frac{1}{2}\int_{-\infty}^\infty \frac{x^2 dx}{(x^2+4)(x^2+9)} = \frac{1}{2} \cdot 2\pi i\left(\frac{i}{5} - \frac{3i}{10}\right) = \frac{\pi}{10} .$$

17.   This is similar to text Example 3. We consider

$$\oint_C \frac{ze^{iz}dz}{z^2 + 4z + 5} ,$$

where C is the contour of text Figure 7.2. We verify that the requirements listed in the text (and of Example 3) are met. Then

$$\int_{-\infty}^{\infty} \frac{xe^{ix}dx}{x^2 + 4x + 5} = 2\pi i \cdot \text{sum of residues of } \frac{ze^{iz}}{z^2 + 4z + 5}$$

in the upper half plane. The singular points are the roots of

$$z^2 + 4z + 5 = 0, \qquad z = \frac{-4 \pm \sqrt{16 - 20}}{2} = -2 \pm i.$$

The only singular point in the upper half plane is at $z = -2 + i$. The residue there is

$$\left. \frac{ze^{iz}}{\frac{d}{dz}(z^2 + 4z + 5)} \right|_{z=-2+i} = \frac{(-2 + i)e^{i(-2+i)}}{2(-2 + i) + 4} = \frac{i - 2}{2i} e^{-1-2i}$$

$$= \tfrac{1}{2}(1 + 2i)e^{-1}(\cos 2 - i \sin 2).$$

Then

$$\int_{-\infty}^{\infty} \frac{x \sin x \, dx}{x^2 + 4x + 5} = \text{Im} \int_{-\infty}^{\infty} \frac{xe^{ix}dx}{x^2 + 4x + 5} =$$

$$= \text{Im}\left[2\pi i \cdot \tfrac{1}{2}(1 + 2i)e^{-1}(\cos 2 - i \sin 2) \right]$$

$$= \frac{\pi}{e}(\cos 2 + 2 \sin 2).$$

21.   (a) We evaluate $\oint_N \frac{f(z)}{z}\,dz$ around the closed contour of text
Figure 7.3. In this problem we shall use N to denote this
closed indented contour (consisting of C, C' and parts of the
real axis) in order to distinguish it from Γ which goes straight
through the pole (see figures). We take the radius r of C'

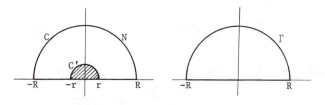

small enough so that there are no singularities of $\frac{f(z)}{z}$ in the
shaded area shown (see figure) except at $z = 0$. Thus the interior
singular points are the same for N and Γ. Along C' we have

$$z = re^{i\theta}, \qquad dz = re^{i\theta}i\,d\theta, \qquad \frac{dz}{z} = i\,d\theta.$$

As $r \to 0$, $z \to 0$, and since $f(z)$ is analytic at $z = 0$, $f(z) \to f(0)$.
Thus

$$\int_{C'} \frac{f(z)}{z}\,dz = \int_{C'} f(z)i\,d\theta \longrightarrow \int_{\pi}^{0} f(0)i\,d\theta = -i\pi f(0).$$

Then the integral around the closed contour N becomes

$$\oint_N \frac{f(z)}{z}\,dz = \int_C \frac{f(z)}{z}\,dz + PV\int_{-R}^{R} \frac{f(x)\,dx}{x} - i\pi f(0).$$

Now the integral around the contour Γ (see figure) is the
integral along C plus the integral from -R to R, so

$$PV\oint_\Gamma \frac{f(z)}{z}\,dz = \oint_N \frac{f(z)}{z}\,dz + i\pi f(0).$$

21. (continued)

We can write the result in terms of residues. By the residue
theorem, the integral along N is $2\pi i$ times the sum of the
residues of $\dfrac{f(z)}{z}$ at interior singular points. Also $f(0)$ is
the residue of $\dfrac{f(z)}{z}$ at $z = 0$. Thus

$$PV \oint_{\Gamma} \frac{f(z)}{z}\, dz = 2\pi i (\textstyle\sum \text{residues at interior points})$$
$$+ i\pi (\text{residue at simple pole on boundary}).$$

(b) The proof is almost identical if the simple pole on the
boundary is at $(a,0)$. Indent
the contour N as shown. Then
along C'

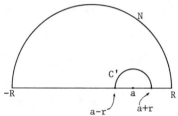

$$z = a + re^{i\theta}, \quad dz = re^{i\theta} i\, d\theta,$$
$$\frac{dz}{z - a} = i\, d\theta.$$

The rest of the proof continues
as above.

In general, consider $\oint F(z)\,dz$ around a closed contour. Suppose that $F(z)$ has several simple poles at points on the boundary
where the contour curve has a tangent; call the residues at these
points R_n'. Also suppose there are residues R_n at interior points.
Then the (PV) integral of $F(z)$ in the counterclockwise direction
around the contour is

$$PV \oint F(z)\,dz = 2\pi i \sum R_n + i\pi \sum R_n' .$$

The theorem can easily be extended to include points on the boundary where the contour turns a corner at a simple pole. You might

21. (continued)

like to show that in this case the contribution to the integral
is (iθ) times the residue, where θ is the interior angle (that
is inside the contour) between the tangents to the contour
curve on the two sides. We have considered the case $\theta = \pi$ above.

24. As in text examples 3 and 4, we consider

$$\oint_C \frac{ze^{i\pi z}}{1 - z^2} \, dz$$

around the upper half plane (text Figure 7.2). The poles at
$z = \pm 1$ are on the contour, so what we will find is the principal
value (PV) of this integral (see text Example 4). As in text
Example 4 and Problem 21 above, the principal value of the
integral is $2\pi i \cdot \frac{1}{2}$(sum of residues at simple poles on the boundary). We find by text equation (6.2):

$$R(-1) = \frac{ze^{i\pi z}}{-2z}\bigg|_{z=-1} = \frac{-e^{-i\pi}}{2} = \frac{1}{2} \, ,$$

$$R(1) = \frac{ze^{i\pi z}}{-2z}\bigg|_{z=1} = \frac{e^{i\pi}}{-2} = \frac{1}{2} \, .$$

Then

$$PV \oint_C \frac{ze^{i\pi z}}{1 - z^2} \, dz = 2\pi i \cdot \frac{1}{2}\left(\frac{1}{2}+\frac{1}{2}\right) = \pi i.$$

As in the text examples (see the theorem stated in the text at
the end of Example 3), the integral around the large semicircle
(text Figure 7.2) tends to zero as the radius tends to infinity.
Thus we are left with the integral along the x axis where $y = 0$
so $z = x$:

24.   (continued)

$$PV \int_{-\infty}^{\infty} \frac{xe^{i\pi x}}{1 - x^2} dx = \pi i.$$

The imaginary part of this equation gives the desired result:

$$PV \int_{-\infty}^{\infty} \frac{x \sin \pi x}{1 - x^2} dx = \pi.$$

26.   As in text examples 3 and 4 we integrate around the upper half plane.   By text page 605,

$$PV \int_{-\infty}^{\infty} \frac{x \, dx}{(x - 1)^4 - 1} = 2\pi i (\text{sum of residues in upper half plane})$$
$$+ i\pi (\text{sum of residues on real axis}).$$

The denominator has zeros at

$$(z - 1)^4 = 1, \qquad z - 1 = \sqrt[4]{1} = -1, \ 1, \ i, \ -i,$$

$$z = 0, \ 2, \ 1 + i, \ 1 - i.$$

The point $z = 0$ is not a pole since the factor z in the numerator cancels the factor $(z - 0)$ in the denominator.   The point $1 - i$ is not in the upper half plane.   We find the residues at $z = 2$ and $z = 1 + i$ using text equation (6.2).

$$R(2) = \frac{z}{4(z - 1)^3} \bigg|_{z=2} = \frac{2}{4} = \frac{1}{2},$$

$$R(1 + i) = \frac{z}{4(z - 1)^3} \bigg|_{z=1+i} = \frac{1 + i}{4i^3} = \frac{1}{4}(i - 1).$$

Thus we find

$$PV \int_{-\infty}^{\infty} \frac{x \, dx}{(x - 1)^4 - 1} = 2\pi i \cdot \frac{1}{4}(i - 1) + i\pi \left(\frac{1}{2}\right) = -\frac{\pi}{2}.$$

33.   We evaluate

$$\oint_C \frac{z^{1/2}\,dz}{1+z^2}$$

around the keyhole contour of text Figure 7.4. Inside C, θ goes
from 0 to 2π, so $\sqrt{z}$ is a single-valued function. By text equation (6.2), the residues of the integrand are the values of

$$\frac{z^{1/2}}{\frac{d}{dz}(1+z^2)} = \frac{z^{1/2}}{2z} = \frac{1}{2}z^{-1/2}$$

at the two poles $z = i = e^{i\pi/2}$ and $z = -i = e^{3\pi i/2}$. We have

$$R(i) = \frac{1}{2}\left(e^{i\pi/2}\right)^{-1/2} = \frac{1}{2}e^{-i\pi/4}.$$

$$R(-i) = \frac{1}{2}\left(e^{3\pi i/2}\right)^{-1/2} = \frac{1}{2}e^{-3\pi i/4}.$$

Then by the residue theorem the integral is

$$\oint_C \frac{z^{1/2}\,dz}{1+z^2} = 2\pi i \cdot \frac{1}{2}(e^{-i\pi/4} + e^{-3\pi i/4})$$

(A) $$= i\pi\left(\frac{1-i}{\sqrt{2}} + \frac{-1-i}{\sqrt{2}}\right) = \pi\sqrt{2}.$$

Now let the radius of the large circle tend to infinity and the
radius of the small circle tend to zero. Along either circle
$z = re^{i\theta}$, $dz = rie^{i\theta}\,d\theta$ so the integral is

$$\int_0^{2\pi} \frac{r^{1/2}e^{i\theta/2}rie^{i\theta}\,d\theta}{1+r^2e^{2i\theta}}.$$

33.   (continued)

As $r \to 0$, the denominator $\to 1$ and the numerator $\to 0$; as $r \to \infty$, the denominator becomes approximately $\left(re^{i\theta}\right)^2$ and the numerator is $i\left(re^{i\theta}\right)^{3/2}$. Thus the integrals along both circles tend to zero and we are left with the two integrals along the positive x axis. Along AB (text Figure 7.4), $\theta = 0$, so $z = r$; along DE, $\theta = 2\pi$, so $z = re^{2\pi i}$. Then these integrals are

$$\int_0^\infty \frac{r^{1/2}\,dr}{1+r^2} + \int_\infty^0 \frac{r^{1/2}e^{\pi i}\,dr}{1+r^2} = 2\int_0^\infty \frac{r^{1/2}\,dr}{1+r^2} = \pi\sqrt{2}$$

by (A) above.   Thus

$$\int_0^\infty \frac{r^{1/2}\,dr}{1+r^2} = \frac{1}{2}\,\pi\sqrt{2} = \int_0^\infty \frac{\sqrt{x}\,dx}{1+x^2}\;.$$

Comment: Another method of evaluating this integral is to let $z^2 = u$. Then the integral is a beta function [text page 463, equation (6.5)], that is, it is a special case of the integral of Example 5, text page 607. However, our purpose here is to learn to find integrals by contour integration.

40. We evaluate $\oint_C \dfrac{z\,dz}{\sinh z}$ around
the contour C shown in the
diagram, and let $a \to \infty$.
Since $\sinh z = 0$ for $z = n\pi i$,
there are no poles inside
the contour. At the pole $z = i\pi$ on the contour, we have by text
page 599, equation (6.2),

$$R(i\pi) = \frac{z}{\dfrac{d}{dz}(\sinh z)}\Bigg|_{z=i\pi} = \frac{i\pi}{\cosh i\pi} = \frac{i\pi}{\cos \pi} = -i\pi.$$

By the theorem stated on text page 605,

(A) $$\oint_C \frac{z\,dz}{\sinh z} = 2\pi i \cdot \frac{1}{2}(-i\pi) = \pi^2.$$

Write the contour integral as a sum of the integrals along the
four sides of the rectangle, carefully putting in the value of
z along each side:

Lower edge:	$z = x$,		x: $-a$ to a,
Right side:	$z = a + iy$,	$dz = i\,dy$,	y: 0 to π,
Top:	$z = x + i\pi$,	$dz = dx$,	x: a to $-a$,
Left side:	$z = -a + iy$,	$dz = i\,dy$,	y: π to 0.

$$\oint_C \frac{z\,dz}{\sinh z} = \int_{-a}^{a} \frac{x\,dx}{\sinh x} + \int_0^\pi \frac{(a+iy)(i\,dy)}{\sinh(a+iy)}$$

$$+ \int_a^{-a} \frac{(x+i\pi)\,dx}{\sinh(x+i\pi)} + \int_\pi^0 \frac{(-a+iy)(i\,dy)}{\sinh(-a+iy)}\ .$$

40.   (continued)

We will show below that the y integrals tend to 0 as $a \to \infty$.

Also $\sinh(x + i\pi) = - \sinh x$ (verify this) so the imaginary

part of the integral from a to -a is zero (integral of an

odd function over a symmetric interval -- see text, page 322).

Thus we have

$$\oint_C \frac{z \, dz}{\sinh z} = \int_{-\infty}^{\infty} \frac{x \, dx}{\sinh x} + \int_{\infty}^{-\infty} \frac{x \, dx}{- \sinh x} = 2 \int_{-\infty}^{\infty} \frac{x \, dx}{\sinh x} = \pi^2$$

by (A) above.   Then

$$\int_{0}^{\infty} \frac{x \, dx}{\sinh x} = \frac{1}{2} \int_{-\infty}^{\infty} \frac{x \, dx}{\sinh x} = \frac{\pi^2}{4}.$$

To show that the y integrals tend to zero, write

$$\sinh(\pm a + iy) = \frac{1}{2}\left(e^{\pm a + iy} - e^{-(\pm a + iy)}\right).$$

As $a \to \infty$ or $a \to -\infty$, one of the two exponentials tends to infinity

so the integrand tends to zero and the integral (over the finite

interval π) tends to zero.

41.   Let $x = u^2$, $dx = 2u\,du$, $du = \frac{1}{2}x^{-1/2}dx$; then

$$\int_0^\infty \sin u^2\,du = \frac{1}{2}\int_0^\infty x^{-1/2}\sin x\,dx,$$

(A)

$$\int_0^\infty \cos u^2\,du = \frac{1}{2}\int_0^\infty x^{-1/2}\cos x\,dx.$$

We evaluate $\oint_C z^{-1/2}e^{iz}\,dz$ around the contour shown in the text. We show below that the integrals along the quarter circles tend to zero as $r \to 0$ and $R \to \infty$. There are no poles inside the contour. Thus the integrals along the x axis and the y axis must add to zero. We choose the branch of $\sqrt{z}$ determined by $0 < \theta < \pi/2$; then along the x axis, $z = x$, and along the y axis, $z = re^{i\pi/2} = ri$. Thus

$$\int_0^\infty x^{-1/2}e^{ix}\,dx + \int_\infty^0 \left(re^{i\pi/2}\right)^{-1/2} e^{i(ri)}i\,dr = 0,$$

so

$$\int_0^\infty x^{-1/2}e^{ix}\,dx = ie^{-i\pi/4}\int_0^\infty r^{-1/2}e^{-r}\,dr = ie^{-i\pi/4}\Gamma\!\left(\frac{1}{2}\right)$$

(B)

$$= i\left(\frac{1-i}{\sqrt{2}}\right)\sqrt{\pi} = \frac{1}{2}(1+i)\sqrt{2\pi}.$$

(For the Γ function, see text, page 461.) Now take real and imaginary parts of (B) and use (A) to get

$$\int_0^\infty \sin u^2\,du = \int_0^\infty \cos u^2\,du = \frac{1}{4}\sqrt{2\pi}.$$

41.   (continued)

We next verify that the integrals along the two quarter circles
tend to zero. Along either circle the integral is

$$\int z^{-1/2} e^{iz}\, dz = \int \left(re^{i\theta}\right)^{-1/2} e^{iz} i r e^{i\theta}\, d\theta.$$

Since $\left|e^{i\theta}\right| = 1$, $|i| = 1$, and $\left|e^{iz}\right| = \left|e^{ir(\cos\theta + i\sin\theta)}\right| = e^{-r\sin\theta}$,
then the absolute value of the integrand is

(C) $r^{1/2} e^{-r\sin\theta}$.

For the small quarter circle this tends to zero as $r \to 0$, so the
integral tends to zero. For the large quarter circle we want
$r \to \infty$. Now the magnitude of the integrand given in (C) does
tend to zero for all θ except $\theta = 0$. Thus the integral from some
arbitrarily small θ up to $\theta = \pi/2$ tends to zero and we only need
consider the integral from $\theta = 0$ to, say, $\theta = \varepsilon > 0$. In this range,
$\cos\theta \cong 1$, so let's multiply (C) by $\cos\theta$ in order to integrate it.
More carefully, since $\cos\varepsilon < \cos\theta$, then $\frac{\cos\theta}{\cos\varepsilon} > 1$; if we multiply
(C) by $\frac{\cos\theta}{\cos\varepsilon}$ and integrate, we get something larger than the
original integral. If this larger result $\to 0$ as $r \to \infty$, then so
does the original integral. We find

$$r^{1/2} \int_0^\varepsilon e^{-r\sin\theta}\, d\theta \leqslant \frac{r^{1/2}}{\cos\varepsilon} \int_0^\varepsilon e^{-r\sin\theta} \cos\theta\, d\theta$$

$$= \frac{r^{1/2}}{\cos\varepsilon}\left(-\frac{1}{r} e^{-r\sin\theta}\right)\Bigg|_0^\varepsilon = r^{-1/2}\, \frac{1 - e^{-r\sin\varepsilon}}{\cos\varepsilon}$$

which tends to zero as $r \to \infty$.

44.   We want to find the number of zeros of the polynomial f(z)
      anywhere in the complex plane by using the argument principle.
      We need to find the increase in the angle θ of f(z) when z
      traverses a circle large enough to inclose all zeros.   For
      $a_n \neq 0$, write

$$f(z) = Re^{i\theta} = a_n z^n + a_{n-1} z^{n-1} + a_{n-2} z^{n-2} + \cdots + a_1 z + a_0$$

$$= a_n z^n \left(1 + \frac{a_{n-1}}{a_n} z^{-1} + \frac{a_{n-2}}{a_n} z^{-2} + \cdots + \frac{a_1}{a_n} z^{-n+1} + \frac{a_0}{a_n} z^{-n} \right)$$

$$= a_n z^n [1 + q(z)].$$

If $z = re^{i\theta}$ with r very large, then

$$f(z) = Re^{i\theta} = a_n r^n e^{in\theta} [1 + q(re^{i\theta})] \cong a_n r^n e^{in\theta}$$

since $|q(z)|$ is small for large r (because of the z^{-1} factors).
As θ increases from 0 to 2π, the angle of $e^{in\theta}$ increases from
0 to $2\pi n$.   By the argument principle [text equation (7.8)], the
number of zeros of f(z) is

$$\frac{1}{2\pi} \cdot 2\pi n = n.$$

Note: We have assumed that there is no change in the angle of
$[1 + q(re^{i\theta})]$; let's see why this is correct.   Since $|q(z)|$ is
small, the complex number $[1 + q(z)]$ stays close to the point 1
in the complex plane and consequently cannot follow a path
which encircles the origin.

46.   The equation

(1) $f(z) = z^3 + 3z^2 + 4z + 2 = 0$

has no positive real roots since all terms are positive. On the
imaginary axis $z = iy$; then

(2) $f(z) = -iy^3 - 3y^2 + 4iy + 2 \neq 0$

(remember that y is real). We now let $z = re^{i\theta}$ go around the first
quadrant (OPQO in the figure) and consider
the change in the angle Θ of $f(z) = Re^{i\Theta}$.

   Along OP: $\theta = 0$, $\Theta = 0$, no change in Θ.

   Along PQ: θ goes from 0 to $\pi/2$, so

                Θ goes from 0 to approximately $3\pi/2$
since $f(z) \cong z^3 = r^3 e^{3i\theta}$ when r is very large.

   Along QO, $f(z)$ is given by equation (2) above; the tangent of
the angle of any complex number is its imaginary part divided by
its real part so

(3) $\tan \Theta = \dfrac{Im\ f(z)}{Re\ f(z)} = \dfrac{4y - y^3}{2 - 3y^2} = \dfrac{y(4 - y^2)}{2 - 3y^2}$.

We now follow $z = iy$ down the y axis and outline what is happening
to $\tan \Theta$ [see equation (3)] and Θ. We use the notation $\sim \infty$ to
mean very large and positive, and $\sim -\infty$ to mean very large magnitude but negative. Remember that we had $\Theta \sim 3\pi/2$ at Q.

46.   (continued)

y	sign of tan θ	tan θ	quadrant of θ	θ
$\sim \infty$ at Q	+	$\sim \dfrac{-y^3}{-3y^2} = \dfrac{y}{3} \sim \infty$	III	$\sim 3\pi/2$
$\sim \infty$ to 2	+	$\sim \infty$ to 0	III	$3\pi/2$ to π
2 to $\sqrt{2/3}$	-	0 to $-\infty$	II	π to $\pi/2$
$\sqrt{2/3}$ to 0	+	∞ to 0	I	$\pi/2$ to 0

The figure shows schematically
the path of the complex number
$f(z) = Re^{i\theta}$. The total change
in the angle θ is zero, so by
the argument principle there are
no roots in the first quadrant.
Since roots of an equation with
real coefficients come in complex conjugate pairs, there are
also no roots in the fourth

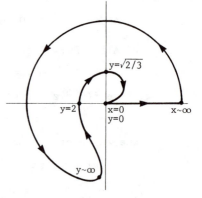

quadrant. Thus the three roots must be either all negative real,
or else one negative real and one in each of quadrants II and
III. Let's find out how many real roots there are. Think of
the graph of f(x); a cubic looks like this $\bigwedge$ or this $\diagup$
If there are three real roots (that is, the graph crosses the

46.   (continued)

x axis three times) then $f(x)$ must have a maximum with $f(x) > 0$
and a minimum with $f(x) < 0$.   We find x for $f'(x) = 0$:

$$f'(x) = 3x^2 + 6x + 4, \qquad x = \frac{-6 \pm \sqrt{36 - 48}}{6}.$$

Since x comes out complex, $f(x)$ does not have a maximum or a
minimum; the graph crosses the x axis only once.   Thus there
is one negative real root, one in quadrant II and one in
quadrant III.

Comment: You may have discovered that one root of the equation
is $z = -1$; thus you can easily find that the other two are $-1 \pm i$
(see Appendix B to these Solutions).   However, most equations
do not have rational roots!

48.   The equation $f(z) = z^4 - z^3 + 6z^2 - 3z + 5 = 0$ has no real positive
roots because for $0 < x < 1$,

$$x^4 + 6x^2 + (5 - x^3 - 3x) > 0$$

and for $x > 1$

$$x^3(x - 1) + 3x(2x - 1) + 5 > 0.$$

On the y axis, set $z = iy$; then

(A) $f(z) = y^4 + iy^3 - 6y^2 - 3iy + 5 \neq 0$

since the imaginary part and the real part of $f(z)$ are not zero
for the same y (check this).   We follow z around the first quad-

48.   (continued)

rant as in Problem 46 and find the change in the angle θ of $f(z) = Re^{i\theta}$ (see the first diagram in the Problem 46 Solution).

   Along OP: $\theta = 0$, $\theta = 0$, no change in θ.

   Along PQ: θ goes from 0 to $\pi/2$, so

               θ goes from 0 to approximately $4 \cdot \frac{\pi}{2} = 2\pi$

since $f(z) \cong z^4 = r^4 e^{4i\theta}$ for large r.

   Along QO, $f(z) = Re^{i\theta}$ is given by (A) above. As in Problem 46, we find

(B) $\tan \theta = \dfrac{\text{Im } f(z)}{\text{Re } f(z)} = \dfrac{y^3 - 3y}{y^4 - 6y^2 + 5} = \dfrac{y(y^2 - 3)}{(y^2 - 1)(y^2 - 5)}$.

Now we outline the change in $\tan \theta$ and in θ as we follow z down the y axis. For $y \backsim \infty$, $\tan \theta$ in (B) is positive so we are in the first quadrant with θ a little over 2π (see the $y \backsim \infty$ point in the diagram below).

y	sign of $\tan \theta$	$\tan \theta$	quadrant of θ	θ
$\backsim \infty$	+	$\backsim y^3/y^4 \backsim 0$	I	$\backsim 2\pi$
$\backsim \infty$ to $\sqrt{5}$	+	0 to ∞	I	$2\pi + \frac{\pi}{2}$
$\sqrt{5}$ to $\sqrt{3}$	-	$-\infty$ to 0	II	3π
$\sqrt{3}$ to 1	+	0 to ∞	III	$3\pi + \frac{\pi}{2}$
1 to 0	-	$-\infty$ to 0	IV	4π

48.   (continued)

The diagram indicates schematically the change in Θ. Since Θ increases by 4π, then by the argument principle [text equation (7.8)], the number of roots in the first quadrant is $\frac{1}{2\pi} \cdot 4\pi = 2$. Since the roots occur in conjugate pairs, the other two roots are in quadrant IV.

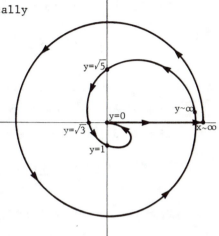

51.   The function

$$f(z) = \frac{z^3(z+1)^2 \sin z}{(z^2+1)^2(z-3)}$$

has a simple pole at $z = 3$, double poles at $z = i$ and $z = -i$, a zero of order 2 at $z = -1$, and a zero of order 4 at $z = 0$ (recall that $\sin z \cong z$ for z near zero). Inside $|z| = 2$, the number of zeros is $N = 6$, and the number of poles is $P = 4$. (Remember that a double pole counts as two poles, and so on.) Thus, by text equation (7.8), the integral around $|z| = 2$ is

$$\oint_C \frac{f'(z)}{f(z)}\,dz = 2\pi i(N - P) = 2\pi i(6 - 4) = 4\pi i.$$

Similarly for the integral around $|z| = 1/2$, find

$$2\pi i(N - P) = 2\pi i(4 - 0) = 8\pi i.$$

52.   Hint: Let $f(z) = 1 + 2z^4$. What is $f'(z)$?

Section 8

7. Replace z by 1/z to get

$$f\left(\frac{1}{z}\right) = \frac{4z^{-3} + 2z^{-1} + 3}{z^{-2}} = \frac{4 + 2z^2 + 3z^3}{z}.$$

Since $f(1/z)$ has a simple pole at $z = 0$, we say that $f(z)$ has a simple pole at $z = \infty$. To find the residue at $z = \infty$ [text equation (8.3)] we find the residue at $z = 0$ of

$$-\frac{1}{z^2} f\left(\frac{1}{z}\right) = -\frac{4 + 2z^2 + 3z^3}{z^3} = -\frac{4}{z^3} - \frac{2}{z} - 3.$$

From the coefficient of $1/z$ we have $R = -2$.

Comment: Note carefully that we must consider two different functions at $z = 0$. To decide the behavior of $f(z)$ at infinity we look at $f(1/z)$ at $z = 0$, but to find the residue of $f(z)$ at infinity we find the residue of $-\frac{1}{z^2} f\left(\frac{1}{z}\right)$ at $z = 0$.

We can also find the residue at infinity directly from $f(z)$. We have

$$f(z) = \frac{4z^3 + 2z + 3}{z^2} = 4z + \frac{2}{z} + \frac{3}{z^2}.$$

Then by Problem 1, the residue at infinity is the coefficient of $1/z$ multiplied by -1. Thus $R(\infty) = -2$, as above.

10.   Replace z by 1/z to get

$$f\left(\frac{1}{z}\right) = \frac{1 + z^{-1}}{1 - z^{-1}} = \frac{z + 1}{z - 1} \,.$$

Since $f(1/z)$ is analytic at $z = 0$, we say that $f(z)$ is analytic at infinity. To find the residue of $f(z)$ at $z = \infty$, we find the residue at $z = 0$ of

(1) $-\dfrac{1}{z^2} f\left(\dfrac{1}{z}\right) = -\dfrac{1}{z^2} \dfrac{z + 1}{z - 1} = \dfrac{1}{z^2} \dfrac{1 + z}{1 - z} \,.$

Using method C, text page 600, we multiply (1) by z^2, differentiate once and evaluate the result at $z = 0$.

$$\frac{d}{dz} \frac{1 + z}{1 - z}\bigg|_{z=0} = \frac{1 - z - (1 + z)(-1)}{(1 - z)^2}\bigg|_{z=0} = 2 = R(\infty) \text{ for } f(z).$$

Alternatively we can expand (1) in the Laurent series which converges near $z = 0$,

$$\frac{1}{z^2} \frac{1 + z}{1 - z} = \frac{1}{z^2}(1 + z)(1 - z)^{-1} = \frac{1}{z^2}(1 + z)(1 + z + z^2 + \cdots)$$

$$= \frac{1}{z^2} + \frac{2}{z} + \cdots,$$

so $R = $ (coefficient of $\frac{1}{z}$) $= 2$.

It is not necessary to find $f\left(\dfrac{1}{z}\right)$; we can find $R(\infty)$ directly from $f(z)$. We show two methods outlined in Problems 1 and 2(a). By Problem 1, we expand $f(z)$ in a Laurent series of powers of $\dfrac{1}{z}$.

10.   (continued)

$$f(z) = \frac{1+z}{1-z} = \frac{\frac{1}{z}+1}{\frac{1}{z}-1} = -\left(1+\frac{1}{z}\right)\left(1-\frac{1}{z}\right)^{-1} = -\left(1+\frac{1}{z}\right)\left(1+\frac{1}{z}+\cdots\right)$$

$$= -\left(1+\frac{2}{z}+\cdots\right).$$

We find $b_1 = $ (coefficient of $\frac{1}{z}$) $= -2$, and by Problem 1,
$R(\infty) = -b_1 = 2$. Alternatively, using Problem 2(a), we see that
$f(z) \to 1$ as $z \to \infty$, so we compute

$$R(\infty) = \lim_{z \to \infty} z^2 f'(z) = \lim_{z \to \infty} z^2 \frac{2}{(1-z)^2} = 2.$$

15.   The integral in the positive direction (counterclockwise) around
$|z| = 5$ is the negative of the integral "around ∞" (clockwise,
see text page 615). Thus we evaluate $2\pi i R(\infty)$ and take its
negative.

$$f(z) = \frac{z^2}{(2z+1)(z^2+9)} \longrightarrow 0 \text{ as } z \longrightarrow \infty$$

so by Problem 2(b),

$$R(\infty) = -\lim_{z \to \infty} z f(z) = -\lim_{z \to \infty} \frac{z^3}{(2z+1)(z^2+9)} = -\frac{1}{2}.$$

Then the integral "around ∞" is $(2\pi i)\left(-\frac{1}{2}\right) = -i\pi$, so the integral
in the positive direction $= i\pi$.

   You might compare the work required to evaluate this integral
if we compute residues at the three finite poles instead of at ∞.
You should find

$$R\left(-\frac{1}{2}\right) = \frac{1}{74}, \qquad R(3i) = \frac{18+3i}{74}, \qquad R(-3i) = \frac{18-3i}{74},$$

so the integral is

$$2\pi i(\text{sum of residues}) = 2\pi i\left(\frac{37}{74}\right) = \pi i.$$

16. In Problem 15 we found

$$R(\infty) = -\frac{1}{2}, \quad \text{sum of residues at finite points} = \frac{37}{74} = \frac{1}{2},$$

and we see that the sum of these is zero. To show this in
general for any function $f(z)$ with a finite number of singular
points, we integrate $f(z)$ around a circle large enough to
inclose all the singular points. If we integrate in the counterclockwise direction, we get $2\pi i$(sum of residues at finite points).
If we integrate in the clockwise direction, we get $2\pi i R(\infty)$.
Since these two integrals are along the same path but in opposite
directions, their sum is zero, so

$$2\pi i[\text{sum of residues at finite points} + R(\infty)] = 0.$$

Section 9

3. Given $w = u + iv = \dfrac{1}{z} = \dfrac{\bar{z}}{z\bar{z}} = \dfrac{x - iy}{x^2 + y^2}$, we have

(1) $u = \dfrac{x}{x^2 + y^2}, \qquad v = \dfrac{-y}{x^2 + y^2}.$

If $u = 0$, then $x = 0$, so the v axis ($u = 0$) maps onto the y axis
($x = 0$). Similarly $v = 0$ (u axis) maps onto $y = 0$ (x axis). If
$u = c \neq 0$, then from (1)

(2) $x^2 + y^2 = \dfrac{x}{c} \qquad \text{or} \qquad \left(x - \dfrac{1}{2c}\right)^2 + y^2 = \dfrac{1}{4c^2}.$

Similarly, if $v = k \neq 0$, then from (1)

(3) $x^2 + y^2 = -\dfrac{y}{k} \qquad \text{or} \qquad x^2 + \left(y + \dfrac{1}{2k}\right)^2 = \dfrac{1}{4k^2}.$

3.   (continued)

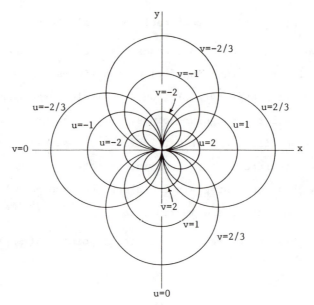

Thus the coordinate grid u = const., v = const., in the w plane
maps onto the axes and two orthogonal sets of circles as shown
in the diagram. However, in order to use the mapping we must
consider this more carefully. Study the diagram to see that
large values of |u| and |v| correspond to small values of |x|
and |y| and vice versa. (This is also clear if we take abso-
lute values in the mapping equation to get |w| = 1/|z|. We
know that the v axis maps to the y axis, but observe from the
diagram the following correspondence:

$y = -\infty$	$-\frac{3}{2}$	-1	$-\frac{1}{2}$	0	$\frac{1}{2}$	1	$\frac{3}{2}$	∞
$v = 0$	$\frac{2}{3}$	1	2	$\pm\infty$	-2	-1	$-\frac{2}{3}$	0

3.   (continued)

You might like to consider similarly the mapping from the u axis
to the x axis. It is also instructive to follow around one of
the circles. Let's start at the origin and go around $v = -\frac{2}{3}$ in
the counterclockwise direction. In the w plane we are following
the line $v = -\frac{2}{3}$; let us observe the values of u. We start at
$u = \infty$ and then go to $u = 2, 1, \frac{2}{3}, 0$ (this is $u = 0$ on the y axis which
is $x = 0$), $-\frac{2}{3}, -1, -2, -\infty$; note the negative direction in the w
plane. It is of interest to consider the mapping of regions
as well as their boundaries. For example, the region between
the circles $u = 1$ and $u = 2$ corresponds to the vertical strip in
the w plane between the lines $u = 1$ and $u = 2$. However, the whole
area inside the circle $u = 1$ corresponds to the infinite w plane
area $u \geqslant 1$. For any applied problem using this mapping (see, for
example, Problem 10.6 below), we must consider an appropriate
mapping of lines and regions.

4.   Given $w = e^{z}$, we find

$$w = u + iv = e^{x+iy} = e^{x}(\cos y + i \sin y),$$

(1) $u = e^{x} \cos y,$ $v = e^{x} \sin y.$

From (1), observe that if $0 < y < \pi/2$, and x takes all values,
then u and v take all positive values. Thus the horizontal
strip $0 < y < \pi/2$ in the z plane is the image of the first quadrant of the w plane. If $\pi/2 < y < \pi$, and x takes all values,
then from (1), u takes all negative values and v takes all
positive values; this is the second quadrant in the w plane.

4.   (continued)

Continue this to convince yourself that horizontal strips of width $\pi/2$ in the z plane are images of quadrants in the w plane where the quadrant is determined by y. Thus a whole w plane maps into a horizontal strip in the z plane of width 2π. We need another w plane to map into the next strip and so on. Then in the w plane we have a Riemann surface consisting of an infinite number of sheets. (We could have seen this at the beginning because $w = e^z = e^{z+2n\pi i}$ so the points z and $z + 2n\pi i$ are images of the same w.)

We sketch a few of the curves u = const. (solid lines) and v = const. (dashed lines). Study the lines and curves and their images and satisfy yourself that they are correctly labeled. Note that $y = -\pi/2$ is the image of the negative v axis and $y = \pi/2$ is the image of the positive v axis. Verify that if you start at A in the figure, go in the direction indicated by the arrow along $y = -\pi/2$ to $x = -\infty$ and then come back to B in the figure along $y = \pi/2$, you have traversed the image points of the v axis from left to right along the v axis. The equation of the v axis is u = 0 and it is convenient to think of this path from A to B as the limit of the curves d, c, b, a as $u \to 0$. (You can map the u axis in a similar way; try it.) Note that all the curves are exactly the same shape. An easy one to plot using your calculator is (b) in the figure:

$$u = e^x \cos y = 1 \quad \text{or} \quad x = -\ln \cos y, \quad -\frac{\pi}{2} < y < \frac{\pi}{2}.$$

4.   (continued)

Then all the curves $u = e^x \cos y = C$ or $x - \ln C = -\ln \cos y$ are simply
shifted in the x direction by $\ln C$.   The curves $v = e^x \sin y = C$
are simply shifted by $\pi/2$ in the y direction.   Remember as you
sketch the curves that they always intersect at right angles
since they are images of the w plane coordinate lines which
intersect at right angles.

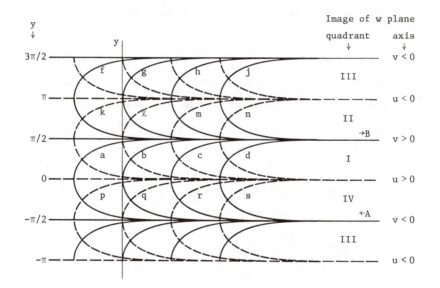

Curves a, b, c, d are images of $u = e^{-2}$, 1, e^2, e^4.

Curves f, g, h, j are images of $u = -e^{-2}$, -1, $-e^2$, $-e^4$.

Curves k, ℓ, m, n are images of $v = e^{-2}$, 1, e^2, e^4.

Curves p, q, r, s are images of $v = -e^{-2}$, -1, $-e^2$, $-e^4$.

7.   Hint: Start by following the hint in Problem 10.10.

9.   In polar coordinates

$$z = re^{i\theta}, \qquad\qquad w = z^3 = r^3 e^{3i\theta}.$$

As θ goes from 0 to $2\pi/3$ in the z plane, the angle of w goes
from 0 to 2π, that is, the whole w plane has been used. We
call this w plane I. As θ goes from $2\pi/3$ to $4\pi/3$, the angle
of w goes from 2π to 4π; we call this w plane II. As θ goes
from $4\pi/3$ to 2π, the angle of w goes from 4π to 6π; this is w
plane III. We have covered one z plane and three w planes.
Now go back to w plane I. We have a Riemann surface of three
sheets in the w plane, since it takes three w planes to plot
the images of the points in one z plane.

14.   Given $w = f(z)$, $f(z)$ analytic, then

(1) $dw = f'(z)dz = \dfrac{dw}{dz}\,dz.$

Recall that the angle of the product of two complex numbers
is the sum of the angles:

$$\left(r_1 e^{i\theta_1} r_2 e^{i\theta_2}\right) = r_1 r_2 e^{i(\theta_1 + \theta_2)}.$$

Then from (1)

      angle of dw = angle of $f'(z)$ + angle of dz.

Thus the rotation angle from dz to dw is the angle of $f'(z)$
or the angle or $\dfrac{dw}{dz}$.

Section 10

4. Hint: Follow the text solution of Example 2 but change text
    Figure 10.1 as follows: In the w plane, draw the $T = 100°$ line
    at $v = \pi/2$; for the z plane use the text diagram in the problem.

5. Hint: Follow the text solution of Examples 2 and 3 with
    Figure 10.2 replaced by the following figure.

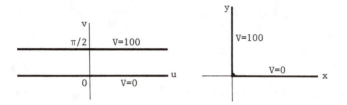

6. As we saw in Problem 9.3, the image of $v = 0$ is $y = 0$, and the
    image of $v = -1$ is the circle

    $$\left(x - \frac{1}{2}\right)^2 + y^2 = \frac{1}{4} .$$

    Look at the figure in the solution of Problem 9.3 to see that
    the region outside the circle and above the x axis corresponds
    to values of v between -1 and 0, and all values of u. Then in
    the w plane we want the temperature to be 0 when $v = 0$ and 100
    when $v = -1$; thus $T = -100v$. We find the temperature in the z
    plane by writing v in terms of x and y from Problem 9.3 above.

    $$T = -100 \frac{-y}{x^2 + y^2} = \frac{100\,y}{x^2 + y^2} .$$

11.   Given $w = \ln \frac{z+1}{z-1}$, we find

$$u + iv = \ln \frac{x + iy + 1}{x + iy - 1},$$

(1) $e^{u+iv} = e^u(\cos v + i \sin v) = \frac{x+1+iy}{x-1+iy} = A,$

(2) $e^{2u} = |A|^2 = \frac{(x+1)^2 + y^2}{(x-1)^2 + y^2},$

(3) $v = \text{angle of } A = \text{angle of } \frac{x^2 + y^2 - 1 - 2iy}{(x-1)^2 + y^2},$

(4) $\cot v = \frac{\text{Re } A}{\text{Im } A} = \frac{x^2 + y^2 - 1}{-2y}.$

From (4) we find the equations of the circles $\cot v = \text{const.}$

(5) $x^2 + (y + \cot v)^2 = 1 + \cot^2 v = \csc^2 v.$

From (2) we find the equations of the circles $u = \text{const.}$

$$(e^{2u} - 1)(x^2 + y^2) - 2(e^{2u} + 1)x = 1 - e^{2u}.$$

$$x^2 + y^2 - 2 \frac{e^{2u} + 1}{e^{2u} - 1} x = -1$$

$$\frac{e^{2u} + 1}{e^{2u} - 1} = \frac{e^u + e^{-u}}{e^u - e^{-u}} = \coth u,$$

(6) $(x - \coth u)^2 + y^2 = \coth^2 u - 1 = \text{csch}^2 u.$

If $v = (2n+1)\pi/2$, then $\cot v = 0$ and (5) is the unit circle. Note, however, that $\cot v = 0$ corresponds to an infinite set of values of v, so much care must be taken [go back to equation (3)] to find the right mapping for a particular problem. See,

11.   (continued)

for example, the hint in Problem 12. We sketch several circles
of each family and note that the two sets are orthogonal as
we know they must be.

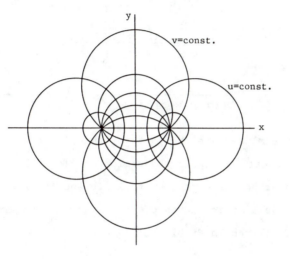

14.   From the definitions, text page 625, and Problem 11 above,
equations (1), (2), (3):

$$\Phi + i\Psi = V_0 w = V_0 [\ell n \ |A| + i(\text{angle of } A)],$$

$$\Phi = \frac{1}{2} V_0 \ \ell n \ \frac{(x+1)^2 + y^2}{(x-1)^2 + y^2},$$

$$\Psi = V_0 \ \tan^{-1} \frac{2y}{1 - x^2 - y^2}, \quad \text{arc tan between } \frac{\pi}{2} \text{ and } \frac{3\pi}{2}.$$

14.   (continued)

We find $\vec{V}$ most easily by using Problem 8 (see text).

$$V_x - iV_y = V_0 \frac{dw}{dz} = V_0 \frac{d}{dz} \ell n \frac{z+1}{z-1}$$

$$= V_0 \frac{-2}{z^2 - 1} = 2V_0 \frac{1}{1 - x^2 + y^2 - 2ixy}$$

$$= 2V_0 \frac{1 - x^2 + y^2 + 2ixy}{(1 - x^2 + y^2)^2 + 4x^2y^2} \, ,$$

$$V_x = 2V_0 \frac{1 - x^2 + y^2}{(1 - x^2 + y^2)^2 + 4x^2y^2} \, , \qquad V_y = \frac{-4xyV_0}{(1 - x^2 + y^2)^2 + 4x^2y^2} \, .$$

The streamlines are the parts of the circles $\Psi = \text{const.}$ or $\cot v = \text{const.}$ lying in the region of flow. Note that the upper and lower halves of the unit circle and the x axis between -1 and 1 are limiting cases of $\cot v = \text{const.}$ and are streamlines. See the sketch in Problem 11.

Section 11

3.   Hint: In this proof you need the fact that

$$(1) \qquad \left| \int_0^{2\pi} F(\theta) \, d\theta \right| \le \int_0^{2\pi} |F(\theta)| \, d\theta$$

where $F(\theta)$ is a complex function of the real variable θ. In words, the absolute value of the integral is less than or equal to the integral of the absolute value of the integrand. To see this recall that an integral is the limit of a sum and consider a sum of complex numbers.

3.   (continued)

From the figure you can see that

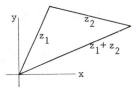

$$|z_1 + z_2| \leqslant |z_1| + |z_2|,$$

that is, the absolute value of a sum

of complex numbers is less than or

equal to the sum of the absolute values.   Replace sum by

integral to get (1).

6.   Hint: See text, page 599, equation (6.2).

12.   We expand $f(z)$ in powers of $1/z$:

$$\sqrt{z^2 - 1} = z\left(1 - z^{-2}\right)^{1/2} = z\left(1 - \frac{1}{2}z^{-2} - \frac{1}{8}z^{-4} \cdots \right)$$

$$= z - \frac{1}{2}z^{-1} - \frac{1}{8}z^{-3} \cdots .$$

(Note that this series gives positive values for large positive

real z.   The other branch of the square root would be the nega-

tive of this series.)   Then by Problem 8.1, text page 616,

$$R(\infty) = -b_1 = -\left(-\frac{1}{2}\right) = \frac{1}{2} .$$

We can check this by using text equation 8.2.

$$R(\infty) = - \text{ residue at 0 of } \frac{1}{z^2}\sqrt{\frac{1}{z^2} - 1}$$

$$= \text{residue at 0 of } -\frac{1}{z^3}\sqrt{1 - z^2}.$$

We can evaluate this by method C, text page 600, or by

Laurent series.

$$-\frac{1}{z^3}\left(1 - z^2\right)^{1/2} = -\frac{1}{z^3}\left(1 - \frac{1}{2}z^2 \cdots\right),$$

so $R = 1/2$ as before.

13.   (a) Warning hint: Multiply by $\left(z - \frac{\pi}{2}\right)^4$, <u>not</u> $(2z - \pi)^4$.

14.   (c) See hint in Problem 6.

18.   We evaluate $\oint_C \dfrac{ze^{i\pi z/2}}{z^4 + 4}\, dz$ around the contour of text Figure 7.2.
       By text, top of page 605, the integral is $2\pi i$ times the sum of
       the residues at the two poles in the upper half plane, namely
       $z = \pm 1 + i$, $z^2 = \pm 2i$. By text equation (6.2) we find

$$R(1 + i) = \left.\frac{ze^{i\pi z/2}}{4z^3}\right|_{z=1+i} = \frac{ie^{-\pi/2}}{8i} = \frac{e^{-\pi/2}}{8}$$

and similarly, $R(-1 + i) = e^{-\pi/2}/8$. (Check it.) Then the contour
integral is $2\pi i \cdot 2e^{-\pi/2}/8 = \pi i e^{-\pi/2}/2$. Thus

$$\int_0^\infty \frac{x \sin(\pi x/2)}{x^4 + 4}\, dx = \frac{1}{2}\,\text{Im}\,\oint_C \frac{ze^{i\pi z/2}}{z^4 + 4}\, dz = \frac{\pi}{4}\, e^{-\pi/2}.$$

25.   We evaluate

$$\text{PV}\,\oint_C \frac{e^{imz}}{z^2 - a^2}\, dz$$

around the contour of text Figure 7.2. By the rule in text
Example 4, page 605, the integral is

$$\frac{1}{2}(2\pi i)\,[R(a) + R(-a)] = \pi i \left(\frac{e^{ima}}{2a} + \frac{e^{-ima}}{-2a}\right)$$

$$= -\frac{\pi}{a}\,\sin ma.$$

Then

$$\text{PV}\int_0^\infty \frac{\cos mx}{x^2 - a^2}\, dx = \frac{1}{2}\,\text{Re}\left(-\frac{\pi}{a}\,\sin ma\right) = -\frac{\pi}{2a}\,\sin ma.$$

29.   We want to evaluate

(1) $\displaystyle\int_0^\infty \frac{(\ln x)^2}{1+x^2}\,dx.$

A quick, careless, and <u>wrong</u> analysis of this problem is to
evaluate the contour integral

(2) $\displaystyle\oint_C \frac{(\ln z)^2}{1+z^2}\,dz = 2\pi i R(i) = -\pi^3/4$

and take half of it $\left(\displaystyle\int_0^\infty = \frac{1}{2}\int_{-\infty}^\infty\right)$.   This is obviously wrong
because the integrand is positive so the integral can't be
negative. Let's look at the problem more carefully. The
integral around the contour of text Figure 7.3 consists of
four parts:

> Integral along the negative x axis,
> Integral along the positive x axis,
> Integral along the large semicircle,
> Integral along the small semicircle.

The integrals along the semicircles tend to zero as $R \to \infty$ and
$r \to 0$. (Show that the integrands tend to zero in both cases.)
Along the negative x axis, $z = re^{i\pi}$, so $\ln z = \ln r + i\pi$, and $x = -\infty$
is $r = \infty$, $\theta = \pi$. Along the positive x axis, $z = r$. The sum of
these two integrals is equal to the contour integral in (2)

$$\int_\infty^0 \frac{(\ln r + i\pi)^2}{1+r^2} e^{i\pi}\,dr + \int_0^\infty \frac{(\ln r)^2}{1+r^2}\,dr = -\pi^3/4$$

(3) $$= \int_0^\infty \frac{(\ln r + i\pi)^2}{1+r^2}\,dr + \int_0^\infty \frac{(\ln r)^2}{1+r^2}\,dr$$

$$= 2\int_0^\infty \frac{(\ln r)^2}{1+r^2}\,dr - \pi^2\int_0^\infty \frac{dr}{1+r^2} + 2\pi i\int_0^\infty \frac{\ln r}{1+r^2}\,dr.$$

29.   (continued)

Now

$$\int_0^\infty \frac{dr}{1+r^2} = \tan^{-1} r \Big|_0^\infty = \pi/2.$$

We take real parts of (3) to get

$$\int_0^\infty \frac{(\ln r)^2}{1+r^2}\, dr = \frac{1}{2}\left(\frac{\pi^3}{2} - \frac{\pi^3}{4}\right) = \frac{\pi^3}{8}.$$

Incidentally, note that we can take imaginary parts of (3) to get

$$\int_0^\infty \frac{\ln r}{1+r^2}\, dr = 0.$$

33.   We follow the method of Example 6, text page 610. Also see
Problems 7.46 and 7.48 in these Solutions. You can easily verify
that $f(z) = z^6 + z^3 + 9z + 64$ is not zero on the positive x axis or
positive y axis. We follow around the contour OPQ of text
Figure 7.5 and find the change in Θ. Along the y axis, $z = iy$, so

$$f(z) = (iy)^6 + (iy)^3 + 9iy + 64 = 64 - y^6 + iy(9 - y^2),$$

$$\tan \Theta = \frac{y(9 - y^2)}{64 - y^6}.$$

Then along OP, Θ remains zero; along PQ, θ increases to $\pi/2$, so
Θ increases to about $6\pi/2 = 3\pi$. Following down the y axis, we
have:

33.   (continued)

y	sign of tan θ	tan θ	quadrant of θ	θ
$\sim \infty$	+	$\sim \dfrac{y^3}{y^6} \sim 0$	III	$\sim 3\pi$
∞ to 3	+	~ 0	III	$\sim 3\pi$
3 to 2	-	0 to $-\infty$	II	$3\pi - \dfrac{\pi}{2}$
2 to 0	+	∞ to 0	I	$3\pi - \pi$

(You might like to sketch a diagram in the w plane as in the solutions of Problems 7.46 and 7.48 showing the approximate path of f(z) in the w plane.) Thus the change in θ is 2π, so f(z) has one zero in quadrant I. Then f(z) has one zero in quadrant IV, and two zeros in each of quadrants II and III.

36.   Let C be the circle $z = a + re^{i\theta}$, $0 \leqslant \theta \leqslant 2\pi$. Then by Cauchy's integral formula [text equation (3.9)]:

(1) $f(a) = \dfrac{1}{2\pi i} \oint_C \dfrac{f(z)}{z - a}\, dz = \dfrac{1}{2\pi i} \int_0^{2\pi} \dfrac{f(z)}{re^{i\theta}}\, rie^{i\theta}\, d\theta$

$= \dfrac{1}{2\pi} \int_0^{2\pi} f(z)\, d\theta = \text{average of } f(z) \text{ on } C.$

Let $f(z) = u(x,y) + iv(x,y)$ and take the real part of equation (1).

$u(x,y)\Big|_{z=a} = \text{average of } u(x,y) \text{ on } C.$

37.   We want to show that a (nonconstant) harmonic function u(x,y)
       cannot take its maximum value or its minimum value at an
       interior point of any region R. We see this by contradiction.
       Suppose it is claimed that u(x,y) takes its maximum value M at
       the point P inside R (see
       diagram). This means that at
       all other points of R, u(x,y) ⩽ M.

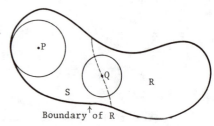

       By Problem 36, the average of
       u(x,y) around any circle (in R)
       centered at P is equal to M.
       This is a contradiction if

       Boundary of R

       u(x,y) < M for any part of any
       such circle. We conclude that u(x,y) = M at all points of the
       largest disk centered at P which lies in R. We can also show
       that u(x,y) = M at <u>all</u> points of R. Suppose this were not true.
       Let S consist of all the points in R where u(x,y) = M; then
       u(x,y) < M at all other points of R. Let Q be a point on the
       boundary of S but not on the boundary of R (see diagram). Since
       u(x,y) is a continuous function, u(x,y) = M at Q but u(x,y) < M on
       part of any circle with center Q. By Problem 36, these assump-
       tions lead to a contradiction. Therefore u(x,y) = M at every
       point of R, that is, u(x,y) = const. The argument can be re-
       peated to show that if u(x,y) takes its minimum value at an in-
       terior point, then u(x,y) = const. Thus a nonconstant harmonic
       function takes its maximum value and its minimum value on the
       boundary of any region R. [We have assumed as in Problem 36 that
       R is contained in the region where f(z) = u(x,y) + iv(x,y) is
       analytic.]

Chapter 15

Section 2

Note that numbers L1 to L35 refer to entries in the Laplace transform table, text pages 636 to 638.

3.   Hint: See text page 70, equation (12.4).

5.   By L4, text page 636,

$$\int_0^\infty e^{-pt} \cos at \, dt = \frac{p}{p^2 + a^2} .$$

We integrate both sides with respect to a.   [Integrate cos at with respect to a to get (sin at)/t.]

$$\int_0^\infty e^{-pt} \frac{\sin at}{t} \, dt = \int \frac{p \, da}{p^2 + a^2} = \text{arc tan} \frac{a}{p} + C .$$

The integration constant C is zero because both sin at and arc tan(a/p) are zero when $a = 0$.   Thus

$$\int_0^\infty e^{-pt} \frac{\sin at}{t} \, dt = \text{arc tan} \frac{a}{p}$$

which is L19.

10.   First write the denominator as it is written in L13 and L14,
      text page 636.

   (A) $\dfrac{2p-1}{p^2-2p+10} = \dfrac{2p-1}{(p-1)^2+9}$.

   In L13 and L14, let $a = -1$ and $b = 3$.   Then

   (B) $L(e^t \sin 3t) = \dfrac{3}{(p-1)^2+3^2}$.

   (C) $L(e^t \cos 3t) = \dfrac{p-1}{(p-1)^2+3^2}$.

   Now we want a linear combination of (B) and (C) so that the
   right-hand sides give (A).   That is, we want constants α and
   β so that

         $3\alpha + (p-1)\beta \equiv 2p - 1$.

   We see that β must be 2 to give the term $2p$; then $3\alpha = 1$, $\alpha = 1/3$.
   Thus the desired inverse transform is

         $\dfrac{1}{3} e^t \sin 3t + 2 e^t \cos 3t$.

15.   Write text equation (2.1) with $f(t)$ and $F(p)$ replaced by $g(t)$
      and $G(p)$:

         $\displaystyle\int_0^\infty g(t)\, e^{-pt}\, dt = G(p)$.

   Differentiate this with respect to p to get

         $\displaystyle\int_0^\infty g(t)(-te^{-pt})\, dt = \dfrac{d}{dp}\, G(p)$.

   Thus the Laplace transform of $t\, g(t)$ is $-\dfrac{d}{dp}\, G(p)$.

15.   (continued)

It is straightforward to prove L32 in general. Differentiating repeatedly with respect to p, we get

$$\int_0^\infty g(t)(-t)^2 e^{-pt}\, dt = \frac{d^2}{dp^2} G(p),$$

$$\cdots$$

$$\int_0^\infty g(t)(-t)^n e^{-pt}\, dt = \frac{d^n}{dp^n} G(p).$$

Thus the Laplace transform of $t^n g(t)$ is $(-1)^n \dfrac{d^n}{dp^n} G(p)$.

18.   Let $g(t) = e^{-at} - e^{-bt}$. Then by L2,

$$G(p) = \frac{1}{p+a} - \frac{1}{p+b} \qquad \text{or} \qquad G(u) = \frac{1}{u+a} - \frac{1}{u+b}.$$

By L31, the Laplace transform of $\dfrac{g(t)}{t}$ is

$$\int_p^\infty G(u)\, du = \int_p^\infty \left(\frac{1}{u+a} - \frac{1}{u+b} \right) du = \ell n\,|u+a| - \ell n\,|u+b| \,\Big|_p^\infty$$

$$= \ell n\,\left|\frac{u+a}{u+b}\right| \Big|_p^\infty = \ell n\, 1 - \ell n\, \frac{p+a}{p+b} = \ell n\, \frac{p+b}{p+a}.$$

21.   By L11,

$$L(t \sin bt) = \frac{2bp}{(p^2 + b^2)^2}.$$

Then by L29, we find the Laplace transform of $e^{-at} t \sin bt$ by replacing p by $(p+a)$. Thus

$$L(e^{-at} t \sin bt) = \frac{2b(p+a)}{[(p+a)^2 + b^2]^2}.$$

27. We write

$$\sin(x - vt) = - \sin v \left(t - \frac{x}{v}\right).$$

Now in L28, let

$$g(t) = - \sin vt \qquad \text{and} \qquad a = x/v.$$

Then

$$g(t - a) = - \sin v \left(t - \frac{x}{v}\right)$$

and by L3

$$G(p) = \{\text{Laplace transform of } g(t)\} = - \frac{v}{p^2 + v^2}.$$

Thus, by L28, the Laplace transform of

$$f(t) = \begin{cases} - \sin v \left(t - \frac{x}{v}\right), & t > \frac{x}{v} \\[2mm] 0, & t < \frac{x}{v} \end{cases}$$

is

$$e^{-pa}G(p) = - \frac{v}{p^2 + v^2} e^{-px/v}.$$

Section 3

3. Take the Laplace transform of each term in the differential
    equation [see L35 and L3 in the text Laplace transform table,
    and remember that $L(y) = Y$.]

$$p^2 Y - py_0 - y_0' + 4(pY - y_0) + 4Y = \frac{1}{p+2}.$$

3.   (continued)

Then substitute the given values for y_0 and y_0', and solve for Y.

$$p^2 Y - 4 + 4pY + 4Y = \frac{1}{p+2},$$

$$(p^2 + 4p + 4) Y = \frac{1}{p+2} + 4,$$

$$(p+2)^2 Y = \frac{1}{p+2} + 4,$$

$$Y = \frac{1}{(p+2)^3} + \frac{4}{(p+2)^2}.$$

Note that we do _not_ combine the fractions on the right-hand side
of the Y equation because we can find the inverse transforms of
each of the terms in the Laplace transform table. We use L6 with
$a = 2$ and $k = 2$ for the first term, and for the second term we use
L6 again with $a = 2$ and $k = 1$. Thus we find

$$y = \frac{1}{2} t^2 e^{-2t} + 4t\, e^{-2t}.$$

7.   Hint: Combine the terms on the right-hand side of the Y equation
and cancel $(p - 2)$. Then use L7.

9.   Using L35 and L4 we take the Laplace transform of each term in
the differential equation and substitute the given initial
conditions.

$$\left(p^2 Y - py_0 - y_0'\right) + 16Y = p^2 Y - 8 + 16Y = \frac{8p}{p^2 + 16}.$$

Solve for Y and then find the inverse transform.

$$(p^2 + 16) Y = \frac{8p}{p^2 + 16} + 8,$$

$$Y = \frac{8p}{(p^2 + 16)^2} + \frac{8}{p^2 + 16}.$$

By L11 and L3 with $a = 4$, we find

$$y = t \sin 4t + 2 \sin 4t.$$

12. Take the Laplace transform of each term of the differential
     equation and substitute the initial conditions to get

$$p^2 Y - p - 2 - Y = \frac{1}{p+1} - 2 \frac{1}{(p+1)^2} .$$

Then

$$(p^2 - 1) Y = \frac{1}{p+1} - \frac{2}{(p+1)^2} + p + 2 ,$$

$$Y = \frac{1}{(p^2 - 1)(p+1)} - \frac{2}{(p^2 - 1)(p+1)^2} + \frac{p+2}{p^2 - 1} .$$

Looking in the text Laplace transform table, we do not find the
first two terms on the right-hand side of the Y equation. Let's
try combining these two terms.

$$\frac{1}{(p^2 - 1)(p+1)} - \frac{2}{(p^2 - 1)(p+1)^2} = \frac{p+1-2}{(p^2 - 1)(p+1)^2} = \frac{1}{(p+1)^3} .$$

Then

$$Y = \frac{1}{(p+1)^3} + \frac{p}{p^2 - 1} + \frac{2}{p^2 - 1} .$$

We find y using L6 with $a = 1$, $k = 2$, for the first term, and
using L10 and L9 for the other terms.

$$y = \frac{1}{2} t^2 e^{-t} + \cosh t + 2 \sinh t$$

$$= \frac{1}{2}(t^2 e^{-t} + 3e^t - e^{-t}) .$$

19. Take the Laplace transform of each term of the differential
     equation, substitute the initial conditions, solve for Y, and
     then find the inverse transform, as follows.

$$p^2 Y - p - 2 + pY - 1 - 5Y = \frac{1}{p-2} \; ,$$

$$(p^2 + p - 5) Y = \frac{1}{p-2} + p + 3 = \frac{1 + p^2 + p - 6}{p-2} \; ,$$

$$Y = \frac{p^2 + p - 5}{(p^2 + p - 5)(p - 2)} = \frac{1}{p-2} \; ,$$

$$y = e^{2t} .$$

Note that we looked ahead and observed that we could not find

the inverse transform of $\dfrac{1}{(p-2)(p^2 + p - 5)}$ in the text table,

so we combined terms on the right-hand side of the Y equation.

20. Hint: $p^3 - 6p^2 + 32 = (p+2)(p-4)^2$ as you can verify. Or use
     partial fractions (see Appendix C to these Solutions).

23. Take the Laplace transform of each term of the differential
     equation, substitute the initial conditions, and solve for Y.

$$p^2 Y - 3 + 2pY + 5Y = \frac{10p}{p^2 + 1} \; ,$$

$$Y(p^2 + 2p + 5) = \frac{10p}{p^2 + 1} + 3 = \frac{3p^2 + 10p + 3}{p^2 + 1} \; ,$$

$$Y = \frac{3p^2 + 10p + 3}{(p^2 + 1)(p^2 + 2p + 5)} \; .$$

Here we see that (using the text Laplace transform table) we
cannot find the inverse transforms of the individual terms on
the right-hand side of the Y equation, and combining the terms

23.   (continued)

does not simplify Y. When this happens, we use the method of
partial fractions to write Y as a sum of two terms as follows.
(Also see Appendix C to these Solutions.)

$$\frac{3p^2 + 10p + 3}{(p^2 + 1)(p^2 + 2p + 5)} \equiv \frac{Ap + B}{p^2 + 1} + \frac{Cp + D}{p^2 + 2p + 5},$$

$$3p^2 + 10p + 3 \equiv (Ap + B)(p^2 + 2p + 5) + (Cp + D)(p^2 + 1).$$

This is an identity; that is, it is true for all values of p
(including complex values). We select some values of p to give
us simple equations for A, B, C, D.

(1) $p = i$: $-3 + 10i + 3 = (Ai + B)(-1 + 2i + 5) + 0$ or

$$5i = -A + 2B + i(2A + B).$$

(2) $p = 0$: $3 = 5B + D.$

(3) $p = 1$: $16 = 8(A + B) + 2(C + D).$

Since A,B,C,D are real, we find from the first equation (take
real and imaginary parts):

$$5 = 2A + B,$$
$$\qquad\qquad so \qquad A = 2B, \qquad B = 1, \qquad A = 2.$$
$$0 = -A + 2B,$$

Then from equation (2), we have $D = -2$, and from equation (3),
$C = -2$.   Thus

$$Y = \frac{2p + 1}{p^2 + 1} - \frac{2p + 2}{p^2 + 2p + 5} = \frac{2p + 1}{p^2 + 1} - 2\frac{p + 1}{(p + 1)^2 + 2^2}.$$

Now using L3, L4, and L14, we find

$$y = 2\cos t + \sin t - 2e^{-t}\cos 2t.$$

26. Hint: Use L29 to obtain a transform which is not in the text
     Laplace transform table.

27. Here we want to take Laplace transforms of both y and z and
     their derivatives. Let Y be the Laplace transform of y as
     usual, and let Z be the Laplace transform of z. Take Laplace
     transforms of both differential equations and substitute the
     initial conditions.

$$\begin{cases} (pY - y_0) + (pZ - z_0) - 3Z = 0, \\ (p^2Y - py_0 - y_0') + (pZ - z_0) = 0, \end{cases}$$

$$\begin{cases} pY + pZ - \dfrac{4}{3} - 3Z = 0, \\ p^2Y + pZ - \dfrac{4}{3} = 0. \end{cases}$$

We can solve these equations simultaneously for Y and Z to find

$$Y = \frac{4}{p^2(4 - p)}\,, \qquad Z = \frac{4}{3}\left(\frac{1}{(p - 4)} - \frac{1}{p(p - 4)}\right).$$

Then, using L2 and L7 to take the inverse transform of Z,
we find z:

$$z = \frac{4}{3}\left(e^{4t} - \frac{1 - e^{4t}}{-4}\right) = \frac{4}{3}e^{4t} + \frac{1}{3}(1 - e^{4t})$$

$$= e^{4t} + \frac{1}{3}\,.$$

We can now either find y from Y, or we can find y' directly from
the original differential equations and integrate to get y. By
the latter method, we find from the first differential equation

27.   (continued)

$$y' = 3z - z' = 3\left(e^{4t} + \frac{1}{3}\right) - 4e^{4t} = 1 - e^{4t},$$

$$y = \int y'\, dt = t - \frac{1}{4}e^{4t} + C.$$

Since $y_0 = 0$, we find $C = 1/4$; then

$$y = t + \frac{1}{4}\left(1 - e^{4t}\right).$$

To find y from Y, write

$$Y = \frac{(4 - p) + p}{p^2(4 - p)} = \frac{1}{p^2} + \frac{1}{p(4 - p)}$$

and use L6 and L7 to get the same y we found above.

29.   Take Laplace transforms of both equations and substitute the
       initial conditions.

$$\begin{cases} pY - 1 + pZ - 1 - 2Y = \dfrac{1}{p}, \\[2mm] Z - (pY - 1) = \dfrac{1}{p^2}, \end{cases} \qquad \text{or}$$

(1) $$\begin{cases} (p - 2)Y + pZ = 2 + \dfrac{1}{p}, \\[2mm] -pY + Z = \dfrac{1}{p^2} - 1. \end{cases}$$

Solving these equations simultaneously, we find

$$Y = \frac{1}{p - 1}$$

so by L2,

(2) $$y = e^t.$$

29.   (continued)

The easiest way to find z is to go back to the given differen-

tial equations. Solving the second equation for z and using (2)

gives

(3) $z = t + y' = t + e^t$.

Alternatively we could find Z from (1).

$$(p^2 + p - 2)Z = p\left(2 + \frac{1}{p}\right) + (p - 2)\left(\frac{1}{p^2} - 1\right) = p + 2 + \frac{p^2 + p - 2}{p^2},$$

$$Z = \frac{p + 2}{(p + 2)(p - 1)} + \frac{p^2 + p - 2}{p^2(p^2 + p - 2)} = \frac{1}{p - 1} + \frac{1}{p^2}.$$

Then by L2 and L5 we find z as in (3).

36.   We compare the given integral with L19 which says that

$$\int_0^\infty \frac{\sin at}{t} e^{-pt} \, dt = \arctan \frac{a}{p}.$$

Let $a = 2$ and $p = 3$; then

$$\int_0^\infty \frac{\sin 2t}{t} e^{-3t} \, dt = \arctan \frac{2}{3}.$$

40.   Hint: L21 says that

$$\int_0^\infty \frac{e^{-(p+a)t} - e^{-(p+b)t}}{t} = \ln \frac{p+b}{p+a}.$$

Note that the value of p is not significant; only $(p + a)$ and

$(p + b)$ are needed. Let $p + a = 2$ and $p + b = 2e$.

Section 4

6.   Using text equations (4.2), we find $g(\alpha)$; this is like finding
     the coefficients c_n in a Fourier series. Then we substitute
     $g(\alpha)$ into the first integral in (4.2) to get $f(x)$; this is like
     substituting the c_n's you have found into a Fourier series.

$$g(\alpha) = \frac{1}{2\pi}\int_{-\infty}^{\infty} f(x)e^{-i\alpha x}\,dx = \frac{1}{2\pi}\int_{-1}^{1} xe^{-i\alpha x}\,dx.$$

[The limits are -1 to 1 because $f(x) = 0$ for $|x| > 1$.]   Then

$$g(\alpha) = \frac{1}{2\pi}\frac{e^{-i\alpha x}}{(-i\alpha)^2}(-i\alpha x - 1)\Big|_{-1}^{1}$$

$$= \frac{1}{2\pi\alpha^2}e^{-i\alpha}(i\alpha - 1) + \frac{1}{2\pi\alpha^2}e^{i\alpha}(i\alpha - 1)$$

$$= \frac{i}{2\pi\alpha}(e^{-i\alpha} + e^{i\alpha}) - \frac{1}{2\pi\alpha^2}(e^{i\alpha} - e^{-i\alpha})$$

$$= \frac{i}{\pi\alpha}\cos\alpha - \frac{i}{\pi\alpha^2}\sin\alpha = \frac{i}{\pi}\frac{\alpha\cos\alpha - \sin\alpha}{\alpha^2},$$

$$f(x) = \int_{-\infty}^{\infty} g(\alpha)e^{i\alpha x}\,dx = \frac{i}{\pi}\int_{-\infty}^{\infty}\frac{\alpha\cos\alpha - \sin\alpha}{\alpha^2}e^{i\alpha x}\,d\alpha.$$

11. We find $g(\alpha)$ using text equation (4.2) and substitute it into
     the integral for $f(x)$.

$$g(\alpha) = \frac{1}{2\pi}\int_{-\pi/2}^{\pi/2} e^{i\alpha x}\cos x\,dx.$$

The limits are $-\pi/2$ to $\pi/2$ because $f(x) = 0$ for $|x| > \pi/2$.   We
write $\cos x$ in complex exponential form (see text page 67) and
integrate to find $g(\alpha)$.

11.   (continued)

$$g(\alpha) = \frac{1}{2\pi}\int_{-\pi/2}^{\pi/2} e^{i\alpha x} \frac{e^{ix} + e^{-ix}}{2}\, dx$$

$$= \frac{1}{4\pi}\left[\frac{e^{i(\alpha+1)x}}{i(\alpha+1)} + \frac{e^{i(\alpha-1)x}}{i(\alpha-1)}\right]_{-\pi/2}^{\pi/2}.$$

In evaluating this, note that $e^{i\pi/2} = i$, so $e^{i(\alpha+1)\pi/2} = ie^{i\alpha\pi/2}$ and similarly for the other terms.   Then

$$g(\alpha) = \frac{ie^{i\alpha\pi/2} - (-i)e^{-i\alpha\pi/2}}{4\pi i(\alpha+1)} + \frac{-ie^{i\alpha\pi/2} - ie^{-i\alpha\pi/2}}{4\pi i(\alpha-1)}$$

$$= \frac{\cos(\alpha\pi/2)}{2\pi}\left(\frac{1}{\alpha+1} - \frac{1}{\alpha-1}\right) = \frac{\cos(\alpha\pi/2)}{\pi(1-\alpha^2)},$$

$$f(x) = \frac{1}{\pi}\int_{-\infty}^{\infty} \frac{\cos(\alpha\pi/2)}{1-\alpha^2}\, e^{i\alpha x}\, d\alpha.$$

16.   We use text (4.15) to find $g_c(\alpha)$ and $f_c(x)$.

$$g_c(\alpha) = \sqrt{\frac{2}{\pi}}\int_{0}^{\infty} f(x)\cos\alpha x\, dx = \sqrt{\frac{2}{\pi}}\int_{0}^{\pi/2}\cos x \cos\alpha x\, dx$$

$$= \sqrt{\frac{2}{\pi}}\frac{1}{2}\left(\frac{\sin(x+\alpha x)}{1+\alpha} + \frac{\sin(x-\alpha x)}{1-\alpha}\right)\bigg|_{0}^{\pi/2} \qquad \text{by tables.}$$

Now $\sin\left(\frac{\pi}{2}+\frac{\alpha\pi}{2}\right) = \cos\frac{\alpha\pi}{2}$ and $\sin\left(\frac{\pi}{2}-\frac{\alpha\pi}{2}\right) = \cos\frac{\alpha\pi}{2}$.

Thus

$$g_c(\alpha) = \sqrt{\frac{2}{\pi}}\frac{1}{2}\left(\cos\frac{\alpha\pi}{2}\right)\left(\frac{1}{1+\alpha} + \frac{1}{1-\alpha}\right) = \sqrt{\frac{2}{\pi}}\frac{\cos(\alpha\pi/2)}{1-\alpha^2},$$

$$f_c(x) = \sqrt{\frac{2}{\pi}}\int_{0}^{\infty} g(\alpha)\cos\alpha x\, dx = \frac{2}{\pi}\int_{0}^{\infty} \frac{\cos(\alpha\pi/2)}{1-\alpha^2}\cos\alpha x\, dx.$$

16.   (continued)

Let us show that this is the same result as in Problem 11.
We write the Problem 11 result using $e^{i\alpha x} = \cos \alpha x + i \sin \alpha x$.

$$f(x) = \frac{1}{\pi} \int_{-\infty}^{\infty} \frac{\cos(\alpha \pi/2)}{1 - \alpha^2} (\cos \alpha x + i \sin \alpha x) \, dx.$$

Now observe that $g(\alpha)$ is even; thus the integral of $g(\alpha) \sin \alpha x$
(over the symmetric interval $-\infty$ to ∞) is zero, and the integral
of $g(\alpha) \cos \alpha x$ is twice the 0 to ∞ integral. We have

$$f(x) = \frac{2}{\pi} \int_0^{\infty} \frac{\cos(\alpha \pi/2)}{1 - \alpha^2} \cos \alpha x \, dx$$

which is the same as f_c above.

18.   We use text equation (4.14) to find $g_s(\alpha)$ and $f_s(x)$.

$$g_s(\alpha) = \sqrt{\frac{2}{\pi}} \int_0^1 x \sin \alpha x \, dx = \sqrt{\frac{2}{\pi}} \frac{1}{\alpha^2} (\sin \alpha x - \alpha x \cos \alpha x) \Big|_0^1$$

$$= \sqrt{\frac{2}{\pi}} \frac{1}{\alpha^2} (\sin \alpha - \alpha \cos \alpha).$$

$$f_s(x) = \frac{2}{\pi} \int_0^{\infty} \frac{\sin \alpha - \alpha \cos \alpha}{\alpha^2} \sin \alpha x \, d\alpha.$$

In Problem 6 we found

$$f(x) = \frac{i}{\pi} \int_{-\infty}^{\infty} \frac{\alpha \cos \alpha - \sin \alpha}{\alpha^2} e^{i\alpha x} \, d\alpha$$

$$= \frac{i}{\pi} \int_{-\infty}^{\infty} \frac{\alpha \cos \alpha - \sin \alpha}{\alpha^2} (\cos \alpha x + i \sin \alpha x) \, d\alpha.$$

We see that $g(\alpha)$ is an odd function (odd times even = odd, and
odd $\div$ even = odd). Thus the integral of $g(\alpha) \cos \alpha x$ is zero and

18.   (continued)

the integral of $g(\alpha)\sin \alpha x$ is twice the 0 to ∞ integral.

$$f(x) = \frac{2i}{\pi} \int_0^\infty \frac{\alpha \cos \alpha - \sin \alpha}{\alpha^2} \, i \sin \alpha x \, d\alpha$$

$$= \frac{2}{\pi} \int_0^\infty \frac{\sin \alpha - \alpha \cos \alpha}{\alpha^2} \sin \alpha x \, d\alpha$$

as above.

22.   From text page 518, we find

$$j_1(x) = x\left(-\frac{1}{x}\frac{d}{dx}\right)\left(\frac{\sin x}{x}\right) = -\frac{d}{dx}\left(\frac{\sin x}{x}\right) = \frac{\sin x - x \cos x}{x^2}$$

Replace x by α and use the result of Problem 18 above to get

$$f(x) = \frac{2}{\pi} \int_0^\infty j_1(\alpha) \sin \alpha x \, d\alpha.$$

Now use the definition of $f(x)$ given in Problem 6 above to get

$$\frac{2}{\pi} \int_0^\infty j_1(\alpha) \sin \alpha x \, d\alpha = f(x) = \begin{cases} x, & |x| < 1, \\ 0, & |x| > 1, \end{cases}$$

which gives the desired result if we multiply by $\frac{\pi}{2}$.

Comment: We also have, with $x = 1$,

$$\int_0^\infty j_1(\alpha) \sin \alpha \, d\alpha = \frac{\pi}{4}$$

since a Fourier integral converges to the midpoint of a jump in $f(x)$ (see text pages 649 and 652).

24.   (a) Hint: Evaluate the integral from $-\infty$ to 0 and the integral

from 0 to ∞ separately:

$$\int_{-\infty}^{\infty} e^{-|x|} e^{-i\alpha x}\, dx = \int_{-\infty}^{0} e^{x} e^{-i\alpha x}\, dx + \int_{0}^{\infty} e^{-x} e^{-i\alpha x}\, dx$$

$$= \int_{-\infty}^{0} e^{(1-i\alpha)x}\, dx + \int_{0}^{\infty} e^{-(1+i\alpha)x}\, dx.$$

28.   (a) We use text equation (4.15) to find the cosine transform of

f(x) and write f(x) as a cosine integral.

$$g_c(\alpha) = \sqrt{\frac{2}{\pi}} \int_{2}^{4} \cos \alpha x\, dx - \sqrt{\frac{2}{\pi}} \left. \frac{\sin \alpha x}{\alpha} \right|_{2}^{4} = \sqrt{\frac{2}{\pi}} \frac{\sin 4\alpha - \sin 2\alpha}{\alpha}\, .$$

$$f_c(x) = \frac{2}{\pi} \int_{0}^{\infty} \frac{\sin 4\alpha - \sin 2\alpha}{\alpha} \cos \alpha x\, d\alpha.$$

We can write this in a simpler form by using the trigonometric

identity

$$\sin x - \sin y = 2 \sin \frac{x-y}{2} \cos \frac{x+y}{2}\, .$$

Then

$$\sin 4\alpha - \sin 2\alpha = 2 \sin \alpha \cos 3\alpha.$$

Thus

$$f_c(x) = \frac{4}{\pi} \int_{0}^{\infty} \frac{\sin \alpha \cos 3\alpha}{\alpha} \cos \alpha x\, d\alpha.$$

(b) We use text (4.14) to find f(x) as a sine integral.

$$g_s(\alpha) = \sqrt{\frac{2}{\pi}} \int_{2}^{4} \sin \alpha x\, dx = \sqrt{\frac{2}{\pi}} \left(-\frac{\cos \alpha x}{\alpha} \right) \Big|_{2}^{4}$$

$$= \sqrt{\frac{2}{\pi}} \frac{\cos 2\alpha - \cos 4\alpha}{\alpha} = \sqrt{\frac{2}{\pi}} \frac{2 \sin 3\alpha \sin \alpha}{\alpha}$$

28.   (continued)

by the identity

$$\cos x - \cos y = -2 \sin \frac{x+y}{2} \sin \frac{x-y}{2}.$$

Then

$$f_s(x) = \frac{4}{\pi} \int_0^\infty \frac{\sin 3\alpha \sin \alpha}{\alpha} \sin \alpha x \, d\alpha.$$

Comment: These two integrals for $f(x)$ in (a) and (b) have the same values when $x > 0$. But for $x < 0$, they are different since $f_c(x)$ is even and $f_s(x)$ is odd. We sketch the functions repre-sented by the two integrals. Note that the functions are <u>not</u> periodic.

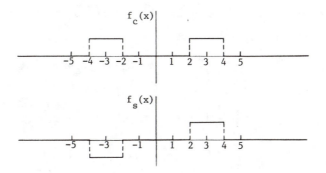

Section 5

5.   Write F(p) as a product

$$F(p) = \frac{p}{(p+a)(p+b)^2} = \frac{p}{(p+a)(p+b)} \cdot \frac{1}{p+b} = G(p)H(p)$$

and use L8 and L2 (see text, page 636) to find

$$g(t) = \frac{ae^{-at} - be^{-bt}}{a-b} \,, \qquad h(t) = e^{-bt}.$$

Then the inverse Laplace transform of F(p) is given by the convolution integral [see text equations (5.6) and (5.7) or L34]:

$$f(t) = g*h = h*g = \int_0^t h(t-\tau)g(\tau)d\tau.$$

[Note from L34 that it doesn't matter whether we use $g(t-\tau)h(\tau)$ or $h(t-\tau)g(\tau)$ but it may be easier to use $(t-\tau)$ in the simpler function. Thus we have

$$f(t) = \frac{1}{a-b}\int_0^t e^{-b(t-\tau)}(ae^{-a\tau} - be^{-b\tau})d\tau$$

$$= \frac{e^{-bt}}{a-b}\int_0^t (ae^{(b-a)\tau} - b)d\tau = \frac{e^{-bt}}{a-b}\left(\frac{ae^{(b-a)\tau}}{b-a} - b\tau\right)\Bigg|_0^t$$

$$= \frac{e^{-bt}}{a-b}\left[\frac{a}{b-a}(e^{(b-a)t} - 1) - bt\right]$$

$$= \frac{a(e^{-bt} - e^{-at}) + b(b-a)te^{-bt}}{(b-a)^2}.$$

12.   Write

$$F(p) = \frac{1}{p(p^2 + a^2)} \cdot \frac{1}{p^2 + b^2} = G(p)H(p).$$

Then by L15 and L3 (text page 636)

$$g(t) = \frac{1}{a^2}(1 - \cos at), \qquad h(t) = \frac{1}{b}\sin bt.$$

By L34 and the text hint:

$$f(t) = \frac{1}{a^2 b}\int_0^t \sin b(t - \tau)(1 - \cos a\tau)d\tau$$

$$= \frac{1}{a^2 b}\int_0^t \left[\sin b(t - \tau) - \frac{1}{2}\sin(bt - b\tau + a\tau)\right.$$
$$\left. - \frac{1}{2}\sin(bt - b\tau - a\tau)\right]d\tau$$

$$= \frac{1}{a^2 b}\left[\frac{\cos b(t - \tau)}{b} - \frac{1}{2}\frac{\cos(bt - b\tau + a\tau)}{b - a}\right.$$
$$\left. - \frac{1}{2}\frac{\cos(bt - b\tau - a\tau)}{b + a}\right]_0^t$$

$$= \frac{1}{a^2 b}\left[\frac{1 - \cos bt}{b} - \frac{1}{2}\frac{\cos at - \cos bt}{b - a} - \frac{1}{2}\frac{\cos at - \cos bt}{b + a}\right]$$

$$= \frac{1 - \cos bt}{a^2 b^2} - \frac{\cos at - \cos bt}{a^2(b^2 - a^2)}$$

$$= \frac{1}{a^2 b^2} - \frac{1}{b^2 - a^2}\left(\frac{\cos at}{a^2} - \frac{\cos bt}{b^2}\right).$$

13. If $f(t) = \int_0^t \sin(t - \tau)e^{-\tau}d\tau$, then by L34, L3 and L2,

$$F(p) = L(\sin t)L(e^{-t}) = \frac{1}{p^2 + 1} \cdot \frac{1}{p+1}.$$

We expand this by partial fractions (see Appendix C to these Solutions) to get

$$F(p) = \frac{1}{2}\left(\frac{1}{p+1} + \frac{1-p}{p^2 + 1}\right) = \frac{1}{2}\left(\frac{1}{p+1} + \frac{1}{p^2 + 1} - \frac{p}{p^2 + 1}\right).$$

Then by L2, L3, and L4:

$$f(t) = \frac{1}{2}(e^{-t} + \sin t - \cos t).$$

17. Take the Laplace transform of each term in the differential equation $y'' - a^2 y = f(t)$ to get (see L35 with $y_0 = y_0' = 0$):

$$p^2 Y - a^2 Y = L(f) \qquad \text{or} \qquad Y = \frac{1}{p^2 - a^2} L(f).$$

By L9, $\dfrac{1}{p^2 - a^2} = L\left(\dfrac{1}{a}\sinh at\right)$. Then $y(t)$ is the convolution of $\dfrac{1}{a}\sinh at$ and $f(t)$. For $t > 0$, $f(t) = 1$, so

$$y(t) = \int_0^t \frac{1}{a}\sinh a(t - \tau) \cdot 1 \cdot d\tau = -\frac{1}{a^2}\cosh a(t - \tau)\Big|_0^t$$

$$= -\frac{1}{a^2}(1 - \cosh at) = \frac{1}{a^2}(\cosh at - 1), \qquad t > 0.$$

For $t < 0$, $f(t) = 0$ so $y(t) = 0$.

20.   From the text example [Figure 4.1 and equation (4.16)] we have
       the Fourier transform pair

$$f(x) = \begin{cases} 1, & -1 < x < 1, \\ 0, & \text{otherwise,} \end{cases} \qquad g(\alpha) = \frac{\sin \alpha}{\pi \alpha}.$$

We want to verify Parseval's theorem, text equation (5.17),
namely

$$\int_{-\infty}^{\infty} |g|^2 \, d\alpha = \frac{1}{2\pi} \int_{-\infty}^{\infty} |f|^2 \, dx.$$

We find

$$\int_{-\infty}^{\infty} |f|^2 \, dx = \int_{-1}^{1} 1 \cdot dx = 2$$

$$\int_{-\infty}^{\infty} |g|^2 \, d\alpha = \int_{-\infty}^{\infty} \frac{\sin^2 \alpha}{\pi^2 \alpha^2} \, d\alpha$$

Integrate by parts:

$$u = \sin^2 \alpha, \qquad\qquad du = 2 \sin \alpha \cos \alpha,$$
$$dv = \frac{d\alpha}{\alpha^2}, \qquad\qquad v = -\frac{1}{\alpha}.$$

Then

$$\int_{-\infty}^{\infty} \frac{\sin^2 \alpha}{\alpha^2} \, d\alpha = -\frac{1}{\alpha} \sin^2 \alpha \Big|_{-\infty}^{\infty} + \int_{-\infty}^{\infty} \frac{2 \sin \alpha \cos \alpha}{\alpha} \, d\alpha$$

$$= 0 + 2 \int_{-\infty}^{\infty} \frac{\sin \alpha \cos \alpha}{\alpha} \, d\alpha = 4 \int_{0}^{\infty} \frac{\sin \alpha \cos \alpha}{\alpha} \, d\alpha$$

since the integrand is an even function. By text equation (4.18)
with $x = 1$, this integral is $4(\pi/4) = \pi$. Thus

$$\int_{-\infty}^{\infty} |g|^2 \, d\alpha = \frac{1}{\pi^2} \cdot \pi = \frac{1}{\pi}, \qquad \text{and} \qquad \frac{1}{2\pi} \int_{-\infty}^{\infty} |f|^2 \, dx = \frac{1}{2\pi} \cdot 2 = \frac{1}{\pi},$$

so these integrals are equal as Parseval's theorem says.

24.   Start with the symmetrized Fourier transform pair

$$f(x) = \frac{1}{\sqrt{2\pi}} \int_{-\infty}^{\infty} g(\alpha) e^{i\alpha x} \, d\alpha,$$

$$g(\alpha) = \frac{1}{\sqrt{2\pi}} \int_{-\infty}^{\infty} f(x) e^{-i\alpha x} \, dx.$$

Let $\alpha = 2\pi p/h$, $f(x) = \psi(x)$, $g(\alpha) = \sqrt{\frac{h}{2\pi}} \phi(p)$; then

$$\psi(x) = f(x) = \frac{1}{\sqrt{2\pi}} \int_{-\infty}^{\infty} \sqrt{\frac{h}{2\pi}} \phi(p) e^{2\pi i p x/h} \frac{2\pi}{h} \, dp$$

$$= \frac{1}{\sqrt{h}} \int_{-\infty}^{\infty} \phi(p) e^{2\pi i p x/h} \, dp,$$

$$\phi(p) = \sqrt{\frac{2\pi}{h}} g(\alpha) = \sqrt{\frac{2\pi}{h}} \frac{1}{\sqrt{2\pi}} \int_{-\infty}^{\infty} \psi(x) e^{-2\pi i p x/h} \, dx$$

$$= \frac{1}{\sqrt{h}} \int_{-\infty}^{\infty} \psi(x) e^{-2\pi i p x/h} \, dx.$$

Section 6

1.   In order to use text equation (6.6) to find $f(t)$, we need the residues at all poles of

$$\frac{z^3 e^{zt}}{z^4 + 4}.$$

There are four simple poles at the four 4th roots of -4, namely at $z = \pm 1 \pm i$ (see text, Chapter 2, Section 10). By text page 599, equation (6.2), the residues are the values at the poles of

$$\frac{z^3 e^{zt}}{\frac{d}{dz}(z^4 + 4)} = \frac{z^3 e^{zt}}{4z^3} = \frac{1}{4} e^{zt}.$$

1.   (continued)

Thus by text equation (6.6),

$$f(t) = \frac{1}{4}\left[e^{(1+i)t} + e^{(1-i)t} + e^{(-1+i)t} + e^{(-1-i)t}\right]$$

$$= \frac{1}{4}e^t\left(e^{it} + e^{-it}\right) + \frac{1}{4}e^{-t}\left(e^{it} + e^{-it}\right)$$

$$= \frac{1}{2}\left(e^t + e^{-t}\right)\cos t = \cosh t \cos t.$$

8.   To find $f(t)$ using text equation (6.6), we need the residues

at $z = 0$ and at $z = -1$ of

$$\frac{(z-1)^2 e^{zt}}{z(z+1)^2}.$$

At $z = 0$, we use text page 599, equation (6.1), to find $R(0) = 1$.

At $z = -1$, there is a double pole. By Method C, text page 600,

we find

$$R(1) = \frac{d}{dz}\frac{(z-1)^2 e^{zt}}{z}\bigg|_{z=-1}$$

$$= \frac{2(z-1)e^{zt}}{z} + \frac{(z-1)^2 t e^{zt}}{z} - \frac{(z-1)^2 e^{zt}}{z^2}\bigg|_{z=-1}$$

$$= 4e^{-t} - 4te^{-t} - 4e^{-t} = -4te^{-t}.$$

Then by text equation (6.6),

$$f(t) = R(0) + R(1) = 1 - 4te^{-t}.$$

9. To use text equation (6.6), we need the residues of $\dfrac{ze^{zt}}{z^4-1}$

at the four poles, $z = \pm 1, \pm i$. As in Problem 1 above, we

find the values of

$$\frac{ze^{zt}}{4z^3} = \frac{1}{4}\frac{e^{zt}}{z^2}$$

at $z = \pm 1, \pm i$. Thus by text equation (6.6),

$$f(t) = \frac{1}{4}\frac{e^t + e^{-t}}{(\pm 1)^2} + \frac{e^{it}+e^{-it}}{(\pm i)^2} = \frac{1}{2}(\cosh t - \cos t).$$

Section 7

2. We want to find $f(t)$ given $F(p) = \dfrac{1}{p} e^{-pa} = G(p)H(p)$.   By L1 and

L27 [or text equation (7.7)]

$$g(t) = 1 \qquad \text{and} \qquad h(t) = \delta(t-a), \qquad a > 0.$$

Then using the convolution integral (L34), we find

$$g(t-\tau) = 1, \qquad h(\tau) = \delta(\tau-a),$$

$$f(t) = \int_0^t g(t-\tau)h(\tau)d\tau = \int_0^t \delta(\tau-a)d\tau = \begin{cases} 1, & t > a, \\ 0, & t < a. \end{cases}$$

To get the last step: The "peak" of the δ function is at $\tau = a$.

If we integrate across $\tau = a$, we get 1; otherwise we get zero.

Thus the integral is 1 if _a_ is between the upper and lower

limits $0 < a < t$. If $a > t$, then the integral is zero.

6a.   As on text page 655, we have

(1) $Y = \dfrac{1}{Ap^2 + Bp + C} F(p) = \dfrac{1}{A(p+a)(p+b)} F(p) = T(p)F(p).$

Let us use $g(t)$ to denote the inverse Laplace transform of $T(p)$
[since $t(t)$ might be confusing].   Then by L7 (text page 636),

(2) $g(t) = \dfrac{1}{A} \dfrac{e^{-at} - e^{-bt}}{b - a},$

and $f(t)$ is one of the functions (from text Figure 7.4)

(3) $f_n(t) = \begin{cases} n, & t_0 < t < t_0 + \dfrac{1}{n}, \\[2mm] 0, & \text{otherwise.} \end{cases}$

We assume that $t_0 > 0$ as shown in text Figure 7.4.   By (1), $y(t)$
is the inverse Laplace transform of $T(p)F(p)$.   We find $y_n(t)$ as
a convolution of $g(t)$ and $f_n(t)$ in (2) and (3) above, using L34
(text page 638).

(4) $y_n(t) = \displaystyle\int_0^t g(t - \tau) f_n(\tau) d\tau$

$ = \dfrac{1}{A(b - a)} \displaystyle\int_0^t \left[e^{-a(t-\tau)} - e^{-b(t-\tau)} \right] f_n(\tau) d\tau.$

Since $f_n(\tau) = 0$ except when $t_0 < \tau < t_0 + \dfrac{1}{n}$, the limits on the
integral are

(5)   from t_0 to $t_0 + \dfrac{1}{n}$ if $t_0 + \dfrac{1}{n} < t,$

(6)   from t_0 to t if $t_0 < t < t_0 + \dfrac{1}{n},$

6a.   (continued)

and the integral is zero if $t_0 > t$. For (5) we get

$$(7) \qquad y_n(t) = \frac{n}{A(b-a)} \int_{t_0}^{t_0 + \frac{1}{n}} \left[e^{-a(t-\tau)} - e^{-b(t-\tau)} \right] d\tau$$

$$= \frac{n}{A(b-a)a} \left[e^{-a(t-t_0-\frac{1}{n})} - e^{-a(t-t_0)} \right]$$

$$- \frac{n}{A(b-a)b} \left[e^{-b(t-t_0-\frac{1}{n})} - e^{-b(t-t_0)} \right].$$

For (6), replace the first exponential in each bracket by 1.
We shall not want this result, however, since, if $t_0 < t$, then
$t_0 + \frac{1}{n} < t$ for sufficiently large n and we want to let $n \to \infty$.
In (7), if we let $x = -a(t - t_0)$ and $\Delta x = \frac{a}{n}$, then as $n \to \infty$,
$\Delta x \to 0$ and we have

$$\frac{n}{a} \left[e^{-a(t-t_0-\frac{1}{n})} - e^{-a(t-t_0)} \right] = \frac{1}{\Delta x} \left(e^{x+\Delta x} - e^x \right) \longrightarrow \frac{d}{dx} e^x = e^x.$$

Similarly the corresponding terms containing b in (7) tend to
$e^{-b(t-t_0)}$, and $y_n(t) \to y(t)$ where

$$(8) \qquad y(t) = \frac{1}{A(b-a)} \left[e^{-a(t-t_0)} - e^{-b(t-t_0)} \right], \qquad t > t_0,$$

and $y(t) = y_n(t) = 0$ if $t < t_0$.

6b.   For the equation

$$Ay'' + By' + Cy = \delta(t - t_0), \quad y_0 = y_0' = 0,$$

we use L27 (text page 638) and (a) above to get

$$Y = \frac{1}{Ap^2 + Bp + C} e^{-pt_0} = \frac{1}{A(p + a)(p + b)} e^{-pt_0}.$$

In L28 (text page 638) replace a by t_0, replace f(t) by y(t) and let

$$G(p) = \frac{1}{A(p + a)(p + b)}.$$

Then g(t) is given by (2) in part (a) above. By L28, f(t) [our y(t)] is g(t) evaluated at $t - a = t - t_0$. Thus

$$y(t) = \begin{cases} \dfrac{1}{A(b - a)}\left[e^{-a(t-t_0)} - e^{-b(t-t_0)}\right], & t > t_0 > 0, \\[2mm] 0, & t < t_0, \end{cases}$$

as we found in (a).

6c.   In equation (2) above, g(t) is the inverse Laplace transform
      of the transfer function T(p). If the impulse is at $t_0 = 0$,
      then y(t) in (8) is the same as g(t) in (2); that is, the
      response of the system is the inverse Laplace transform of
      the transfer function for $t > 0$ and 0 for $t < 0$. If $t_0 > 0$,
      then the response is $g(t - t_0)$ for $t > t_0$ and zero for $t < t_0$
      (that is, before the impulse).

9.   Take the Laplace transform of each term in the differential
      equation assuming $y_0 = y_0' = 0$ (see Problem 6 and L35):

$$p^2 Y + 2pY + 10Y = e^{-pt_0},$$

$$Y = \frac{1}{p^2 + 2p + 10} e^{-pt_0} = \frac{1}{(p+1)^2 + 9} e^{-pt_0}.$$

We find y using L13 and L28. By L13, the inverse Laplace
transform of $\dfrac{1}{(p+1)^2 + 9}$ is $\frac{1}{3} e^{-t} \sin 3t$. Then by L28,

$$y = \begin{cases} \frac{1}{3} e^{-(t-t_0)} \sin 3(t - t_0), & t > t_0, \\ 0, & t < t_0. \end{cases}$$

Section 8

2.   Hint: Write $\sin \omega(t - t') \sin \omega t'$ as a difference of cosines
      by using the trigonometric identity

$$\sin x \sin y = \frac{1}{2} [\cos(x - y) - \cos(x + y)].$$

4. We are given

$$(1) \qquad f(t) = \begin{cases} 1, & 0 < t < a, \\ 0, & \text{otherwise}. \end{cases}$$

Then by text equation (8.6)

$$(2) \qquad y(t) = \int_0^t \frac{1}{\omega} \sin \omega(t - t') f(t') dt'.$$

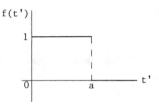

Now $f(t') = 1$ for $0 < t' < a$ (see figure).
If we put $f(t') = 1$ in (2), then the
upper limit is t if $t < a$, and the upper
limit is a if $t > a$.   Thus

$$y(t) = \begin{cases} \dfrac{1}{\omega}\displaystyle\int_0^t \sin \omega(t - t') dt', & t < a, \\[2ex] \dfrac{1}{\omega}\displaystyle\int_0^a \sin \omega(t - t') dt', & t > a, \end{cases}$$

$$= \begin{cases} \dfrac{1}{\omega^2} \cos \omega(t - t') \Big|_0^t = \dfrac{1}{\omega^2}(1 - \cos \omega t), & t < a, \\[2ex] \dfrac{1}{\omega^2} \cos \omega(t - t') \Big|_0^a = \dfrac{1}{\omega^2}[\cos \omega(t - a) - \cos \omega t], & t > a. \end{cases}$$

6. Hint: Solve by Laplace transforms the differential equation

$$G'' - a^2 G = \delta(t - t'), \qquad G_0 = G_0' = 0.$$

13.   By text equation (8.17)

(1) $y = -(\cos x) \int_0^x (\sin x') f(x') \, dx'$

$$- (\sin x) \int_x^{\pi/2} (\cos x') f(x') \, dx'.$$

We sketch the given function $f(x')$

(2) $f(x') = \begin{cases} x', & 0 < x' < \pi/4, \\[2mm] \dfrac{\pi}{2} - x', & \dfrac{\pi}{4} < x' < \dfrac{\pi}{2}. \end{cases}$

Now we use the graph to see what
$f(x')$ is in each of the integrals
in (1). This depends on whether
$x < \dfrac{\pi}{4}$ or $x > \dfrac{\pi}{4}$.

For $x < \dfrac{\pi}{4}$:

In the integral from 0 to x, we see from the graph that
$f(x') = x'$. But the integral from $x' = x$ to $x' = \dfrac{\pi}{2}$ must be written
in two parts. For x' between x and $\dfrac{\pi}{4}$, we have $f(x') = x'$, but
for x' between $\dfrac{\pi}{4}$ and $\dfrac{\pi}{2}$, we have $f(x') = \dfrac{\pi}{2} - x'$. Thus the integrals in (1) are:

$$\int_0^x (\sin x') f(x') \, dx' = \int_0^x (\sin x') x' \, dx' = \sin x - x \cos x,$$

$$\int_x^{\pi/2} (\cos x') f(x') \, dx' = \int_x^{\pi/4} x' \cos x' \, dx'$$
$$+ \int_{\pi/4}^{\pi/2} \left(\frac{\pi}{2} - x' \right) \cos x' \, dx'$$

$$= \cos x' + x' \sin x' \Big|_x^{\pi/4} + \frac{\pi}{2} \sin x' - (\cos x' + x' \sin x') \Big|_{\pi/4}^{\pi/2}$$

$$= -\cos x - x \sin x + \sqrt{2}$$

13.   (continued)

(after some algebra). Substitute these results into (1) to
find $y(x)$ when $x < \frac{\pi}{4}$.

$$y(x) = -(\cos x)(\sin x - x \cos x) - (\sin x)(-\cos x - x \sin x + \sqrt{2})$$

$$= x - \sqrt{2} \sin x, \quad x < \frac{\pi}{4} .$$

For $x > \frac{\pi}{4}$:

This time the integral from 0 to x must be written as a sum
of two integrals. For x' between 0 and $\pi/4$, we have $f(x') = x'$
(see graph), and for x' between $\pi/4$ and x, we have $f(x') = \frac{\pi}{2} - x'$.
For the integral from x to $\pi/2$, we have $f(x') = \frac{\pi}{2} - x'$. Substitute these into the integrals in (1) and evaluate as above to
get $y(x)$ for $x > \frac{\pi}{4}$. Thus find

$$y(x) = \begin{cases} x - \sqrt{2} \sin x, & x < \frac{\pi}{4} , \\[2mm] \frac{\pi}{2} - x - \sqrt{2} \cos x, & x > \frac{\pi}{4} . \end{cases}$$

It is straightforward to verify that $y'' + y = f(x)$ and that
$y(0) = y(\pi/2) = 0$ (check these).

14.   See comments in Problem 16 below.

16.   Write the differential equation in the form of text Equation
(8.19):

(1) $y'' - \frac{2}{x} y' + \frac{2}{x^2} y = \frac{1}{x} \ln x.$

We want to use text equation (8.21) to find a particular solution
of this equation given the solutions $y_1 = x$ and $y_2 = x^2$ of the

16.   (continued)

corresponding homogeneous equation. We first find the Wronskian

of y_1 and y_2.

$$W = \begin{vmatrix} x & x^2 \\ 1 & 2x \end{vmatrix} = x^2 .$$

Then we find $y_p(x)$ from text equation (8.21) with $y_1 = x$, $y_2 = x^2$,

$W = x^2$, $f(x) = \frac{1}{x} \ell n\, x$ [compare equation (1) above and text equation

(8.19)].

$$y_p = x^2 \int \frac{x}{x^2} \frac{\ell n\, x}{x} \, dx - x \int \frac{x^2}{x^2} \frac{\ell n\, x}{x} \, dx = x^2 \int \frac{\ell n\, x}{x^2} \, dx - x \int \frac{\ell n\, x}{x} \, dx$$

$$= x^2 \left(-\frac{\ell n\, x}{x} - \frac{1}{x} \right) - x \left[\frac{1}{2} (\ell n\, x)^2 \right] = - x\, \ell n\, x - x - \frac{1}{2} x (\ell n\, x)^2 .$$

Comment: You might wonder here whether it is correct to use

text equation (8.21) when the functions $y_1 = x$ and $y_2 = x^2$ do not

satisfy $y_1(a) = 0$, $y_2(b) = 0$ for any a and b except $a = b = 0$. Yes,

it is correct, but let us see why. The significant point to

consider is whether we want a particular solution which satisfies

certain boundary conditions or whether we are satisfied with just

any particular solution. Green functions are most often used to

obtain solutions of a differential equation which satisfy speci-

fied boundary conditions, and text equations (8.17) and (8.20)

give such solutions. [In (8.17), $y = 0$ at $x = 0$ and at $x = \frac{\pi}{2}$.   In

(8.20), $y = 0$ at $x = a$ and at $x = b$.] However, in text equations

(8.18) and (8.21) we no longer have solutions satisfying these

boundary conditions, but we do still have solutions of the given

differential equations [(8.18) is a solution of (8.7) and (8.21)

16.   (continued)

is a solution of (8.19)]. This is what is meant by a particular

solution y_p when no conditions are specified. Since we are not

specifying boundary conditions for the solution y_p, there is no

need to specify them for the Green function. Thus if we wanted

to do this problem without using text equation (8.21) we could

proceed as follows. For the Green function we use

$$G(x,x') = \begin{cases} Cx'^2x, & x < x', \\ Cx'x^2, & x > x'. \end{cases}$$

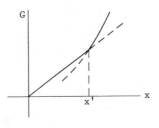

The factor C is the same in both cases

to make G continuous at x'. To make

the change in slope equal to 1 at $x = x'$

(see text, page 672), we have

$$Cx' \frac{d}{dx}x^2 - Cx'^2 \frac{dx}{dx}\bigg|_{x=x'} = 1,$$

$$C = \frac{1}{W} \qquad \text{where} \qquad W = \begin{vmatrix} x' & x'^2 \\ 1 & 2x' \end{vmatrix} = x'^2.$$

Then a solution of the differential equation is given by

$$y_p = \int G(x,x')f(x')dx'$$

with any limits we like as long as we cross the point $x' = x$.

Thus

$$y_p = \int^x Cx'x^2f(x')dx' + \int_x Cx'^2xf(x')dx'$$

$$= x^2 \int^x \frac{x'f(x')}{W}dx' - x \int^x \frac{x'^2f(x')}{W}dx'$$

as above.

22.   Green's second identity (text page 281) is

(1) $\int (\phi \nabla^2 \psi - \psi \nabla^2 \phi) d\tau = \oint (\phi \nabla \psi - \psi \nabla \phi) \cdot \vec{n} \, d\sigma,$

where the τ integral is over a volume and the σ integral is
over the closed surface bounding the volume. Let

$$\psi = G(\vec{r}, \vec{r}') = -\frac{1}{4\pi |\vec{r} - \vec{r}'|} + F(\vec{r}, \vec{r}'),$$

$$\phi = u(\vec{r}),$$

where, by text equations (8.22), (8.23), and (8.28)

$$\nabla^2 \phi = \nabla^2 u = f(\vec{r}),$$

$$\nabla^2 F = 0,$$

$$\nabla^2 \psi = \nabla^2 G = \delta(\vec{r} - \vec{r}'),$$

$$\psi = G = 0 \text{ on the surface } \sigma.$$

We substitute these equations into (1) and integrate with
respect to the primed variables.

$$\int [u(\vec{r}') \delta(\vec{r} - \vec{r}') - G(\vec{r}, \vec{r}') f(\vec{r}')] d\tau' = \oint [u(\vec{r}') \nabla G - 0] \cdot \vec{n}' \, d\sigma'$$

or

$$u(\vec{r}) = \int G(\vec{r}, \vec{r}') f(\vec{r}') d\tau' + \oint u(\vec{r}') \frac{\partial G}{\partial n'} \, d\sigma'.$$

Section 9

3.   Take the t Laplace transform of the heat flow equation for $u(x,t)$,

(1) $\dfrac{\partial^2 u}{\partial x^2} = \dfrac{1}{\alpha^2} \dfrac{\partial u}{\partial t}$

with $u_0 = u(x,0) = 100x/\ell$. This gives the differential equation
for $U(x,p)$:

3.   (continued)

(2) $\dfrac{\partial^2 U}{\partial x^2} = \dfrac{1}{\alpha^2}(pU - u_0) = \dfrac{1}{\alpha^2} pU - \dfrac{100x}{\alpha^2 \ell}$.

The solutions of $\dfrac{\partial^2 U}{\partial x^2} = \dfrac{p}{\alpha^2} U$ are $\sinh\left(x\sqrt{\dfrac{p}{\alpha^2}}\right)$ and $\cosh\left(x\sqrt{\dfrac{p}{\alpha^2}}\right)$.

If $U = Kx$, then $\dfrac{\partial^2 U}{\partial x^2} = 0$, and $\dfrac{p}{\alpha^2} U = \dfrac{100x}{\alpha^2 \ell}$ if $K = \dfrac{100}{p\ell}$.

Thus the general solution of (2) is

(3) $U(x,p) = A \sinh(xp^{1/2}/\alpha) + B \cosh(xp^{1/2}/\alpha) + \dfrac{100x}{p\ell}$.

Since the temperature at $x = 0$ is held at $0°$ for all t, we have

$u(0,t) = 0$ so

$\qquad U(0,p) = [\text{Laplace transform of } u(0,t)] = 0$.

Similarly

$\qquad u(\ell,t) = 0,\qquad U(\ell,p) = 0$.

Substitute these values into (3) to get

$\qquad B = 0,\quad A \sinh(\ell p^{1/2}/\alpha) + \dfrac{100}{p} = 0,\quad A = \dfrac{-100}{p \sinh(\ell p^{1/2}/\alpha)}$.

Thus (3) becomes

$\qquad U(x,p) = - \dfrac{100 \sinh(xp^{1/2}/\alpha)}{p \sinh(\ell p^{1/2}/\alpha)} + \dfrac{100x}{p\ell}$.

We assume the expansion given:

$\qquad \dfrac{100 \sinh(xp^{1/2}/\alpha)}{p \sinh(\ell p^{1/2}/\alpha)} = \dfrac{100x}{p\ell} - \dfrac{200}{\pi} \sum_{n=1}^{\infty} \dfrac{(-1)^{n+1} \sin(n\pi x/\ell)}{n[p + (n\pi\alpha/\ell)^2]}$.

Take inverse Laplace transforms of each of the terms using L2

(text page 636) to get

$\qquad u(x,t) = \dfrac{200}{\pi} \sum_{n=1}^{\infty} \dfrac{(-1)^{n+1}}{n} e^{-(n\pi\alpha/\ell)^2 t} \sin(n\pi x/\ell)$

as on text page 552, equation (3.15).

6.   Change the order of integration to get

$$u(x,y) = \frac{200}{\pi} \int_0^1 dt \int_0^\infty e^{-ky} \sin kx \sin kt \, dk.$$

Use identity

$$\sin kx \sin kt = \frac{1}{2}[\cos k(x - t) - \cos k(x + t)].$$

The easiest way to evaluate integrals of the form $\int_0^\infty e^{-au} \cos bu \, du$
is to use a table of Laplace transforms. From L4 on text page
636, we have

$$\int_0^\infty e^{-au} \cos bu \, du = a/(a^2 + b^2).$$

Use this result with $u = k$, $a = y$, $b = (x - t)$, to find

$$\int_0^\infty e^{-ky} \cos k(x - t) \, dk = \frac{y}{y^2 + (x - t)^2},$$

and similarly for the $\cos k(x + t)$ integral.   Thus

$$u(x,y) = \frac{200}{\pi} \int_0^1 \frac{1}{2}\left[\frac{y}{y^2 + (x - t)^2} - \frac{y}{y^2 + (x + t)^2}\right] dt$$

$$= \frac{100}{\pi}\left[- \tan^{-1}\frac{x - t}{y} - \tan^{-1}\frac{x + t}{y}\right]_{t=0}^1$$

$$= \frac{100}{\pi}\left[2 \tan^{-1}\frac{x}{y} - \tan^{-1}\frac{x - 1}{y} - \tan^{-1}\frac{x + 1}{y}\right]$$

as in text equation (9.14).

Section 10

3.   By L13 (text page 636),

$$L(e^{-at} \sin at) = \frac{a}{(p + a)^2 + a^2},$$

$$L(e^{at} \sin at) = \frac{a}{(p - a)^2 + a^2}.$$

3. (continued)

Recall that $\sinh at = \frac{1}{2}(e^{at} - e^{-at})$. Thus

$$L(\sinh at \sin at) = \frac{1}{2}\left[\frac{a}{(p-a)^2 + a^2} - \frac{a}{(p+a)^2 + a^2}\right]$$

$$= \frac{a}{2}\frac{(p+a)^2 + a^2 - (p-a)^2 - a^2}{(p-a)^2(p+a)^2 + a^2(2p^2 + 2a^2) + a^4} = \frac{2a^2 p}{p^4 + 4a^4}.$$

Let's verify this result by finding the inverse transform using
text equation (6.6), page 663. We find the residues of

$$\frac{2a^2 ze^{zt}}{z^4 + 4a^4}$$

at the four poles. The easiest way to do this is to use text
equation (6.2), page 599.

$$\frac{2a^2 ze^{zt}}{\frac{d}{dz}(z^4 + 4a^4)} = \frac{2a^2 ze^{zt}}{4z^3} = \frac{e^{zt}}{2z^2/a^2}.$$

We need the values of z and of $(z/a)^2$ at the poles. These are
(see text, Chapter 2, Section 10, if needed)

$$z = a(-4)^{1/4} = a\sqrt{2}\,e^{i\pi/4}\,e^{in\pi/2}$$

$$= a(1+i),\ a(-1+i),\ a(-1-i),\ a(1-i).$$

$$(z/a)^2 = \quad 2i, \qquad -2i, \qquad 2i, \qquad -2i.$$

Then the inverse transform is (text, page 663)

$$\frac{1}{2}\frac{1}{2i}\left[e^{a(1+i)t} - e^{a(-1+i)t} + e^{a(-1-i)t} - e^{a(1-i)t}\right]$$

$$= \frac{1}{2}(e^{at} - e^{-at})\frac{1}{2i}(e^{iat} - e^{-iat}) = \sinh at \sin at.$$

5. Hint: Use L26 and L32.

6,7,8.Hint: Try several methods and see which is simpler (it depends
          on the problem). Methods include the convolution integral, the
          Bromwich integral, combining transforms in the tables, differen-
          tiating or integrating with respect to a parameter.

13.   From Problem 23.20 of Chapter 12, with $n = 0$, we have

$$J_0(x) = \frac{1}{\pi}\int_0^\pi \cos(x \sin \theta)\,d\theta.$$

Since $\sin \theta$ (or any function of $\sin \theta$) is symmetric about $\theta = \pi/2$,
we can write $\int_0^\pi f(\sin \theta)\,d\theta = 2\int_0^{\pi/2} f(\sin \theta)\,d\theta.$ Thus

$$J_0(x) = \frac{2}{\pi}\int_0^{\pi/2} \cos(x \sin \theta)\,d\theta.$$

Now let $\alpha = \sin \theta$.   Then $d\alpha = \cos \theta\, d\theta$, so

$$d\theta = \frac{d\alpha}{\cos \theta} = \frac{d\alpha}{\sqrt{1 - \alpha^2}}\,, \qquad \alpha \text{ limits are 0 to 1,}$$

$$J_0(x) = \frac{2}{\pi}\int_0^1 \frac{\cos \alpha x}{\sqrt{1 - \alpha^2}}\,d\alpha.$$

This says that J_0 is the Fourier cosine transform of the function

$$g(\alpha) = \sqrt{\frac{2}{\pi}}\begin{cases} \dfrac{1}{\sqrt{1 - \alpha^2}}\,, & 0 \leqslant \alpha < 1, \\[2ex] 0, & \alpha > 1. \end{cases}$$

Then $g(\alpha)$ is the inverse transform of $J_0(x)$, that is

$$g(\alpha) = \sqrt{\frac{2}{\pi}}\int_0^\infty J_0(x)\cos \alpha x\, dx \qquad \text{so}$$

13.   (continued)

$$\int_0^\infty J_0(x)\cos \alpha x\, dx = \begin{cases} \dfrac{1}{\sqrt{1-\alpha^2}}\,, & 0 \leqslant \alpha < 1, \\[2mm] 0, & \alpha > 1. \end{cases}$$

For $\alpha = 0$ we have

$$\int_0^\infty J_0(x)\, dx = 1.$$

Also see Problem 8, text page 515.

17.   You might guess that for the symmetrized sine and cosine
integrals given in text equations (4.14) and (4.15), the
integrals of $|f|^2$ and $|g|^2$ would be equal. This is correct;
let us verify it for the cosine transforms. We have

$$f_c(x) = \sqrt{\frac{2}{\pi}} \int_0^\infty g_c(\alpha)\cos \alpha x\, d\alpha,$$

$$g_c(\alpha) = \sqrt{\frac{2}{\pi}} \int_0^\infty f_c(x)\cos \alpha x\, dx.$$

Since $\cos \alpha x$, $g_c(\alpha)$, and $f_c(x)$ are all even, we can write the
integrals as one-half the $-\infty$ to ∞ integrals. We can also
replace $\cos \alpha x$ by $e^{i\alpha x}$ or $e^{-i\alpha x}$ in the $-\infty$ to ∞ integrals since
the extra $\sin \alpha x$ is odd. Thus we can write

$$f_c(x) = \sqrt{\frac{1}{2\pi}} \int_{-\infty}^\infty g_c(\alpha)\, e^{i\alpha x}\, d\alpha,$$

$$g_c(\alpha) = \sqrt{\frac{1}{2\pi}} \int_{-\infty}^\infty f_c(x)\, e^{-i\alpha x}\, dx,$$

or

17.   (continued)

$$\sqrt{2\pi}\, f_c(x) = \int_{-\infty}^{\infty} g_c(\alpha) e^{i\alpha x}\, d\alpha,$$

$$g_c(\alpha) = \frac{1}{2\pi} \int_{-\infty}^{\infty} \sqrt{2\pi}\, f_c(x) e^{-i\alpha x}\, dx.$$

These equations are now of the form of text equation (4.2) if
we let $g(\alpha) = g_c(\alpha)$ and $f(x) = \sqrt{2\pi}\, f_c(x)$.   Thus text equation
(5.17) applies so we have

$$\int_{-\infty}^{\infty} |g_c(\alpha)|^2 d\alpha = \frac{1}{2\pi} \int_{-\infty}^{\infty} |\sqrt{2\pi}\, f_c(x)|^2 dx = \int_{-\infty}^{\infty} |f_c(x)|^2 dx.$$

Since $|g|^2$ and $|f|^2$ are even functions, we can write

$$\int_{0}^{\infty} |g_c(\alpha)|^2 d\alpha = \int_{0}^{\infty} |f_c(x)|^2 dx.$$

19.   (a) In Problem 4.11, we had

$$f(x) = \begin{cases} \cos x, & -\pi/2 < x < \pi/2, \\ 0, & |x| > \pi/2, \end{cases}$$

$$g(\alpha) = \frac{\cos(\alpha\pi/2)}{\pi(1 - \alpha^2)}.$$

Thus by text equation (5.17)

$$\int_{-\infty}^{\infty} \frac{\cos^2(\alpha\pi/2)}{\pi^2(1 - \alpha^2)^2}\, d\alpha = \frac{1}{2\pi} \int_{-\pi/2}^{\pi/2} \cos^2 x\, dx.$$

The integral of $\cos^2 x$ is $\pi/2$ (see text, page 306) and the $-\infty$
to ∞ integral is twice the 0 to ∞ integral of an even function.
Thus

$$\int_{0}^{\infty} \frac{\cos^2(\alpha\pi/2)}{(1 - \alpha^2)^2}\, d\alpha = \frac{1}{2}\, \pi^2\, \frac{1}{2\pi}\, \frac{\pi}{2} = \frac{\pi^2}{8}.$$

19.   (continued)

(b) We consider

$$\oint_C \frac{1+e^{i\pi z}}{(z-1)^2(z+1)^2} \, dz$$

around the upper half plane. We evaluate this integral as discussed on text page 605. We first find the residues at the poles. Since $e^{i\pi} = e^{-i\pi} = -1$, they are simple poles.

$$R(1) = \lim_{z\to 1} \frac{1+e^{i\pi z}}{(z-1)(z+1)^2} = \frac{1}{4}\lim_{z\to 1}\frac{i\pi e^{i\pi z}}{1} = -\frac{i\pi}{4}\,.$$

Similarly (check it),

$$R(-1) = -\frac{i\pi}{4}\,.$$

Then the contour integral is

$$\frac{1}{2}(2\pi i)(2)\left(-\frac{i\pi}{4}\right) = \pi^2/2.$$

As discussed on text page 605, the integral along the large semicircle tends to zero and we have

$$\int_{-\infty}^{\infty} \frac{1+e^{i\pi x}}{(x-1)^2(x+1)^2}\,dx = \frac{\pi^2}{2} = \int_{-\infty}^{\infty}\frac{1+\cos \pi x}{(x^2-1)^2}\,dx$$

(take the real part). The integrand is even so we can write the 0 to ∞ integral. We also use the identity

$$1+\cos \pi x = 2\cos^2(\pi x/2).$$

Thus we get

$$\int_0^{\infty} \frac{\cos^2(\pi x/2)}{(x^2-1)^2}\,dx = \frac{\pi^2}{8}$$

as in part (a).

21.   (a) Following the text outline, we write:

$$h(x) = \sum_{k=-\infty}^{\infty} f(x + 2k\pi) = \sum_{n=-\infty}^{\infty} c_n e^{inx},$$

$$c_n = \frac{1}{2\pi} \int_0^{\pi} h(x) e^{-inx} \, dx = \sum_{k=-\infty}^{\infty} \frac{1}{2\pi} \int_0^{2\pi} f(x + 2k\pi) e^{-inx} \, dx.$$

Let $u = x + 2k\pi$; then $e^{-inx} = e^{-in(u-2k\pi)} = e^{-inu}$, so

$$c_n = \sum_{k=-\infty}^{\infty} \frac{1}{2\pi} \int_{2k\pi}^{2(k+1)\pi} f(u) e^{-inu} \, du = \frac{1}{2\pi} \int_{-\infty}^{\infty} f(u) e^{-inu} \, du.$$

Now by text equation (4.2)

$$g(\alpha) = \frac{1}{2\pi} \int_{-\infty}^{\infty} f(u) e^{-i\alpha u} \, du, \qquad \text{so}$$

$$c_n = g(n).$$

(b) From (a), we have

$$\sum_{k=-\infty}^{\infty} f(x + 2k\pi) = \sum_{n=-\infty}^{\infty} g(n) e^{inx}.$$

Put $x = 0$ to get Poisson's summation formula,

$$\sum_{k=-\infty}^{\infty} f(2k\pi) = \sum_{n=-\infty}^{\infty} g(n).$$

Chapter 16

2. We can think of tossing three coins one at a time, or of tossing
    one coin three times in a row, or of tossing three coins labeled
    1, 2, 3, all at once. These are equivalent experiments and there
    are eight possible outcomes, namely

hhh	hht	hth	thh
ttt	tth	tht	htt

    These outcomes are equally likely, mutually exclusive, and
    collectively exhaustive. By actual count of the number of
    favorable cases, we see that:

    Probability of two heads and one tail = 3/8.

    Probability of hht is 1/8.

    If we know that there are at least two heads, then there are
    just four possible outcomes, namely the first line of the table
    above. Of these, one outcome is three heads. Thus:

    (Probability of three heads
    assuming at least two heads) = 1/4.

5. Hint: Compare the discussion of text equation (1.1).

7. Hint: How many different letters are there in the word
    "probability"?

10. Caution: The problem says "on the same side of the mall", not
    necessarily the same entrance. Call the entrances N, S, E_1, E_2;
    "at random" means these four are equally likely.

Section 2

2.   The sample space is text equation (2.3) or the table in Prob-
     lem 1.2 above, with probability 1/8 for each point. We find
     three sample points corresponding to two heads and one tail
     (hht, hth, thh) so the probability of this is $3 \cdot \frac{1}{8} = 3/8$. The
     probability that the first two are heads and the third tails
     (sample point hht) is 1/8. If at least two are heads, the
     sample space is

          hhh   hht   hth   thh

     with probability 1/4 attached to each point. Then the proba-
     bility of all heads is 1/4.

5.   Hint: For the third part, see Problem 13. For the last part,
     consider an appropriate part of the sample space for a three
     child family [compare text equation (2.3)].

10. Hint: See Problem 1.10. Set up a sample space like text equa-
     tion (2.4) using symbols such as E_2,N meaning that you enter
     through the second entrance on the East, and exit on the North.

13. Here are the uniform sample spaces and resulting answers for
     three different problems:
     (a) If the older child is a girl, what is the probability that
         both are girls? Sample space: gg   gb,   p = 1/2.
     (b) If the younger child is a girl, what is the probability that
         both are girls? Sample space: gg   bg,   p = 1/2.
     (c) If at least one child is a girl, what is the probability
         that both are girls? Sample space: gg   bg   gb,   p = 1/3.

16.   By text definition (1.2)

$$p_1 = N_1/N, \quad p_2 = N_2/N, \quad \cdots$$

and by text definition, page 688,

$$p_1 = N_1 \cdot \frac{1}{N}, \quad p_2 = N_2 \cdot \frac{1}{N}, \quad \cdots.$$

These are the same.   Then

$$p_1 + p_2 + \cdots + p_n = (N_1 + N_2 + \cdots + N_n)/N = 1$$

since

$$N_1 + N_2 + \cdots + N_n = N.$$

By text equation (1.2)

$$p = (N_i + N_j + N_k)/N.$$

By text definition, page 688,

$$p = p_i + p_j + p_k = (N_i + N_j + N_k)/N.$$

Thus the two definitions give the same result.

Section 3

2.   Using the notation given in the problem, we have

$$P(AB) = p_{n+1} + p_{n+2} + \cdots + p_{n+k},$$

AB: n+1 to n+k

$$P(A) = p_1 + p_2 + \cdots + p_{n+k}.$$

A: 1 to n+k   AB   B: n+1 to n+k+ℓ

Now to find $P_A(B)$, we must first find the new value P_i to be associated with the sample points in A when we remove the rest of the sample space [compare the text discussion of equation (1.1) where we crossed off the sample point tt]. We want the values of P_i to be proportional to the values of p_i, and we also want

$$P_1 + P_2 + \cdots + P_{n+k} = 1$$

since A is now the whole sample space. These conditions are satisfied if we take

$$P_i = p_i / (p_1 + p_2 + \cdots + p_{n+k}) = p_i / P(A).$$

Then

$$P_A(B) = P_{n+1} + P_{n+2} + \cdots + P_{n+k} = \frac{1}{P(A)} (p_{n+1} + p_{n+2} + \cdots + p_{n+k})$$

so

$$P(AB) = P(A) P_A(B)$$

as claimed in text equation (3.8).

9.   Let S mean "spade" and N mean "not spade". Then the symbol $N_1 S_2 S_3$ means: First card drawn is not a spade, second card drawn is a spade, third card drawn is a spade. In this notation, the possible ways of discarding two cards and then drawing a spade are:

9.   (continued)

(1) $S_1 S_2 S_3$, $S_1 N_2 S_3$, $N_1 S_2 S_3$, $N_1 N_2 S_3$.

These events are mutually exclusive so by text equation (3.7)
we add their probabilities. Now to compute the probability of,
say, $S_1 N_2 S_3$ we use text equation (3.5) repeatedly. The probability of S_1 is 13/52. Then the probability of N_2 is 39/51
since there are now 51 cards left and 39 of them are not spades.
Now the probability of S_3 is 12/50 since there are 12 spades
left and 50 cards left. Thus the probability of $S_1 N_2 S_3$ is

$$\frac{13}{52} \cdot \frac{39}{51} \cdot \frac{12}{50} .$$

Similarly we compute all four of the probabilities in (1) (you
should verify the other three) and add them to get

$$\frac{13}{52} \cdot \frac{12}{51} \cdot \frac{11}{50} + \frac{13}{52} \cdot \frac{39}{51} \cdot \frac{12}{50} + \frac{39}{52} \cdot \frac{13}{51} \cdot \frac{12}{50} + \frac{39}{52} \cdot \frac{38}{51} \cdot \frac{13}{50}$$

$$= \frac{13}{52} \left[\frac{12(11+39) + 39(12+38)}{51 \cdot 50} \right] = \frac{1}{4} .$$

10.   Hint: Set up the sample space like this:

	A	B	C
(1)	c	b	a
(2)	a	c	b

and so on. Lines (1) and (2) are two sample points. Set up the
rest of the sample space; how many sample points are there?

13.   (a) Hints: Set up the sample space like this:

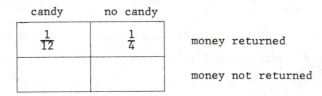

The probability that you get your money back (given $= \frac{1}{3}$) is the sum of the probabilities of "money + candy" and "money, no candy", so you can find the probability of "money, no candy" $= \frac{1}{3} - \frac{1}{12} = \frac{1}{4}$. Similarly, complete the sample space (remember that $\sum p_i = 1$ for any sample space).

16.   Let A mean "nickel" and B mean "right pocket".   Then we want the conditional probability $P_A(B)$.   See Bayes's formula, text equation (3.8).   Let us find $P(A)$ and $P(AB)$. $P(AB)$ is the probability that you select a nickel from your right pocket.   This is

$$P(AB) = \text{probability of selecting the right pocket}$$
$$\text{and then selecting a nickel from it}$$

$$= \frac{1}{2} \cdot \frac{3}{7} = \frac{3}{14}$$

since there are two equally likely pockets and then 3 nickels out of 7 coins in your right pocket.   To find $P(A)$ we add the probabilities of the mutually exclusive events of selecting a nickel from the right pocket or from the left pocket:

$$P(A) = \frac{1}{2} \cdot \frac{3}{7} + \frac{1}{2} \cdot \frac{2}{3} = \frac{23}{42}.$$

Then by text equation (3.8),

$$P_A(B) = \frac{3}{14} \div \frac{23}{42} = \frac{9}{23}.$$

22. From text equation (2.4), we see that the probability of a double when two dice are tossed is $\frac{1}{6}$ (since there are 6 favorable sample points out of 36). Let D mean "double" and N mean "not double"; then the probability of D is $\frac{1}{6}$ and the probability of N is $\frac{5}{6}$. Let A be the first player and B the second player. In the table below, A_1, A_2, ... mean first, second, ... toss for the first player, and B_1, B_2, ... mean first, second, ... toss for the second player. We outline the events which give a win for player A at A_1, A_2, ..., and their probabilities. Note that if A is to win at, say, A_3, all previous tosses for both A and B must have been N (not double).

$$A_1 \quad B_1 \quad A_2 \quad B_2 \quad A_3 \quad \cdots$$

A wins at A_1:	D					$p_1 = 1/6$
A wins at A_2:	N	N	D			$p_2 = (5/6)^2 (1/6)$
A wins at A_3:	N	N	N	N	D	$p_3 = (5/6)^4 (1/6)$

$\cdots$ $\cdots$

The total probability is an infinite series (geometric, see text, page 3) which we can sum to get

$$p = \frac{1}{6}\left[1 + \left(\frac{5}{6}\right)^2 + \left(\frac{5}{6}\right)^4 + \cdots\right] = \frac{1}{6}\frac{1}{1 - (5/6)^2} = \frac{6}{11}.$$

This is the probability of a win for the first player. Then the probability of a win for the second player is $1 - p = \frac{5}{11}$.

Section 4

4.   Hints: The uniform sample space consists of all the possible
     hands of 5 cards chosen from 52 cards. How many of these hands
     are all spades, that is, how many sets of 5 cards can you choose
     from the 13 spades? How many hands are all face cards, etc.?

5.   Hint: Compare text, page 718, Example 1.

6.   Hint: Include the completely off position in your calculation
     and then subtract 1. Use text equation (4.1).

8.   For the first part, the uniform sample space consists of all
     possible sets of 2 cards chosen from 52 cards. The number of
     sample points is $C(52,2)$. There are 4 aces, so $C(4,2)$ is the
     number of sample points corresponding to a pair of aces. Thus
     the probability of drawing 2 aces when 2 cards are drawn from
     a shuffled deck is

$$\frac{C(4,2)}{C(52,2)} = \frac{4!50!2!}{2!2!52!} = \frac{12}{52 \cdot 51} = \frac{1}{221} \, .$$

     If at least one of the 2 cards is an ace then the sample
     space consists of all sets of "one ace, one not ace" (there are
     $4 \cdot 48$ of these) plus all sets of "2 aces" [there are $C(4,2)$ of
     these]. Thus the probability of 2 aces if at least one is an
     ace is

$$\frac{C(4,2)}{4 \cdot 48 + C(4,2)} = \frac{6}{4 \cdot 48 + 6} = \frac{1}{33} \, .$$

     Similarly, we find the probability of 2 aces if one is the
     ace of spades:

$$\frac{3}{48 + 3} = \frac{1}{17} \, .$$

8.   (continued)

Students often ask why knowing that one card is a particular ace gives a higher probability of 2 aces than just knowing that at least one card is an ace. Notice that the sample space is much smaller when one card is the ace of spades; there are only 51 two-card sets containing the ace of spades, but there are 198 two-card sets containing at least one ace. Thus the probability associated with each sample point is greater when one card is the ace of spades $\left(\frac{1}{51} > \frac{1}{198}\right)$.

These last two parts of the problem are conditional probability problems and so can be done by Bayes's formula [text equation (3.8)]. You might like to solve them that way.

10.   The probability that two people have different birthdays is the probability that the second person was born on one of the 364 days other than the birthday of the first person; this probability is $\frac{364}{365}$. The probability that the third person has a different birthday from either of the first two is $\frac{363}{365}$, and so on. Now by text equation (3.1), the probability that the first two people have different birthdays, and then that the third person has a still different birthday, is the product

$$P(AB) = \frac{364}{365} \cdot \frac{363}{365} = \left(1 - \frac{1}{365}\right)\left(1 - \frac{2}{365}\right).$$

Continuing in this way for n people, we have

$$p = \left(1 - \frac{1}{365}\right)\left(1 - \frac{2}{365}\right) \cdots \left(1 - \frac{n-1}{365}\right).$$

10.   (continued)

By text, page 24, equation (13.4), we have for $n - 1 \ll 365$,

$$\ell n\ p = -\frac{1}{365} - \frac{2}{365} - \cdots - \frac{n-1}{365} = -\frac{1+2+\cdots+(n-1)}{365}$$

$$= -\frac{n(n-1)}{2(365)} .$$

If $p < \frac{1}{2}$, $\ell n\ p < \ell n\ \frac{1}{2} = -\ell n\ 2$, so we want

$$-\frac{n(n-1)}{2(365)} < -\ell n\ 2 \qquad \text{or} \qquad \frac{n(n-1)}{2(365)} > \ell n\ 2 .$$

We find $n \geqslant 23$. Thus in a group of 23 people, the probability is slightly over $\frac{1}{2}$ that two will have the same birthday. For 50 people, the probability of a coincidence is about 0.97.

15.   For Maxwell-Boltzmann statistics, the particles are distinguish-able (we label them 1 and 2) and there are $3^2 = 9$ possible ways of putting 2 particles in 3 boxes. The probability of each arrangement is $\frac{1}{9}$.

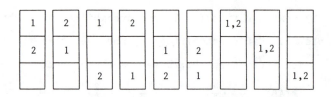

15.   (continued)

For Fermi-Dirac statistics, the particles are indistinguishable
(we draw them as circles o) and only one particle is allowed
per box. The number of ways of putting 2 particles in 3 boxes
is $C(3,2) = 3$. Each arrangement has probability $\frac{1}{3}$.

For Bose-Einstein statistics, the particles are indistinguishable,
but there may be more than one particle per box. To find the number of distinguishable arrangements of 2 particles in 3 boxes, we
sketch a picture as on page 704 of the text: |o| |o| . The 4
lines are the sides of the 3 boxes and the circles are the 2 particles. The outside lines must remain where they are but the 2
circles and the 2 interior lines may be rearranged in any way and
still picture an allowed arrangement. This can be done in
$C(4,2) = 6$ ways as we also see by actually drawing the arrangements. Each arrangement has probability $\frac{1}{6}$.

Section 5

2.   The uniform sample space is text equation (2.4). The sample
     space in which the points correspond to different values of
     x is, by text equation (2.5):

$$x = 2 \quad 3 \quad 4 \quad 5 \quad 6 \quad 7 \quad 8 \quad 9 \quad 10 \quad 11 \quad 12$$

$$p = \frac{1}{36} \quad \frac{2}{36} \quad \frac{3}{36} \quad \frac{4}{36} \quad \frac{5}{36} \quad \frac{6}{36} \quad \frac{5}{36} \quad \frac{4}{36} \quad \frac{3}{36} \quad \frac{2}{36} \quad \frac{1}{36}$$

Then by text equation (5.2)

$$\mu = \sum p_i x_i$$

$$= \frac{1}{36}\left[(2+12) + 2(3+11) + 3(4+10) + 4(5+9) + 5(6+8) + 6 \cdot 7\right]$$

$$= \frac{252}{36} = 7.$$

To find Var(x) we subtract μ from each x value and square
the result:

$$(x - \mu)^2 = 25 \quad 16 \quad 9 \quad 4 \quad 1 \quad 0 \quad 1 \quad 4 \quad 9 \quad 16 \quad 25$$

$$p \quad = \frac{1}{36} \quad \frac{2}{36} \quad \frac{3}{36} \quad \frac{4}{36} \quad \frac{5}{36} \quad \frac{6}{36} \quad \frac{5}{36} \quad \frac{4}{36} \quad \frac{3}{36} \quad \frac{2}{36} \quad \frac{1}{36}$$

Then by text equation (5.3),

$$\text{Var}(x) = \sum p_i (x_i - \mu)^2$$

$$= \frac{1}{36}\left[25 + 25 + 2(16+16) + 3(9+9) + 4(4+4) + 5(1+1) + 6 \cdot 0\right]$$

$$= \frac{210}{36} = \frac{35}{6}.$$

2.   (continued)

By text equation (5.4),

$$\sigma_x = \sqrt{35/6} \; .$$

For other ways of finding $\sigma^2 = \text{Var}(x)$, see Problem 17 and Problem 7.13.

We can find μ more simply by using Problem 9 below. Let x be the number on the first die and y the number on the second die. Then the sample space for either x or y is

x or y: 1   2   3   4   5   6

p: $\frac{1}{6}$ $\frac{1}{6}$ $\frac{1}{6}$ $\frac{1}{6}$ $\frac{1}{6}$ $\frac{1}{6}$

By text equation (5.2) or (5.5),

$$\bar{x} = \bar{y} = \frac{1}{6}(1 + 2 + 3 + 4 + 5 + 6) = \frac{21}{6} = \frac{7}{2} \; .$$

Then by Problem 9,

$$\mu = \bar{x} + \bar{y} = \frac{7}{2} + \frac{7}{2} = 7$$

as above.

3.   Hint: You will need to find $\displaystyle\sum_{n=1}^{\infty} n\left(\frac{1}{2}\right)^n$ and $\displaystyle\sum_{n=1}^{\infty} n^2\left(\frac{1}{2}\right)^n$. To do this first find [see text, page 3, equation (1.8)]

$$\sum_{n=1}^{\infty} x^n = \frac{x}{1 - x} \; .$$

Differentiate both sides of this equation and multiply by x [label the equation you get as equation (1)]. Differentiate equation (1) and multiply by x [call the result equation (2)]. Now put $x = 1/2$ in your equations (1) and (2).

7. The uniform sample space is given in text equation (2.3) but a
    simpler sample space for this problem is:

	3h	2h,1t	2t,1h	3t
x_i:	3	1	-1	-3
P_i:	p^3	$3p^2(1-p)$	$3p(1-p)^2$	$(1-p)^3$

Note that if p is the probability of a head, then $(1-p)$ is the
probability of a tail. Then the probability of hht is $p^2(1-p)$.
We have combined the three sample points hht, hth, thh to find
a total probability for 2h,1t of $3p^2(1-p)$. Similarly the probability of 2t,1h is $3p(1-p)^2$. Now by text equation (5.2)

$$\mu = \sum x_i P_i = 3p^3 + 3p^2(1-p) - 3p(1-p)^2 - 3(1-p)^3$$

$$= 3\left[p^2(\cancel{p+1-p}) - (1-p)^2(\cancel{p+1-p})\right]$$

$$= 3(p^2 - 1 + 2p - p^2) = 3(2p-1).$$

A simpler way of finding this result is given in Problem 15 below.
We can find Var(x) by text equation (5.3):

7.   (continued)

$$\text{Var}(x) = \sum p_i (x_i - \mu)^2$$

$$= p^3 (3 - \mu)^2 + 3p^2 (1 - p)(1 - \mu)^2$$

$$+ 3p(1 - p)^2 (-1 - \mu)^2 + (1 - p)^3 (-3 - \mu)^2$$

$$= p^3 (6 - 6p)^2 + 3p^2 (1 - p)(4 - 6p)^2$$

$$+ 3p(1 - p)^2 (2 - 6p)^2 + (1 - p)^3 (6p)^2$$

$$= 12p(1 - p).$$

The algebra in the last step is straightforward but long.
Fortunately there are easier ways of making this calculation
(see Problem 17 below and also Problem 7.13). Then by text
equation (5.4) we have

$$\sigma = \sqrt{\text{Var}(x)} = \sqrt{12p(1 - p)} = 2\sqrt{3p(1 - p)}.$$

9.   We are assuming that we have a sample space [such as in text
equations (2.1) to (2.4)] with probabilities p_i corresponding
to the sample points. Now let x and y be two random variables
with values x_i and y_i corresponding to the sample points.
(See example below.) By text equation (5.2) or (5.5):

$$E(x) = \bar{x} = \mu_x = \sum p_i x_i$$

$$E(y) = \bar{y} = \mu_y = \sum p_i y_i$$

$$E(x + y) = \overline{x + y} = \sum p_i (x_i + y_i) = \sum p_i x_i + \sum p_i y_i$$

$$= \bar{x} + \bar{y} = E(x) + E(y).$$

9.   (continued)

Example and further comment: Suppose 3 coins are tossed.
Let x be the number of heads, and let y be 1 if the first coin
is a head and -1 if it is a tail. The sample space and values
of p_i, x_i, and y_i look like this:

	hhh	hht	hth	thh	htt	tht	tth	ttt
p_i:	$\frac{1}{8}$	$\frac{1}{8}$	$\frac{1}{8}$	$\frac{1}{8}$	$\frac{1}{8}$	$\frac{1}{8}$	$\frac{1}{8}$	$\frac{1}{8}$
x_i:	3	2	2	2	1	1	1	0
y_i:	1	1	1	-1	1	-1	-1	-1

We can combine sample points corresponding to each possible pair
(x,y) and list the probabilities $f(x_i,y_j)$ that $x = x_i$ and $y = y_j$.

y \ x	3	2	1	0
1	$\frac{1}{8}$	$\frac{2}{8}$	$\frac{1}{8}$	0
-1	0	$\frac{1}{8}$	$\frac{2}{8}$	$\frac{1}{8}$

Then

$$E(x) = 3 \cdot \frac{1}{8} + 2\left(\frac{2}{8} + \frac{1}{8}\right) + 1\left(\frac{1}{8} + \frac{2}{8}\right) + 0\left(\frac{1}{8}\right),$$

$$E(y) = 1\left(\frac{1}{8} + \frac{2}{8} + \frac{1}{8}\right) + (-1)\left(\frac{1}{8} + \frac{2}{8} + \frac{1}{8}\right),$$

$$E(x+y) = (3+1)\frac{1}{8} + (3-1)\cdot 0 + (2+1)\frac{2}{8} + (2-1)\frac{1}{8} + (1+1)\frac{1}{8}$$

$$+ (1-1)\frac{2}{8} + (0+1)\cdot 0 + (0-1)\frac{1}{8} = E(x) + E(y).$$

9.   (continued)

In general

$$E(x) = \sum_i \sum_j x_i f(x_i, y_j),$$

$$E(y) = \sum_i \sum_j y_j f(x_i, y_j),$$

$$E(x+y) = \sum_i \sum_j (x_i + y_j) f(x_i, y_j) = E(x) + E(y).$$

Compare this with Problem 6.9.

15.   We let

> $x =$ number of heads minus number of tails in toss 1,
>
> $y =$ number of heads minus number of tails in toss 2,
>
> $z =$ number of heads minus number of tails in toss 3.

Then by Problem 9,

$$E(x+y+z) = E(x) + E(y) + E(z).$$

The sample space for one toss of the coin in Problem 7 is

	h	t
$p_i:$	p	1 - p
x_i or y_i or $z_i:$	1	-1

Then

$$\bar{x} = p - (1 - p) = 2p - 1,$$

and similarly,

$$\bar{y} = \bar{z} = 2p - 1.$$

Thus, for three tosses, the average of "number of heads minus number of tails" is

$$E(x+y+z) = 3(2p - 1)$$

as we found in Problem 7.

16. Hint: Be careful to distinguish between $\mu^2 = \bar{x}^2$ and $E(x^2) = \overline{x^2}$; they are very different. For example, in Problem 15 above with $p = 1/2$, we find $\bar{x} = 0$ so $\bar{x}^2 = 0$, but $\overline{x^2} = 1$.

17. Let us use Problem 16 to find $\sigma^2 = \text{Var}(x)$ in Problem 7. Using the sample space given in Problem 7, we have

$$E(x^2) = 9p^3 + 3p^2(1 - p) + 3p(1 - p)^2 + 9(1 - p)^3$$

so by Problem 16 and the value of μ from Problem 7 or Problem 15, we have

$$\sigma^2 = E(x^2) - \mu^2$$
$$= 9p^3 + 3p^2(1 - p) + 3p(1 - p)^2 + 9(1 - p)^3 - 9(2p - 1)^2.$$

The algebra is simpler if we let $1 - p = q$; then $2p - 1 = p - q$ and $p + q = 1$. We find

$$\sigma^2 = 9[p^3 + q^3 - (p - q)^2] + 3p^2q + 3pq^2$$
$$= 9[(\cancel{p+q})(p^2 - pq + q^2) - (p^2 - 2pq + q^2)] + 3pq(\cancel{p+q})$$
$$= 9pq + 3pq = 12pq$$

as in Problem 7. For a still easier way of finding σ^2, see Problem 7.13.

Section 6

1. Hints: For simple harmonic motion

$$x = a \cos \omega t,$$

$$v = \frac{dx}{dt} = -a\omega \sin \omega t.$$

The probability of finding the particle in a given dx is proportional to the time dt it spends there:

1.   (continued)

$$dt = \left|\frac{dx}{v}\right| = \left|\frac{dx}{a\omega \sin \omega t}\right| = \frac{dx}{\omega \sqrt{a^2(1 - \cos^2\omega t)}} = \frac{dx}{\omega \sqrt{a^2 - x^2}}.$$

Thus the probability function for the position of the particle is

$$f(x) = \frac{K}{\sqrt{a^2 - x^2}}$$

where K must be found so that $\int_{-a}^{a} f(x)\, dx = 1$.   Evaluate K and
continue the problem.

2.   Hint: $f(x) = Ke^{-x/\lambda}$. Find K so that $\int_{0}^{\infty} f(x)\, dx = 1$.   Note that
$\int f(x)\, dx = 1$ is always a requirement for a probability function
since it says that the probability that x has some value is 1.
The integral is over all values of x in the given problem.

8.   Since the particle has equal probabilities of being found in any
two volume elements of the same size, the probability of finding
it in a volume V must be proportional to V.   Then the probability
F(r) of finding it inside a sphere of radius r is proportional to
the volume of a sphere of radius r, that is, to $\frac{4}{3}\pi r^3$ or proportional to r^3.   Thus $F(r) = kr^3$.   But F(1) is the probability that
the particle is inside the sphere of radius 1 (where we know it
is) so F(1) = 1, and k = 1.   Then we have

$$F(r) = r^3, \qquad \text{so} \qquad f(r) = \frac{dF}{dr} = 3r^2,$$

$$\bar{r} = \int_{0}^{1} rf(r)\, dr = \int_{0}^{1} 3r^3\, dr = 3/4,$$

$$\overline{r^2} = \int_{0}^{1} r^2 f(r)\, dr = \int_{0}^{1} 3r^4\, dr = 3/5,$$

$$\sigma^2 = \overline{r^2} - (\bar{r})^2 = \frac{3}{5} - \frac{9}{16} = \frac{3}{80}, \qquad \sigma = \sqrt{3/80}.$$

(See Problem 5.16.)

9. Using text equations (6.6), we have

$$\overline{x+y} = \int\int (x+y) f(x,y)\cdot dx\, dy$$

$$= \int\int xf(x,y)\, dx\, dy + \int\int yf(x,y)\, dx\, dy = \overline{x}+\overline{y}.$$

Compare the equation at the end of the solution of Problem 5.9.

12. Hints: We want K so that $K\displaystyle\int_{r=0}^{\infty}\int_{\theta=0}^{\pi}\int_{\phi=0}^{2\pi} e^{-2r/a}\, r^2 \sin\theta\, dr\, d\theta\, d\phi = 1.$
You should find that the integration over angles gives 4π and
that $K = 1/(\pi a^3)$. Then the probability that the electron is at
a distance between r and $r+dr$ is

$$f(r)\, dr = \frac{1}{\pi a^3}\int_{\theta=0}^{\pi}\int_{\phi=0}^{2\pi} e^{-2r/a}\, r^2 \sin\theta\, dr\, d\theta\, d\phi = \frac{4}{a^3} e^{-2r/a}\, r^2\, dr.$$

$$f(r) = \frac{4}{a^3} r^2 e^{-2r/a}.$$

Use this to complete the problem. To evaluate the r integrals,
see a table of Laplace transforms, for example, text, page 636,
entry L5.

Section 7

As an illustration of Problems 1 to 6, we continue the case $n = 8$
discussed in the text.

(a) $f(x) = C(8,x)(1/2)^8$. Values of $f(x)$ (multiplied by 2^8) for
$x = 0$ to $x = 8$ are tabulated in part (c) below. The graph of $f(x)$
is text figure 7.1.

(b) The graph of $nf(x)$ as a function of x/n is text figure 7.4.

Case n = 8, (continued)

(c) By definition, the distribution function is

$$F(x) = f(0) + f(1) + \cdots + f(x),$$

that is, given the values of $f(x)$ for each x, we find

$$F(0) = f(0), \quad F(1) = f(0) + f(1), \quad F(2) = f(0) + f(1) + f(2),$$

and so on. It is convenient to tabulate first the numerators
of $f(x)$; these are the binomial coefficients $C(8,x)$, namely
1, 8, 8·7/2, 8·7·6/3!, etc. Similarly we have tabulated $2^8 F(x)$.
Then to find $f(x)$ or $F(x)$, we divide by $2^8 = 256$. The function
$F(x)$ (see graph) is called the distribution function or cumulative distribution. Note that $F(x)$ increases to 1.

x	0	1	2	3	4	5	6	7	8
$2^8 f(x)$	1	8	28	56	70	56	28	8	1
$2^8 F(x)$	1	9	37	93	163	219	247	255	256
$F(x)$	.004	.035	.145	.363	.637	.855	.965	.996	1

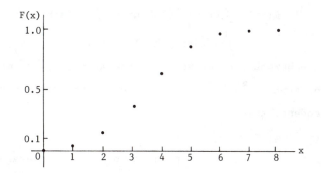

Case n = 8, (continued)

(d) We can use either f(x) or F(x) (table of values or graphs) to answer the questions asked in the instructions:

Probability of exactly 3 heads = $f(3) = \frac{56}{256} = \frac{7}{32} = .219$.

Probability of at most 3 heads = $\sum_{x=0}^{3} f(x) = F(3) = .363$.

Probability of at least 3 heads = $\sum_{x=3}^{8} f(x)$

$$= 1 - \sum_{x=0}^{2} f(x) = 1 - F(2) = 1 - .145 = .855.$$

Most probable number of heads from Figure 7.1 is 4 since f(x) is largest at x = 4.

Expected number of heads = $\sum_{x=0}^{8} xf(x)$ which we could compute, but it is easier to use Problem 5.11 which gives n/2 = 8/2 = 4.

12. If the probability that $x = x_i$ and $y = y_j$ is $f(x_i)g(y_j)$ or $p_i p_j$, that is, if x and y are independent, then

$$E(xy) = \sum_i \sum_j x_i y_j p_i p_j = \left(\sum_i x_i p_i\right)\left(\sum_j y_j p_j\right) = E(x)E(y).$$

In the continuous case, if f(x,y) = g(x)h(y), then

$$E(xy) = \iint xy f(x,y)\, dx\, dy = \iint xy g(x)h(y)\, dx\, dy =$$

$$= \int xg(x)\, dx \int yh(y)\, dy = E(x)E(y) = \mu_x \mu_y.$$

Remember that expected value and average value mean the same thing $[E(x) = \mu_x = \bar{x}]$ so we could say that if x and y are independent, then

$$\overline{xy} = \bar{x}\,\bar{y}.$$

In words, "average of product = product of averages" for independent random variables.

13. We first show that $E[(x - \mu_x)(y - \mu_y)] = 0$ if x and y are independent, that is (see Problem 12 above), if $E(xy) = \mu_x\mu_y$. We have

$$E[(x - \mu_x)(y - \mu_y)] = E(xy - \mu_x y - \mu_y x + \mu_x\mu_y)$$

$$= E(xy) - E(\mu_x y) - E(\mu_y x) + E(\mu_x\mu_y)$$

by Problem 5.9 (average of sum = sum of averages). By Problem 12 above, $E(xy) = \mu_x\mu_y$ if x and y are independent. Now μ_x is a constant, so $E(\mu_x y) = \mu_x E(y)$ by Problem 5.13, that is,

(average of Ky) = (K times average of y).

Similarly $E(\mu_y x) = \mu_y E(x) = \mu_y\mu_x$. Finally $\mu_x\mu_y$ is a constant and the average of a constant is just that constant so $E(\mu_x\mu_y) = \mu_x\mu_y$. Thus (1) gives

$$E[(x - \mu_x)(y - \mu_y)] = \mu_x\mu_y - \mu_y\mu_x - \mu_x\mu_y + \mu_x\mu_y = 0.$$

We use this result to find Var$(x + y)$ when x and y are independent. By definition,

$$Var(x + y) = E\left\{\left[(x + y) - (\mu_x + \mu_y)\right]^2\right\}$$

$$= E\left\{\left[(x - \mu_x) + (y - \mu_y)\right]^2\right\}$$

$$= E\left[(x - \mu_x)^2 + 2(x - \mu_x)(y - \mu_y) + (y - \mu_y)^2\right]$$

$$= E\left[(x - \mu_x)^2\right] + 2E\left[(x - \mu_x)(y - \mu_y)\right] + E\left[(y - \mu_y)^2\right]$$

$$= Var(x) + 2 \cdot 0 + Var(y) = Var(x) + Var(y).$$

Let us use this result to find the variance in Problems 5.7, 5.15, and 5.17. Define x, y, z as in Problem 5.15 as number of heads minus number of tails in tosses 1, 2, 3. The three tosses are independent, so

13.   (continued)

$$\text{Var}(x + y + z) = \text{Var}(x) + \text{Var}(y) + \text{Var}(z).$$

The sample space for one toss is, as in Problem 5.15,

$$h \qquad t$$

$$p_i: \quad p \quad 1-p$$

$$x_i: \quad 1 \quad -1$$

and by Problem 5.15, $\mu_x = 2p - 1$. Then by Problem 5.16,

$$\text{Var}(x) = E(x^2) - \mu_x^2 = 1^2 p + (-1)^2(1 - p) - (2p - 1)^2$$

$$= p + 1 - p - (4p^2 - 4p + 1) = 4p(1 - p).$$

Similarly

$$\text{Var}(y) = \text{Var}(z) = 4p(1 - p), \qquad \text{so}$$

$$\text{Var}(x + y + z) = 3 \cdot 4p(1 - p) = 12p(1 - p)$$

as in Problems 5.7 and 5.17, but with less algebra.

You might like to find the variance in Problem 5.2 by this method. The two tosses of a die are independent so find the variance for one toss and multiply by 2. Also use Problem 5.16.

15. Let x = number of successes in one Bernoulli trial. Then the sample space for x is

$$x: \quad 1 \qquad 0$$

$$\text{probability:} \quad p \qquad q = 1 - p$$

Then

$$\bar{x} = 1 \cdot p + 0 \cdot q = p,$$

$$\overline{x^2} = 1^2 p + 0^2 q = p,$$

so by Problem 5.16

$$\text{Var}(x) = \overline{x^2} - (\bar{x})^2 = p - p^2 = pq.$$

Now for n trials, let x_1, x_2, $\cdots$, x_n be the number of successes in trials $1, 2, \cdots, n$, and let

$$x = x_1 + x_2 + \cdots + x_n$$

be the total number of successes in n trials. Then, since the trials are independent,

$$\text{Var}(x) = \text{Var}(x_1 + x_2 + \cdots + x_n)$$

$$= \text{Var}(x_1) + \text{Var}(x_2) + \cdots + \text{Var}(x_n)$$

$$= pq + pq + \cdots + pq = npq,$$

and

$$\sigma = \sqrt{\text{Var}(x)} = \sqrt{npq}.$$

Section 8

3.   We want to compute μ and σ for the normal distribution given in
text equation (8.1). By definition

$$\mu = \int xf(x)\, dx \qquad \text{and} \qquad \sigma^2 = \int (x - \mu)^2 f(x)\, dx.$$

We evaluate these integrals using

$$f(x) = \frac{1}{\sqrt{2\pi npq}}\, e^{-(x-np)^2/(2npq)}.$$

Then

$$\mu = \frac{1}{\sqrt{2\pi npq}}\int_{-\infty}^{\infty} x e^{-(x-np)^2/(2npq)}\, dx,$$

$$\sigma^2 = \frac{1}{\sqrt{2\pi npq}}\int_{-\infty}^{\infty} (x - \mu)^2 e^{-(x-np)^2/(2npq)}\, dx.$$

Make the change of variables

$$t = \frac{x - np}{\sqrt{2npq}}, \qquad x = np + t\sqrt{2npq}, \qquad dx = \sqrt{2npq}\, dt.$$

We find

$$\mu = \frac{1}{\sqrt{2\pi npq}}\int_{-\infty}^{\infty} \left(np + t\sqrt{2npq}\right) e^{-t^2} \sqrt{2npq}\, dt.$$

Now $\int_{-\infty}^{\infty} t e^{-t^2}\, dt = 0$ since the integrand is an odd function (see
text, page 322) and we are integrating over a symmetric interval.
Cancel $\sqrt{2npq}$ to get

$$\mu = np\,\frac{1}{\sqrt{\pi}}\int_{-\infty}^{\infty} e^{-t^2}\, dt = np$$

(see text page 468 and remember that e^{-t^2} is an even function).

3.   (continued)

To find σ, put $\mu = np$ in σ^2 and make the same change of variables as in the μ integral:

$$\sigma^2 = \frac{1}{\sqrt{2\pi npq}} \int_{-\infty}^{\infty} (x - np)^2 e^{-(x-np)^2/(2npq)} \, dx$$

$$= \frac{1}{\sqrt{2\pi npq}} \int_{-\infty}^{\infty} (t\sqrt{2npq})^2 e^{-t^2} \sqrt{2npq} \, dt$$

$$= \frac{2npq}{\sqrt{\pi}} \int_{-\infty}^{\infty} t^2 e^{-t^2} \, dt = \frac{2npq}{\sqrt{\pi}} \frac{\sqrt{\pi}}{2} = npq.$$

Thus $\mu = np$ and $\sigma = \sqrt{npq}$ for the normal distribution.

10.   Exactly 31 4's in 180 tosses of a die means, in the notation of the text:

$$n = 180, \quad p = 1/6, \quad q = 5/6, \quad x = 31.$$

Then

$$\mu = np = 30, \quad \sigma = \sqrt{npq} = \sqrt{180 \cdot \frac{1}{6} \cdot \frac{5}{6}} = 5,$$

$$t = \frac{x - np}{\sigma} = \frac{31 - 30}{5} = \frac{1}{5} = 0.2.$$

By text equations (8.8) and (8.6) the probability is

$$\frac{1}{\sigma} \phi(t) = \frac{1}{5} \frac{1}{\sqrt{2\pi}} e^{-t^2/2} = \frac{1}{5\sqrt{2\pi}} e^{-0.02} = 0.0782$$

by calculator. Alternatively the value of $\phi(t)$ can be found in a table of ordinates of the normal distribution. Then you must divide by σ to get

$$\frac{1}{\sigma} \phi(0.2) = \frac{1}{5}(0.3910) = 0.0782.$$

11.   The values of μ and σ are the same as in Problem 10 above. Here we have $x_1 = 29$ and $x_2 = 33$, so

$$t_1 = \frac{x_1 - \mu}{\sigma} = -\frac{1}{5} = -0.2 \qquad \text{and} \qquad t_2 = \frac{x_2 - \mu}{\sigma} = \frac{3}{5} = 0.6.$$

The probability that x is between 29 and 33 is, by text equation (8.12),

$$P(t_1, t_2) = P(-0.2, 0.6) = P(0, 0.2) + P(0, 0.6)$$
$$= 0.0793 + 0.2258 = 0.305$$

from a table of $P(0,x)$. (Look in the area column of a table that gives area $= 0$ at $t = 0$.) If your table reads area $= 0.5000$ at $t = 0$, it is a table of $P(-\infty, x)$; subtract 0.5 from each entry to get the result above.

   A more accurate result (see comment in the text at the end of Section 8) is

$$P(0, 0.3) + P(0, 0.7) = 0.1179 + 0.2580 = 0.376.$$

(Comment: The sum of the terms of the binomial distribution from $x = 29$ to $x = 33$ is 0.37160.)

14.   See solution of Problem 9.2 in Chapter 11.

20.   Hint: If P is the probability that x is between $\mu - \sigma$ and $\mu + \sigma$, then $1 - P$ is the probability that $|x| > \sigma$. Find the probability P as in Problem 11 above or in text Example 2. Here $x_1 = \mu - \sigma$, $x_2 = \mu + \sigma$, so $t_1 = -1$ and $t_2 = 1$. Thus the probability that $|x| > \sigma$ is

$$1 - 2P(0, 1) = 0.32.$$

Similarly find the probability that $|x| > 2\sigma$, etc.

21.   Hint: What is the probability that x is between $\mu - \frac{1}{2}\sigma$ and
      $\mu + \frac{1}{2}\sigma$?   (See Problem 20 above.)   Note that you get something
      near 40% but not exactly 40%.   Look at the table to see how
      you could adjust x slightly to get 40%; this gives you the
      new C borderlines.   Similarly consider the other borderlines.

Section 9

1.   Hint: See text, page 347.   Note that $P_n(0) = 0$ for $n > 0$.

2.   For the Poisson distribution, $P_n = \mu^n e^{-\mu}/n!$.   The average and
      variance of n are (see definitions, text, page 710)

(1) $\bar{n} = \sum n P_n = e^{-\mu} \sum_{1}^{\infty} \frac{n\mu^n}{n!}$,

(2) $\sigma^2 = e^{-\mu} \sum_{1}^{\infty} \frac{(n-\mu)^2 \mu^n}{n!} = e^{-\mu} \sum_{1}^{\infty} (n^2 - 2\mu n + \mu^2)\frac{\mu^n}{n!}$.

We evaluate these sums as follows.   By text page 24,

$$e^x = \sum_{0}^{\infty} \frac{x^n}{n!} .$$

Differentiate this with respect to x and multiply by x to get

$$e^x = \sum_{1}^{\infty} \frac{nx^{n-1}}{n!} ,$$

(3) $xe^x = \sum_{1}^{\infty} \frac{nx^n}{n!}.$

Now differentiate (3) with respect to x and multiply by x:

2. (continued)

$$xe^x + e^x = \sum_1^\infty \frac{n^2 x^{n-1}}{n!} \ ,$$

(4) $(x^2 + x) e^x = \sum_1^\infty \frac{n^2 x^n}{n!} \ .$

Let $x = \mu$ in (3) and (4):

(5) $\sum_1^\infty \frac{n\mu^n}{n!} = \mu e^\mu \ ,$

(6) $\sum_1^\infty \frac{n^2\mu^n}{n!} = (\mu^2 + \mu) e^\mu .$

Substitute (5) into (1), and use (5) and (6) in (2) to get

$$\bar{n} = e^{-\mu}\mu e^\mu = \mu,$$

$$\sigma^2 = e^{-\mu}\left[(\mu^2 + \mu) e^\mu - 2\mu(\mu e^\mu) + \mu^2 e^\mu\right] = \mu.$$

Thus we have

$$\bar{n} = \mu, \qquad \sigma^2 = \mu, \qquad \sigma = \sqrt{\mu}.$$

7. Hint: This is a binomial distribution, but since $p = 1/365$ is small, you expect that the Poisson approximation [text equation (9.9)] is better than the normal approximation. Write all three formulas and then use your calculator to compare the results.

8. The average number of misprints per page is $\mu = 100/40 = 2.5$. The probability of no misprints on a given page is then [by text equation (9.8)]

$$P_0 = \frac{\mu^0 e^{-\mu}}{0!} = e^{-2.5} = 0.082.$$

The expected number of pages with no misprints is

$$40P_0 = 3.3.$$

8.   (continued)

Similarly the expected number of pages with two misprints is

$$40P_2 = 40 \frac{(2.5)^2 e^{-2.5}}{2!} = 10.3,$$

and the expected number of pages with five misprints is

$$40P_5 = 40 \frac{(2.5)^5 e^{-2.5}}{5!} = 2.7.$$

(Remember that "expected" is a technical term in probability meaning "average." Although a number of pages must be an integer, an average number may not be.)

9.   Hint: The probability of more than 10 defects is equal to $1 - \sum_0^{10} P_n$. A simpler way to evaluate this is to use a table of cumulative sums of the Poisson distribution.

Section 10

2.   We want $E(\bar{x})$ where $\bar{x} = \frac{1}{n} \sum_1^n x_i$, given that $E(x_i) = \mu$ for each x_i. By Problem 5.13,

$$E\left(\frac{1}{n} \sum_1^n x_i\right) = \frac{1}{n} E\left(\sum_1^n x_i\right),$$

and by Problem 5.9,

$$E(x_1 + x_2 + \cdots + x_n) = E(x_1) + E(x_2) + \cdots + E(x_n).$$

Thus

$$E(\bar{x}) = \frac{1}{n}\left[E(x_1) + E(x_2) + \cdots + E(x_n)\right] = \frac{1}{n} n\mu = \mu.$$

Next we want $Var(\bar{x})$ given that $Var(x_i) = \sigma^2$ for each x_i. By Problem 5.13 $[Var(Kx) = K^2 Var(x)]$,

$$Var(\bar{x}) = Var\left(\frac{1}{n} \sum_1^n x_i\right) = \frac{1}{n^2} Var\left(\sum_1^n x_i\right).$$

2.   (continued)

We assume that the measurements x_1, x_2, $\cdots$, x_n are independent. Then by Problem 7.13, the variance of the sum is the sum of the variances. The variance of each x_i is σ^2, so

$$\operatorname{Var}(\bar{x}) = \frac{1}{n^2}\left[\operatorname{Var}(x_1) + \operatorname{Var}(x_2) + \cdots + \operatorname{Var}(x_n)\right] = \frac{1}{n^2} n\sigma^2 = \frac{\sigma^2}{n}.$$

The corresponding standard deviation in $\bar{x}$ is

$$\sigma_m = \sigma/\sqrt{n}.$$

This means that the probability function for $\bar{x}$ is much narrower than the probability function for the x_i. In other words an average of n measurements is more likely to be near μ than a single measurement is.

3.   We want to find $E(s^2)$ where

$$s^2 = \frac{1}{n}\sum_1^n (x_i - \bar{x})^2.$$

By Problems 5.13 and 5.9, we find as in the first two steps of Problem 2 above

$$E(s^2) = \frac{1}{n} E\left[\sum_1^n (x_i - \bar{x})^2\right] = \frac{1}{n}\sum_1^n E\left[(x_i - \bar{x})^2\right].$$

Since all the x_i have the same probability function [that is, $f(x)$ of the parent population], all n terms have the same value. Thus

$$(1) \quad E(s^2) = \frac{1}{n} nE\left[(x_i - \bar{x})^2\right] = E\left[(x_i - \bar{x})^2\right]$$

3.   (continued)

where x_i is any one of the n measurements. We write

(2) $(x_i - \bar{x})^2 = [(x_i - \mu) - (\bar{x} - \mu)]^2$

$$= (x_i - \mu)^2 - 2(x_i - \mu)(\bar{x} - \mu) + (\bar{x} - \mu)^2$$

and find the expected value (that is, the population average) of each of the three terms. By the definition of σ^2 (population variance)

(3) $E\left[(x_i - \mu)^2\right] = \sigma^2.$

By Problem 2 above [Var $\bar{x} = \sigma^2/n$],

(4) $E\left[(\bar{x} - \mu)^2\right] = \sigma^2/n.$

To find the expected value of the middle term in (2), write

(5) $\bar{x} - \mu = \left(\dfrac{1}{n}\sum_1^n x_i\right) - \mu = \dfrac{x_1 + x_2 + \cdots + x_n}{n} - \dfrac{n\mu}{n}$

$$= \frac{1}{n}[(x_1 - \mu) + (x_2 - \mu) + \cdots + (x_n - \mu)].$$

Since the x_i are independent, then for $i \neq j$,

(6) $E[(x_i - \mu)(x_j - \mu)] = E(x_i - \mu)E(x_j - \mu) = 0 \cdot 0 = 0.$

Multiply (5) by $(x_i - \mu)$; the terms on the right are then of the form $(x_i - \mu)(x_j - \mu)$ except for the one term which is $(x_i - \mu)^2$. The expected values of all the product terms are zero by (6) and the expected value of $(x_i - \mu)^2$ is σ^2 as in (3). Then from (5)

(7) $E[(x_i - \mu)(\bar{x} - \mu)] = \dfrac{1}{n}E\left[(x_i - \mu)^2\right] = \dfrac{1}{n}\sigma^2.$

Now collect the results in (3), (7), and (4) to find the expected value of $(x_i - \bar{x})^2$ in (2) which is $E(s^2)$ in (1):

$$E(s^2) = E\left[(x_i - \bar{x})^2\right] = \sigma^2 - \frac{2}{n}\sigma^2 + \frac{\sigma^2}{n} = \sigma^2\left(1 - \frac{1}{n}\right) = \frac{n-1}{n}\sigma^2.$$

8. For x, we find $\bar{x} = 5$, and by text equation (10.1), (10.2), and
    (10.3):

$$s_x^2 = \frac{1}{8} \sum_1^8 (x_i - 5)^2 = 0.015, \quad s_x = 0.122;$$

$$\sigma_x^2 = \frac{1}{7} \sum_1^8 (x_i - 5)^2 = 0.017, \quad \sigma_x = 0.131;$$

$$\sigma_{mx} = \sigma_x / \sqrt{8} = 0.046;$$

$$r_x = 0.67 \sigma_{mx} = 0.031.$$

If we just want the probable error r, we could find it directly
using text equation (10.4) and the definition $r = 0.67 \sigma_m$:

$$r_x = 0.67 \sqrt{\frac{\sum (x_i - 5)^2}{8 \cdot 7}} = 0.031.$$

(You may have a calculator which will give you the values of
$\bar{x}$, s, and σ when you enter the x_i values. In that case you will
only need to find σ_m and r from formulas.) Repeat the calcula-
tions above for the given y values to find $\bar{y} = 1$ and
$r_y = 6.4 \times 10^{-3}$.

Now we want to combine the values we have found to compute
average values and probable errors of various functions of x
and y. We use text equations (10.12) and (10.14), and the values
found above, namely,

$$\bar{x} = 5, \quad \bar{y} = 1, \quad r_x = 3.1 \times 10^{-2}, \quad r_y = 6.4 \times 10^{-3}.$$

8.   (continued)

For $w = x + y$,

$$\bar{w} = 5 + 1 = 6, \quad r_w = \sqrt{r_x^2 + r_y^2} = 0.03.$$

For $w = xy$,

$$\bar{w} = 5 \cdot 1 = 5, \quad r_w = \sqrt{\bar{y}^2 r_x^2 + \bar{x}^2 r_y^2} = 0.04.$$

For $w = x^3 \sin y$,

$$\bar{w} = 105, \quad r_w = \sqrt{\left(3\bar{x}^2 \sin \bar{y}\right)^2 r_x^2 + \left(\bar{x}^3 \cos \bar{y}\right)^2 r_y^2} = 2.$$

(If you didn't get this, be sure your calculator is in radian
mode.)

For $w = \ln x$,

$$\bar{w} = 1.61, \quad r_w = \left|\frac{\partial w}{\partial x}\right| r_x = \frac{1}{\bar{x}} r_x = 0.006.$$

Section 11

5.   Let A mean that the first pair of cards match (one card from each

deck); let B mean that the second pair match, C that the third

pair match, etc.   Then we want $P(A + B + C + D + \cdots)$.   By a general-

ization of Problem 3.8 this is

$$P(A) + P(B) + \cdots - [P(AB) + P(AC) + \cdots]$$
$$+ [P(ABC) + P(ABD) + \cdots] - [P(ABCD) + \cdots] + \cdots.$$

Now the probability that the first pair match is $\frac{1}{52}$ and there are

52 terms of the form $P(A)$ so the first group of terms gives the

sum 1.   But we have included twice the cases where two cards

match so we must subtract terms of the form $P(AB)$.   The probabil-

5.   (continued)

ity that the first two pairs match is $\dfrac{1}{52\cdot51}=\dfrac{50!}{2!}$ and the number
of sets of two pairs is $C(52,2)$ so the sum of terms of the form
$P(AB)$ is

$$\frac{50!}{52!}\frac{52!}{50!2!}=\frac{1}{2!}.$$

Similarly, the next group of terms of the form $P(ABC)$ adds up to

$$\frac{1}{52\cdot51\cdot50}\,C(52,3)=\frac{49!}{52!}\frac{52!}{49!3!}=\frac{1}{3!}.$$

Satisfy yourself that the next groups of terms add to $\dfrac{1}{4!}$, $\dfrac{1}{5!}$,
and so on. Thus the probability of at least one match is

$$1-\frac{1}{2}+\frac{1}{3!}-\frac{1}{4!}+\frac{1}{5!}-\cdots-\frac{1}{52!}.$$

The series for $1-e^{-x}$ is

$$1-\left(1-x+\frac{x^2}{2!}-\frac{x^3}{3!}+\frac{x^4}{4!}-\cdots\right)=x-\frac{x^2}{2!}+\frac{x^3}{3!}-\frac{x^4}{4!}\cdots.$$

Let $x=1$ to see that the probability we found above is nearly
$1-e^{-1}$. [Since $1/53!=2.3\times10^{-70}$, the answer is $1-e^{-1}$ to 69
decimal places!]

Note that this analysis applies to matching of any two sets
of n things, say students and their exam papers, or couples at
a party. Note also that the series converges so rapidly that
the probability of a match is practically the same for 10 cards
or 100 cards. In fact $p=0.63$ to two decimal places for $n\geqslant5$.

7.   Hint: Don't forget that 0 is divisible by 3.

9.   (a) The possible values of x are 0,1,2. We compute the associated probabilities.

$$P(\text{two 3's}) = P(\text{box 2})P(3,3, \text{ in toss of 2 dice})$$

$$= \frac{1}{2} \cdot \frac{1}{36} = \frac{1}{72} .$$

$$P(\text{one 3}) = P(\text{box 1})P(3 \text{ in toss of die}) +$$

$$P(\text{box 2})P(\text{exactly one 3 in toss of 2 dice})$$

$$= \frac{1}{2} \cdot \frac{1}{6} + \frac{1}{2} \cdot \frac{10}{36} = \frac{16}{72} .$$

(If you don't understand the $\frac{10}{36}$, see text, page 689; there are 10 sample points corresponding to exactly one 3.)

$$P(\text{no 3's}) = \frac{1}{2} \cdot \frac{5}{6} + \frac{1}{2} \cdot \frac{25}{36} = \frac{55}{72} .$$

Verify that the three probabilities add to 1 as they must. Then the sample space is

x:	0	1	2
p:	$\frac{55}{72}$	$\frac{16}{72}$	$\frac{1}{72}$

(b) From the sample space, we see that the probability of at least one 3 is $\frac{17}{72}$.

(c) Let A mean "at least one 3" and B mean "first box". Then we want

$$P_A(B) = \frac{P(AB)}{P(A)} .$$

We found P(A) in part (b). P(AB) is the probability of picking the first box and getting a 3. This is

$$P(AB) = \frac{1}{2} \cdot \frac{1}{6} = \frac{1}{12} .$$

Thus

$$P_A(B) = \frac{1}{12} \div \frac{17}{72} = \frac{6}{17} .$$

9. (continued)

(d) $\quad \bar{x} = 0 \cdot \frac{55}{72} + 1 \cdot \frac{16}{72} + 2 \cdot \frac{1}{72} = \frac{18}{72} = \frac{1}{4}$,

$\overline{x^2} = 0 + 1^2 \cdot \frac{16}{72} + 2^2 \cdot \frac{1}{72} = \frac{20}{72} = \frac{5}{18}$,

$\sigma^2 = \overline{x^2} - (\bar{x})^2 = \frac{5}{18} - \frac{1}{16} = \frac{31}{144}$.

15. This is a binomial distribution with $n = 1095$, $p = \frac{1}{365}$, and $x = 2$. Thus the binomial probability is

$$C(1095, 2) \left(\frac{1}{365}\right)^2 \left(\frac{364}{365}\right)^{1093} = 0.224.$$

To find the Poisson approximation, we need $\mu = np = 3$; this is the average number of people born per day. Then the probability that exactly 2 are born on any one day (say January 1) is

$$P_2 = \frac{3^2 e^{-3}}{2!} = 0.224.$$

Caution: Don't confuse n in the binomial distribution and n in the Poisson distribution. The correspondence is

binomial: p   n $\mu = np$ x

Poisson: μ n

We could also try using the normal approximation to the binomial although we expect it to be poor because

$$\mu = np = 3 \ll n = 1095.$$

15.   (continued)

We find

$$\sigma = \sqrt{(1095)\left(\frac{1}{365}\right)\left(\frac{364}{365}\right)} = 1.73$$

$$t = \frac{2 - 3}{\sigma} , \qquad t^2 = \frac{1}{\sigma^2}$$

$$\frac{1}{\sigma}\,\phi(t) = \frac{1}{\sigma\sqrt{2\pi}}\,e^{-1/(2\sigma^2)} = 0.195.$$

We observe that, as expected, this is not as accurate as the Poisson approximation.

Appendix A

Integrals and Integration

The purpose of this appendix is twofold. Section 1 is a list of some integrals which are needed frequently in the problems and also some integrals which are not always listed in standard tables. Section 2 offers some general advice about evaluation of integrals as well as specific suggestions for easy ways of evaluating certain kinds of integrals.

<u>Section 1</u>   Some useful integrals.

(a)   The following indefinite integrals can easily be found in integral tables. However, they occur so often in Fourier series problems that it is convenient to have them listed here.

$$\int x \sin ax \, dx = (\sin ax - ax \cos ax)/a^2$$

$$\int x \cos ax \, dx = (\cos ax + ax \sin ax)/a^2$$

$$\int x \, e^{ax} \, dx = e^{ax}(ax - 1)/a^2$$

$$\int x^2 \sin ax \, dx = [2ax \sin ax - (a^2x^2 - 2)\cos ax]/a^3$$

$$\int x^2 \cos ax \, dx = [2ax \cos ax + (a^2x^2 - 2)\sin ax]/a^3$$

$$\int x^2 e^{ax} \, dx = e^{ax}(a^2x^2 - 2ax + 2)/a^3$$

(b)   Integrals in terms of inverse hyperbolic functions.

You may find that the following integrals are given in your tables only in terms of logarithms. It is often simpler to use inverse hyperbolic functions. For a discussion of the relation between the two forms see the text, page 75. Two answers con-

(b)   (continued)

nected by "or" are not equal but differ by a constant of integration. Assume $x > 0$ and $a > 0$. Since the integrands are even
functions, an integral over negative x is equal to the corres-
ponding integral over positive x. For example, $\displaystyle\int_{-2}^{3} = \int_{0}^{2} + \int_{0}^{3}$
for an even function.

$$\int \frac{dx}{\sqrt{x^2 + a^2}} = \sinh^{-1}\frac{x}{a} \qquad \text{or} \qquad \ell n\left(x + \sqrt{x^2 + a^2}\right).$$

$$\int \frac{dx}{\sqrt{x^2 - a^2}} = \cosh^{-1}\frac{x}{a} \qquad \text{or} \qquad \ell n\left(x + \sqrt{x^2 - a^2}\right), \qquad x \leqslant a.$$

Also note that you can find other related <u>indefinite</u> integrals
in terms of inverse hyperbolic functions by replacing a logarithm given in your tables by the corresponding inverse hyperbolic function; that is,

$$\ell n\left(x + \sqrt{x^2 + a^2}\right) + C = \sinh^{-1}\frac{x}{a} + C',$$
$$\ell n\left(x + \sqrt{x^2 - a^2}\right) + C = \cosh^{-1}\frac{x}{a} + C'.$$

<u>Section 2</u>   Evaluation of integrals.

In order to solve the text problems, it is essential to have
a good table of integrals, both indefinite and definite. However, there are sometimes easier ways to evaluate integrals.
For example, you should be able without tables or calculation
to find that

$$\int_{-10}^{10} x^3 e^{-x^2} dx = 0 \qquad \text{and} \qquad \int_{0}^{\pi/2} \sin^2 3x \, dx = \pi/4,$$

Section 2   (continued)

and you should know that a good way to evaluate

$$\int_0^\infty x \cos ax \, e^{-bx} \, dx$$

is to turn to a table of Laplace transforms. Here is an outline
of suggestions about evaluating integrals.

(a)   There are a number of places in the text where methods of evalu-
        ating integrals are discussed. Look in the text index, page 783,
        under "Integrals". In particular, see the integrals in Chapter
        11, and the solutions of Chapter 11 problems.

(b)   If you are evaluating a definite integral, try the section on
        definite integrals in your integral table. The integral you want
        may be listed as a definite integral and not as an indefinite
        integral since some definite integrals [for example $\int_0^\infty e^{-x^2} \, dx$,
        see (g) below] can be evaluated although the corresponding in-
        definite integral cannot be found in terms of elementary functions.

(c)   The method of differentiating with respect to a parameter is
        often useful. See text, pages 194, 196 (Problems 14 to 16), 640
        (text and problems). Also see the solution of Chapter 4 Problem
        12.14 and Chapter 15 Problem 2.15.

(d)   It is often trivial to evaluate a definite integral of $\sin^2 ax$ or
        $\cos^2 ax$. See these Solutions, page 56, Problem 12, and text,
        pages 306 and 307 (problems).

(e)   The integral of an odd function (text page 322) over a symmetric
      interval is zero and you should not waste time evaluating it!
      (The integral of $x^3 e^{-x^2}$ mentioned above is an example.)   Learn
      to recognize odd and even functions.   The integral of an even
      function from -a to a is twice the integral from 0 to a.

(f)   Any integral of the form $\int_0^\infty f(x) e^{-ax}\, dx$ is a Laplace transform.
      Use a table of Laplace transforms to evaluate it.   See text,
      pages 636-638 and Problems 34 to 43 on page 647.   Also see these
      Solutions, page 531, Problems 36 and 40.

(g)   Integrals of powers of x times e^{-ax} or $e^{-a^2 x^2}$ occur at several
      places in the text.   See text pages 194, 196, 457 to 461, 467 to
      472, 723 to 726.

(h)   Always watch for a substitution which reduces a given integral
      to a simpler form.   For example, you are not likely to find in
      tables

$$\int \frac{e^{\arcsin\sqrt{1-x^2}}}{\sqrt{1-x^2}}\, dx$$

but the substitution $u = \arcsin\sqrt{1-x^2}$ reduces it to $-\int e^u\, du$
(try it).   Even more important, be aware of simpler integrals
for which you don't need tables, such as:

$$\int e^{\sin x} \cos x\, dx = \int e^u\, du = e^u = e^{\sin x},$$

$$\int (1-x^2)^3 x\, dx = \int u^3 \frac{du}{-2} = -\frac{1}{2}\frac{u^4}{4} = -\frac{1}{8}(1-x^2)^4,$$

(h)   (continued)

and so on. You may find it useful to think of integrals like
these with u replaced by a "box". Thus

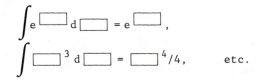

with the "box" replaced by any function you like.

Synthetic Division

We are going to show the meaning and use of synthetic division in two ways.

1. Suppose you want to find values of

$$f(x) = 5x^3 - 17x^2 - 14x + 8$$

for various x values. Verify that

$$f(x) = x[x(5x - 17) - 14] + 8.$$

To evaluate this, we start with the inside parenthesis and work our way out, as follows:

(a) $\begin{cases} \text{Multiply 5 times x and add to } -17. \\ \text{Multiply the result times x and add to } -14. \\ \text{Multiply the result times x and add to 8.} \end{cases}$

(Incidentally, this is the recommended method of programming a computer to evaluate a polynomial.) Follow the arrows in the following table to see that it shows a systematic way of carrying out the steps in (1) to find f(x) if x = -2:

(b)

5		-17	a d		-14	a d		8	a d
↓		-10	d		54	d		-80	d
5	times x	-27		times x	40		times x	-72 = f(-2)	

1.   (continued)

Notice that the first row is just the set of coefficients (of x^3, x^2, x, const.) in f(x). You must be careful to include any zero coefficients. For example, if $f(x) = 2x^3 - 5x + 6$, the first row must read 2 0 -5 6; if you write 2 -5 6, you have $2x^2 - 5x + 6$ and not $2x^3$ as you wanted.

Now we can condense the scheme in (b) if we remember the diagonal "times x" steps; for convenience we put the x value in a box on the right:

```
        5   -17   -14    8   |-2
(c)          -10   54   -80
        ─────────────────────
        5   -27    40   -72
```

Compare (c) with (b) to see what the steps are. Method (c) is called <u>synthetic division</u>.

Suppose that it turns out that the value of f(x) is zero; this means that x satisfies the equation f(x) = 0. Using the f(x) above, try x = 4:

```
        5   -17   -14    8   |4
(d)           20    12   -8
        ─────────────────────
        5    3    -2     0
```

Thus x = 4 is a root of $f(x) = 5x^3 - 17x^2 - 14x + 8 = 0$, or (x - 4) is a factor of f(x).

2.   It is also true that if we divide $f(x)$ by $(x-4)$, the quotient
     is $5x^2 + 3x - 2$ with coefficients given by the first 3 numbers in
     the third row of (d). To see why this is true, do the long
     division:

(e)

$$
\begin{array}{r}
5x^2 + 3x - 2 \\
x - 4 \enclose{longdiv}{5x^3 - 17x^2 - 14x + 8} \\
5x^3 - 20x^2 \\[-0.3em]
\hline
3x^2 - 14x \\
3x^2 - 12x \\[-0.3em]
\hline
-2x + 8 \\
-2x + 8 \\[-0.3em]
\hline
0
\end{array}
$$

so $5x^3 - 17x^2 - 14x + 8 = (x - 4)(5x^2 + 3x - 2)$.

Now compare columns in (d) and (e). The first coefficient [of x^3
in $f(x)$ and of x^2 in the quotient] is 5 in both cases. The next
coefficient in the quotient in (e) is $-17 - (-20) = 3$; in (d) we
wrote $-17 + 20 = 3$. The next coefficient in (e) is $-14 - (-12) = -2$;
in (d) we wrote $-14 + 12 = -2$. The final number 0 is $8 - 8$ in (e)
and $8 + (-8)$ in (d). Thus synthetic division (d) is really the
same as long division (e) but much condensed and therefore
quicker to use.

     You can use synthetic division when you are searching for roots
of a polynomial equation. Try a possible root; if the last number
in the third row is zero, you have found a root and also factored
the polynomial. For some examples of the use of synthetic division, see pages 307 and 313 of these Solutions.

Appendix C

Partial Fractions

It is often convenient to write a fraction as a sum of two simpler fractions; for example $\dfrac{6x - 2}{(x + 3)(x - 1)} = \dfrac{A}{x + 3} + \dfrac{B}{x - 1}$.

To find A and B, we first clear the fractions:

$$6x - 2 = A(x - 1) + B(x + 3).$$

Now this is not just an equation; it is an identity. That is, it is to be true for <u>all</u> values of x. We use $x = 1$ and $x = -3$:

For $x = 1$: $\quad 4 = B(4)$, $\quad\quad B = 1$.

For $x = -3$: $\quad -20 = A(-4)$, $\quad A = 5$.

Thus $\qquad \dfrac{6x - 2}{(x + 3)(x - 1)} = \dfrac{5}{x + 3} + \dfrac{1}{x - 1}$.

The method just shown is valid if all the denominator factors are linear and the degree of the polynomial in the numerator is at least one less than the degree of the denominator. Next consider a fraction with a squared factor in the denominator. For example,

$$\frac{2x^2 - 5x + 9}{(x - 1)^2(x + 2)} = \frac{A}{(x - 1)^2} + \frac{B}{x - 1} + \frac{C}{x + 2}.$$

Note that for a squared factor we need two terms on the right; for a cubed factor we would need three, etc. Again clearing fractions:

$$2x^2 - 5x + 9 = A(x + 2) + B(x - 1)(x + 2) + C(x - 1)^2.$$

For $x = 1$: $\quad 6 = A(3)$, $\quad\quad A = 2$.

For $x = -2$: $\quad 27 = C(3)^2$, $\quad C = 3$.

We need one more equation to find B. We can use any other value of x, but perhaps $x = 0$ is the simplest. For $x = 0$,

$$9 = 2A - 2B + C = 7 - 2B, \qquad B = -1.$$

Then
$$\frac{2x^2 - 5x + 9}{(x+1)^2 (x+2)} = \frac{2}{(x-1)^2} - \frac{1}{x-1} + \frac{3}{x+2} \ ..$$

If one of the denominator factors is quadratic and we do not want to factor it (because the factors involve square roots or complex numbers), then we can proceed as follows:

$$\frac{5x^2 - 4x + 11}{(x-1)(x^2 + x + 2)} = \frac{A}{x-1} + \frac{Bx + C}{x^2 + x + 2} \ .$$

Clear fractions:

$$5x^2 - 4x + 11 = A(x^2 + x + 2) + (Bx + C)(x - 1).$$

For $x = 1$: $12 = A(4)$, $A = 3$.

To find B and C, one method is to use two other values of x, say:

For $x = 0$: $11 = 2A - C$, $C = -5$.

For $x = -1$: $20 = A(2) + (-B + C)(-2) = 16 + 2B$, $B = 2$.

Another method for finding B (or B and C, or even all three) is to collect powers of x on the right:

$$5x^2 - 4x + 11 = (A + B)x^2 + (A - B + C)x + 2A - C.$$

Since this is an identity we must have
$$\begin{aligned} A + B &= 5, \\ A - B + C &= -4, \\ 2A \quad - C &= 11. \end{aligned}$$

From these equations we can find A, B, C. Or, if we use the previously found $A = 3$, then the first equation gives $B = 2$, and the third equation gives $C = -5$. The second equation provides a check: $3 - 2 - 5 = -4$. Thus

$$\frac{5x^2 - 4x + 11}{(x-1)(x^2 + x + 2)} = \frac{3}{x-1} + \frac{2x - 5}{x^2 + x + 2} \ .$$

For other examples of the use of partial fractions, see pages 473 and 528 of these Solutions.